T0141950

Springer Water

Series Editor

Andrey G. Kostianoy, Russian Academy of Sciences, P.P. Shirshov Institute of Oceanology, Moscow, Russia

The book series Springer Water comprises a broad portfolio of multi- and interdisciplinary scientific books, aiming at researchers, students, and everyone interested in water-related science. The series includes peer-reviewed monographs, edited volumes, textbooks, and conference proceedings. Its volumes combine all kinds of water-related research areas, such as: the movement, distribution and quality of freshwater; water resources; the quality and pollution of water and its influence on health; the water industry including drinking water, wastewater, and desalination services and technologies; water history; as well as water management and the governmental, political, developmental, and ethical aspects of water.

More information about this series at http://www.springer.com/series/13419

Abdelazim M. Negm · Gheorghe Romanescu ·
Martina Zeleňáková
Editors

Water Resources
Management in Romania

Editors
Abdelazim M. Negm
Faculty of Engineering
Zagazig University
Zagazig, Egypt

Gheorghe Romanescu
Faculty of Geography and Geology
University "Alexandru Ioan Cuza"
Iaşi, Romania

Martina Zeleňáková
Technical University of Kosice
Košice, Slovakia

ISSN 2364-6934   ISSN 2364-8198   (electronic)
Springer Water
ISBN 978-3-030-22322-9   ISBN 978-3-030-22320-5   (eBook)
https://doi.org/10.1007/978-3-030-22320-5

This Springer imprint is published by the registered company Springer Nature Switzerland AG
The registered company address is: Gewerbestrasse 11, 6330 Cham, Switzerland

Preface

The interest in water resources management of hydrographical basins in Romania makes the emphasis on its exploitation very strong. The subject of water resources management is very wide, and we chose the most topical issues for Romania to be covered in this volume. It is obvious that a number of good books are available on specific parts of the topic, but this book intent to cover much more breadth and depth of the subject, and it was the idea of water resources management in Romania book came about. The book has been treated as the product of teamwork of more than 40 distinguished researchers and scientists from different institutions, academic, and research centers with major concerns regarding water management.

This book presents current knowledge on water resources management in Romania mainly from hydrological point of view. It will attract researchers, experts, scientists, practitioners as well as graduate or anybody interested in water resources management. Sustainable development of water management is based on the principle that water as a natural resource may be utilized only to that extent which ensures future generations sufficient usable supplies of water in the seas, rivers, lakes, and reservoirs, and that reserves contained in porous environments below the surface of the land remain preserved in the same quantity and quality. For this reason, it is necessary to devote all the attention to the knowledge and protection of water resources. The book is focused on a wide variety of water resources issues, from hydrology, climate change, water quantity, water quality, water supply, flood protection, hydrological hazard, and ecosystems. The book presents state-of-the-art knowledge that can be effectively used for solving a variety of problems in integrated water resources management as well as the latest developments in the research area mainly in Romania.

The Water Resources Management in Romania volume consists of 17 chapters and is divided into 7 parts. Part I: "**Introducing the book**" was prepared by editors Abdelazim M. Negm from Water and Water Structures Engineering Department, Faculty of Engineering, Zagazig University, Martina Zeleňáková from Department of Environmental Engineering, Faculty of Civil Engineering, Technical University of Košice, and Ionut Minea from Department of Geography, Faculty of Geography and Geology, Alexandru Ioan Cuza University of Iași.

v

Part II of the book is devoted to: "**Water Quality**" issues. Chapter 2 of the book "**Implementation of EU Water Framework Directive (2000/60/EC) in Romania —European Qualitative Requirements**" outlines the contemporary reality of water resources and the stage of their quality in Romania. It was written by editor of this book Professor Gheorghe Romanescu and his colleagues Cristian Constantin Stoleriu from Department of Geography, Faculty of Geography and Geology, and Alin Mihu-Pintilie from Interdisciplinary Research Department, Field Science, all from Alexandru Ioan Cuza University of Iași. Chapter 3 "**Causes and Effects of Water Pollution in Romania**" pays attention to water quality in Romania. It states that pressures to water are generated by (i) human agglomeration (lack of con- nection to the sewerage system and to the sewage treatment plants); (ii) industry (wastewater discharges); (iii) agriculture (nutrient and pesticide emissions); (iv) hydrotechnical works (dams, dikes, sills, weirs, diversions); and (v) other anthropic activities. This chapter was written by Iuliana Gabriela Breaban from Department of Geography and CERNESIM, Faculty of Geography and Geology, Alexandru Ioan Cuza University of Iași, and Ana Ioana Breaban from Faculty of Hydrotechnical Engineering, Geodesy and Environmental Engineering, Technical University "Gheorghe Asachi" of Iași. In Chap. 4 "**Management of Surfaces Water Resources—Ecological Status of the Mureș Waterbody (Superior Mureș Sector), Romania,**" the authors assess the ecological status of the water over three sections, Tîrgu Mureș, Ungheni/Mureș, and Iernut, based on the state of the physicochemical and biological elements. The authors of this chapters are Florica Morar from Industrial Engineering and Management Department, Faculty of Engineering "Petru Maior" University of Tîrgu Mureș, Dana Rus Department of Electrical Engineering and Computers, Faculty of Engineering, "Petru Maior" University of Tîrgu Mureș, and Petru-Dragoș Morar—Ph.D. student at Technical University Bucharest.

Part III of the book deals with "**Water Supply**" topics. Chapter 5 "**Water Supply Challenges and Achievements in Constanta County**" introduces drink- able water supply systems for which is used mainly groundwater, but also surface water from the Danube. It presents two case studies, one regarding a drinkable water system and the other an irrigation water pumping station. It was prepared by Anca Constantin and Claudiu Ștefan Nițescu from Faculty of Civil Engineering, Ovidius University from Constanta. Chapter 6 "**Drinking Water Supply Systems—Evolution Towards Efficiency**" is oriented to past, present, and future of water supply systems, with emphasis to water supply systems in Cluj City. It was prepared by Ciprian Bacotiu, Cristina Iacob from Building Services Engineering Faculty, Technical University of Cluj-Napoca, and Peter Kapalo from Institute of Building and Environmental Engineering, Faculty of Civil Engineering, Technical University of Košice.

Part IV of the book titled "**Antropic Influence to Water Resources**" includes three chapters. Chapter 7 "**The Vulnerability of Water Resources from Eastern Romania to Anthropic Impact and Climate Change**" identifies the main issues associated with the anthropic impact and climate change upon the water resources of Eastern Romania, taking into account the social and natural characteristics of the region. This chapter was written by Ionut Minea from Department of Geography, Faculty of Geography and Geology, "Alexandru Ioan Cuza" University of Iaşi. Chapter 8 "**Romanian Danube River Floodplain Functionality Assessment**" identified three types of areas that belong to Romanian Danube River floodplain: The first type represents the areas with only agricultural potential, the second type are the areas with potential to be ecologically restored, and the third one is a combination of the first two. The chapter was written by Cristian Trifanov from Informational System and Geomatics Department, Danube Delta National Institute for Research and Development, Tulcea; Alin Mihu-Pintilie from Interdisciplinary Research Department, Alexandru Ioan Cuza University of Iaşi; Marian Tudor from Management Department, Marian Mierla from Informational System and Geomatics Department, Mihai Doroftei from Biodiversity Conservation and Sustainable Use of Natural Resources Department and Silviu Covaliov from Ecological Restoration and Species Recovery Department of the Danube Delta National Institute for Research and Development, Tulcea. Chapter 9 "**Deforestation and Frequency of Floods in Romania**" assesses deforested areas in the period 2000–2016, as well as the frequency of floods at the level of each administrative unit from Romania for the same time. It was prepared by Daniel Peptenatu, Alexandra Grecu, Adrian Gabriel Simion, Karina Andreea Gruia, Ion Andronache, Cristian Constantin Draghici, Daniel Constantin Diaconu from Department of Meteorology and Hydrology and Research Center for Integrated Analysis and Territorial Management, Faculty of Geography, University of Bucharest.

Part V is devoted to the topic "**Hydrology.**" Chapter 10 "**Hydrological Impacts of Climate Changes in Romania**" provides a comprehensive synthesis of studies on hydroclimatic changes in Romania and presents some original results on hydrological responses to climate changes in Valea Cerbului River basin (area of 26 km^2) located in the Carpathian Mountains, based on the analysis of historical data and hydrological simulations. The chapter presents the results of the study of authors Liliana Zaharia and Gabriela Ioana-Toroimac from Faculty of Geography, University of Bucharest, and Elena-Ruth Perju from National Institute of Hydrology and Water Management. Chapter 11 "**Monitoring and Management of Water in the Siret River Basin (Romania)**" approaches the methodology of water resources management, and also the purpose of performing different types of measurements (both quantitative and qualitative) and measures to be taken in order to avoid the negative effects that can result from the exploitation of these resources. Chapter 12 "**Water Resources from Romanian Upper Tisa Basin**" analyzes and assesses the hydrological regime and the water resources of Upper Tisa Basin. It was prepared by Gheorghe Şerban and Răzvan Bătinaş from Faculty of Geography, Babeş-Bolyai University, Cluj-Napoca, Daniel Sabău from Romanian Waters National Administration—Someş-Tisa Regional Water Branch, Cluj-Napoca, Petre Breţcan from Department of Geography, Faculty of Humanities, Valahia University in Târgovişte, Elena Ignat from Coţofăneşti Secondary School in Bacău County, and Simion Nacu from Romanian Waters, National Administration in Bucharest.

Part VI presents the "**Case Studies.**" Chapter 13 "**Particularities of Drain Liquid in the Small Wetland of Braila Natural Park, Romania**" highlights the particularities of the hydrological regime and, especially that of the liquid leakage on the lower course of Danube, within one of the many wetlands along the river to correctly quantify the water intake critical periods, for various uses. The chapter was written by Daniel Constantin Diaconu from Department of Meteorology and Hydrology and Research Center for Integrated Analysis and Territorial Management, Faculty of Geography, University of Bucharest. Chapter 14 "**Assessment of Some Diurnal Streamwater Profiles in Western and Northern Romania in Relation to Meteorological Data**" detects the shapes of diurnal profiles and their spatial variations by water and air measurements that were conducted in river valleys of Romania. It was prepared by Andrei-Emil Briciu, Dinu Iulian Oprea, Dumitru Mihăilă, Liliana Gina Lazurca (Andrei), Luciana-Alexandra Costan (Briciu) from Department of Geography, Ştefan cel Mare University of Suceava and Petruţ-Ionel Bistricean from Suceava Weather Station, National Meteorological Administration. Chapter 15 "**Drought and Insolvency: Case Study of the Producer-Buyer Conflict (Romania, the Period Between the Years 2011–2012)**" presents the impact of the drought recorded in the autumn of the year 2011 and the spring of the year 2012 which entailed a drastic reduction in power production provided by hydropower plants, reasons for which the company Hidroelectrica S.A. became unable to distribute power to beneficiaries. The chapter was prepared by Gheorghe Romanescu and Ionuţ Minea from Department of Geography, Faculty of Geography and Geology, "Alexandru Ioan Cuza" University of Iaşi. Chapter 16 "**Water Resources from Apuseni Mountains —Major Coordinates**" presents aspects related to the general organization of the available water resources, followed by an assessment of the factors determining the water flow, followed the observations related to the water flow parameters. It was prepared by Răzvan Bătinaş and Gheorghe Şerban from Faculty of Geography, Babeş-Bolyai University, Cluj-Napoca and Daniel Sabău from Romanian Waters National Administration—Someş-Tisa Regional Water Branch, Cluj-Napoca.

The last, seventh, part of this book presents "**Conclusion**" and was prepared by editors of this book.

We would like express special thanks to all who contributed to this high-quality volume which presents a real source of knowledge and latest findings in the field of Water Resources Management of Romania. We would like to thank all the authors for their contributions. Without their patience and effort in writing and revising the different versions to satisfy the high-quality standards of Springer, it would not have been possible to produce this volume and make it a reality. Much appreciation and great thanks are also owed to the editors of the Environmental Earth Science book series at Springer for the constructive comments, advices, and the critical reviews. Acknowledgments must be extended to include all members of the Springer team who have worked long and hard to produce this volume. We believe that the book will be a valuable source of information, knowledge and experiences for the academics, practitioners, researchers, graduate students, and scientists not only in Romania.

The volume editor would be happy to receive any comments to improve future editions. Comments, feedback, suggestions for improvement, or new chapters for next editions are welcomed and should be sent directly to the volume editors. The emails of the editors can be found inside the books at the footnote of their chapters.

The book is especially devoted to university Prof. Gheorghe Romanescu, editor of this book, an eminent teacher and researcher in field of water geography, who unexpectedly has left us on October 3, 2018, during the processing of this volume. We appreciate his great effort in invitation of the authors because without his contribution and hard work, the book would not arouse.

We would close the preface by the statement of Heraclitus "No man ever steps in the same river twice, for it is not the same river and he is not the same man." and statement of Cehov "Everything must be beautiful in man, from the face to the clothing, from the soul to the thought."

Zagazig, Egypt
Iași, Romania
Košice, Slovakia
March 2019

Abdelazim M. Negm
Ionut Minea
Martina Zeleňáková

Contents

Part I
Introducing the Book

Chapter 1
Introduction to "Water Resources Management in Romania"

Abdelazim M. Negm, Martina Zeleňáková and Ionut Minea

Abstract This chapter presents the main features of the book titled "Water Resources Management in Romania". The book consisted of 7 parts including the introduction (this chapter) and the conclusions (the closing chapter). The main body of the book consists of five themes to cover the water quality, the water supply, the antropic influences to water resources, hydrology and several case studies. The main technical elements of each chapter are presented under its relevant theme.

Keywords Romania · Management · Quality · Quantity · Modelling · Water resources · Pollution · Hydropower · Danube River · Antropic · Hydrology

1.1 Romania: A Brief Background

Figure 1.1 shows the main features of Romania country that are associated with the contents of the book. Romania benefits from different types of aquatic units including rivers, lakes, underground waters, marine waters. The particular hydrographic and hydrological characteristics of Romania, determined mainly by the geographic position of the country in the temperate-continental climate and by the presence of the Carpathian Mountain arch condition the spatial and temporal distribution of water resources.

A. M. Negm (✉)
Water and Water Structures Engineering Department, Faculty of Engineering, Zagazig University, Zagazig 44519, Egypt
e-mail: amnegm85@yahoo.com; amnegm@zu.edu.eg

M. Zeleňáková
Department of Environmental Engineering, Faculty of Civil Engineering, Technical University of Košice, Košice, Slovakia
e-mail: martina.zelenakova@tuke.sk

I. Minea
Department of Geography, Faculty of Geography and Geology, "Alexandru Ioan Cuza" University of Iasi, Iasi, Romania
e-mail: ionutminea1979@yahoo.com

© Springer Nature Switzerland AG 2020
A. M. Negm et al. (eds.), *Water Resources Management in Romania*, Springer Water,
https://doi.org/10.1007/978-3-030-22320-5_1

3

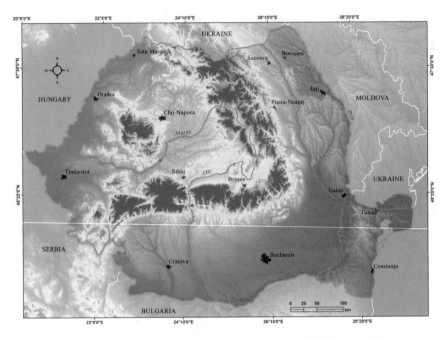

Fig. 1.1 The main feature of Romania country (https://www.naturalearthdata.com/downloads/)

The anthropic factor has contributed with some modifications of these peculiarities through the multiple hydro-technical constructions (accumulation lakes, derivations, catchments) made on the course of natural rivers or lakes. Romania's water resources are relatively poor and unevenly distributed over time and space. Global assessments estimate a volume of 134.6 billion m^3 of water (consisting of surface waters—rivers, lakes, the Danube River—and underground waters).

Only 30% of this volume of water (about 40 billion m^3) can be considered the usable resource, according to the degree of river basin planning. If we relate to the number of inhabitants, the specific water resources are 1894 m^3/year/place. From this point of view, Romania is considered one of the countries with the lowest water resources in Europe, ranking the 21st among the European states. The water requirements for the population, and economic activities supply (industry, agriculture, fish farming, tourism) have increased from 14.4 billion m^3/year in 1975 to 35 billion m^3/year in 1990 and to 46 billion m^3/year within the period 2010–2015 [4].

At the same time, Romania depends to a great extent on the water resources coming from different upstream countries (representing the contribution of the rivers that form on the territory of other countries and then enters the territory of Romania. This includes the Danube case and the watercourses in the upper basin of the Siret rivers and Prut in the north-eastern and eastern border). Note that there is a slight compensation between the surface and underground water resources. Thus, if in the mountainous area predominates the surface resource, in the plains, the underground resource compensates for the necessary (Romanian Plain and Dobrogea). With all this compensation, there are areas along the Danube River or rivers that have their source on the territory of other states where water resources are insufficient [2].

The temporal and spatial distribution of water resources in Romania are unequal. The Carpathians area with a surface of 28% of the territory of Romania generates over 65% of the surface water resources. The essential volume of the rivers flow is registered during the spring (almost 50% from the entire year). As a result, it was necessary to prevalently build storage lakes. In 2018, lakes were counting over 1450 units, and the most important—400—were storing around 15 billion m^3 of water [5]. They mitigate the variability of the hydrological regime of water resources in Romania, determined by the large volumes accumulated or discharged in the rainwater overflow periods (spring), and the drought-induced deficits (when the flow rate drops to very low values in summer and autumn). Therefore, unlike the countries of Western and Northern Europe, the lack of sufficient water resources risks becoming a limiting factor of economic development if a strict policy of rational water use by the responsible actors is not promoted [1].

Anthropic measures to regulate flows through accumulating lakes aim at restraining maximum flows during surplus periods to make them available during periods of drought. In order to satisfy the water requirement, some accumulation requirements are imposed by the overflows in the mountain area, transfers of water from the surplus river basins to the deficient ones, injection of suitable transfer volumes (sands, gravels) of the transfer volumes or the phases with maximum drainage be subsequently extracted with appropriate health protection.

It is worth mentioning that in 1976, the National Programme for the Romanian Hydrographic Basins Management had been promoted, for the hydrographic basins and even for the territorial-administrative units (counties) to assure a good knowledge and management of the water resources. In the framework of this programme, the building of storage lakes, damming in, water transfers between the hydrographic basins, irrigation systems extension, water quality monitoring was achieved. Most of these objectives had been partially stopped after 1989, and restarted after Romania integration into European Union, under other terms and parameters introduced by Water Framework Directive 2000/60/EC [6, 7].

1.2 Main Themes of the Book

The book covers the following themes: water quality in three chapters numbers 2, 3 and 4, water supply in two chapters numbers 5 and 6, anthropic influence to water resources in three chapters numbers 7, 8 and 9, hydrology in three chapters numbers 10, 11 and 12 and four cases studies in Chaps. 13, 14, 15 and 16.

1.3 Water Quality

Chapter 2 addresses an overview of water quality issues in line with the environmental objectives of the Water Framework Directive 2000/60/EC. Romania's accession to the EU in 2007 led to the generation of river basin management plans that monitor and manage the water resource and its quality. The first management plans that are in line with the environmental objectives of the Water Framework Directive 2000/60/EC date back to 2004. Through the studies that led to the design of the management plans, it was emphasized that the industrial and mining activities, as well as the lack of sewerage systems and treatment plants (in the area of rural and urban small towns), represent the main sources of degradation of water quality. The methodology consisted of the analysis of water management plans within river basins, data from the National Institute of Hydrology and Water Management, and data obtained on field studies in eastern Romania. The sources identified as diminishing water quality are those belonging to old industries, rural settlements, agriculture, and new smaller industries. The Danube River Basin extends to the neighbouring countries of Bulgaria, Serbia, Hungary, Ukraine and the Republic of Moldova in the area of Romania. Thus in the border area with these countries, there is a cross-border influence on water quality. Hydrographic basin management plans contain management measures to increase the surface and underground water quality. In order to reduce surface and underground water pollution, the implementation of the measures mentioned in the hydrographic basin management plans started in 2009, in order to reduce water pollution.

Chapter 3 shows that Romanian legislation correctly reflects the environmental requirements agreed at EU level, but the real challenge is effective field enforcement, due to lack of adequate planning, coordination, and funding. The distribution on hydrographic basins shows that the accidental pollution cases are ranging between 67 and 3, most accidental pollutions occurring in Mures (67), Arges-Vedea (61), Olt (43), Siret (40) and Prut (36), a few cases being recorded in Banat (3) and Crisuri (6). By year, most cases have been noticed in 2017 (70), 2012 (65), 2015 (63). Groundwater quality monitoring revealed that in 2016, about 16% (23/143) underground water bodies were outside the chemical parameters imposed by the standards. In 2016, there were reported 47 accidental pollutions, the distribution of which according to the nature of the pollutant is as follows: untreated wastewater 18 events—accounting for 38.30% of the total accidents; petroleum products 14 events—29.79%; semi-solid waste 6 events—12.76%; pollution of other nature 4 events—8.51%; unidentified substances 3 events—6.38%; mine water 1 event—2.13%; chemicals that attract low oxygenation 1 event—2.13%. It has been noticed a particular situation where almost half of the residents are not connected to the sewage system, therefore wastewater is directly transferred to the natural emissary. In addition, it is highlighted a slow continuous increase regarding the beneficiaries of treated wastewater from 40.2% (2005) to 47.8% (2015), from which currently only 1.9% do not treat the collected wastewaters.

The chapter number 4 brings to the reader's attention some aspects regarding both the quality of surface waters in general (Sects. 4.1–4.3) and the theoretical information on the environmental factor—WATER, and a case study is presented in Sect. 4.4. The water quality status of the Mureş water body is presented and analyzed on three sections (Târgu-Mureş socket, Ungheni socket, and Iernut socket) for the period 2015–2017. It is necessary to know the quality of surface water at any moment, because on the basis of the data obtained and analyzed it is possible to make a prognosis of the water quality evolution on the monitored sections. The water quality in Romania is monitored according to the structure and methodological principles of the Integrated Water Monitoring System in Romania (S.M.I.A.R.), restructured according to the requirements of the European Directives. The study showed an improvement in the process of monitoring and evaluation of the waters of the Mures River basin (referring to the area under study) in 2017 compared to previous years.

1.4 Water Supply

Chapter 5 indicates that a more flexible interconnected drinkable water supply system is needed in Dobrogea, as well as modern and energy efficient irrigation water systems. The energy efficiency along with the environmental protection have to be among the main goals of the Romanian authorities, and designers. Dobrogea is a beautiful but dry-climate region in Romania, where the drinkable water has to be abstracted from deep aquifers, and the irrigation water supply systems have to deal with high pumping heads resulting in considerable energy consumption. The main challenges related to water are water quality, energy efficiency, water loss mitigation and above all, a more responsible attitude of man towards waters and environmental protection. New vast on-going projects aim to modernize the water supply systems. A new interconnected drinkable water supply system is going to meet either the water quality standards in small villages or the variable demand of the towns on the Black Sea coast. The modernization of the irrigation water supply systems aims to support agriculture with higher efficiency. Numerical simulation of water systems operation is a real key to better engineering design and to a more energy efficient use of the water resources. The case studies presented in this chapter refer to the possible energy saving emerging from an optimal operation of a drinkable water distribution system and respectively an irrigation water pumping station. The environmental impact of the new construction sites has to be continuously assessed in order to preserve the existing habitats.

Chapter 6 presents the history and evolution of drinking water supply systems in the major city of Transylvania, Cluj-Napoca, focusing on the achievements made during the last 30 years under the ruling concept of efficiency. The accelerated urban development of Cluj-Napoca and the political events of the last century had a strong influence on the evolution of drinking water supply systems and the faith of the local water operator. The communist regime, following its industrialization plans, wanted to build quickly and cheaply. After 1989, the centralized economy died, and this quantitative approach had to be abandoned. The last 30 years represent a period of profound transformations in the field of water supply systems. The market economy imposed new standards of efficiency and quality. Water companies had to survive inflation, re-organize themselves, access European funds, adopt new materials and technologies, reduce water losses and find new clients. Thus, since 1997, Cluj-Napoca's local water company has conducted four major investment projects for the expansion, modernization, and rehabilitation of water and sewerage infrastructure in the serviced area. New technologies like GIS and SCADA were implemented, water metering was finalized and new clients were found in a nearby county. The Operational Program for Large Infrastructure in the period 2014–2020 is promoting integrated water and wastewater systems in a regional approach, in order to reduce the costs and obtain maximum efficiency for such investments.

1.5 Antropic Influences to Water Resources

Chapter 7 shows that water resources from Eastern part of Romania are becoming increasingly vulnerable as human impacts increase and regional climate change. With a surface of 20,569 km^2 and a population of 2.2 million, the Eastern part of Romania is classified as vulnerable from an economic and social point of view, but also as far as the impact of human activity and climate change on water resources. Taking into account the natural condition the anthropic impact was evaluated using WEI+ index (developed by the European Environmental Agency). According to this index, the mean value calculated for this region (15.7) indicated limited pressure upon water resources, but with the potential of increasing in severity over dry years (like 2000, 2007, 2012). More than half of the rivers that drain the eastern part of Romania display a trend toward a rise in mean annual and seasonal flow, in agreement with the regional trend associated with climate change. However, this trend does not significantly reduce the anthropogenic impact on water resources in this region.

Chapter 8 describes the extensive interdisciplinary studies done along the Romanian Danube River Floodplain within the ecological and economic resizing program of the Danube floodplain and delta. The main aspect that was followed in this program was the reconsideration of the flood line of defense for the localities along the river, by creating water reserve polders, ecological restoration of some areas, etc. In order to determine which areas of the floodplain could be changed from one use to another, various surveys were done along the river to evaluate the floodplain's functionality. To determine the precise state of the dykes, the topology of the area, LiDAR measurements were conducted on 1075 km stretch with a variable width of 80 km. With the results of the LiDAR and bathymetrical measurements, hydraulic modeling was possible in order to develop flood scenarios and to determine the feasible areas where the land use could be changed (taking into consideration numerous variables). The chapter stressed that the shift to sustainable development is a process of a large scale, achievable at the same time by the cooperation of all countries because the effects on the environment do not know limits imposed by the borders.

Chapter 9 indicates that reducing forest areas under pressure from economic activities that require this type of resource, is one of the most important topics of debate, both at the level of the scientific world, as well as for the affected communities in the short, medium and long term. Therefore, the chapter aimed to evaluate the forest areas in the period 2000–2016 and their correlation with the frequency of flooding. To highlight the correlation between the forest cuts and the increase of the flood frequency, detailed analyzes were carried out on the main mountain groups from Romania, where the growth of deforested areas has led to an obvious increase in the frequency of floods. A database of deforested areas has been made by extracting 654,178 Landsat 7 ETM+ satellite images, characterizing the areas covered with forest and changes from 2000 to 2016. For spatial quantification at the local level, a spatial join was made, for each administrative-territorial unit, the newly created file containing the attribute table, including the column that sums the pixels in each administrative-territorial unit. The database was completed with the frequency of floods at the level of each territorial administrative unit. Encouraging exports of woody raw materials and low incomes of the population have made that illegal cuts get a great extent, and the lack of the general cadastre made the evidence of forest areas uncertain, favouring illegalities in the management of forest resources. Under these circumstances, cutting forests on nearly 400,000 ha has led to major imbalances in ecosystems, changing the drainage of water on the slopes is the immediate consequence. Researching the relationship between deforested areas and frequency of floods can open a new research area, on identifying types of deforestation by fractal and non-fractal analysis.

1.6 Hydrology

Chapter 10 presents a review on the main research findings on hydro-climatic changes in Romania and provides new results on hydrological responses to climate changes in a Carpathian watershed, based on the analysis of historical data and hydrological modeling. The hydrological changes in Romania are integrated into the context of the streamflow changes identified at global and European spatial scales, which are synthesized in the first part of the chapter. Further, an overview of observed and projected changes in main climate parameters controlling the river flow in Romania is presented, followed by a review of studies on streamflow trends and hydrological impacts (observed and projected) of climate changes in Romania. In the last part of the chapter, the authors investigated the hydrological responses to climate changes in a small mountain catchment (Valea Cerbului River) located in the Carpathians. Based on recorded data, changes in the magnitude and frequency of floods, as well as in the frequency of low flows were analyzed. The possible future streamflow changes in Valea Cerbului River catchment were estimated by simulations with WaSiM-Eth model, under B1, A2 and A1B climatic scenario, for the period 2001–2065 relative to 1961–2000 period. In summary, the chapter shows that after 1960, a general decreasing trend of the mean annual streamflow was detected in Romania.

For the future (by 2050–2065), the projections indicate a decline of the annual runoff, increases of discharges in winter and decreases in late summer and autumn.

Chapter 11 comprises detailed information about how Siret Water Basin Administration works. Both the structure of the administration and the mechanisms used for measuring, analyzing and finding the correct solutions are presented. The hydrological activity of the Siret Water Basin Administration is carried out on an area of 28,878 km^2, managed by four subunits called Water Management Systems (WMS), and each WMS has one or several hydrological stations subordinated to it. In their turn, the hydrological stations manage several hydrometric stations. The main activities ongoing at the hydrometric stations are level recordings, water and air temperature recordings, precipitation, meteorological elements recordings, determination of flows, alluvial sampling, water quality sampling, etc. In addition to river hydrometry, the Siret Basin Water Administration also monitors the lakes. Each hydrological station has under administration a number of lakes. In addition to the quantitative monitoring of water bodies in the Siret Water Basin Administration, there is also qualitative monitoring for the 353 water bodies: 343 water bodies—natural rivers; 10 water bodies—lakes. Out of the total number of monitored waterbodies, 50 bodies of water are natural, and 5 bodies of water are heavily modified. The natural and heavily modified water bodies in the Siret River Basin are monitored qualitatively through 87 sections.

The aims of Chap. 12 are to offer an integrated image on the condition factors (geological, morphological, climatic, hydrographic and phyto-pedogeografic cover) and the most important, on hydrological regime and balance, which allows the assessment of the water resources of the analysed area, and to compare the raw natural water quantity with the water quantity accessed by the population. It is analyzed the actual water resource, being mainly pursued the differences of river runoff, between the two slopes of the basin, one exposed to the east, and the other to the west. Among the various parameters, used to emphasize the resource of water in a territory have been analysed: the volume of average flow, the average drained layer, the average specific runoff, the coefficient of average river-flow. There are identified high rates of flow, according to pluvial quantities—over 1100 mm in high mountainous space. The analysis of how the population benefits of water resources in terms of the quantity and of the failures existing in the water supply systems is also important. A flow layer composite raster structure was built, for comparing the surface natural resource with used by population water resource, and showing the areas where the amount of water available for the population is in deficit. To sum up, Romanian Upper Tisa Basin, an rainwater surplus area, with a moderate offer concerning the water resource in the majority of the basin space, due to the geological layer and anthropogenic pressure. The surface water resource is very consistent in all Romanian Upper Tisa Basin. The underground water resource is slightly less represented on the Maramureş Basin, and better emphasized in the Tisa Corridor or the lower plain of the West.

1.7 Case Studies

Chapter 13 aims to highlight the role of water resources in the extinction of vegetation fires in protected or hardly accessible areas, as well as the specific methodology for calculating the water resource available in periods of low humidity. The study area is the Danube River, in the lower sector, more precisely Balta Mică a Brăilei. This wetland is a characteristic area of the Danube before its sinking into the Black Sea. The research theme consisted of determining the characteristics of the water resource in this hydrographic area, to be used in dry periods to extinguish potential plant fires. Having thus organized a GIS database characteristic for the studied area, using the ArcSwat application, there have been created layers representing: the basins and the hydrographical network of this area, soil features, land-use characteristics, as well as the land slopes. ArcSWAT, which is the GIS interface for SWAT, was used in this study. ArcSWAT requires information on topography, soils, land use, slope categories, and weather data from the study region. Actual data was gathered to calibrate and validate the SWAT for the study area, resulting in a series of specific maps. The area is one that raises many challenges and problems in running the informational application (low slopes of land and water, difficult determination of reception areas). The research demonstrates the usefulness of GIS hydrological methods (in this case ArcSWAT), in assessing the characteristics of the hydrological regime of the rivers. The aggregation of hydrological information with other parameters leads to the identification of the optimal solutions to solving the tasks. Therefore, minimal anthropic intervention on the environment, protection of water resources and the environment, are objectives that lead to sustainable development of the current human society.

Chapter 14 shows mean diurnal cycles of various streamwater parameters in different catchments, river sections and time intervals in western and northern Romania. The study area is located in the Romanian Carpathians, and the Moldavian Plateau and 21 sites were used for measuring rivers. Water parameters used in this study are temperature, electrical conductivity, and pressure/level. Air parameters are temperature, relative humidity, and atmospheric pressure. The studied time interval ranges from 2011 to 2018 and is represented by compact measuring campaigns of a few weeks or months of continuous monitoring. The measurements were done every 60, 30, 15 or 10 min, depending on site and campaign. Authors applied a detrending to the time series, that was obtained by, firstly, calculating a relevant trend of the entire time series through an additive seasonal decomposition which uses a period equal to the length of a day and by, secondly, removing this trend from the raw time series. The water monitoring sites have revealed similar diurnal shapes of the temperature, level and electrical conductivity in areas with the natural flow: afternoon maximum water level and temperature and minimum electrical conductivity; the opposite events occur early in the morning. For both water and air parameters, the moments when minima occur are less temporally variable than those of maxima, at all sites. To sum up, chapter indicated that when using the detrended time series, the water temperature recorded diurnal cycles with amplitudes, ranging from 0.095 °C (site 21, Şugău River) to 2.978 °C (site 5, Bobîlna River).

Chapter 15 indicates that the regional climate is increasingly unpredictable, and the hydrological risk events (droughts and tidal waves) occur more frequently. In this case, infrastructures capable of water intake are required (either to compensate for the drought periods or to mitigate floods). For the past 25 years, drought periods have increased in length and severity, though, the mean amount of precipitations has augmented. The drought recorded in the autumn of the year 2011 and the spring of the year 2012 entailed a drastic reduction in power production provided by hydropower plants, reasons for which the company Hidroelectrica S.A. became unable to distribute power to beneficiaries. Hidroelectrica S.A. declared insolvency and made a change in the Energetic Strategy of Romania for the period between the years 2011–2035. In this case, there is the possibility of compromising the objectives assumed by Romania according to Annex I of the Directive, 2009/28/EC, which stipulates that the objective set for Romania for the year 2020, is to get 24% of the energy from renewable sources. Authorities are allowed to construct hydropower plants with an installed capacity of 1400 MW by the year 2035 when the national hydroelectric power potential will have been used up by 67% (59% by the year 2020). In such a scenario, the Hidroelectrica S.A. will consider increasing by 2035 the production capacity from 17.33 TWh (in a normal hydrological year) to 32.20 TWh. Several future projects aim to raise the capacity by 7.99 TWh. These include the 2.28 TWh representing ongoing projects that are to be finalized by 2015 (according to the Development Strategy of Hidroelectrica S.A.), a number of units on the Danube (totalling 4.66 TWh), and others with a micro-potential of 2.22 TWh.

To sum up, the proved that the drought recorded in the autumn of the year 2011, and in the spring of the year 2012, entailed a drastic reduction in power production provided by the hydropower plants and triggered an energy crisis at the level of entire Romania. Therefore, in the National Strategic Energy Project from the period 2020–2035 it is important to take into account scenarios similar to those produced by the hydrological drought in 2011–2012. Also it is necessary to increase the production capacity such that until 2035 the national hydroelectric power potential to be used at over 65%.

In Chap. 16, the authors organize in a concentrated way the very complex and extended thematic of Apuseni Mountains water resource. Being the pole of precipitations in Romania, this mountainous area represents a veritable natural obstacle against the air masses and a barrier for Transylvanian road and railway infrastructure. The petrographic composite is one of the most complex in Carpathian area, practically a mosaic of rocks, from the volcanic and metamorphic rocks who built the heart of this space, to limestones and sandstones who comes to complicate this mountainous structure. This substrate represents the adequate structure for a very important land surface drainage of the water and less for the groundwater accumulation. A big difference can be observed between the western and eastern slopes of the Apuseni Mountains. On the western slopes, the quantities of precipitations exceed easily 1000 mm when on the eastern slope this represents only over 700 mm. The liquid runoff follows this trend and presents the same discrepance: 800–900 mm versus 300–600 mm. Among the various parameters, used to emphasize the resource of water in a territory the volume of average flow, the average drained layer, etc. have been analyzed. Very important are the stocks of raw water, the grace of reservoirs built in the brain of the mountain, where the rocks offer practically the perfect substrate with a high rate of impermeability. The study closes with a water balance of this mountainous area, who prove that this area is a rainwater surplus one, regarding the water resources.

The book ends with the conclusions and recommendations chapter numbered 17.

Acknowledgements The writers of this chapter would like to acknowledge the efforts of all authors of the book chapters during the different phases of the book preparation including their inputs in this chapter.

References

1. Batinas RH, Sorocovschi V (2011) Resurse de apa, potential si valorificare turistica. Presa Universitara Clujeana, Cluj-Napoca
2. Bretcan P, Tampu MF (2008) The Razim-Sinoie lacustrine complex. Protection, resources, valorization. Lakes Reservoirs Ponds 1–2:99–112
3. European Commission (2012) Draft of the commission services. Member state: Romania accompanying the report document of the commission towards the European Parliament and the council on the implementation of the Water Framework Directive (2000/60/CE). Management plans of the hydrographical basins. http://ec.europa.eu/environment/water/water-framework/pdf/CWD-2012-379_EN-Vol25_RO_ro.pdf
4. Gâștescu P (2010) Resursele de apă din România, Potențial, calitate, distribuție teritorială, management. In: Water resources from Romania. Vulnerability to the pressure of man's activities, conference proceedings, pp 10–30
5. Romanescu G, Sandu I, Stoleriu CC, Sandu IG (2014) Water resources in Romania and their quality in the main Lacustrine Basins. Rev Chim (Bucharest) 63(3):344–349
6. Romanian Waters National Administration (2013) The national plan for the development of the hydrographical basins in Romania. Synthesis, revised version, Feb 2013, Bucharest
7. Tuchiu E (2011) Directiva Cadru Apa – directii de dezvoltare ale Planurilor de Management ale Bazinelor Hidrografice. In: Institutul National de Hidrologie si Gospodarire a Apelor, annual scientific conference, 1–3 Nov 2011, pp 55–68

Part II
Water Quality

Chapter 2
Implementation of EU Water Framework Directive (2000/60/EC) in Romania—European Qualitative Requirements

Gheorghe Romanescu, Cristian Constantin Stoleriu and Alin Mihu-Pintilie

Abstract The EC (European Community) enacted the Water Framework Directive (2000/60/EC) on September 23rd, 2000. This was aimed at maintaining and improving the biological and chemical status of natural waters until 2015. The main water resources in Romania are represented by surface waters such as rivers, lakes, and ponds, with the Danube River as main hydrographical axis and as a transboundary river. These water resources are influenced by the values of overland flows. About 94% of the drainage basins in Romania are located within the Danube catchment area (29% of its surface). With its water resources, Romania is ranked the 13th in Europe according to the yearly mean of water quantity distributed per resident (1840 m^3/person/year). In sum, there are several categories of waters in Romania, such as: 55,535 km of permanent rivers (representing 70% of all water courses), 23,370 km of perennial rivers (30%); 117 natural lakes larger than 0.5 km^2, with 52% of them located in Danube Delta; 255 reservoirs larger than 0.5 km^2, 174 km of transitional waters and 164 km of coastal waters. The ecological status of Romania's surface waters has been assessed as good and least good for about 64% of them, and only 2% have been classified as poor or bad. As regards the chemical status, about 93% of surface bodies of water have been declared good and only 7% in a bad status. The total length of rivers overlapping protected areas represents 15.3% of the total water courses length. There are 216 areas with protected habitats and species (occupying 14,437.26 km^2 and 6.1% of Romania's territory), where water is a major factor.

Keywords Anthropic impact · International Hydrographical District of Danube · Pollution · Surface bodies · Water quality

G. Romanescu · C. C. Stoleriu (✉)
Department of Geography, Faculty of Geography and Geology, Alexandru Ioan Cuza University of Iasi, Bd. Carol I 20A, 700505 Iasi, Romania
e-mail: cristoan@gmail.com

A. Mihu-Pintilie
Institute for Interdisciplinary Research, Science Research Department, Alexandru Ioan Cuza University of Iasi, St. Lascar Catargi 54, 700107 Iasi, Romania

© Springer Nature Switzerland AG 2020
A. M. Negm et al. (eds.), *Water Resources Management in Romania*, Springer Water,
https://doi.org/10.1007/978-3-030-22320-5_2

2.1 Introduction

The rational water references in the context of sustainable development show major importance concerning this topic, which was considered an inexhaustible resource for a long time. Today, unfortunately, it has become a limitative factor in the socioeconomic development of all States, including of the economically developed ones. Has water become a critical raw material? It appears that at the global level the answer is yes. However, the issue of the local management also includes cases of proper management: Norway, Finland, Canada, Russia, etc. The "lack" of water makes the studies related to resources and quality extremely numerous at international level [1–21] and national [22–52].

The European Community adopted on 23.10.2000 the 2000/60/EC Directive, which instituted the communitarian water policy. We take into account the achieving of a good biological and chemical status of natural waters (coastal ones included) until the end of the year 2015. Romania has been making great efforts for the water—an indispensable natural resource—to meet the standards imposed by the European Community. The old mining exploitations, the big iron and steel and the chemical factories continually polluted the water of most big rivers: Siret, Mures, Somes, Olt, Dambovita, etc. Nowadays, the water resources in Romania mainly comprise surface waters (rivers, lakes, to which we add the Danube as a main, but transboundary water artery) and they depend on the value of surface runoff. Unfortunately, the mean multiannual discharge of all Romanian rivers (except the Danube) is 1200 m^3/s. The Danube has a mean multiannual discharge of 6450 m^3/s, and it compensates for the water deficit within the south of the country. Climate and landforms impose the quantitative and qualitative value of water resources [53]. The temperate continental climate of transition with excessive nuances, specific to Romania, makes the precipitations insufficient for around 60% of the Romanian territory: Oltenia, Muntenia, Moldova, Dobrogea, Banat, etc. For the east of the country, the value of evapotranspiration is significantly higher than the average amount of precipitations. For this reason, some hydrographical arteries dry out.

The European Union promotes legislative instruments for the protection and sustainable management of water resources. The Framework Directive 2000/60/EC that defines the water as a heritage to be protected, treated and preserved represent the most important instrument from this point of view. The Directive ensures the framework necessary for the sustainable water management, and it involves the quantitative and qualitative water and aquatic ecosystem management. To achieve a "good status" of waters, we must ensure equal environmental living standards for all the inhabitants of Europe. The implementation instrument of the Framework Directive—regulated through Article 13 and Annex VII—is represented by the Management Plan of the Basin or the Hydrographical District. Based on knowing the status of bodies of water, target objectives are determined for a period of six years and measures are proposed for reaching a "good status" of waters [54].

The Framework Directive 2000/60/EC for the Danubian States outlined the elaboration of the Management Plan of the Danube Hydrographical District. The States that signed the International Commission for the Protection of the Danube River (ICPDR) established for the Management Plan of the Danube Hydrographical District to comprise:

- part A—the general plan that comprises basin problems with transboundary effects (the main streams of the rivers with hydrographical basins >4000 km^2; lakes measuring >100 km^2; the main channels; transboundary aquifers measuring >4000 km^2; the Danube Delta and the coastal waters);
- part B—the national management plan of the Danubian countries.

The groundwater resources in Romania are insufficiently exploited (4–6%), and they are mainly destined to the supplying of big urban centres (Iasi, Ploiesti, Craiova, Brasov, Sibiu, etc.). The most important groundwater resources are affected by mineralization (saltification—the Moldavian Plateau, the Transylvanian Depression); they are hard waters (Dobrudja, the Mehedinti Plateau, the Moldavian Plateau, etc.) or they are at great depths (Baragan, Oltenia).

The faulty character in water resources of hydrographical basins in Romania makes the accent on exploitation be very strong. The post-Communist period led to the fall of the energy- and water-consuming industrial branches: iron and steel, chemistry, agriculture, etc. In this case, the water demands decreased visibly. Recently, an invigoration has been recorded for the agricultural sector, where irrigated fields have increased significantly, mostly in Baragan, the Moldavian Plateau (the county of Vaslui), Oltenia and Dobrudja. Romania has passed from 3 million ha in the Communist period to 300,000 irrigated hectares today. However, the trend is rapidly increasing. This evolution may give the illusion of better security in the water supply. Nonetheless, numerous challenges subject the Romanian economy to securing the water supply of cities and villages that have been neglected for a long time. For example, most villages are not connected to the centralized water supply system. The perennial character of a global water supply threatened by increased aridity and the invigoration of big water-consuming industrial branches (agriculture, chemistry); meeting the environment and quality standards, etc.

In what terms are these challenges expressed? What are the environmental constraints and to what extent is a new approach to water resources management necessary? Is Romania capable of meeting the future need for water? The topic related to water management and to the development of hydrographical basins is destined to researches in the field of water resources and quality, of territorial planning, of hydrotechnical constructions, of environmental geography, etc. This study outlines the contemporary reality of water resources and the stage of their quality.

2.2 Methodology

The hydrological data were provided by the National Institute of Hydrology and Water Management in Bucharest and by the local Water Basin Administrations (Siret, Prut-Barlad, Jiu, Arges-Vedea, Mures, Banat, Crisuri, Somes-Tisa, Olt, Buzau-

Ialomita, Littoral Dobrudja). The entire data row was completed with the information collected from the scientific literature published in the national and international journals [53, 55–91].

The graphic materials were edited using licensed software programs. The field trips were conducted in the period 2010–1017, and they concerned the big hydrographical arteries and the most important lacustrine basins (natural and artificial). The database of Siret and Prut (Moldova) hydrographic basins was obtained by the authors based on field work. Most of the measurements were focused on the water bodies with risk of pollution. The results were published in the Romanian and foreign literature.

The samples were collected in conformity with the manual of water quality monitoring system elaborated by the specialists of the National Institute of Hydrology and Water Management in Bucharest. The chemical and biological analyses were conducted in the Geoarchaeology Laboratory within the Faculty of Geography and Geology Iasi and within the CERNESIM Laboratory of Alexandru Ioan Cuza University in Iasi. For expeditionary measurements HACH Drel/2010 multi-parameter was used. Through the devising of management plans for each attachment, in agreement with the EC Water Framework Directive 2000/60, an abiotic typology of rivers, based on natural conditions (altitude, mean depth, geology, retention time), was established. For each river, the pollution-related pressures (punctiform, diffuse and hydromorphological) are identified. An assessment of the anthropic impact is carried out through the analysis of the groups of chemical indicators based on which the quality classes in which each sector of the river is included, along with the ecological state of the water, are established. The global analysis of water chemistry and water quality may be the result of the interaction between two study models:

– according to the Normative on the establishment of quality classes for surface water bodies so as to determine their ecological state (Ord. MMGA no. 161/2006), which states that a river is a static ecosystem, therefore it operates with the absolute values of the results of the analysis, five water quality classes are carefully delineated based on the stipulated limits;
– the river is regarded as a dynamic ecosystem, therefore it operates with values that are average, maximum or minimum, compared with the quality indicators of the current period (2017) and those of 2010 [92].

2.3   Results and Discussions

The hydrographical district of Romania was graded using RO1000 (Figs. 2.1 and 2.2) [93]. 97.4% of the surface of Romania is part of the International Hydrographical District of the Danube (IHDD) accounting for 29% of its surface. The Romanian part of the IHDD comprises 11 sub-hydrographical basins and 4 categories of surface

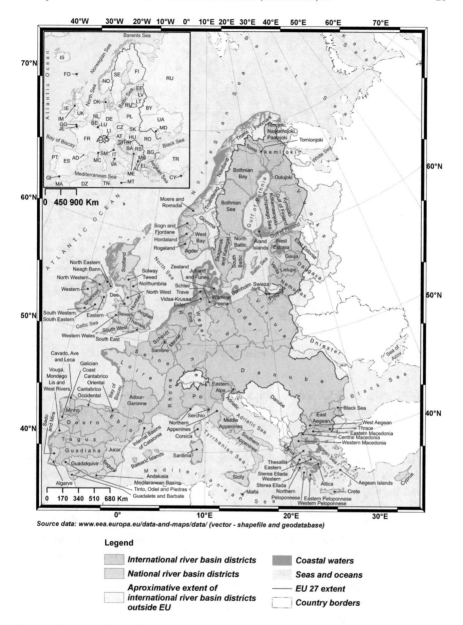

Source data: www.eea.europa.eu/data-and-maps/data/ (vector - shapefile and geodatabase)

Legend

International river basin districts	Coastal waters
National river basin districts	Seas and oceans
Aproximative extent of international river basin districts outside EU	—— EU 27 extent
	Country borders

Fig. 2.1 The map of the national and international basin districts in Europe

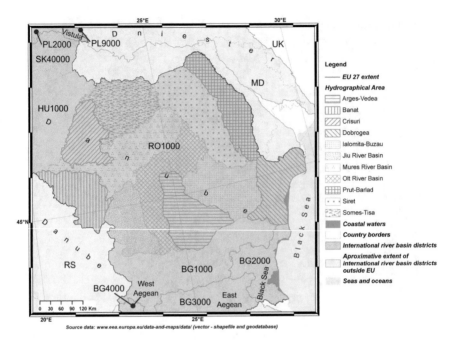

Fig. 2.2 The map of the hydrological district RO1000

bodies of water: rivers, lakes, transitory waters and coastal waters (Table 2.1). For each category of water, there are rivers—20 types of streams (4 temporary); lakes—18 types of natural lakes, 14 types of reservoirs; transitory waters—2 types; coastal waters—2 types.

In Romania, 3399 surface bodies of water were identified, of which 3262 water river bodies. The delimitation of bodies of water is based on the category, the typology and the physical characteristics of surface waters. In the category of additional criteria, one may include the status of waters and the hydromorphological alterations. After the year 2008, the biological criteria were also added. The small bodies of water (rivers with a capturing area <10 km², lakes smaller than 50 ha) were grouped into a single body of water because they are subjected to the same pressures, they are in the same state, and they have the same type. In the small hydrographical basins, the entire river can be considered one body of water if it is not affected or influenced by a certain type of pressure (e.g., hydroenergy, water capturing, agriculture, industry, etc.) (Table 2.2) [93].

The following pressures on the bodies of water were identified: point and diffuse pollution sources; land use; hydromorphological alterations; infrastructure projects; sources with high accidental pollution risk; pisciculture/aquaculture activities; sand and ballast from riverbeds; woodland exploitations, etc. (Table 2.3; Fig. 2.3) [93].

The following are significant diffuse pollution sources: the human agglomerations without wastewater collection systems or proper mud collection and elimination systems from the treatment stations; the agro-zootechnical farms without

Table 2.1 Transboundary hydrographical basins by category and weighting in Romania, expressed in percentages (the category 1: cooperation agreement, cooperation body, existing PMBH; the category 2: cooperation agreement, existing cooperation body; the category 3: existing cooperation agreement; the category 4: there is no official collaboration) [93]

Hydrographical district/hydrographical sub-basin	Name	Surface (km^2)	% of the Romanian territory	Countries with common borders
RO1000	The Danube	238,391	–	BG, HU, MD, RS, UA
Hydrographical sub-basin				
SO	Somes-Tisa	22,380	9.4	HU, UA
CR	Crisuri	14,860	6.3	HU
MU	Mures	28,310	11.9	HU
BA	Banat	18,393	7.7	RS
JI	Jiu	16,676	47.1	–
OT	Olt	24,050	10.1	–
AG	Arges-Vedea	21,479	9	–
IL	Buzau-Ialomita	24,699	10.1	–
SI	Siret	28,116	11.9	UA
PR	Prut-Barlad	20,267	8.5	MD, UA
DL	Dobrudja-Litoral	19,161	8	BG, MD, RS, UA

proper dejection storage/usage systems; the localities identified as vulnerable zones to pollution from nitrates from agricultural sources; the industrial deposits; the non-complying waste storage systems; the abandoned industrial sites, etc. The point and diffuse pollution sources contribute to chemical pollution. The main pressures are represented by the human agglomerations, industry, agriculture and land use manner. The types and size of the hydromorphological pressures were defined based on the recommendations within the CIS orientation Document No. 3—Pressures and impacts (IMPRESS) and on the criteria of the regional UNDP-GEF project of the Danube, which take into account the hydrotechnical works, the magnitude of pressure and their effects on the ecosystems. The hydromorphological pressures are as follows: dams; spillways; regulations and damming; derivations; bank defenses; navigable paths/channels; water captures/restitutions; navigation; future infrastructure projects, etc. [93]. On the Romanian territory, over 2500 protected areas were identified, of which 1879 areas are destined for drinking water capturing (Table 2.4).

The surface and groundwaters are monitored within surveillance programs and separate operational programs, drafted up in conformity with the demands of the Framework-directive regarding the water (Fig. 2.4; Table 2.5). The monitoring program is permanent. All the monitoring points were defined as surveillance points. The networks for the surveillance monitoring and for operational monitoring are identical. A monitoring point may belong to several programs for the surface waters,

Table 2.2 Size of surface and groundwater bodies [93]

| Hydrographical district | Surface waters | | | | | | | | Groundwaters | |
| | Rivers | | Lakes | | Transitory | | Coastal | | | |
	No.	Average length (km)	No.	Average surface (km^2)	No.	Average surface (km^2)	No.	Average surface (km^2)	No.	Average surface (km^2)
RO1000	3262	23	131	8	2	391	4	143	142	1857

Table 2.3 Number and percentage of surface bodies of water affected by significant pressures [93]

Hydrographical district	Without pressure		Point source		Diffuse source		Water capturing		Discharge regulations and morphological alterations		River management		Transitory and coastal water management		Other morphological alterations		Other Pressures	
	No.	%	No.	%	No.	%	No.	%	No.	%	No.	%	No.	%	No.	%	No.	%
RO1000	1914	56.31	260	7.65	1105	32.51	49	1.44	445	13.09	115	3.38	2	0.06	6	0.18	78	2.29

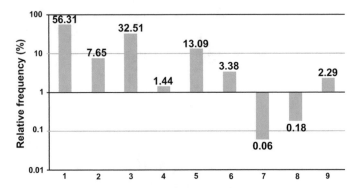

Fig. 2.3 Surface bodies of water affected by significant pressures. 1 = without pressure; 2 = point source; 3 = diffuse source; 4 = water capturing; 5 = discharge regulations and morphological alterations; 6 = river management; 7 = transitory and coastal water management; 8 = other morphological alterations; 9 = other pressures

and it includes several monitoring sub-points. A body of water may have one or more monitoring stations, but the quality elements are monitored at the representative station [93].

The surface waters in Romania were evaluated as having a good or a better ecological status in a proportion of 64% (Fig. 2.5). Only 2% of the surface bodies of water are in a low or bad ecological status. The chemical status of surface bodies of water in Romania is good in a proportion of 93% (Fig. 2.5). Only 7% is evaluated as having a low chemical status [93].

Measurements can be taken based on assessing the status, and they are submitted to the approach of the International Commission for the Protection of the Danube River (ICPDR). The common program comprises measures with importance at basin level, and it represents more than a list of national measures. The common program includes specific measures for restoring the continuity of rivers, for reducing the level of pollution from nutrients, organic substances, and hazardous substances. The measures at the level of hydrographical sub-basin or body of water are featured in 11 management plans specific to sub-basins. The Romanian Waters National Administration is in charge of monitoring the implementation of the program of measures (PM), as well as with reporting the stage of its implementation [93].

Agriculture is a source of point and diffuse source of pollution from nutrients, organic substances, and hazardous substances. Agriculture pollutes through the animal farms that do not benefit from dejection recycling/storage facilities, through the households without a sewerage system, through the use of fertilizers or pesticides, etc. The capturing of the water for irrigations may represent a basin pressure.

The hydromorphological measures focus on restoring lateral and longitudinal connectivity:

– reconnecting the lateral branches for the Lower Danube;
– the construction of retention basins;
– operational modifications for the pulsating effect of the waves;

Table 2.4 The number of protected areas in each hydrographical district and at the level of the entire country for the surface and groundwaters [93]

Hydrographical district	Number of protected areas										
	Art. 7 drinking water capturing	Bathing	Birds	Others (European)	Fish	Habitats	At local level	At national level	Nitrates	Molluscs	UWWT
RO1000	1879	35	106	–	12	213	–	381	42	4	–

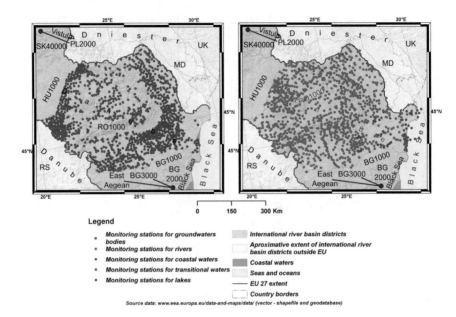

Fig. 2.4 Monitoring stations of surface waters (left) and of groundwaters (right) [93]

- dredging reduction or modification;
- cross-basin adductions;
- restoring floodplains; the improvement of the hydrological conditions of channels;
- restoring habitats;
- diversifying the structure of banks;
- creating buffer bands along river sectors in order to reduce diffuse pollution;
- building passages for the ichthyofauna to pass;
- hydrotechnical works for improving the circulation of the water along certain channels within the Danube Delta;
- the improvement of the hydrological regime downstream from the reservoirs, etc. (Fig. 2.6) [93].

For the groundwaters, preventative measures have been set in place:

- it is prohibited to evacuate pollutants directly in the groundwaters;
- regulation of evacuations from point sources that may produce pollution;
- measures for preventing the significant pollutant leaking from technical installations;
- the prevention/reduction of the impact of accidental pollutions;
- the modernization of the wastewater treatment stations through the introduction of a new stage;
- extending the aeration procedure in the biological stage;
- improving the quality of the effluents;
- installing an automatic system of monitoring the quality of wastewater;

Table 2.5 The number of monitoring points by water category [93]

Hydrographical district	Rivers		Lakes		Transitory		Coastal		Underground		
	Surveillance	Operational	Surveillance	Operational	Surveillance	Operational	Surveillance	Operational	Surveillance	Operational	Quantitative
RO1000	1263	547	434	228	12	12	42	42	2365	1224	3338
Total per type of site	1263	547	434	228	12	12	42	42	2365	1224	3338
Total number of monitoring points	1263		434		12		42		3397		

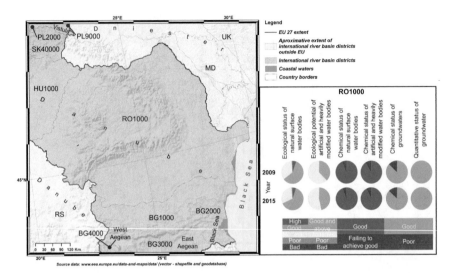

Fig. 2.5 Ecological status, ecological potential, chemical status and quantitative status of natural, anthropic and heavily modified surface bodies of water and of groundwaters in 2009 and foreseen for 2015 (according to the European Commission [93])

Fig. 2.6 The Crisan channel as a bridge between the Sulina branch and the Rosu-Lumina lacustrine complex

– neutralizing the mining wastewater before spilling;
– closing the mining areas;
– applying the code of agricultural good practices to limit the pollution from pesticides, etc. [93].

There are no basic or additional measures for the groundwaters because the quantitative status of bodies of water was good. The general additional measures refer to 19 groundwater bodies with low chemical status:

– sewerage systems for agglomerations <2000 inhabitants;
– the use of action plans and of the code of agricultural good practices in the non-vulnerable areas;
– the use of ecological agriculture, the additional monitoring of the lists of hazardous/priority substances in the surface waters, the groundwaters and the wastewaters;
– studies regarding the ecological reconstruction of the former mining areas;
– the evaluation of the annual amount of contaminants and of their impact on the quality of the water;
– the rehabilitation of the wastewater treatment stations;
– the ecological reconstruction of the polluted areas, etc.

For the transboundary groundwater bodies, there are agreements with Hungary, Bulgaria, Serbia and the Republic of Moldova. There are no agreements with Ukraine. The international coordination was ensured by the International Commission for the Protection of the Danube River (ICPDR) for 8 groundwater bodies considered to be of transboundary importance [93].

The measures concerning chemical pollution refer to: (a) the reduction of pollution from priority substances and other substances in the surface waters; (b) the preliminary regulation of the evacuations from point sources that may produce pollution; measures to prevent pollution from diffuse sources (agriculture, industry, households); (c) prohibiting of the direct input of pollutants into the groundwaters; preventing the significant losses of pollutants from the technical installations and preventing/reducing the impact of accidental pollutions; (d) building wastewater treatment facilities (e) sensitizing the public opinion; (f) restoring habitats to consolidate the purification function of the natural ecosystems; preventing/reducing the significant losses of contaminants from the technical installations; (g) improving the sewerage system connection of human agglomerations; (h) restoring floodplains; (i) creating buffer bands along rivers; the construction of retention basins or the modernisation of the wastewater treatment stations; (j) extending the sewerage system in order to connect it to the industrial one; and (m) stabilizing the waste deposits, etc. [93].

The lack of water is the most serious global issue of humankind in an early 21st century. At the Dublin Conference on Water and the Environment (January 1992) and at the United Nations Conference on Environment and Development (July 1992) the following principles regarding water were applied: the basin principle; the principle of unitary quantity-quality management; the principle of solidarity; the principle of the pollutant paying; the economic principle—the beneficiary pays; the principle of access to water. The listed principles are the foundation for the concept of integrated management of water resources, which combine the use of water and the protection of natural ecosystems through integration at the level of the basin of water usages.

The integrated management of water resources (IMWR) promotes the development and coordinated management of the water, the land and the resources, to optimize a balanced social and economic development, without compromising the sustainability of ecosystems. The concept of integrated management of water resources implies—in opposition to the traditional management of water resources—their integrated approach, on both a physical and a technical level, and at the level of planning and management. The integration level is represented by the hydrographical basin, a natural unity of the formation of water resources.

The system of water resources—formed by both the natural system and the water management infrastructure and the administrative structure—ensures goods and services for water usages, as well as the preservation of aquatic ecosystems. The system of water resources ensures the water depending on the quantitative and qualitative requirements expressed by usages that pay for the service ensuring the water.

The system of water resources and that of water usages interact with the economic and social system, but also with the environment. From this point of view, the connections are twofold:

- the system of water resources and the water usages have an impact upon the development of economic and social activities and upon the quality of the environment;
- society and the environment react to this impact through legislative, technical, economic and administrative measures in order to reduce the negative effects to ensure a durable economic development and the conservation of the environment.

The planning and management of water resources are based on the technical and the economic requirements, meant to ensure the water supply of the usages and protection of aquatic ecosystems. The most important economic regulations concern the fees, rates, penalties, and bonuses.

The sustainable development of the systems of water resources comprises the following aspects:

- the durability of physical aspects—maintaining the natural circuit of water and nutrients;
- the technical durability—equilibrating the water demand-supply balance;
- the environmental durability—"zero tolerance" for the pollution that exceeds the self-treatment capacity of the environment. In this case, there are no long-term effects or irreversible effects upon the environment;
- the social durability—preserving water demands and the will to pay services for ensuring water resources;
- the economic durability—the economic support of the measures ensuring a high living standard in the field of water;
- the institutional durability—maintaining the capacity of planning, management, and operation of the system of water resources.

The sustainable management of water resources is based on their integrated management. The services performed by the system of water resources satisfy the current objectives of the society without compromising the ability of the system [94].

The integrated management of water resources has several components:

a. The integration of the system of natural water resources.
    It is represented by the hydrological cycle and its components: precipitations, evaporation, surface runoff, groundwater runoff. The maintaining of the hydrological assessment and of the relations between its components is based upon the biophysical links between the forest, the land, and the water resources within a hydrographical basin.
b. The integration of the infrastructure of water resources management within the natural capital and the elaboration of an environment-friendly infrastructure of water management. It ensures the optimal water supply of the usages; it reduces the risk of flood occurrence; it preserves and maintains the increase in the biodiversity of aquatic ecosystems.
c. The integration of water usages.
    The water supply of the household field, the agriculture and the industry, as well as the preservation of aquatic ecosystems, is approached from a sectoral perspective, traditionally. Most water usages solicit water resources in increasing amounts and of very good quality. The solution to the resources-water demand equation and to water protection resources requires the analysis of the usages at the level of the hydrographical basin. The decisions and actions within the field of the integrated management of water resources must be taken by all parties affected (the principle of subsidiarity).
d. The upstream-downstream integration.
    The usages upstream must acknowledge the rights of the usages downstream regarding the use of water resources of higher quality and in sufficient amounts. The excessive pollution of water resources by the upstream usages leads to increased costs and discomfort for the downstream usages.
e. The integration of water resources in the planning policies.
    Water can be considered a limiting factor in the development of the society.

The integrated management of water resources is based upon the Hydrographical Basin Management Plan (in conformity with the provisions of the Framework Directive 2000/60 of the European Union). Based on learning the status of bodies of water, the Management Plan determines the target objectives for six years, and it proposes, at the level of the hydrographical basin, measures for attaining a higher quality of water with the purpose of sustainable usage [94].

2.3.1 The Hydrographical Basin Management Plan

In accordance with the provisions of the Law on Waters No. 107/1996, with the subsequent amendments and supplements, guidelines were elaborated by basins or groups of hydrographic basins until 22nd of December 2009. The aim was to establish the fundamental orientations regarding the sustainable, unitary, balanced and complex management of water resources and aquatic ecosystems, as well in order to protect the wetlands [45, 46, 76, 95, 96].

The guidelines set the water quantity and quality objectives, meant to ensure:

- a good status of surface waters or, for the artificial or highly modified bodies of water, a good ecological potential and a good chemical status of surface waters;
- a good chemical status and a balance between abstraction and recharge of groundwater resources;
- the optimal water supply for the usages;
- the reduction of the negative effects of waters due to floods, droughts and accidental pollutions.

The Guideline of Development and Management of the hydrographical basin is the planning tool in the field of waters on the hydrographical basin, and it comprises two parts: the development plan of the Hydrographical Basin (DPHB) and the Hydrographical Basin Management Plan (MPHB).

Due to the increased pressure on water resources, the European Union has promoted legislative instruments for their protection and sustainable management. All the European inhabitants will attain the same living standards regarding the aquatic environment [97]. The implementation tool of the Framework Directive is the hydrographical basin (District) Management Plan that—based on knowing the status of bodies of water—establishes the target objectives and proposes the measures stipulated for attaining a "good status" of waters.

The Romanian territory is 97.4% situated in the hydrographical basin of the Danube, meaning 29% of its surface. The specific resource within the domestic rivers is 1840 m^3/in. year, which ranks Romania the 13th in Europe. Romania has the following categories of waters: permanent rivers 55,535 km, accounting for 70% of all streams; non-permanent rivers 23,370 km, accounting for 30% of all streams; natural lakes 117 with a surface exceeding 0.5 km^2, of which 52% are in the Danube Delta; reservoirs 255 with a surface exceeding 0.5 km^2; transitory waters 174 km; coastal waters 116 km [54]. The transitory waters are surface bodies of water from the river mouths, partially salinized by the seawaters, but heavily influenced by freshwaters (Table 2.6).

The Romanian hydrographical network comprises 78,905 km in length and the water resources within the domestic rivers are 40 billion m^3, namely 20% of the water resources in the Danube. The specific resource within the domestic rivers is 1840 m^3/in. year (the 13th in Europe). Water management by hydrographical basins debuted on the Romanian territory as late as 1959. The hydrographical basin of the Danube comprises a surface of 232,193 km^2 of the Romanian surface, namely 97.4%. The same basin also includes the coastal waters as well as the basins of the tributaries discharging in the Black Sea (5198 km^2), thus forming the Danube Hydrographical District (in conformity with the provisions of the Framework Directive 2000/60/EC). The coastal waters, also influenced by those of the Danube, are delimited to a nautical mile (1852 m) to the shoreline.

Table 2.6 Basins/Hydrographical spaces and coastal waters on which the Management Plans are elaborated

No.	Basins/Hydrographical spaces	Surface (km^2)	% of the Romanian surface
1	Somes-Tisa	22,380	9.42
2	Crisuri	14,860	6.26
3	Mures	28,310	11.93
4	Banat	18,394	7.74
5	Jiu	16,713	7.05
6	Olt	24,050	10.14
7	Arges-Vedea	21,479	9.04
8	Buzau-Ialomita	23,874	10.05
9	Siret	28,116	11.85
10	Prut-Barlad	20,267	8.53
11	Dobrudja-Littoral (Danube, Delta, Hydrographical space Dobrudja + Coastal waters)	18,949 + 1130	7.98
Total	Romania	237,391 + 1130	100

The elaboration of the management plans is done in phases:

– phase 2004: the elaboration of Chaps. 1–5 and Sects. 6.1 and 6.1.2 of the Hydrographical Basin Management Plan;
– phase 2007: elaboration of the integrated monitoring system of waters (Chap. 7), in conformity with the provisions of the Framework Directive and the other European directives;
– phase 2008: the elaboration of the Hydrographical Basin Management Plan through Chaps. 1–13 based on the data provided by the new integrated monitoring system of waters. The publication of the first draft of the Plan and its submission for information and consultation by the public;
– phase 2009: the publication of the Hydrographical Basin Management Plan and its approval by the Romanian Government;
– phase 2009–2015: the implementation of the Hydrographical Basin Management Plan to attain the good status of waters in Romania;
– phase 2013: the analysis of the elaboration of the Hydrographical Basin Management Plan and the beginning of the elaboration of the second Hydrographical Basin Management Plan to be implemented until 2021.

The cycle of planning and of elaboration and implementation of strategy in the field of water management continues every six years, each cycle with specific objectives. The first Hydrographical Basin National Management Plan was completed in 2009. The Framework Directive Water has a profoundly scientific character. Its implementation creates a legal and operational framework of integrated management of water resources at the level of the hydrographical basin. The Basin Management Plan is in close correlation with the socioeconomic development and it features the starting point for the measures corresponding to anthropic activities, including the measures of water management at the basin and local level [90, 91]. The Framework Directive Water in Romania respects the timescale and the implementation phases at European level. The following were analyzed: the characterization of hydrographical basins; the impact of human activities; the economic activity. The following were identified, analyzed and evaluated: the categories of surface water; the eco-regions, the typology and reference conditions for the surface waters; the significant anthropic pressures and their impact on the water resources; the surface bodies of water; the groundwater bodies; the protected areas [91].

The typology of streams was redefined and synthesized, leading to the reduction of the number of types. Twenty types of streams were defined nationwide, with sub-types differentiated depending on the geological factor (Fig. 2.7). The number of types of natural lakes is 18. For the reservoirs, 14 types were defined, with sub-types differentiated depending on retention time. The transitory and coastal waters were differentiated into two specific types each, depending on the typological character-istics. The application of delimiting criteria for the bodies of water has led to the identification of 3399 surface bodies of water [3004 bodies of water-rivers (1184 non-permanent bodies of water), 164 bodies of water-reservoirs and 130 bodies of water-natural lakes, 95 artificial bodies of water, 2 transitory bodies of water—one lacustrine and one marine] and 4 coastal bodies of water (Fig. 2.8) [91].

Many (1764) point sources of pollution were analyzed, which led to 947 signif-icant point sources (436 urban, 325 industrials, 181 agricultural and 5 belonging

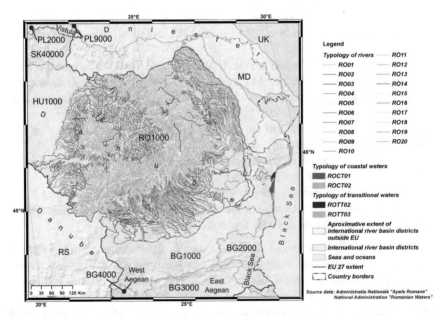

Fig. 2.7 The typology of streams, of transitory waters and of coastal waters (processing after Tuchiu [91])

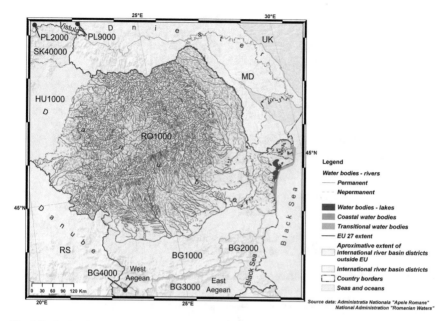

Fig. 2.8 Surface bodies of water in Romania (processing after Tuchiu [91])

to other types). The human agglomerations counting over 2000 inhabitants are the most important significant sources of pollution (2605 human agglomerations). The diffuse sources, generally the human agglomerations and the agricultural activities, contribute to the pollution of surface waters. Specific azote and phosphorus emissions were determined: 5 kg N/ha and 0.4 kg P/ha. The hydromorphological alterations, mostly the hydrotechnical works of transverse damming and those along riverbeds, affect significantly the ecological status of the bodies of water [91].

In order to assess the impact and the risk of failing to achieve the environmental objectives, the following categories of risk were taken into account: the pollution from organic substances; the pollution from nutrients; the pollution from hazardous substances; hydromorphological alterations. On the Romanian territory, there are 1241 bodies of water, accounting for 36.5% of all bodies of water, which involve the risk of failing to attend the environmental objectives in the year 2015. A number of 3399 bodies of water were analyzed and characterized from the perspective of the global status, of which 2008 bodies have very good status/maximum potential and good status/good potential (59.08%) (Figs. 2.9, 2.10 and 2.11). From the ecological perspective, on the Romanian territory, there are natural bodies of water (82%), heavily modified bodies of water (15%) and artificial bodies of water (3%) [91].

The pressures acting upon the bodies of water are the same as for surface bodies of water. Thus, 142 groundwater bodies were identified and delimited, of which 86.6% achieve the good qualitative status and 13.4% fail to achieve the good status from the

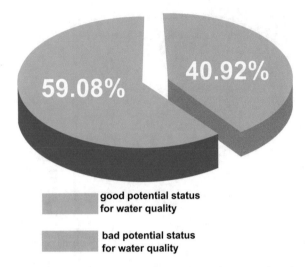

Fig. 2.9 The global status of surface bodies of water in Romania (processing after Tuchiu [91])

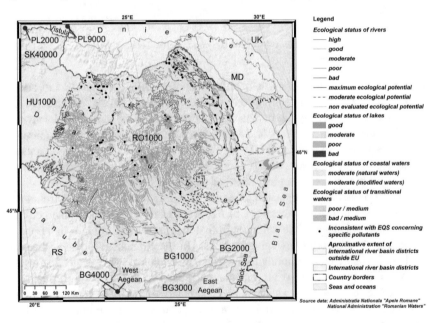

Fig. 2.10 Ecological status/ecological potential of surface bodies of water in Romania (processing after Tuchiu [91])

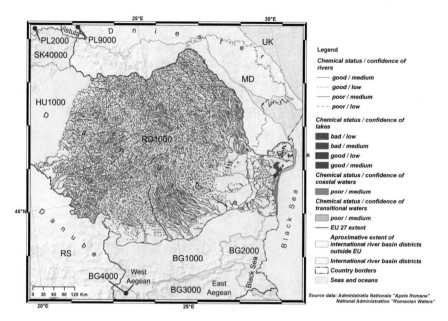

Fig. 2.11 Chemical status of surface bodies of water in Romania (processing after Tuchiu [91])

qualitative (chemical) perspective; there are also exceptions, of course (Figs. 2.12 and 2.13). For 15 groundwater bodies, the trends were evaluated as upward, while for four bodies of water, the trend is downward.

The following categories of protected areas were identified and mapped:

– protection areas for water captures for drinking water;
– areas for protection of the important aquatic species from the economic perspective;
– areas for the protection of habitats and of the species where maintaining or improving water status is an important factor;
– areas sensitive to nutrients and areas vulnerable to nitrates;
– areas for bathing [91].

There are 269 surface water captures: 60% have protection areas ensured. For groundwaters, there are 1617 water captures: 85% have protection areas ensured. The length of the areas for the protection of the important species is 1795 km. For the Black Sea, four areas were appointed for the increase and exploitation of the molluscs, with a total surface of 567 Mm^2. The areas destined for the protection of habitats and species, where maintaining or improving the status of water is an important factor, accounting for around 20% of the Romanian territory. Also, the Romanian territory was identified as an area sensitive to pollution from nutrients. The areas vulnerable to the pollution from nitrates accounts for about 58% of the Romanian territory. Fifteen bathing areas (35 sectors) were identified, monitored and evaluated from the perspective of the water quality for the Black Sea [91].

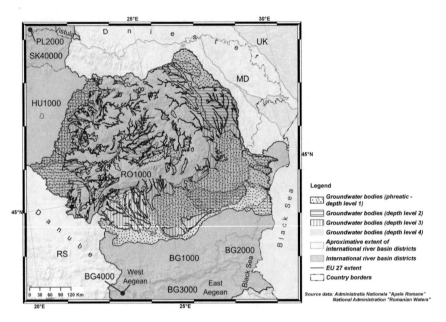

Fig. 2.12 The groundwater bodies in Romania (processing after Tuchiu [91])

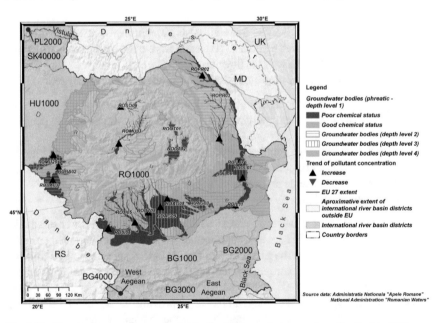

Fig. 2.13 Chemical status of groundwater bodies in Romania (processing after Tuchiu [91])

The bodies of water are represented by the water mass belonging to certain lakes (natural or artificial), streams (river, channel, river sector), transitory waters and coastal waters. They are differentiated through natural or anthropic characteristics. The criteria used for the delimitation of bodies of water were the following: categories of surface water; the typology of surface waters; the physical-geographical and hydromorphological characteristics of the basin; the pressures and the state of surface waters; the limits of the protected areas. 3717 bodies of water were delimited, of which eight for the Danube Delta. The average length of bodies of water is 21.3 km. Out of the total, 37% are accounted for non-permanent bodies of water. The average length of a body of water in Romania is 21.3 km, comparable with the lengths within other Danubian countries: 26 km in Germany, 12 km in Austria, 17.9 km in Hungary, etc. [97].

A body of water must belong to only one category of water and, thus, it cannot include a river and a lake. A body of water does not criss-cross the limits of another body of water, and it must have the same quality. Confluence may represent the limit of a surface body of water. The body of water must have the same quality status (very good, good, moderate, satisfactory, unsatisfactory). The chemical and hydromorphological pressures section the body of water into smaller bodies. The river sectors on which hydromorphological alternations are recorded are the heavily modified bodies of water. The anthropic intervention creates artificial bodies of water. The protected areas of an aquatic nature represent independent bodies of water.

The environmental objectives for the surface waters, in conformity with Article 4 of the Water Framework Directive 2000/60 of the European Union are as follows:

– preventing the deterioration; protecting and restoring until achieving a good status of waters;
– protecting and restructuring the artificial and heavily modified bodies of water in order to achieve a good ecological potential and a good chemical status of waters;
– progressively reducing pollution.

The monitoring of surface waters is provided by specific methodologies (Table 2.7).

The monitoring frequency for the protected areas that ensure >100 m^3/day is determined by the number of inhabitants ascribed to the centralized supply system (Table 2.8).

The groundwater bodies are represented by an aquifer, several aquifers or a part of an aquifer. They are monitored using the multiparameter electronic sensors. They are identified based on the following criteria: geological (nature of the deposits, age, structure); hydrodynamic (extension of the geological limit of the hydrographical basin—for the phreatic); status (quantitative and qualitative). The bodies of water can also be transboundary. The delimitation is only done for the aquifers with economic importance (discharges exceeding 10 m^3/day). The other aquifer deposits do not represent bodies of water.

The process of identifying highly modified bodies of water is based on the biological data. Where biological data were not available, abiotic criteria were also used for temporary identification of highly modified bodies of water. These are based

Table 2.7 The minimum frequencies of monitoring the characteristics of surface waters

Elements	Rivers	Lakes	Transitory waters	Coastal waters
Biological				
Phytoplankton	Biannually	Biannually	Biannually	Biannually
Other aquatic flora	Every 3 years	Every 3 years	Every 3 years	Every 3 years
Macro-invertebrates	Every 3 years	Every 3 years	Every 3 years	Every 3 years
Ichthyofauna	Every 3 years	Every 3 years	Every 3 years	
Hydromorphology				
Longitudinal profile	Every 6 years			
Hydrological elements	Continually	Monthly		
Morphological elements	Every 6 years	Every 6 years	Every 6 years	Every 6 years
Physical-chemical				
Temperature	3 times a year	3 times a year	3 times a year	3 times a year
Oxygen regime	3 times a year	3 times a year	3 times a year	3 times a year
Salinity	3 times a year	3 times a year	3 times a year	3 times a year
Nutrient regime	3 times a year	3 times a year	3 times a year	3 times a year
pH, alkalinity, bicarbonates, carbonates, hardness	3 times a year	3 times a year		
Other pollutants	3 times a year	3 times a year	3 times a year	3 times a year
Priority substances	Monthly	Monthly	Monthly	Monthly

Table 2.8 The monitoring frequency for the water sources in the protected areas

Number of inhabitants	Frequency
<10,000	4 times/year
10,000–30,000	8 times/year
>30,000	12 times/year

on types of hydrotechnical works and their effects on aquatic ecosystems. For the basins/hydrographical spaces in Romania, the following resulted: 413 highly modified bodies of water (17.6%); 365 bodies of water candidate for highly modified (15.5%); 1492 bodies of water that are not highly modified (63.6%); 77 artificial bodies of water (3.3%) [54, 98].

Among the major effects of the impact of human activities on water resources with significant economic and social implications, the following are worth mentioning:

– The degradation of the quality of waters: 43.7% of the bodies of water risk not achieving the environmental objectives in conformity with the provisions of the Framework Directive 2000/60.

The eutrophication of certain lakes situated on the interior streams and especially of the bodies of water in the Danube Delta and of the coastal ones, due to the important amount of nutrients carried by the Danube. The eutrophication of coastal waters is also determined by certain sources on the Romanian Black Sea coast (The Petrochemical Factory at Navodari, the Naval Sites of Constanta-Agigea, Mangalia, etc., the dejections from coastal localities, etc.), which are reduced and which have strictly local influence. The eutrophication phenomenon was frequent in the period 1970–1990 and more rarely after this period.

– The decrease in the biodiversity of aquatic flora and fauna due to: the pollution of waters and the alteration of habitats because of the hydromorphological pressures; the change of structure by species, mostly of the ichthyofauna within the Danube Delta and the Black Sea, currently dominated by crucian (a species with low economic value); the reduction of migrating fish species.
– Shore erosion on around 127 km (57% of the length of the Romanian Black Sea coast). It is caused by the increased sea level and by the decreased amount of alluvia transported by the Danube. To them, we add the reduction of biogenic sand due to a decrease in the population of molluscs due to the pollution of waters.

The intensity of erosion process at the interface between the sea and the land was the following in the period 1980–2003:

– in the deltaic accumulative sector (Sulina-Vadu), the highest withdrawal rate was observed: 4–7 m/year for the sector between the Imputita brook and the north of the Saraturile sand bank, the Sahalin island, Zaton, Portita north—Portita lighthouse, Chituc north of the sand bank [42, 99];
– in the sector with waterfronts and beach (Navodari-Vama Veche), erosion is low. Low values, of 2–3 m/year, are found on the sectors with beaches (the central sector of the Mamaia beach, the southern sector of the Techirghiol sandbar, the Mangalia beach, etc.) and very low, of 0.5–1 m/year, on those with waterfronts and capes (the Tuzla cape, Costinesti, the north of the Olimp resort, etc.) [54, 98].

For the following period, it is necessary to implement the following measures:

• Elaborating the integrated water monitoring: of the areas and mediums of investigation and of the monitored elements/components (biological, hydromorphological, physical-chemical). The great number mare of at-risk bodies of water leads to special monitoring that involves the surveillance of the entire body of water and the exact determination of all pressures and of its states [97].
• Extending the role of international relations with the neighbouring countries, in the field of waters, mostly with those within the basin of the Danube and of the Black Sea, in order to reduce the anthropic impact.
• Extending, modernizing and elaborating new treatment stations in conformity with the provisions of the position paper concerning the Directive 91/271/EEC concerning the urban wastewater treatment. The rhythm of achieving the objectives is set in the position Paper concluded between Romania and the European Community.

- Modernizing and extending of the wastewater treatment stations within the industrial units, taking into account the provisions of the Position Paper concerning the Directive 76/464/EEC concerning the priority/priority hazardous substances. The attainment rhythm is provided in the Position Paper.
- Applying the Code of Good Agricultural Practices in the areas vulnerable to nitrates in conformity with the provisions of the Directive 91/676/EEC and of the Position Paper.
- Reconstructing the rivers by applying the new defense concept that ensures both the reduction of risks due to floods and droughts, by creating lakes and dams, and the increase in the biodiversity of aquatic flora and fauna by elaborating habitats, mostly of wetlands [45, 46, 77, 78, 95].
- Elaborating on the areas of sanitary protection for water captures for drinking water.
- The imposition by the Danube Commission of the following regulations in order to reduce of accidental water pollution:

  - financial guarantee (Certificate of Financial Availability) for the ships transporting polluting substances on the Danube, which facilitated the retrieval of operational depolluting costs ("the pollutant pays");
  - endowing the fluvial ships in order to separate bilge waters, with centrifugal separators, and the use of the residues remained in the auxiliary boiler from the board of the ship, which led to the reduction of the amounts of polluting waters by about 1000 times and the elimination of the danger of accidental spills;
  - the existence at the board of the fluvial ships of record journals, manipulation and reporting of hydrocarbons (Oil Record Book), similar to the journals for sea ships, pursuant to the provisions of the MARPOL 73/78 Convention.

- Elaborating the protection works against littoral erosion by accessing Danubian funds (reducing the amount of alluvia transported by the Danube).
- Elaborating studies for determining the background pollution of waters situated in the mining areas, where the mother rock is at the surface.
- Promoting Danubian and European researches to establish the environmental objectives for the rivers that dry out [54, 98].

The chemical status of streams was based on five groups of indicators: oxygen regime: O_2, CCO-Mn, CCO-Cr, CBO_5; general ions, salinity: fixed residue, chlorides, sulphates, Ca, Mg, Na, Fe, Mn, Ba; nutrients regime: $N-NO_3$, $N-NO_2$, $N-NH_4$, N_{total}, $P-PO_4$, P_{total}; metals: Zn, Cu, Cr, As (and other heavy metals if they are monitored); specific pollutants: phenols, active anionic detergents, cyanides (and other substances if they are monitored) [54, 98]. The quality class is established by comparing the determined values with the limit values within the Order 1146/2002 concerning the reference objectives for the classification of the quality of surface waters, for each indicator. The physical-chemical characterization of the quality of surface waters (the global quality class) is provided by the least favourable class, determined at one of the groups of indicators. The length of the river sectors with water of a certain quality—related to the total length of the rivers monitored from the physical-chemical perspective (21,924 km), corresponding to the classes of quality, in conformity with the Order 1146/2002—is the following: 3547 km (16.2%)—very good status; 6492.5 km (29.6%) are 2nd class—good status; 7072.5 km (32.3%) are 3rd class—moderate status; 3141 km (14.3%) are 4th class—satisfactory status; 1671 km (7.6%) are 5th class—unsatisfactory status [54, 98].

The classification of rivers from the biological perspective is done based of the saprobic index of the macroinvertebrates (the Pantle–Buck method). The evaluation is based on the Order 1146/2002 concerning the reference objectives for the classification of the quality of surface waters. Out of the total length of 20,877 km of streams monitored from the biological perspective, the following are worth outlining [54, 98]:

- 2293 km (11%) very good status (oligosaprobic);
- 11,072.2 km (53%) good status (N-mesosaprobic);
- 5585.8 km (26.8%) moderate status (N-O-mesosaprobic);
- 1440 km (6.9%) satisfactory status (O-mesosaprobic);
- 486 km (2.3%) unsatisfactory status (polysaprobic).

The quality of water within the 22 natural lakes studied is the following: 2 (9.1%)—very good quality; 2 (9.1%)—good quality; 7 (31.8%)—moderate quality; 3 (13.6%)—satisfactory quality; 8 (36.4%)—unsatisfactory quality [54, 98]. The quality of water within the 96 reservoirs is the following: 17 (17.7%)—very good quality; 36 (37.5%)—good quality; 35 (36.5%)—moderate quality; 7 (7.3%)—satisfactory quality; 1 (1%)—unsatisfactory quality [54, 98].

The trophicity degree of lakes depends on the nutrients (total azote, total phosphorus) and on phytoplankton biomass: 0 (0%)—ultraoligotrophic; 1 (4.6%)—oligotrophic; 3 (13.6%)—mesotrophic; 9 (40.9%)—eutrophic; 9 (40.9%)—hypertrophic [54, 98]. Within the category of reservoirs: 16 (16.7%)—ultraoligotrophic; 28 (29.2%)—oligotrophic; 21 (21.9)—mesotrophic; 11 (11.4%)—eutrophic; 20 (20.8%)—hypertrophic [54, 98].

The quality of coastal waters depends on the quality of Danubian waters because the latter contributes with 99.53% nutrients; 99% total mineral azote; 91.83% phosphorus from orthophosphates. The littoral currents with a N-S direction favours the dispersion of Danubian pollutants in the coastal waters [54, 98, 100]. The eutrophication of waters was frequently manifested in the period 1970–1990 and more rarely after the year 1990 when the Romanian economy lapsed.

The criteria of the European guideline were the basis for identifying surface bodies of water on the Romanian territory. The criteria used for the delimitation of bodies of water are as follows: categories of surface water; the typology of surface waters; the physical-geographic and the hydromorphological characteristics of the basin; the pressures and the status of surface waters; the limits of the protected areas [54, 98]. From this point of view, 3715 freshwater bodies were delimited, of which eight for the Danube and the Danube Delta, including the Razim-Sinoie lagoon complex. The average length of bodies of water is 21.3 km (with the Danube). Of all bodies of water, 1368 (37%) represent non-permanent bodies of water. Six transitory bodies of water were delimited (three fluvial, on the Chilia, Sulina and St George branch), two lacustrine (Sinoie and Mangalia) and one marine. For the coastal waters, three bodies of water were identified.

In order to identify highly modified bodies of water, biological and abiotic data are taken into account. The criteria are based on types of hydrotechnical works and on their effects on aquatic ecosystems. For the permanent freshwater bodies (2347) within the hydrographical basins and spaces, the following are delimited: 415 (17.6%)—highly modified; 364 (15.5%)—candidate to "highly modified"; 1491 (63.6%)—without hydromorphological alterations; 77 (3.3%)—artificial bodies of water. For the transitory waters, two highly modified bodies of water were identified: one fluvial; one lacustrine. For the coastal waters, one highly modified body of water and one body of water candidate to "highly modified" were identified [54, 98].

For the evaluation of the risk of failing to achieve the environmental objectives for the bodies of water, the following criteria were taken into account: pollution from organic substances; pollution from nutrients; pollution by priority hazardous substances; hydromorphological alterations. A body of water is "at risk" if one of the criteria concerning pressure or impact is met. If none of the criteria is met, the body of water is "not at risk". If data for the evaluation of the risk are not available, the body is considered "possibly at risk". The 2347 freshwater bodies with permanent character (including the Danube) have the following characteristics:

– organic substances: 224 (9.5%) at risk; 128 (5.5%) possibly at risk; 1995 (85%) not at risk;
– nutrients: 290 (12.3%) at risk; 171 (7.3%) possibly at risk; 1886 (80.4%) not at risk;
– priority or priority hazardous substances: 56 (2.4%) at risk; 77 (3.3%) possibly at risk; 2214 (94.3%) not at risk;
– hydromorphological alterations: 492 (20.9%) at risk; 364 (15.5%) possibly at risk; 1491 (63.6%) not at risk;
– all the categories of risk: 639 (27.2%) at risk; 370 (15.8%) possibly at risk; 1338 (57.0%) not at risk.

For the Jiu river, there are four bodies of water "possibly at risk" by the total matters in suspension. For the six transitory bodies of water, the following resulted:

– organic matter—five bodies of water—possibly at risk; one body of water—at risk;
– nutrients—six bodies at risk;
– priority hazardous substances—three bodies at risk; two bodies possibly at risk;
– hydromorphological alterations—two bodies of water at risk.

For the coastal waters, with three bodies, the following result:

– organic matter—one body of water is possibly at risk;
– nutrients—three bodies of water at risk;
– priority hazardous substances—two bodies of water possibly at risk;
– hydromorphological alterations—one body of the water at risk; one body of water possibly at risk.

The identification and the delimitation of groundwater bodies were done based on the following criteria: geologic; hydrodynamic; the qualitative and quantitative status of the bodies of water. The delimitation of groundwater bodies was done only for the areas where there are significant aquifers for water supplies: exploitable discharges >10 m^3/day. Were delimited 129 groundwater bodies, of which 19 are transboundary. Twenty bodies of water are at risk due to historical sources represented by agro-zootechnical units or complexes that have ceased or reduced their activity, but also due to the current sources usually situated in the vulnerable areas [54, 98].

The wetlands and the deep waters within the mountainous units of the Carpathians have suffered small transformations compared to those situated in the lower sectors of the Romanian landforms. The drainages and agricultural re-usage of wetlands took place only in the large depressions, where the soil allowed the cultivation of plants necessary to the local productions. In the high sectors, some wetlands disappeared due to natural evolution, while the hydrotechnical works created other surfaces, adding to the current ones.

Nationwide, the morphometrical characteristics and the climatic conditions allowed the installation of a great variety of wetlands and their preservation was supported by conditions improper for drainages or by low habitation density. The most important argument in the preservation and rehabilitation of wetlands is represented by the role played by them: regulating the hydrological cycle, increasing biodiversity, reducing soil erosion, improving the microclimate, mitigating the floods, ensuring the natural life conditions for numerous species of plants and animals, a tourist destination for recreation, etc. [76–78].

The hydroclimatic events within the past years bring back into the discussion the preservation and the reconstruction of the former wetlands within the riparian sectors of the main hydrographical arteries that cause important damages during overflows. The drainage of wetlands was not specific only to rich countries. The listing of wetlands represents only a temporary phase of evaluation. The natural transformations are very quick and often radical. In order to highlight the evolutionary trends of humid environments, listings must be repeated at certain time intervals [51, 76, 77, 101].

The terrestrial, aquatic and aerial compartments constitute a global system, while the intervention on one of the elements has repercussions on the functionality of the others. The systemic analysis, most of the interactions between the elements of a system and the study of changes within systems, becomes the strong point of current scientific research, known as global or holistic research. This is the foundation for another management policy, more capable of understanding conflicts and thus more capable of solving them in terms of sustainable land development. The global view has the advantage of highlighting the role of the interfaces between the compartments of the systems or between the systems, a role often neglected due to the theme gaps. In this case, the issue of water is symptomatic: is it possible to monitor the functionality of an aquatic system and to guarantee water resources without intervening in soil processing by developing the hydrographical basin? The analysis of the "earth-water" system functioning outlines that the transfers are regulated through the existing areas

Fig. 2.14 The Braila pond was turned into the most fertile agricultural field in Romania

of transition between the terrestrial and the aquatic mediums, generally known as wetlands (Figs. 2.14 and 2.15).

After the campaign of the great damming works of the 70s, some wetlands in Romania were completely dried out. The new surfaces lost their role of protection against flood waves or reloading aquifers and as habitat for the local fauna and flora. In the hydrographical basins and spaces in Romania, 250 surface water captures were identified, of which 179 (71.6%) have areas of protection and 1223 have groundwater captures, of which 980 (80.1%) comprise areas of protection. There are no areas assigned for the protection of aquatic species from the economic perspective [54, 98].

The total length of rivers with protected areas represents 15.3% of the total length of streams in Romania. The areas of habitat or species protection where the water is an important factor are in a number of 216, and they account for a surface of 14,437.26 km^2, namely 6.1% of the surface of Romania [54, 98].

2.4 Conclusions

The environmental objectives for the surface and groundwaters—in conformity with the Framework Directive for Water 2000/60 of the European Union—have been

Fig. 2.15 The Danube floodplain with a role of wetland within the Danube Delta Biosphere Reservation

achieved almost integrally. The status of waters is relatively good and good for about 60%. The point sources of pollution—belonging to old industrial units from the Communist period or to new and small ones—bring significant damage to the small streams. The most important problems are related to the surface and groundwaters within rural settlements because they are polluted by animal and human dejections that end up straight in the bodies of water. By connecting all the localities to the sewerage system and to the water supply system, pollution will cease almost automatically nationwide.

2.5   Recommendations

The pollution sources from the old plants storage sites are becoming more environmentally friendly and therefore pollution is substantially diminished. However, it is necessary to ecologize by afforestation or erosion of tailings dumps that belong to the mining companies in the northern Carpathians, the Apuseni Mountains and Oltenia. Also, blocking emissions from abandoned quarries by coal, copper or other non-ferrous or ferrous ores in southern and western Romania, must become a priority.

Fig. 2.16 Garbage accumulated behind the Bicaz dam on the Bistrita River in July 2018 (Eastern Carpathians)

The most important source of pollution of surface and underground waters is the household, especially the rural ones. The lack of septic tanks and the storage of rubbish at the edge of the villages make the water supply network seriously affected. It is necessary to urge the generalized water supply of all villages, to conduct sewage disposal and to educate the population. Active management is also imperative for mitigating and collecting garbage in recreational areas. Random garbage is often found behind large dams and water cleaning involves additional financial efforts (Fig. 2.16).

Acknowledgements This work was made possible by the financial support of the Ministry of National Education (Exploratory research project PN-III-P4-ID-PCE-2016-0759, the Ethnoarchaeology of Salt in the Inner Carpathian area of Romania, no 151/2017).

References

1. Araoye PA (2009) The seasonal variation of pH and dissolved oxygen (DO_2) concentration in Asa lake Ilorin, Nigeria. Int J Phys Sci 4(5):271–274
2. Chapman LJ, Chapman CA, Srisman TL, Nordlie FG (1998) Dissolved oxygen and thermal regimes of a Ugandan crater lake. Hydrobiologia 385:201–211
3. Cummins KW, Sedell JR, Swanson FJ, Nunshall GW, Fisher SG, Cushing CE, Petersen RC, Vannote RL (1983) Organic matter budgets for stream ecosystems: problems in their

evaluation. In: Barnes JR, Nunshall GW (eds) Stream ecology. Plenum Publ. Corp., New York

4. Felföldi T, Ramganesh S, Somogyi B, Krett G, Jurecska L, Szabó A, Vörös L, Márialigeti K, Máthé I (2015) Winter planktonic microbial communities in highland aquatic habitats. Geomicrobiol J 1–36

5. Gržetič I, Čamprag N (2010) The evolution of the tropic state of the Palić Lake (Serbia). J Serb Chem Soc 75(5):717–732

6. Kominkova D, Nabeikova J, Vitvar T (2016) Effects of combined sewer overflows and storm water drains on metal bioavailability in small urban streams (Prague metropolitan area, Czech Republic). J Soils Sediments 16(5):1569–1583

7. Kowalewski GA (2013) Changes in Lake Rotcze catchment over the last 200 years: implications for lake development reconstruction. Limnol Rev 13(4):197–207

8. Marusic G, Marusic D, Putuntica A (2016) RiverPrut-software for determination and management of water quality. Meridian Ingineresc 2:45–48

9. Mebirouk H, Boubendir-Mebirouk F, Hamma W (2017) Main sources of pollution and its effects on health and the environment in Annaba. Urbanism Architect Constr 9(2):167–182

10. Mirecki N, Agič R, Šunić L, Milenković L, Ilić ZS (2015) Transfer factor as indicator of heavy metals content in plants. Fresenius Environ Bull 24(11c):4212–4219

11. Millot C, Touchart L, Bartout P, Azaroual A, Aldomany M (2014) Ponds: conservation through a better management of environmental impacts. Diversion: an optimal management? In: The case of Loire headwater catchment. Water resources and wetlands. Conference proceedings 11–13 Sept 2014, Tulcea, Romania, pp 110–115

12. Mishra GP, Yadav AK (1978) A comparative study of physico-chemical characteristics of river and lake water in Central India. Hydrobiologia 59(3):275–278

13. Ouattara I, Kamagate B, Dao A, Noufe D, Savane I (2016) Groundwaters mineralisation process and transfers of flow within fissured aquifers: case of transboundary basin of Comoe (Cote d'Ivoire, Burkina Faso, Ghana, Mali). Int J Innov Appl Stud 17(1):57–69

14. Radevski I, Gorin S (2017) Floodplain analysis for different return periods of river Vardar in Tikvesh valley (Republic of Macedonia). Carpathian J Earth Environ Sci 12(1):179–187

15. Rudic Z, Raicevic V, Bozic M, Nikolic G, Obradovic V, Petrovic JJ (2014) Microbiological indicators in the water and sediment of the shallow Panonnian lake-Lake Palic. In: Water resources and wetlands. Conference proceedings 11–13 Sept 2014, Tulcea, Romania, pp 124–129

16. Sakcali MS, Ylmaz R, Gucel S, Yarci C, Ozturk M (2009) Water pollution studies in the rivers of the Edirne Region-Turkey. Aquat Ecosyst Health Manage 12(3):313–319

17. Skowron R, Piasecki A (2014) Overgrowth of lakes as an indicator of their disappearance - on the example of the lakes of north-western Poland. In: Water resources and wetlands. Conference proceedings 11–13 Sept 2014, Tulcea, Romania, pp 94–101

18. Su X, Nilsson C, Pilotto F, Liu S, Shi S, Zeng B (2017) Soil erosion and deposition in the new shorelines of the Three Gorges Reservoir. Sci Total Environ 599–600:1485–1492

19. Trivedi RC (2010) Water quality of the Ganga River – an overview. Aquat Ecosyst Health Manage 13(4):347–351

20. Wrzesiński D, Choiński A, Ptak M, Skowron R (2015) Effect of the North Atlantic Oscillation on the pattern of lake ice phenology in Poland. Acta Geophys 63(6):1664–1684

21. Zelenáková M, Fijko R, Diaconu DC, Remenáková I (2018) Environmental impact of small hydro power plant—a case study. Environments 5(12):1–10

22. Alexe M, Serban G (2008) Considerations regarding the salinity and water temperature of salty lakes of Sovata and Ocna Sibiului. Studia Universitatis "Vasile Goldis" Seria Stiintele Vietii 18:305–311

23. Banaduc D, Panzar C, Bogorin P, Hoza O, Curtean-Banaduc A (2016) Human impact on Tarnava Mare river and its effects on aquatic biodiversity. Acta Oecol Carp IX:187–197

24. Banaduc D, Rey S, Trichkova T, Lenhardt M, Curtean-Banaduc A (2016) The Lower Danube River-Danube Delta–North West Black Sea: a pivotal area of major interest for the past, present and future of its fish fauna—a short review. Sci Total Environ 545–546:137–151

25. Batinas RH (2009) The lakes from Rosia Montana. Lakes Reservoirs Ponds 3:85–93
26. Batinas RH (2010) Studiul calitatii apelor de suprafata din bazinul Ariesului. Presa Universi-
    tara Clujeana, Cluj-Napoca
27. Batuca DG, Jordaan JM (2000) Silting and desilting of reservoirs. CRC Press, Rotterdam
28. Bretcan P, Tampu MF (2008) The Razim-Sinoie lacustrine complex. Protection, resources,
    valorization. Lakes Reservoirs Ponds 1–2:99–112
29. Cical E, Mihali C, Mecea M, Dumuta A, Dippong T (2016) Considerations on the relative
    efficacy of aluminium sulphates versus polyaluminium chloride for improving drinking water
    quality. Stud Univ Babes-Bolyai Chem 61(2):225–238
30. Cirtina D, Capatina C (2017) Quality issues regarding the watercourses from Middle Basin
    of Jiu River. Rev Chim (Bucharest) 68(1):72–76
31. Cirtina D, Capatina C (2017) Preliminary study on assessment of mineralization degree and
    nutrient content of groundwater bodies in Gorj County. Rev Chim (Bucharest) 68(2):221–225
32. Contiu HV (2007) Culoarul Muresului dintre Reghin si confluenta cu Ariesul. Casa Cartii de
    Stiinta, Cluj-Napoca
33. Coops H, Buijse L, Buijse ADT, Constantinescu A, Covaliov S, Hanganu J, Ibelings BW,
    Menting F, Navodaru I, Oosterberg W, Staras M, Torok L (2008) Trophic gradients in a large-
    river Delta: ecological structure determined by connectivity gradients in the Danube Delta
    (Romania). River Res Appl 24(5):698–709
34. Curtean-Banaduc A, Banaduc D (2016) *Thymallus thymallus* (Linnaeus, 1758), ecological sta-
    tus in Maramures Mountains Nature Park (Romania). Transylv Rev Syst Ecol Res 18(2):71–84
    ("The Wetlands Diversity")
35. Diaconu DC (2008) The Siriu reservoir, Buzau River (Romania). Lakes Reservoirs Ponds
    1–2:141–149
36. Dirtu D, Pancu M, Minea ML, Chirazi M, Sandu I, Dirtu AC (2016) Study of the quality
    indicators for the indoor swimming pool water samples in Romania. Rev Chim (Bucharest)
    67(6):1167–1171
37. Dumitran GE, Vuta LI (2011) The Eutrophication phenomenon in Golesti Lake - Romania. In:
    Air and water components of the environment, 18–19 Martie 2011, Cluj-Napoca, pp 155–162
38. Ion LC (2010) Calitatea apei lacurilor din bazinul hidrografic al Ialomitei pana la confluenta
    cu Prahova. Resursele de apa din Romania. Vulnerabilitate la presiunile antropice. In: The
    proceedings of the first national symposium, 11–13 June 2010, Targoviste, pp 386–389
39. Mititelu LA (2010) Vidraru reservoir, Romania. Environmental impact of the hydrotechnical
    constructions on the upper course of Arges River. Lakes Reservoirs Ponds 4(1–2):152–166
40. Pantea I, Ferechide D, Barbilian A, Lupusoru M, Lupusoru GE, Moga M, Vilcu ME, Ionescu
    T, Brezean I (2017) Drinking water quality assessment among rural areas supplied by a
    centralized water system in Brasov county. Univ Politeh Bucharest Sci Bull Ser C Electr Eng
    Comput Sci Ser B 79(1):61–70
41. Raischi MC, Oprea L, Deak G, Badilita A, Tudor M (2016) Comparative study on the use
    of new sturgeon migration monitoring systems on the lower Danube. Environ Eng Manage J
    15(5):1081–1085
42. Romanescu G (2006) Complexul lagunar Razim-Sinoie. Studiu morfohidrografic. "Alexandru
    Ioan Cuza" University Press, Iasi
43. Romanescu G (2006) Hidrologia uscatului. Editura Terra Nostra, Iasi
44. Romanescu G (2008) The ecological characteristics of the romanian littoral lakes - the sector
    Midia Cape -VamaVeche. Lakes Reservoirs Ponds 1–2:49–60
45. Romanescu G (2009) Trophicity of lacustrine waters (lacustrine wetlands) on the territory of
    Romania. Lakes Reservoirs Ponds 3:62–72
46. Romanescu G (2009) The physical and chemical characteristics of the lake wetlands in the
    central group of the east Carpathian Mountains. Lakes Reservoirs Ponds 4:94–108
47. Romanescu G (2010) Trophicity of lacustrine waters (lacustrine wetlands) on the Romanian
    territory. Lakes Reservoirs Ponds 4(1–2):41–51
48. Romanescu G, Stoleriu C (2014) Seasonal variation of temperature, pH and dissolved oxygen
    concentration in Lake Rosu, Romania. CLEAN Soil Air Water 42(3):236–242

49. Serban G (2008) Anthropo-saline and karsto-saline lakes from Ocna Sugatag - Maramures (Romania). Lakes Reservoirs Ponds 1–2:80–89
50. Stefan DS, Neacsu N, Pincovschi I, Stefan M (2017) Water quality and self-purification capacity assessment of Snagov Lake. Rev Chim (Bucharest) 68(1):60–64
51. Telteu CE (2012) Hydrochemical features of the south Dobrogea's lakes and impact of the climatic conditions on these features. In: Water resources and wetlands. Conference proceedings, 14–16 Sept 2012, Tulcea, Romania, pp 163–168
52. Timofti D, Doltu C, Trofin M (2011) Eutrophication phenomena in reservoirs. In: Air and water components of the environment, 18–19 Martie 2011, Cluj-Napoca, pp 473–479
53. Romanescu G, Lasserre F (2006) Le potentiel hydraulique et sa mise en valeur en Moldavie Roumaine. In: Brun A, Lassere F (eds) Politiques de l'eau. Grands principes et realites locales. Presses de l'Universite du Quebec, pp 325–346
54. Romanian Waters National Administration (2004) The hydrographical basins management plans. National Report 2004 – Romania, Bucharest
55. Barbulescu A (2016) Analysis and Models for surface water quality. Stud Time Ser Appl Environ Sci 103:145–151
56. Batinas RH, Sorocovschi V (2011) Resurse de apa, potential si valorificare turistica. Presa Universitara Clujeana, Cluj-Napoca
57. Bretcan P, Tuchiu E, Bretcan S (2012) Importance of water circulation in lakes Razim-Golovita. Status and quality of ecosystems. In: Water resources and wetlands. Conference proceedings, 14–16 Sept 2012, Tulcea, Romania, pp 100–105
58. Capatina C, Cirtina D (2017) Comparative study regarding heavy metals content in air from Targu Jiu and Rovinari. Rev Chim (Bucharest) 68(12):2839–2844
59. Diaconu DC, Mailat E (2010) Studiu complex al ecosistemelor lacustre din cadrul Tinovului Mohos. Resursele de apa din Romania. Vulnerabilitate la presiunile antropice. In: The proceedings of the first national symposium, 11–13 iunie 2010, Targoviste, pp 431–435
60. Diaconu DC, Mailat E (2012) Puiulet Lake Danube delta - bathymetrical characteristics. In: Water resources and wetlands. Conference proceedings, 14–16 Sept 2012, Tulcea, Romania, pp 151–154
61. Diaconu DC, Peptenatu D, Simion AG, Pintilii RD, Draghici CC, Teodorescu C, Grecu A, Gruia AK, Ilie AM (2017) The restrictions imposed upon the urban development by the piezometric level. Case study: Otopeni-Tunari-Corbeanca. Urbanism Architect Constr 8(1):27–36
62. Diaconu DC, Andronache I, Ahammer H, Ciobotaru AM, Zelenakova M, Dinescu R, Pozdnyakov AV, Chupikova SA (2017) Fractal drainage model - a new approach to determinate the complexity of watershed. Acta Montanist Slovaca 22(1):12–21
63. Fodorean I (2008) Considerations regarding the integrated management of freshwater lakes in Transylvanian Plain. Lakes Reservoirs Ponds 1–2:113–126
64. Fodorean I (2010) Lacurile dulci din Podisul Transilvaniei. Presa Universitara Clujeana, Cluj-Napoca
65. Mihu-Pintilie A, Romanescu G, Stoleriu C (2014) The seasonal changes of the temperature, pH and dissolved oxygen in the Cuejdel Lake, Romania. Carpathian J Earth Environ Sci 9(2):113–123
66. Mihu-Pintilie A, Paiu M, Breaban IG, Romanescu G (2014) Status of water quality in Cuejdi hydrographic basin from Eastern Carpathian, Romania. In: 14th SGEM geoconference on water resources. Forest, marine and ocean ecosystems, SGEM2014 conference proceedings, 19–25 June 2014, vol 1, pp 639–646. www.sgem.org
67. Mihu-Pintilie A, Romanescu G, Stoleriu C, Stoleriu O (2014) Ecological features and conservation proposal for the largest natural dam lake in the Romanian Carpathians – Cuejdel Lake. Int J Conserv Sci 5(2):243–252
68. Muraresu O, Druga M, Puscoi B (2008) The anthropic lakes from the hydrographic basin of upper Ialomita river (Romania). Lakes Reservoirs Ponds 1–2:150–157
69. Navodaru I, Staras M, Cernisencu I (2008) The challenge of sustainable use of the Danube Delta Fisheries, Romania. Fish Manage Ecol 8(45):323–332

70. Omer I (2016) Water quality assessment of the groundwater body RODL01 from North Dobrogea. Rev Chim (Bucharest) 67(12):2405–2408
71. Panaitescu E (2007) Acviferul freatic si de adancime din bazinul hidrografic Barlad. Casa Editoriala Demiurg, Iasi
72. Patroescu V, Ionescu I, Tiron O, Bumbac C, Mares MA, Jinescu G (2016) Nitrification front evolution in a biological filter using expanded clay as a filter media. Rev Chim (Bucharest) 67(5):958–961
73. Patroescu IV, Dinu LR, Constantin LA, Alexie M, Jinescu G (2016) Impact of temperature on groundwater nitrification in an up-flow biological aerated filter using expanded clay as filter media. Rev Chim (Bucharest) 67(8):1433–1435
74. Popescu LR, Iordache M, Buica GO, Ungureanu EM, Pascu LF, Lehr C (2015) Evolution of groundwater quality in the area of chemical platform. Rev Chim (Bucharest) 66(12):2060–2064
75. Romanescu G, Romanescu G, Minea I, Ursu A, Margarint MC, Stoleriu C (2005) Inventarierea si tipologia zonelor umede din Podisul Moldovei. Studiu de caz pentru judetele Iasi si Botosani. Editura Didactica si Pedagogica, Bucharest
76. Romanescu G, Romanescu G, Stoleriu C, Ursu A (2008) Inventarierea si tipologia zonelor umede si apelor adanci din Podisul Moldovei. Editura Terra Nostra, Iasi
77. Romanescu G, Lupascu A, Stoleriu C, Raduianu D, Lesenciuc D, Vasiliniuc I, Romanescu G (2009) Inventarierea si tipologia zonelor umede si apelor adanci din Grupa Centrala a Carpatilor Orientali. Editura Universitatii "Alexandru Ioan Cuza", Iasi
78. Romanescu G, Dinu C, Radu A, Torok L (2010) Ecologic characterization of the fluviatile limans in the south-west Dobrudja and their economic implications (Romania). Carpathian J Earth Environ Sci 5(2):25–38
79. Romanescu G, Mihu-Pintilie A, Stoleriu C, Romanescu AM (2012) Present state of trophic parameters of the main lakes from Siret and Pruth watersheds. In: Water resources and wetlands. Conference proceedings, 14–16 Sept 2012, Tulcea, Romania, pp 33–38
80. Romanescu G, Dinu C, Radu A, Stoleriu C, Romanescu AM, Purice C (2013) Water qualitative parameters of fluviatile limans located in the south-west of Dobrogea (Romania). Int J Conserv Sci 4(2):223–236
81. Romanescu G, Stoleriu CC, Enea A (2013) Limnology of the Red Lake, Romania. An interdisciplinary study. Springer, Dordrecht
82. Romanescu G, Cretu MA, Sandu IG, Paun E, Sandu I (2013) Chemism of streams within the Siret and Prut Drainage Basins: water resources and management. Rev Chim (Bucharest) 64(12):1416–1421
83. Romanescu G, Sandu I, Stoleriu C, Sandu IG (2014) Water resources in Romania and their quality in the main Lacustrine Basins. Rev Chim (Bucharest) 63(3):344–349
84. Romanescu G, Jora I, Panaitescu E, Alexianu M (2015) Calcium and magnesium in the groundwaters of the Moldavian Plateau (Romania) - distribution and managerial and medical significance. In: International multidisciplinary scientific geoconference SGEM 2015, water resources. Forest, marine and ocean ecosystem. Conference proceedings, vol I, Hydrology & water resources, pp 103–112
85. Serban G, Antonie M, Roman C (2009) Remanent lakes formed through the work of kaolin exploiting from Aghiresu (Cluj County). Lakes Reservoirs Ponds 3:40–52
86. Serban G, Mirisan B, Danciu D (2010) Functiile acumularilor din zona montana si din zona colinara - studiu comparativ, amenajarile Somesul Cald si Crasna Superioara. Resursele de apa din Romania. Vulnerabilitate la presiunile antropice. In: The proceedings of the first national symposium, 11–13 iunie 2010, Targoviste, pp 51–58
87. Sorocovschi V (2008) The lakes of the Transylvanian Plain: genesis, evolution and territorial repartition. Lakes Reservoirs Ponds 1–2:36–48
88. Sorocovschi V (2009) The mineralisation and chemical composition of the lakes in the Transylvanian Plain. Lakes Reservoirs Ponds 3:13–24
89. Tokar A, Negoitescu A, Hamat C, Rosu S (2016) The chemical and ecological state evaluation of a storage lake. Rev Chim (Bucharest) 67(9):1860–1863

90. Tuchiu E (2010) Characteristics of the lower Danube water bodies between Portile de Fier (Iron Gates) and Isaccea. Lakes Reservoirs Ponds 4(1–2):109–118
91. Tuchiu E (2011) Directiva Cadru Apa – directii de dezvoltare ale Planurilor de Management ale Bazinelor Hidrografice. Institutul National de Hidrologie si Gospodarire a Apelor. In: Annual scientific conference, 1–3 Nov 2011, pp 55–68
92. Romanescu G, Pascal M, Mihu-Pintilie A, Stoleriu CC, Sandu I, Moisii M (2017) Water quality analysis in wetlands freshwater: common floodplain of Jijia-Prut Rivers. Rev Chim (Bucharest) 68(3):553–561
93. European Commission (2012) Draft of the commission services. Member state: Romania accompanying the report document of the commission towards the European Parliament and the council on the implementation of the Water Framework Directive (2000/60/CE). Management plans of the hydrographical basins. http://ec.europa.eu/environment/water/water-framework/pdf/CWD-2012-379_EN-Vol25_RO_ro.pdf
94. Serban P, Galie A (2006) Managementul apelor. Principii si reglementari europene. Editura Tipored, Bucharest
95. Romanescu G (2009) Evaluarea riscurilor hidrologice. Editura Terra Nostra, Iasi
96. Romanescu G (2009) The risk of wetlands disappearance in the Moldavian Plateau under the conditions of rudimentary agricultural techniques use. Riscuri si Catastrofe 8(7):1179–1192
97. Mihailovici M, Serban P (2006) Planul de Management pentru bazinul hidrografic al Dunarii. In: The 4th conference of Romanian hydro-energeticians, 1
98. Romanian Waters National Administration (2013) The national plan for the development of the hydrographical basins in Romania. Synthesis, revised version, Feb 2013, Bucharest
99. Romanescu G, Romanescu AM, Romanescu G (2014) History of building the main dams and reservoirs. In: Water resources and wetlands, conference proceedings, 11–13 Sept 2014, Tulcea, Romania, pp 485–492
100. Romanescu G, Stefan M (2012) Geografa Marii Negre. Editura Transversal, Targoviste
101. Zelenáková M, Zvijaková L (2017) Risk analysis within environmental impact assessment of proposed construction activity. Environ Impact Assess Rev 62:76–89

Chapter 3
Causes and Effects of Water Pollution in Romania

Iuliana Gabriela Breaban and Ana Ioana Breaban

Abstract After 2007, Romania adopted and implemented EU legislation on water management. Romanian groundwaters are polluted locally from point sources (landfills) and from diffuse sources of pollution (fertilizers, pesticides and canals, in urban areas). Over-exploitation of underground water wells can lead to indirect diffuse pollution, causing salty water intrusion, exploitation of mineral aggregates. In order to establish water quality, Romania applies a monitoring flow in accordance with the WFD, both for surface water bodies as well as for groundwater bodies. Nowadays are 3.027 surface water bodies from which 2470 are natural water bodies, 488 heavily modified water bodies and 69 artificial water bodies. 50% of all bodies of surface water have been classified as heavily modified water bodies (35%) or artificial water (15%) while 16% (23/143) of underground water bodies were outside the chemical parameters imposed by the rules. Pressures are generated by (i) human agglomeration (lack of connection to the sewerage system and to the sewage treatment plants); (ii) industry (wastewater discharges); (iii) agriculture (nutrient and pesticide emissions); (iv) hydrotechnical works (dams, dikes, sills, weirs, diversions); (v) other anthropic activities.

Keywords Pollution · Water quality · Surface water · Groundwater · Sources

3.1 Introduction

Water is the source of life, the fresh water being a precious resource. The quality of water resources is a matter of necessity and attractiveness, of continuous relevance given that the freshwater resources of the planet represent only 2.5% of the water on

I. G. Breaban (✉)
Department of Geography, Faculty of Geography and Geology,
Alexandru Ioan Cuza University of Iasi, Bd. Carol I 20A, 700505 Iasi, Romania
e-mail: iulianab2001@gmail.com

Alexandru Ioan Cuza University of Iasi, CERNESIM, 11 Carol I Bld., 700506 Iasi, Romania

A. I. Breaban
Faculty of Hydrotechnical Engineering, Geodesy and Environmental Engineering, Technical University "Gheorghe Asachi" of Iasi, Bd. D. Mangeron 65, 700050 Iasi, Romania

© Springer Nature Switzerland AG 2020
A. M. Negm et al. (eds.), *Water Resources Management in Romania*, Springer Water,
https://doi.org/10.1007/978-3-030-22320-5_3

earth, of which only 30.8% is found in groundwater systems, and 0.3% is directly available in rivers and lakes. According to the Directive *2000/60/EC* of the European Parliament and of the Council establishing a framework for Community action in the field of water policy the *"Water is not a commercial product like any other but, rather, a heritage which must be protected, defended and treated as such"*. Why is water quality so important? The main point of view is the fact that the fresh water is scarce (only 3% of the world's water is suitable for domestic use (drinking, irrigation, …etc.) of which 2% is found in ice and glaciers. Consequently, only 1% of the earth's water is accessible and drinkable!). From other point of view, nowadays a large number of chemicals are used in our everyday lives and in commerce, making their way into our waters. Another difficult task to achieve is the assessment of long-term changes in water quality over the last decades, with increasing demand for water quality monitoring in many rivers through regular measurements of different physicochemical and biological quality of water. Monitoring activities have led to the gradual accumulation of valuable long-term information on water quality, and the study of these data has led to the development of future scenarios of the long-term evolution of freshwater quality.

Water is vital to life, promoting life in many ways. At first, there is no need to clarify further how important it is, but nevertheless, water pollution is one of the most serious environmental threats that humanity faces today. Protecting fresh water resources requires the diagnosis of threats on a large scale, from the local to the global level.

In accordance with the requirements of the Water Framework Directive (WFD), to identify the quality of water bodies, it is necessary to know the anthropic pressure created on water resources in order to adopt appropriate measures for the protection and conservation of water in Romania, which has multiple cross-border boundaries in the form of rivers. A growing literature on the water quality is driven due to increasing challenges related to both water scarcity and poor water distribution, the quantity and quality of water being the cause of concern in the last period of time (Fig. 3.1). It was found that 378 articles dealt with the use of water pollution in Romania between 1992 and 2018, after 2005 the number of papers increased significantly from 9 to 51 in 2015. Although 250 journals have published studies using water pollution Romania, 10 journals published them much more often, with 52% of all publications. The International Multidisciplinary Scientific Geoconference: SGEM published the most (47 papers), followed by Environmental Engineering and Management Journal (42 papers), Revista de Chimie and Journal of Environmental Protection and Ecology (29 papers each), Carpathian Journal of Earth and Environmental Sciences (16 papers) and other 5 journals that published between 4 and 10 papers.

More and more nations are beginning to face difficulties in water supply due to the increased water needs of agriculture and industry, along with poor management or a lack of water management. Pollution caused by various industrial activities, the development of urban areas as well as the presence of non-point sources generated by agriculture and cattle farms, has favoured the deterioration of water quality. Given the increasingly serious concerns about water quality, scientific research has focused on this topic at all levels of importance: global, regional or local (Fig. 3.2).

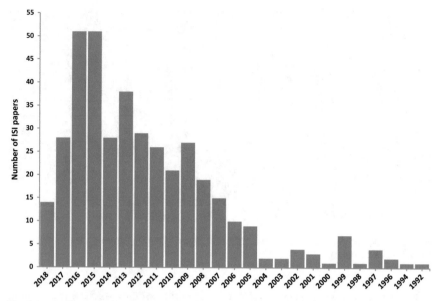

Fig. 3.1 The dynamic of ISI papers record between 1992 and 2018 concerning "water pollution Romania" (webofknowledge.com)

| 202 ENVIRONMENTAL SCIENCES ECOLOGY | 58 WATER RESOURCES | 14 PHYSICS | 9 TOXICOLOGY | 8 MINING MINERAL PROCESSING | 7 ENERGY FUELS | 7 OCEANOGRAPHY |

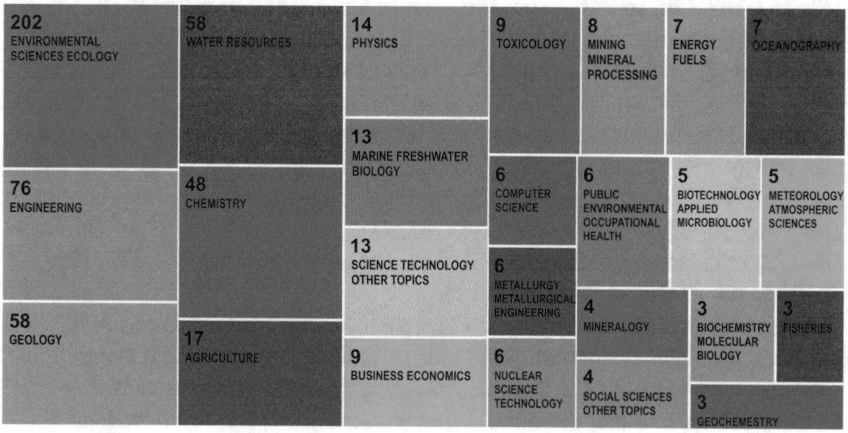

Fig. 3.2 The main 25 research area of ISI papers record between 1992 and 2018 concerning "water pollution Romania" (webofknowledge.com)

Fig. 3.3 The origin of contributors of ISI papers record between 1992 and 2018 (webofknowledge.com)

From 378 recorded papers, 62.46% were scientific articles, 42.59% proceedings paper, the other being book chapters, editorial materials, meeting abstracts and notes. Concerning the origin of contributors, mostly are from Romania, Romanescu et al. [165, 157, 169, 170, 162, 163, 166, 156, 171, 167, 158–161], Romanescu and Stoleriu [164] and Teodosiu et al. [187–189] having more than 10 papers each, followed by authors from Hungary, Germany, U.K., Switzerland, USA (Fig. 3.3).

The high production of articles on water pollution in Romania can be explained by changes throughout the last 30 years in sources of human water consumption, pressures exerted by the increasing demand for good water quality resources has led to action in order to manage water and establish limits for some priority substances [1–3, 5–16, 18–30, 33–42, 44–49, 51–57, 65, 67–69, 71–75, 77, 85–93, 95, 96, 66, 99, 101–112, 114, 115, 117–122, 128–138, 141, 145, 147–153, 155, 172, 173, 175–179, 181–184, 186, 190–193, 195–198, 201].

Water pollution and contamination of water bodies worldwide is one of the environmental problems facing our world today. While it is known that human-induced pollution exceeds by far the pollution induced by natural causes, sometimes they prefer eye closure in this direction. What we sometimes fail to understand is that besides humanity, affected by contamination are also various living organisms on the planet.

3.1.1 Water Pollution and Types

Water pollution can be characterized in several ways. It usually indicates that one or more substances have increased their concentration in water to such an extent that they may lead to perturbations and even damage the health of animals or humans.

Water sources of all kinds can be cleaned naturally by dispersing certain quantities of pollutants. Water pollution specifically refers to quantities: firstly, the amount of pollutant released and, secondly, how large is the volume of water it delivers. A low amount of toxic chemical can have a low impact if it flows into the ocean. But the same amount of the same chemical can have a much greater impact if pumped into a lake or river, with less clean water to ensure its dispersion.

Water pollution almost always means that a water source suffered a series of damages. By happy chance the Earth is kindly, and the damage induced by water pollution is often reversible.

3.1.1.1 Point-Source, Nonpoint Source, Cross-Border

When we say the water resources of the planet, we refer initially to oceans, lakes and rivers, in other words to the surface waters that are first affected by pollution processes (a spill of oil products creates a hydrophobic oil layer that can affect a vast area of an ocean/watercourse). But not all the water on the planet is on the surface, a substantial amount of water is found in underground rock structures known as aquifers or underground waters. Aquifers feed the rivers, supplying much of the drinking water. These, as well as surface waters, are susceptible to pollution (the leaching of pesticides used in agriculture), their pollution being much less obvious than that of the surface waters, which is a matter of particular importance Hill [76] (Fig. 3.4).

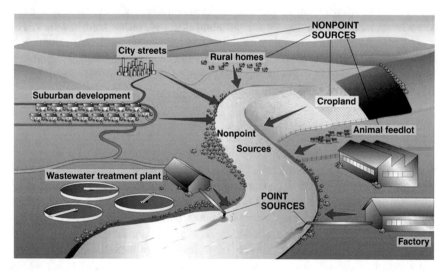

Fig. 3.4 Point-source and nonpoint source pollutants (processing after Wright and Boorse [200])

Table 3.1 Comparison of water point-source pollution and nonpoint-source pollution

Sources of point pollution	Sources of nonpoint pollution
• Leaking storage lagoons for polluted waste • Unlined landfills • Leaking underground storage tanks that contain chemicals • Water discharged from industry • Public and industrial waste treatment plants • Polluted water from abandoned and active mines	• Precipitation containing air pollutants • Soil runoff from farms and building sites • Fertilizer, pesticides from farmland and residential lawns • Water runoff from city and suburban streets • Chemicals added to the road (salt and deicing agents) • Feces and agricultural chemicals from livestock feedlots • Oil and gasoline from personal watercraft

Both, surface and groundwater bodies are the two types of water resources affected by pollution that can occur in two different ways. If pollution originates from a single source, such as an exhaust pipe, it is point-source pollution (oil spill from a tank, an industrial chimney outlet, accidental leakage car oil into a drainage channel), being often difficult to identify (Table 3.1).

Nonpoint or diffuse pollution is associated with land-use activities. Most water pollution situations occur from several diffuse sources known as nonpoint-source pollution (chemicals add to the road surface, water runoff from city and suburban streets) and not from point-sources. When pollution originating from point-sources comes into contact with the environment, the most affected area is usually the area adjacent to the source. In the case of a terrestrial accident involving leakage of organic liquids from a tank, the pollutant layer is concentrated around the tank itself while in an aquatic environment the pollutant diffuses further from the tank. For non-point source pollutants, this is less likely to happen because contamination is made from several places simultaneously. Particular situations are also distinguished when pollution entering in the environment in one place has an effect thousands of miles away, like cross-border pollution.

3.1.1.2   Sources of Pollution

Water can be polluted by a multitude of sources. It is possible to distinguish: wastewaters (domestic and industrial); runoff (agriculture and stormwater); bathing, cloth washing, etc., which is discharged into water bodies. Thus, many organic pollutants come from wastewater and waste from both livestock farms and food processing plants consuming oxygen from water, choking fish and other aquatic animals. Nitrates and phosphates are a category of widespread pollutants from agriculture and domestic activities, ranging from organic fertilizers to farms and to household detergents, being able to "over-fertilize" the water that leads to the growth of algae, some of which are toxic. When algae die, it sinks at the bottom of the water, breaks down, consumes oxygen and affects the aquatic ecosystems (Fig. 3.5).

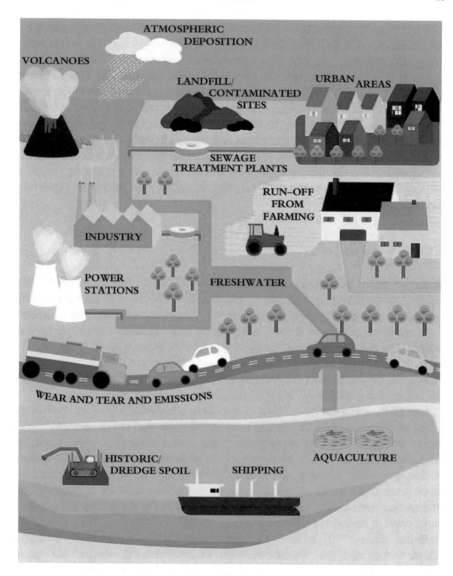

Fig. 3.5 Sources of water pollution (processing after EEA [64])

Fecal contamination of household sewage contributes to deteriorating water quality for recreational activities (swimming, boat trips or fishing), the water being unpleasant and uncertain. Chemical contaminants, such as heavy metals, pesticides and some industrial chemicals, can also pose significant pressures on the life and health of living organisms. Sediment leaching on the land contributes to the formation of water mud films, blocking sunlight and, consequently, the disappearance of living creatures. By irrigating agricultural land with water from inappropriate sources can increase the content of salts, nutrients or other pollutants in water and soils [198].

Natural water resources can be contaminated by the discharge of residual effluents from multiple industrial or agricultural activities. They may contain one or more substances of inorganic nature (heavy metals, acids), organic, microporous plastics, nutrients, etc. at the same time.

3.1.1.3   Pollutant Types

Water pollution occurs when waste is released into water bodies. Polluted water can lead to the destruction of existing plants and organisms in or near the aquatic ecosystem, affecting humans, plants and animals that consume it. In the literature there are different classification concerning the water pollutants based on various criteria. For the present study the following classification was chose taking into account the specific of the pollution processes at regional scale: (1) Organic, (2) Pathogens, (3) Inorganic, (4) Nutrients, (5) Macroscopic, (6) Radioactive, and (7) Other forms of pollution—plastics, alien species, accidental pollution.

Europe has recognized the importance of owning high-quality water resources by acting in this regard by adopting and implementing the WFD in December 2000 to ensure that these resources are properly managed to achieve good water quality in all European waters by 2027 by managing water bodies (rivers, lakes, groundwater bodies, transitional and coastal waters). The WFD aims to implement harmonized water protection regulations within the European Union, with the monitoring and management unit being the "body of water" considered as a significant element of surface water, uniform in type and status.

Eurobarometer [60], following the September 2017 survey of 28 EU Member States, reconfirmed Europe's concern for water resources, from the first ten threats of the environment, four concerns water, such as pollution of rivers, lakes and groundwater (36%), marine pollution (33%), shortage of drinking water (30%), frequent droughts or floods (25%).

Legislative instruments have been promoted at EU level the protection and sustainable management of water resources. The WFD considers that citizens participation in EU is very important for the achievement of the proposed goals (Table 3.2).

Table 3.2   Major water pollutants and its contribution to water and biota

Pollutant category	General contribution/examples	Relative contribution		Sources	
		Water supplies	Biota	Natural	Anthropic
Organic	Add toxics; Increased oxygen demand *Plant matter, livestock excreta, pesticides; solvents*	Increased need of treatment	Tolerated in moderate quantities if is not released too quickly, serious if DO drops too quickly	Run-off and seepage through soil	Domestic sewage, food processing, animal wastes, industry
Immiscible liquids	Formation of a layer at the water surface that could prevent O_2/CO_2 interchange *Oil*		Reduced dissolved oxygen; insect breeding affected	Unlikely	Oil-related activity
Microorganisms	Pathogenic to humans Cause disease Bacteria; viruses; protozoa; parasites *Escherichia coli, total coliforms, faecal coliforms and enterococci*		None	Animal excrement	Human and animal wastes
Endocrine disruptors	Alteration of ecology *Drug residues, hormones and feed additives*	Can be present in water sold in plastic bottles	May adversely affect health and reproduction of both	*Fusarium* species of fungus	Chemical manufacture, intensive farming
Nutrients	Excessive growth of plants and other species *Nitrates and phosphates*		Demand on dissolved oxygen	Natural degradative processes	Sewages Animal wastes, fertilisers, detergents

(continued)

Table 3.2 (continued)

Pollutant category	General contribution/examples	Relative contribution		Sources	
		Water supplies	Biota	Natural	Anthropic
Suspended solids	Reduction in light penetration, blanketing, introduction of colour Soil, silt	Obstruction of filters; increased need of treatment	Photosynthesis reduced; blanketing of benthic plants and animals; obstruction of gills of fish	Land erosion, storms, floods	Pulp mills, quarrying, building or development work involving ground displacement
Toxic chemicals (e.g. heavy metals, pesticides, phenols, PCBs)	Toxic to humans, animals and plants Disrupt immune and endocrine system Copper, mercury, arsenic, selenium and manganese		Could be lethal	Rare	Detergents, pesticides, tanneries, pharmaceuticals, refineries, mining unlined landfills
Acids/alkalis	Lowering/raising of pH; acids can dissolve heavy metals	Corrosion	Only narrow range of pH tolerable; heavy metals toxic	Naturally acid or alkaline rock	Battery, steel, chemical and textile industry; coal mining
Heat	Decrease in dissolved oxygen; increase in metabolic rate of aquatic organisms	None	Possible reduced breeding or growth of aquatic organisms	Unlikely	Power plants, steel mills
Taste-, odour- and colour-forming compounds	Taste, malodour, colour		Tainting of fish	Peat	Chemical manufacture or processing

Adapted from Nesaratnam [126]

In order to evaluate the status and control the effectiveness of water protection measures applied after the adoption of the WFD, a comprehensive monitoring plan was developed to obtain resilient and reproducible information across all EU member states [59]. The WFD Implementation Instrument, regulated by Article 13 and Annex VII, is the River Basin Management Plan (RBMP) which, based on knowledge of the state of water bodies, sets the targets for a period of six years, involving successive stages of planning and implementation, including public participation to achieve them. Initially, to obtain a good environmental status for all waters, 2015 was set, but this proved to be too ambitious, setting new deadlines for RBMP implementation, the second (2016–2021) and third cycle (2022–2027), which will allow for the identification of new programs of measures to be prepared for each updated RBMP.

3.2 Methodology

The overall status of water can be quantified through two distinctive methods. The first method assesses the chemical status by collecting water samples and determining the concentrations of the chemical parameters. Based on the liaison between obtained concentration values and current normative can be asses the water quality. The other method involves studying aquatic biota (fish or other invertebrates) to determine water quality. A richer biota of species and species corresponds to a higher quality water compared to situations where the biota is extremely low reaching the total absence of a situation characterized by a reduced water quality (Fig. 3.6).

By 2015, most EU countries have failed to achieve the objective of good environmental status but are not yet formally non-compliant because the WFD allows for temporary relief for specific water bodies requested by the Member States. Analysing the latest results, experts have concluded that for some polluted or heavily modified water bodies, good environmental status may be impossible, and after a further decade of exemptions, they could be classified as falling into less environmental objectives provided that the country can demonstrate that it has acted with all possible means to achieve good environmental status. Romania is almost entirely located (97.4%) in the Danube River Basin and has 29% of the area and 21.7% of the population of the entire basin. Romania is relatively poor in water resources, disposing of only 1870 m^3 water/inch/year, confronted by the average of 4000 m^3 water/inch/year in Europe. It has a total area of 238,391 km^2 and a total population of 22.3 million inhabitants of whom only 19.9 million are permanent residents [125, 123]. Romania is still largely a rural country, with 46% of the population living in rural areas (the highest proportion amongst all EU countries), and this has a significant impact on water management.

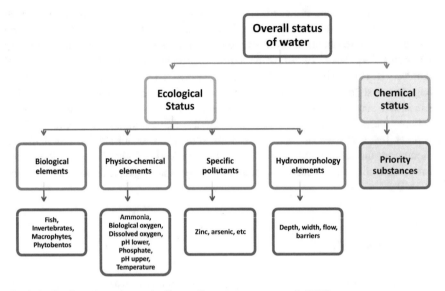

Fig. 3.6 The flow chart of water bodies quality status assessment in WFD

National Administration "Romanian Waters" (ANAR) is the national operational arm of the water sector, being in charge of managing all large water resources infrastructure (except dams dedicated to hydropower generation). In order to ensure consistent monitoring programs across the country, the national working team from the ANAR has developed a conceptual framework for the development of monitoring programs and the assessment of the status of surface and underground waters. Essential elements of this framework include environmental quality standards for certain substances, outline monitoring programs, determination of the mode and the frequency of sampling, the specification of sampling locations in different water categories and to establish water-based assessment rules based on the monitoring results. The results obtained from the monitoring are reported to the European Commission as a management plan and, given the failure to meet the original WFD objectives, such as "good environmental status"/"good environmental potential" and "good chemical status", are planned and implemented then remedial measures.

National assessments of the ecological status or ecological potential of the bodies of water are based on the results obtained in the annual monitoring programs.

Ecological status is based on biological quality elements and supporting physico-chemical and hydromorphological quality elements. Thus, surface water bodies, ecological status, or ecological potential are assessed according to biological quality elements using different methods (BQE): fish, fauna, benthic invertebrates, macroalgae, phytobenthos and phytoplankton. The final result is based on the concept that the worst result of a BQE assessment determines the outcome of the global assessment. To classify the ecological status of water bodies, five status classes are distinguished:

(1) high; (2) good; (3) moderate; (4) poor and (5) bad. For classes 3–5 imposing actions are mandatory to improve the state to what is stipulated by WFD.

In WFD, heavily modified and artificial waters differ from natural water bodies; those in the first category either have been artificially created or have undergone changes to a level that no longer allows for a "good environmental status" without significantly affecting the use of those waters. In this case, the environmental objective of a "good environmental potential" requires improvements in hydromorphological pressures without affecting the use of non-substitutable water.

Another component of the WFD objective, "good *chemical status*", applies to all types of water bodies (natural, artificial and heavily modified), being determined by compliance with environmental quality standards for a series of pollutants with significant impact at European level. Moreover, along with BQE, QE are assessed: River Basin Pollutants (RBSP), Physical-Chemical QE (temperature, oxygen, pH, nutrient conditions) and Hydrological QE. These supplementary QEs are classified as "good" or "less good", according to well-defined limits that have been defined for most of these QEs for different types of water bodies in each water category. The requirements for obtaining a "good environmental status" are:

(i)   that all BQEs must achieve "good status" (Class 2 or better);
(ii)   all ("good") quality values should not be exceeded; and
(iii) for other physicochemical substances, support for QE and hydromorphology must fit in a field that allows good ecosystem functionality (Fig. 3.7).

In accordance with the requirements of the WFD, are considered significant pressures those that result in failure to achieve the environmental objectives for studied water bodies. Depending on how the water body system is operating, it can know if pressure can cause an impact. This approach linked list of pressures and the particular characteristics of the receiving basin led to the identification of significant pressures. European directives present the limits over which the pressures can be potentially significant and the substances and the groups of substances taken into account. In the view of the European Commission's new reporting requirements for the Management Plan, the methodology for identifying significant pressures and assessing the impact on bodies of surface water for the updated RBMP has been revised. The

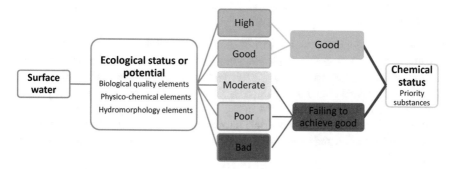

Fig. 3.7 Assessment of the ecological and chemical status of surface water bodies (after WFD)

pressures were compiled in accordance with the EU Guidance on Reporting of the Updated Management Plan, like punctual, diffuse pressures, hydromorphological alterations (including water leakage), quantitative pressures for groundwater, other anthropogenic pressures, unknown pressures.

3.2.1 Significant Point Sources of Significant Pollution in Romania

In establishing the potentially significant pressures—point sources, a set of criteria has been applied which has led to the identification of potentially significant punctual pressures, taking into account the discharges of treated or impure waters into surface water resources.

For determining the diffuse potential significant pressures, the following are considered the main categories of diffuse pollution sources:

- human agglomerations/localities that do not have wastewater collection systems or adequate sludge collection and disposal systems in sewage treatment plants as well as localities with non-compliant household waste dumps;
  In Romania, there are 1917 human agglomerations (>2000 p.e.) with a total organic load of 19,912,886 p.e., considered as potentially significant pressures.
- agriculture: agro-zootechnical farms that do not have adequate systems of storage/ utilization of manure, localities that do not have collection systems centralized individual with manure platforms, units using pesticides that do not comply with the legislation in force, other agricultural units/activities that may lead to significant diffuse emissions;
  Romania is now one the few EU countries that have designated its entire territory a Nutrient Vulnerable Zone. The Code of Good Agricultural Practices was duly developed, establishing the rules for proper management of manure coming from animal farms and for maximum quantities of N, P and K in crops fertilizers allowed by the Nitrates Directive [100].
- industry: warehouses for raw materials, finished products, auxiliary products, non-compliant waste storage, units producing diffuse accidental pollution, abandoned industrial sites.

This chapter includes an analysis of the most recent data on the status of water quality in WFD implementation in Romania as it was presented in the interim report on compliance as of December 31, 2016, sent to the EC, validated and published in WISE WFD database in July 2018 [63]. These data have been validated by those from scientific literature published in the ISI journals, related to water pollution in Romania. For the graphic materials have been used licensed software programs, based on information and cartographic materials from www.eea.europa.eu/data-and-maps/ [62].

3.3 Results and Discussions

Regarding the surface water bodies at EU level, the main significant pressures are *hydromorphological pressures* (affecting 40% of water bodies), *diffuse sources* (38%), particularly from agriculture, and atmospheric deposition, mercury, followed by *point sources* (18%) and *water abstraction* (7%). The main impacts on surface water bodies are nutrient enrichment, chemical pollution and altered habitats due to morphological changes.

In the updated RBMP, 3027 water bodies were redeemed (compared to 3399 water bodies, identified in the first RBMP), the average length of water bodies located on the hydrographic network being 25.5 km.

Water management in Romania has been organized around river basins into 11 river basin authorities, the Siret River Basin having the largest area (42,890 km^2) and the greatest water resource (RO1-Banat; RO2-Jiu; RO3-Olt; RO4-Arges-Vedea; RO5-Buzau-Ialomita; RO6-Dobrogea-Litoral; RO7-Mures; RO8-Crisuri; RO9-Somes-Tisa; RO10-Siret; RO11-Prut-Barlad).

Geological formations in Romania are very different in terms of petrography and morphology. From a geological point of view, Romania is characterized by silica (predominantly), limestone and organic substrates.

Land use at national level, according to official data, includes agricultural land (61.39%), forests (26.65%), water and ponds (3.5%), built surfaces (3.06%), roads and railways (1.63%) and other surfaces (3.77%).

According to WISE WFD database, in Romania are the following categories of surface water [63]:

- rivers (natural, strong modified and artificial)—75.486 km (2891 cadastral rivers);
- natural lakes—130 (1009 km^2);
- transitional waters—2 (383 km^2);
- coastal waters—4 (252 km^2) (Figs. 3.8, 3.9 and 3.10).

3.3.1 Surface Water Bodies (SWB)

In Romania, 50% of all bodies of surface water have been classified as heavily modified water bodies (35%) or artificial water (15%) (Figs. 3.11, 3.12 and 3.13).

Detectability of Cd, Cu, Pb and Zn levels in Maramureş and Satu Mare counties in NW Romania has been found to reflect not only metal loadings from local geology and anthropogenic sources but also to be influenced by water chemistry, including pH and Eh. In general, Cu and Zn were found to be the most widely detectable metals, the enrichment above intervention values is restricted to Cd and Zn, two relatively mobile elements. Cd is by far the most toxic and would appear to present a greater threat to human health; however, Zn enrichment is more widespread. In RO9 Somes-Tisa is located the Aurul tailings pond where the groundwater is polluted with heavy metal elements (Fe, Mn, Ni, Zn, Cd, Pb) [42].

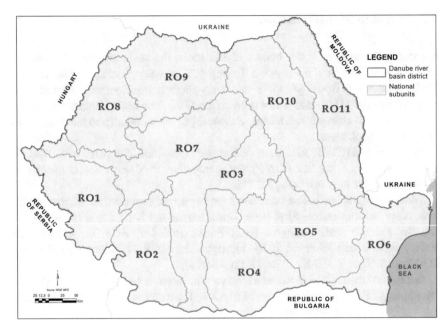

Fig. 3.8 Romanian water basin authorities' distribution

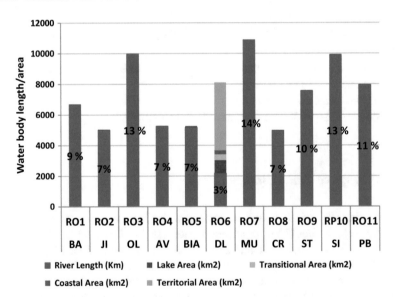

Fig. 3.9 Types of Romanian surface water bodies

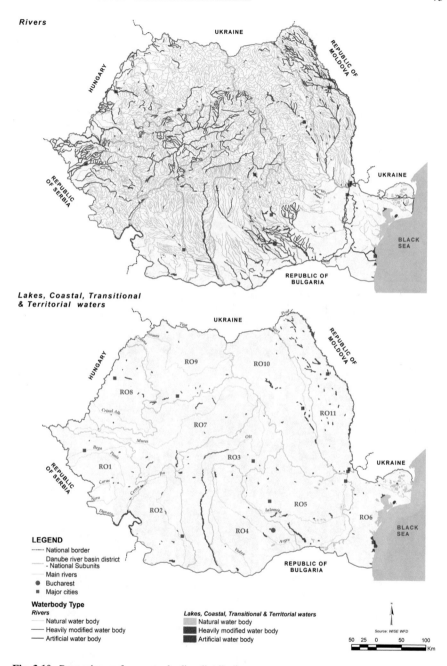

Fig. 3.10 Romanian surface water bodies distribution

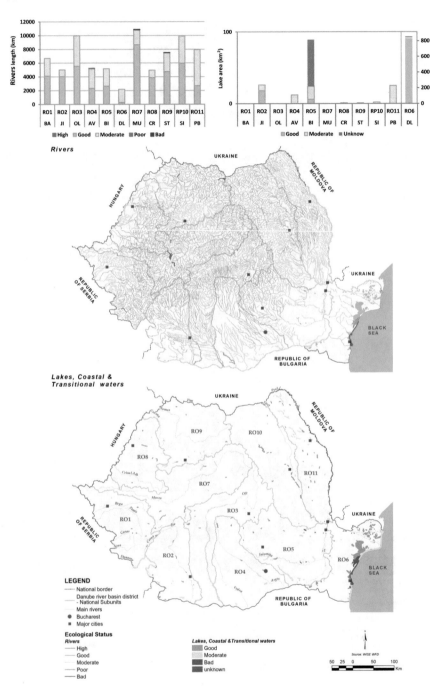

Fig. 3.11 Surface water body: waterbody category and ecological status or potential

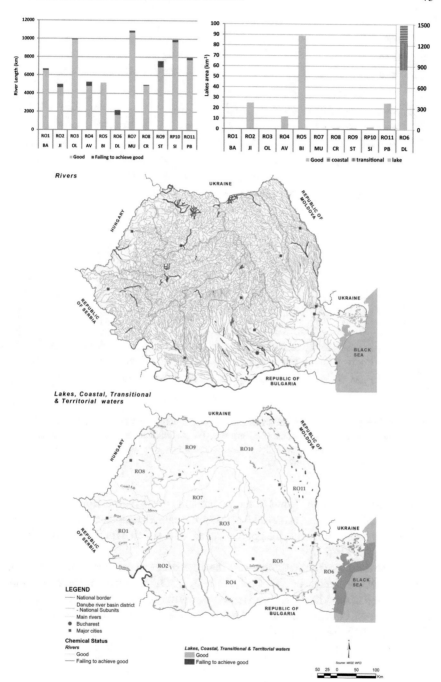

Fig. 3.12 Surface water body: waterbody category and chemical status

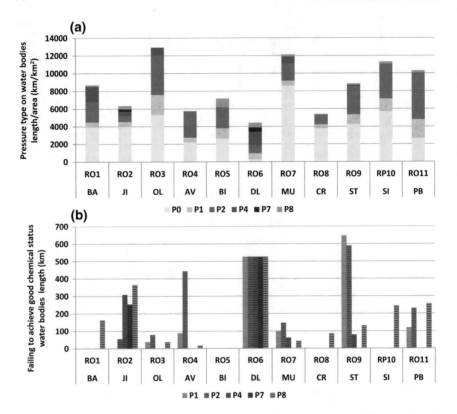

Fig. 3.13 Significant pressures on surface water bodies length (**a**) and the failing to achieved good chemical status (**b**) on National subdistricts (P0—no significant anthropogenic sources; P1—point sources; P2—diffuse sources; P4—hydromorphology; P7—anthropogenic pressure-other; P8—anthropogenic pressure-unknown)

The examination of the Tamiš River water quality reveal that certain improvements in water quality have been achieved, but this is not enough to update the classification of Tamiš along its course into the good class, which is suitable for bathing, recreation and water sports [4].

In some rural area for drinking water, the exceedings were recorded in the case of pH, TDS, DO, turbidity, Na^+, K^+, Ca^{2+}, F^-, NO_2^-, SO_4^{2-}, Fe and Pb. The continuous consumption of drinking water from these sources may be associated with symptoms such as nausea, cramps, diarrhea and headaches [50, 112, 114, 115, 153, 155, 164] (Figs. 3.14, 3.15, 3.16, 3.17 and 3.18).

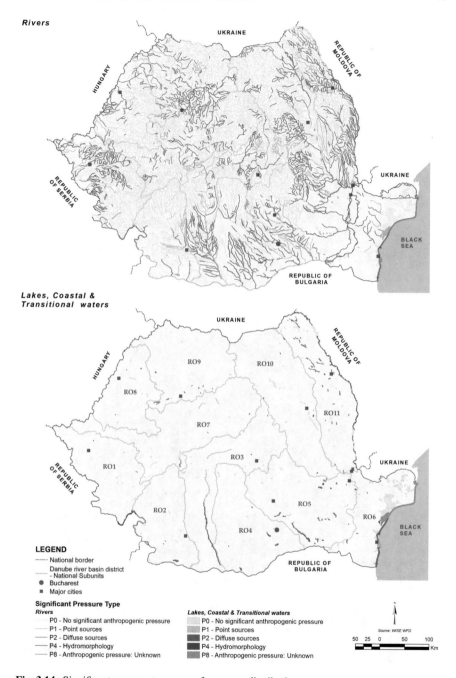

Fig. 3.14 Significant pressure type on surface water distribution

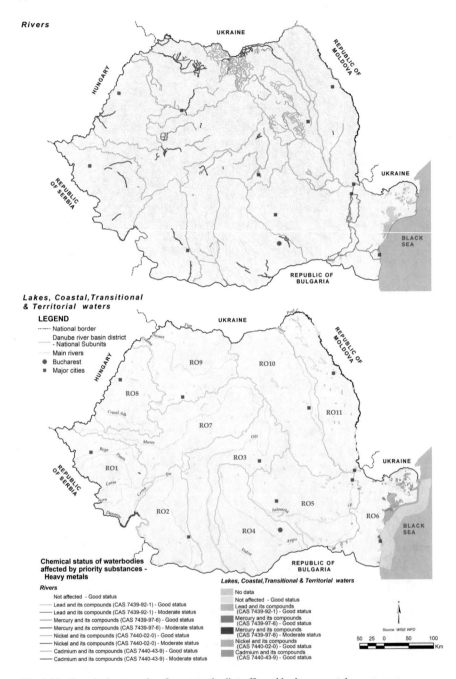

Fig. 3.15 Chemical status of surface water bodies affected by heavy metals

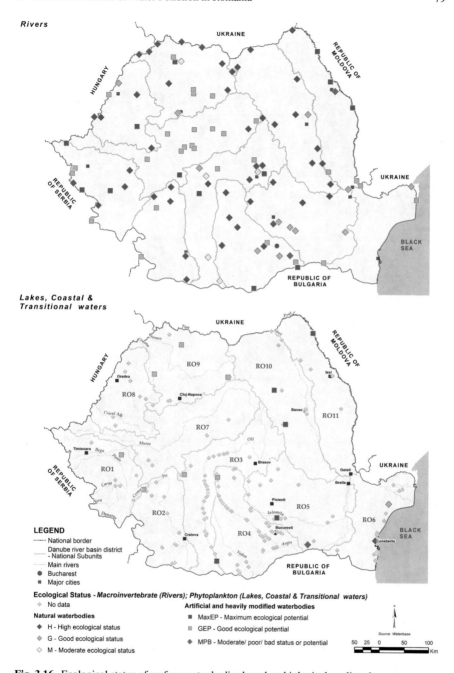

Fig. 3.16 Ecological status of surface water bodies based on biological quality elements

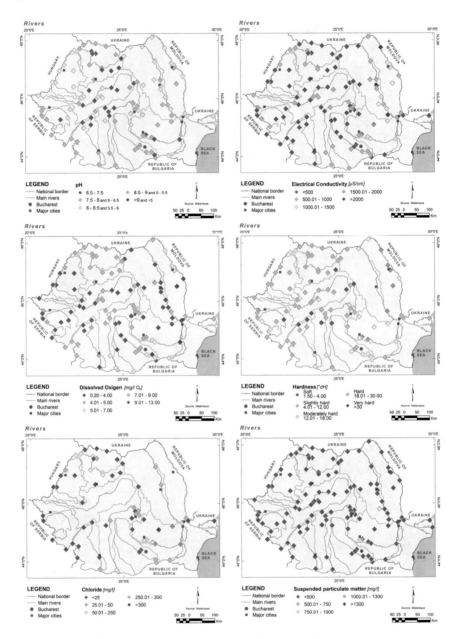

Fig. 3.17 Distribution of main chemical parameters of Romanian rivers

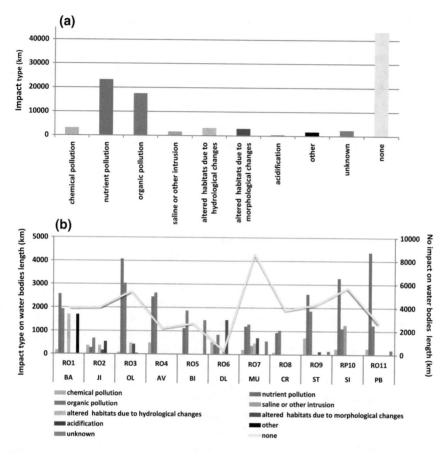

Fig. 3.18 Significant impacts on surface water bodies (**a**) at national level, (**b**) on national subdistricts

Higher content of nitrite and ammonium ions indicates the presence of reducing conditions in the groundwater in spring, than in other seasons [1, 14, 43, 97, 98, 127, 132]. Nitrate ions dominate the other species, but also four samples situated close to the garbage dumps, have a large content of ammonium ions.

3.3.1.1   Organic Pollutants

When an excess of organic matter, such as manure or sewage, comes into contact with different types of water, an organic pollution occurs. In the case of a vapor by organic matter input, the number of decomposition products increases, the decomposition yield being directly proportional to the amount of oxygen required, reaching in some situations the depletion of dissolved oxygen and implicitly the damage to aquatic organisms. As they die, they are broken down by aerobic degradation mechanisms, which lead to further depletion of the oxygen level. When nitrogen and/or phosphorus-containing pollutants are accumulated in aquatic ecosystems, there is an over-development of plants and algae, which after the end of their lifetime is transformed into organic oxygen-consuming matter, the process being known as eutrophication (Fig. 3.19).

Organic pollutants can be divided into following categories:

(a) *Oxygen Demanding Wastes* The waste water, regardless of their type (domestic sewage, industrial: slaughterhouses, food canning, textile, leather, paper and pulp, etc.), are loaded with varying amounts of biodegradable organic compounds, in various forms: dissolved, suspended, colloidal [10, 32, 67, 75, 114, 163, 161, 174, 188, 190, 191]. These, through microorganisms, are aerobically biodegraded, degradation and decomposition taking place simultaneously with the consumption of dissolved oxygen available in water sources. Aquatic ecosystems are adversely affected if the amount of dissolved oxygen decreases below 4 mg/L, dissolved oxygen lowering the outlines of an indicator of pollution in waters.

(b) *Synthetic Organic Compounds* As a result of various anthropogenic activities, synthetic organic compounds are likely to enter aquatic ecosystems, such as their production processes, discharges during transport, and their use in various applications. These compounds include synthesis products such as volatile organic compounds (VOCs), pesticides, insecticides, detergents, food additives, pharmaceuticals, paints, synthetic fibers, plastics, solvents, which even in low concentrations are not suitable for various uses, most resistant to microbial degradation (detergents form foams and volatile substances that can cause explosion in sewers). Persistent organic pollutants (POPs) are compounds organic, non-polar, and hydrophobic, with extremely low solubility in water, lipophilic, which causes accumulation in the fatty tissues of the organisms. In aquatic ecosystems, these compounds are adsorbed on sediment particles or particles suspended in the mass of water, particularly on organic particles, and accumulate in the tissues of end-of-trophic chain (fish, mammal, bird and human species) and may be much higher than the concentration in the environment [70, 66, 111, 147].

(c) *Oil* is a natural product, a complex mixture of hydrocarbons, degradable under bacterial action [143, 175, 176]. It enters in aquatic ecosystems through leaks from oil wells, transport pipelines and wastewater from production and refining, oil tanker spills. Having lower densities than water, they have the ability to dissipate at the surface of the water, thus blocking water contact with air, implicitly decreasing the amount of dissolved oxygen, as well as reducing the photosynthetic activity by obstructing light transmission. For different oils the biodegradation yield is distinct, the tars being some of the slowest.

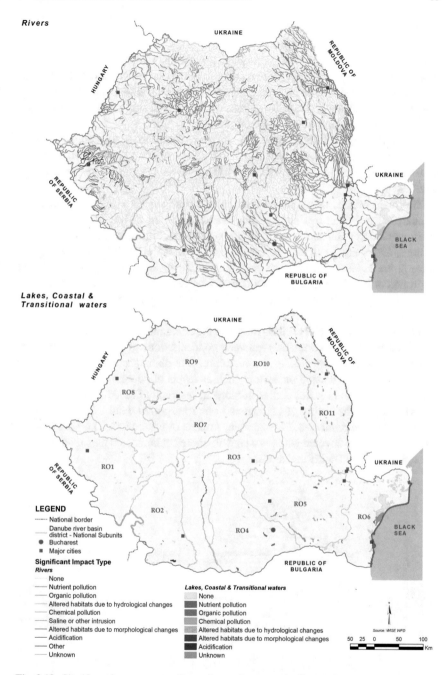

Fig. 3.19 Significant impact type on Romanian surface water bodies

3.3.1.2 Pathogens

Pathogens are microbes of very small size that cause diseases, bacteria, viruses, fungi and parasites being the main representatives. Pathogenic contamination of water and associated diseases is a major concern for water quality worldwide. Contaminant infection is a serious problem for almost all types of water bodies, which makes recognition and understanding essential [194].

Microbial contamination of water is often from feces related to humans (wastewater treatment plants, non-collective sewage systems, combined wastewater overflow sewers), domestic animals (manure spreading, garbage overloading) or wildlife [88].

The main origins of microbial contamination of natural aquatic resources are discharges from water treatment plants, decontamination stations, hospitals, industries regarded as point sources, etc. On the other hand, diffuse sources (suspensions, manure, and sludge) can also be considered. Two of the most known pathogenic bacteria are *coliform* and *E. coli* bacteria. *Coliforms* are normally present in the environment in safety rank and can actually be used to detect other pathogens in water. If *coliforms* increase in number, become dangerous for the environmental health (Fig. 3.20).

The existence of *E. coli* bacteria indicates the contamination of water with human or animal waste. The correlation between concentrations of pathogens and urban activities is well represented in the literature. The abundance and importance of pathogens in water depend on factors such as levels of contamination, the persistence of pathogens in water bodies, biological reservoirs (including aquatic plants and sediments), the ability of pathogens to be transported, land use management, the hydrographic basin size. For example, streams of watercourses passing through partially or wholly covered with grassland are more contaminated than those flowing through forests and cultivated areas.

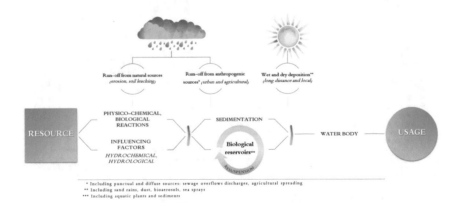

Fig. 3.20 Schematic synthesis of source and fate of allochthonous microorganisms in water (processing after Jung et al. [94])

Agriculture leads to water pollution through three components: pastures, vegetable and animal production. Diffuse pollution resulting from agricultural crops is mainly caused by nutrients, heavy metals and salts from chemical fertilizers and organic chemicals used to protect crops. Animal growth generates both increased amounts of faeces, urine and grassland degradation through erosion processes, the main agents threatening water quality are nutrients, often accompanied by heavy metals and minerals, their environmental impact being very negative [16, 18, 23, 31, 137].

The optimum workflow of the waste water treatment starts with the design and implementation of the sewage collection system (Fig. 3.21 orange bars), after which the collected waters have to pass through the primary and secondary treatments (Fig. 3.21 yellow respectively, green bars). The first treatments implies physical processes for retaining the large residues by filtering out the contaminants while the secondary is characterised by chemical and/or bio-chemical mechanisms for water purification through oxidizing/ aeration. More stringent 'tertiary' treatment (Fig. 3.21 blue bars) can then be applied to remove mainly nutrients such as phosphorous and nitrates [46, 178, 187].

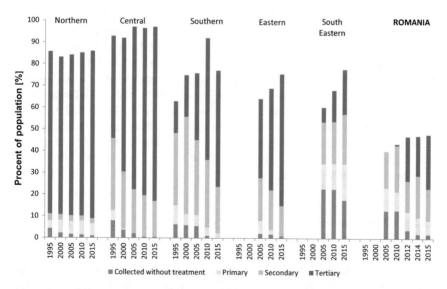

Fig. 3.21 Changes in urban waste water treatment in Europe and Romania (*Northern Europe* Norway, Sweden, Finland and Iceland; *Central Europe* Austria, Belgium, Denmark, Netherlands, Germany, Switzerland, Luxembourg and United Kingdom; *Southern Europe* Greece, Italy, Malta and Spain; *Eastern Europe* Czech Republic, Estonia, Hungary, Latvia, Lithuania, Poland and Slovenia; *South-eastern Europe* Bulgaria, Romania and Turkey) https://www.eea.europa.eu/data-and-maps/indicators/urban-waste-water-treatment/urban-waste-water-treatment-assessment-4

From an european perspective, the treatment of urban waste water registered an increased trend, the connectivity to waste water treatment plants being different for each region. Despite the case of the northern region, where from 1995 above 80% of the population has access to treatment plants which receive in proportion of 70% tertiary treatment, in the central countries a large part of the population (97%) is currently integrated into the local waste water treatment system, from which 75% benefits of tertiary treated water. In southern, south-eastern and eastern Europe the proportion of the population connected to urban waste water treatment is generally lower than in other parts of Europe but has increased over the last 10 years being at a level of approximately 70% [61].

For Romania has been noticed a particular situation where almost half of the residents are not connected to sewage system, therefore waste water is directly transferred to natural emissary. In addition, it is highlighted a slow continuous increase regarding the beneficiaries of treated waste water from 40.2% (2005) to 47.8% (2015), from which currently only 1.9% do not treat the collected waste waters. After the EU accession in 2007, the primary/secondary/tertiary ratios are changed in the favour of the last stage of treatment, prevailing the removing of the mainly nutrients (39:61:0 in 2005 to 14:32:54 in 2015).

3.3.1.3 Inorganic Pollutants

Inorganic pollutants are naturally present in the environment, but due to human activities, the amount of pollutants released into water bodies has increased rapidly over the past decades. The most common inorganic pollutants in the water are ions derived from nutrients (ammonium, nitrate, phosphorus) or heavy metals (arsenic, cadmium, zinc, lead, mercury, etc.) characterized by high toxicity to both the environment and humans (Table 3.3). Pollutants like ammonia and phosphorus potentiate the eutrophication process, excessive amounts of pollutants released into the water can accelerate this phenomenon in a few years, while heavy metals such as arsenic can cause skin, lung, bladder and kidney cancer, as well as changes in pigmentation, skin thickening (hyperkeratosis) neurological disorders, muscle weakness, loss of appetite and nausea (Fig. 3.22).

3.3.1.4 Nutrients

Agricultural run-off, wastewater from the fertilizer and wastewater industry contains a substantial concentration of nutrients such as nitrogen and phosphorus. These drinking water supplies plants can have effects stimulating the growth of algae and other undesirable aquatic plants in the natural receptors, leading to the degradation of these bodies of water. In the long run, the body of water reduces the amount of dissolved oxygen, experiencing eutrophication and dead water [25, 43, 104, 105, 120, 132].

Table 3.3 Main inorganic pollutants studied in Romanian waters, sediments and biota

Pollutant	Area	Water type	References
Copper, cadmium, chromium, lead, nickel	Stanca Costesti lake	Freshwater reservoir, sediment, biota	Strungaru et al. [185]
Mercury	Babeni reservoir—Olt river	Sediment	Bravo et al. [12, 13]
Mercury + Hg-methylating		Water	
Total form of iron, manganese, cobalt, zinc, chromium + nutrients	Suceava river	Surface	Briciu et al. [15, 16]
Arsenic	Mine		Kim et al. [95]
Nutrients	Cuejdi river	Surface	Mihu-Pintilie et al. [113]
Nutrients	Rebricea river	Surface	Moisii et al. [116]
Arsenic +	Aries river	Surface	Marin et al. [106]
	Middle Aries river mine	Surface	Ozunu et al. [131]
	Zlatna Gold mine	Surface and ground	Papp et al. [135, 134]
Nitrate	Balaesti, Gorj	Ground	Popa et al. [141]
	Galda river	Surface	Popa et al. [142]
	Blaj	Surface	Popa et al. [140]
Mercury		Soil, surface, sediment	Popescu et al. [145]
Lead, cadmium, zinc, arsenic, copper	Eastern Carpathian mines	Acid mine drainage	Stumbea [186]
Arsenic	Bega Timis	Well	Senila et al. [177]
Gold	Rosia Montana		Stoica et al. [184]
Mercury +	Black sea	Seawater, algae	Trifan et al. [192]
	Danube delta	Surface water	Vosniakos et al. [197]
	Vardar river		

As to human health dangers, people who swim in eutrophic waters that contain blue-green algae may have skin and eye irritation, gastroenteritis and vomiting, and high levels of nitrogen in the water supply cause a potential risk, especially in infants six months when methaemoglobin causes a decrease in blood oxygen transport capacity because nitrate ions in the blood slightly oxidize ferrous ions to hemoglobin (Fig. 3.23).

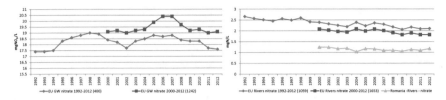

Fig. 3.22 The evolution of nitrate content in ground and river waters in EU and Romania between 1992 and 2012 (data source: EEA WISE-WFD [63])

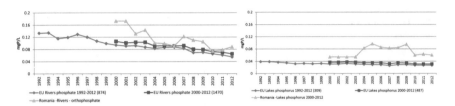

Fig. 3.23 The evolution of phosphate content in river and lakes waters from EU and Romania between 1992 and 2012 (data source: EEA WISE-WFD [63])

3.3.1.5 Macroscopic Pollutants

One of the main contributors to pollution are humans through their daily activity that impact the environment. Water pollution comes in very diverse forms and macroscopic pollutants are one of the most recognisable. Defined as large and visible objects that are found floating in water, are mainly consist of trash, especially plastic waste. Moreover, other identified types of polluting materials are nurdles, debris, shipwrecks, municipal solid waste (toilet paper, textiles, rubber, tin, plastic bags, glass, metals, aluminium, polystyrene foam, cigarettes, diapers) and even oil and grease [180]. Furthermore, the smoke released by anthropic activity (traffic, industry, house heating) returns from the atmosphere as wet and dry depositions causing significant damage to the aquatic environment [78]. In several situations, the discharge of macroscopic items is done accidentally or during severe weather events (storms) or natural hazards such as floods.

3.3.1.6 Radioactive Waste

People consider radioactive waste to be particularly dangerous both for the environment and for humanity [20, 36, 35, 68, 103, 121, 144, 146, 198]. At fairly high concentrations they are fatal, while at lower concentrations they can cause cancers and other diseases. The largest sources of radioactive contamination in Europe are two factories reprocessing spent fuel from nuclear power plants: Sellafield on the northwest coast of Britain and Cap La Hague on the northern coast of France, both discharging radioactive waste into the sea.

Radionuclides, as well as natural constituents of the aquatic environment, come from the erosion of rocks with radionuclides. Natural radionuclides are incorporated as elements in the crystal lattice of minerals and are present in sediments, their activities being dependent on the type of existing minerals. Existing radioactive materials in soil and sediments could pose a potential health risk, especially if they are assisted by natural processes such as atmospheric deposition and erosive activities.

As a result of natural processes, minerals containing radionuclides are transported to the lake environment and are eventually incorporated into the sediment in the water bodies [202].

Radioactive materials come from the following:

- exploitation and processing of ores; use in research, agriculture, medical and industrial activities such as I^{131}, P^{32}, Co^{60}, Ca^{45}, S^{35}, C^{14} etc.; radioactive discharges from nuclear power plants and nuclear reactors, e.g. Sr^{90}, Cs^{137}, Pu^{248}.
- the radioactive isotopes are toxic to life forms; they accumulate in bones, teeth, and can cause serious disorders; the safe use for lifelong use is 1×10^{-7} μCi/mL.

A large number of studies on the behavior of radionuclides in European lakes have been carried out over the last decades. The accidental release of radioactive substances into the environment leads to the need to apply appropriate countermeasures to restore the polluted environment. However, despite their obvious benefits, such interventions can have harmful economic, ecological and social effects, which need to be carefully assessed. Human significant doses may occur after radioactive contamination of water bodies such as rivers, lakes and reservoirs.

3.3.1.7 Other Forms of Pollution

Plastics

Plastic pollution is one of the biggest threats facing our oceans, which range between whole plastic bottles and small microplastics found in seas around the world. Almost all researches on plastic contamination of water systems focus on the oceans. But the biggest problem is plastic that is found in freshwater ecosystems. Preoccupied citizens read about all the plastic contaminants found in the ocean—from the whole plastic waste to the dead whales, with plastic-filled bellies. "But not the plastic in the oceans, we should worry most about it," but from the rivers and lakes around us. Most people find this a worrying situation and believe that this should be prevented, but we live in a plastic age where we are surrounded by plastic products that are easy to obtain and easy to throw. In the life cycle of plastic products, a multitude of actors are involved, ranging from the chemical industry and plastic producers to garbage collectors and regulators [56] (Figs. 3.24 and 3.25).

Fig. 3.24 Cleaning action by ANAR of plastic wastes from the Izvorul Muntelui lake in July 2018 (https://www.green-report.ro/administatia-nationala-apele-romane-a-inceput-o-operatiune-de-ecologizare-a-lacului-bicaz/)

Fig. 3.25 Plastic in Black Sea (https://www.newsinlevels.com/products/plastics-in-the-black-sea-level-3/)

One of the main sources is packing with our clothing. Very small acrylic, nylon, spandex and polyester fibers are removed and discarded each time we wash clothes and are transported to wastewater treatment plants or discharged into the open environment.

According to a recent study, more than 700,000 microscopic plastic fibers could be released into the environment during each cycle of a washing machine. This has not yet been studied in handwashing, which is more common in developing regions, but the effects could also be significant there. In order to influence the physical effects of the micro-plastic itself, ingested plastic remains can act as a concentration medium and transfer chemicals, persistent, bioaccumulative and toxic (PBT) substances, such as polychlorinated biphenyls, PCBs to organisms [124].

Microplastics have been detected in Asia, the EU and North America. They are in remote and protected areas (e.g. Hovsgol Lake, Mongolia) and in quantities large enough to overcome natural particles (e.g. the Danube River, Austria). It is considered to be a significant share of landfill trash traded to the oceans, for example, it is estimated that 1533 tones of plastic trash per year enter the Black Sea on the Danube. Initial studies suggest that both invertebrates in freshwater and fish consume microplastics by ingestion resulting in physical effects including physiological stress responses and even signs of tumor formation [139].

Alien Species

Invasive alien species refer to all species and subspecies introduced outside their natural habitat, both past and present, from all taxonomic groups. This includes any part of the organism: gametes, seeds, eggs or propagules that could survive and reproduce later. Invasive alien species refers to any foreign species of which introduction causes or is likely to cause economic or environmental damage or harm human health. The invasive state is usually assessed by exponential population growth and/or gamma expansion. In recent years, invasive species have become a policy with high levels of proliferation worldwide. Represent a major threat requiring international cooperation and a multidisciplinary approach at different levels: academic, administrative and local communities. The current list of freshwater alien species reported in Romania includes 44 species: 28 fishes and 16 invertebrate species, most of them from North America and Southeast Asia. Along the Black Sea coast of Romania, a total of 58 species of foreign animals were reported, of which one was a fish species, the rest being invertebrates. The largest group is crustaceans, easily transported with ballast water or body load. Very few species have been deliberately introduced, e.g. *Crassostrea virginica*, for aquaculture. Most of the foreign species reported in the Black Sea come from the Atlantic Ocean or are cosmopolitan [2, 44].

Accidental Pollution

The distribution on hydrographic basins shows that most accidental pollutions occurred in the Mures (67), Arges-Vedea (61), Olt (43), Siret (40) and Prut (36), basins with accidental pollution cases ranging between 67 and 3, a few cases being recorded in Banat (3) and Crisuri (6) (Fig. 3.26). By year, most cases have been noticed in 2017 (70), 2012 (65), 2015 (63). The cause of oil pollution is largely due to damage to oil and saltwater pipelines due to their spillage. In 2015, the decreased oxygen concentration due to high summer temperatures led to 9.5% fish mortality,

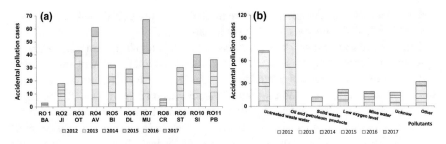

Fig. 3.26 Distribution of accidental pollution on national sub basins (**a**) and pollutants type (**b**) between 2012 and 2017

Fig. 3.27 Pollution of the Lapus River with mine waters from the Breitner mine (March 2018) *from Viorel Coroian*

much lower than in the previous year (18%), a significant amount being collected from the Jijia River and Bahlui in June (Figs. 3.27, 3.28, 3.29, 3.30, 3.31, 3.32, 3.33, 3.34 and 3.35).

Fig. 3.28 Pollution of Constanta port water with petroleum products (http://oildepol.ro)

Fig. 3.29 Pollution of the Ampoi River, in the Zlatna area of the Apuseni Mountains (July 2017) http://www.radiocluj.ro

Fig. 3.30 Dead fish, because of heat wave, in a pond in Sannicolau Mare—Timis (July 2017)
https://www.mediafax.ro

3.3.2 Groundwater Bodies

As far as the categories of groundwater bodies in Romania were concerned, identified, delineated and characterized 143 bodies of groundwater, compared to 142 existing in the first RBMP, 17 of which have a cross-border character. With regard to groundwater body categories, out of the total of 143 water bodies, 115 are phreatic groundwater bodies (horizon 1), the remaining 28 are groundwater bodies of depth (horizons 2, 3, 4).

The pressure type analysis was performed for each groundwater body, taking into account only significant pressures that exert an impact on the chemical and quantitative status of the groundwater body. At the national level, diffuse and point pressures caused by human agglomerations, agricultural sources, as well as those from the agricultural sources, have been considered as significant sources of pollution, which may have an impact on the chemical status of groundwater bodies from the industry.

Characterization of groundwater status, i.e. quantitative status and chemical status, is based on a classification system consisting of 2 classes: good and other than good (poor).

Fig. 3.31 The Dumitrelu sediment retention reservoir (2008)

Fig. 3.32 Acid mine drainage in Valea Vinului mine from Rodnei Mountain and Valea Sesei tailing ponds from Apuseni Mountain (2015)

Fig. 3.33 Pollution of the Sasar River, in Baia Mare area and leaks of $Fe(OH)_3$ from Aurul and Bozânta tailing ponds (2012) [115]

Fig. 3.34 The central part of Aurul Tailing pond (2014) [17]

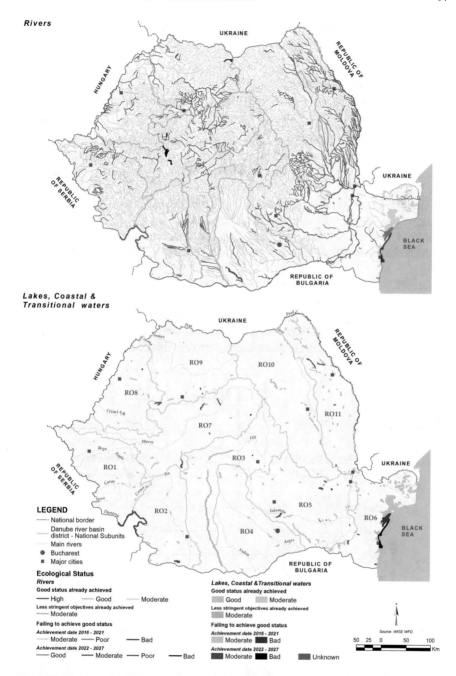

Fig. 3.35   The date of achievement the good ecological status of surface water bodies

The following criteria were used to assess the quantitative status of groundwater bodies:

– water balance;
– connection with surface water;
– influence on groundwater ecosystems dependent on groundwater;
– intrusion of saline water or other intrusions.

The freatic pollution is often an almost irreversible phenomenon, with significant consequences on the use of the underground water supply for drinking purposes, and the de-pollution of groundwater sources is a very difficult process (Figs. 3.36, 3.37 and 3.38).

The groundwater bodies are characterized by the slow response (unsaturated zone, recharge process, aquifer conditions, underground flow conditions, in part merely characteristic for a flow line), the hydrodynamic situation being essential for monitoring actions due to the long time period from recharge to discharge (Figs. 3.39 and 3.40).

Groundwater quality monitoring revealed that in 2016, about 16% (23/143) underground water bodies were outside the chemical parameters imposed by the rules. In 2016 there were reported 47 accidental pollution, the distribution of which according to the nature of the pollutant is as follows: untreated wastewater 18 (events)—accounting for 38.30% of the total accidents; petroleum products 14 events—29.79%;

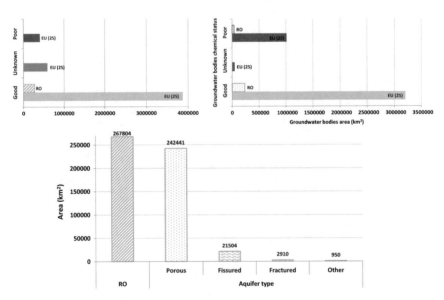

Fig. 3.36 Groundwater basin—chemical status of aquifers in correlation with aquifer type in EU and Romania [62]

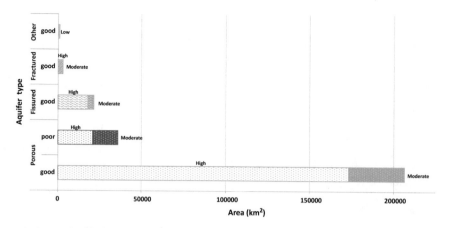

Fig. 3.37 Groundwater basin—chemical status of aquifers in correlation with productivity [62]

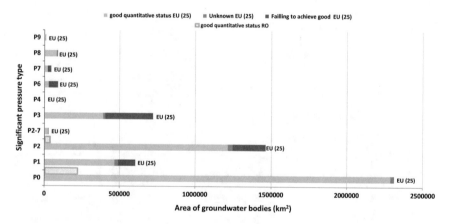

Fig. 3.38 The level of good quantitative status and failing to achieved status in EU and Romanian groundwaters based on pressure type [62]

semi-solid waste 6 events—12.76%; pollution of other nature 4 events—8.51%; unidentified substances 3 events—6.38%; mine water 1 event—2.13%; chemicals that attract low oxygenation 1 event—2.13%.

Based on frequency and type of the pollution of phreatic aquifers in urban areas events, a standard procedure was developed. For a hydrostructure polluted with petroleum products, a 3D conceptual model was created based on geostatistical tools taking into account: the spatial model, the parametric model and the hydrodynamic model [173, 175], observe the physical processes involved in water and conservative tracer flow in an aquifer system surrounded by two lakes [107]. Due to ground-

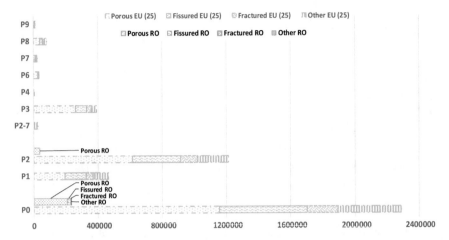

Fig. 3.39 Types of aquifer in EU (25) and Romania with good quantitative status

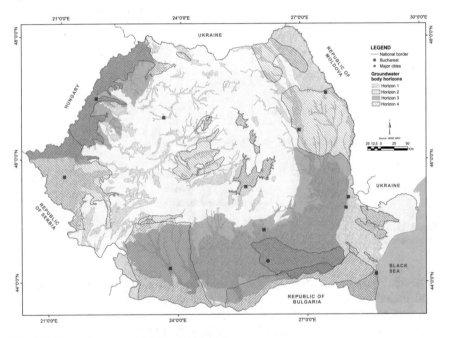

Fig. 3.40 Romanian groundwater bodies horizons

water reservoir uses for human consumption, the presence of microorganisms with pathogenic potential has been a significant health hazard, mostly in rural area, highly contaminated with petroleum products and phenolic compounds due to the groundwater aquifer from the Prahova-Teleajen alluvial cone. A positive correlation was found between the density of total coliform bacteria, *Escherichia coli*, enterococci and phosphate concentrations at sampling site. The main cause is the infiltration of residual waters from surrounding households, rainfall water leaching from fertilized agricultural crops (gardens) or manure storage [198].

From the analysis of the published works for the underground water can be noticed that the most studied indicators with a special impact on the quality of the environment and population health are: nitrates, ammonium, chlorides, sulphates, toxic metals for the environment, orthophosphates, phenols, arsenic. The causes of pollution of the aquatic groundwater with nitrogen are multiple and have a cumulative character. The main sources of nitrate pollution are permanent washing of soil impregnated with nitrogen compounds from the application of chemical fertilizers on some arable land categories, by atmospheric precipitation and irrigation water; the lack of wastewater collection systems, especially in rural agglomerations. Exceedances of the quality standard for nitrates are mainly recorded in areas where the soil is affected by the application of chemical fertilizers, large present and former chemical compound areas, but these overflows, which may pose a danger of pollution of the aquifers in the area, given the hydrodynamic character and the hydraulic conductivity of the water (Table 3.4). According to a report on the implementation of the Nitrates Directive, Southern Romania is one of the areas in the European Union where groundwater contains a high concentration of nitrates (Figs. 3.41, 3.42, 3.43, 3.44, 3.45, 3.46, 3.47, 3.48, 3.49, 3.50, 3.51, 3.52, 3.53, 3.54, 3.55, 3.56, 3.57, 3.58 and 3.59).

Table 3.4 Main pollutants studied in Romanian groundwaters

Location	Elements	Water type	References
Moldovian plateau	Ca^{2+}, Mg^{2+}	Ground	Romanescu et al. [171]
Vaslui river		Ground	Romanescu et al. [170]
Suha basin		Surface Ground	Romanescu et al. [167, 168]
NE Romania	Heavy metal	Ground	Bird et al. [9]
Birlad	Water quality index	Ground	Breaban et al. [14]
N Romania—tailing mining ponds		Ground	Cuciureanu et al. [42]
S Romania		Ground	Macalet et al. [101]
Lake	Modeling the aquifer pollution	Ground	Marinov et al. [107]
Rural settlements	Nitrate		Paiu and Breaban [132]
Zlatna river	Gold		Papp et al. [134]

(continued)

Table 3.4 (continued)

Location	Elements	Water type	References
Gorj county	Nitrate		Popa et al. [141]
W Romania	Phenol		Preda et al. [147]
Gorj county mining			Prunisoara et al. [149]
Targoviste Plain	Salt, cations, Fe, Mn		Radulescu et al. [151]
Buzau county	Cations, anions, heavy metals		Roba et al. [154]
Tillaberi region	Nitrates, cations, anions		Salihou Djari et al. [172]
Focsani	Oil	Vadose zone	Scradeanu et al. [175]
Khenchela	Cations, anions		Sedrati et al. [176]
Prahova county	Organic, nutrients, heavy metals		Stoica et al. [183]
Romania	Nutrients	Drinking water	Tevi et al. [190]

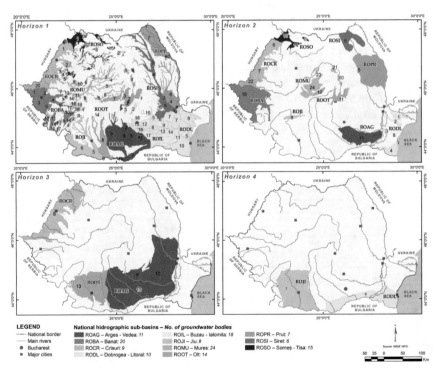

Fig. 3.41 Romanian hydrographic sub-basin

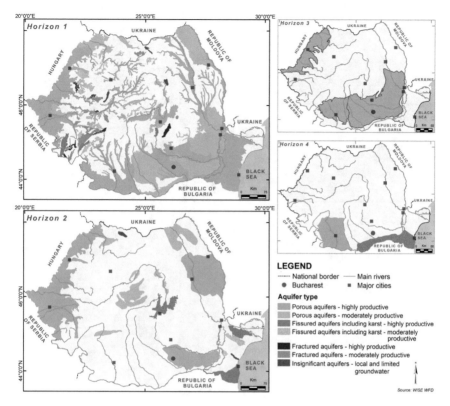

Fig. 3.42 Romanian aquifer types

3.4 Conclusions

Nowadays, the effects of climate change have caused disruptions in water cycle performance, the forecasting of the quantitative and qualitative spatio-temporal distribution of the water resource is becoming a difficult task as well as its management and planing. In a modern society characterized by technological contrasts together with education, access to resources, human health and natural environment, the development of appropriate policies should be based on both the results of systematic and complex analyzes of the specialists in the field and on the active feedback achieved trough public participation.

Since joining the EU in 2007, Romania has significantly improved its environmental performance, according to the World Bank Group [199]. Romanian legislation correctly reflects the environmental requirements agreed at EU level, but the real challenge is effective field enforcement, due to lack of adequate planning, coordination and funding.

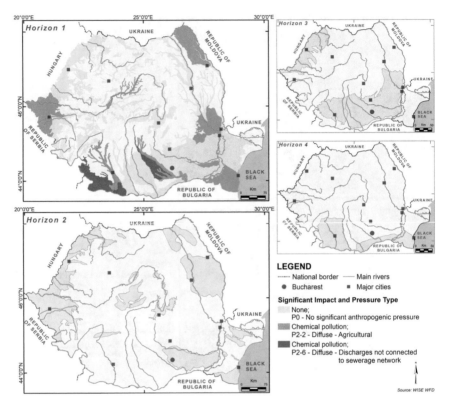

Fig. 3.43 Romanian significant impact and pressure type on groundwater bodies

Following the completion of the second River Basin Management Plan (2010–2015), the WFD results indicate at the EU level a water ecosystems crisis, the ecological status chart of 89,000 river water bodies indicating: 6% bad; 14% poor; 42% moderate; 30% good and only 8% high. After nearly a century, with all the technological and scientific advances recorded, emerging pollutants still present unknown issues that affect our waters through chemical pollution, leading to difficulties in correlating pressures, status and impact, while analytical performance is still limited for determination of mixtures effects of pollutants and atypical modes of action.

Among the EU countries, Romania is one with the lowest water availability per capita, vulnerable to the variability of precipitation and with significant variations between wet and dry years. In Romania, water management has been organized for almost a century (1920) based on hydrographic basins in 11 river basin authorities.

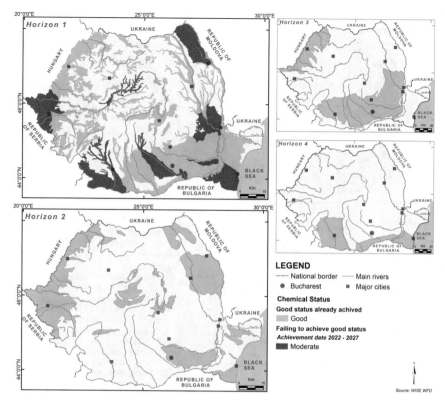

Fig. 3.44 Romanian groundwater chemical status

The environmental objectives for the surface and groundwaters—in conformity with the Framework Directive for Water 2000/60 of the European Union—have been achieved almost integrally. According to the WISE 2018 database, in Romania, 66% of water bodies have a good or high status, 31% are exceptions and only 3% cannot be classified. This situation highlights the demand to improve the implementation of environmental policies aimed at protecting water and expanding synergies between them. The point sources of pollution—belonging to old industrial units from the Communist period or to new and small ones—bring significant damage to the small streams.

Economic domains with the most significant contributions to the potential for chemical pollution of wastewater (in descending order) are: (i) water capture and processing for the supply of the population; (ii) chemical processing; (iii) industries producing electricity or heat; (iv) extractive industry.

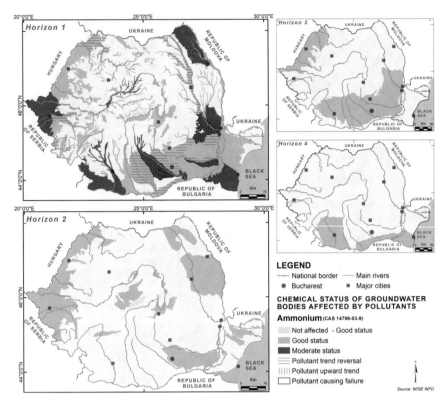

Fig. 3.45 Chemical status of Romanian groundwater bodies affected by ammonium

The most important problems are related to the surface and groundwaters within rural settlements because they are polluted by animal and human dejections that end up straight in the bodies of water. By connecting all the localities to the sewerage system and to the water supply system, pollution will cease almost automatically nationwide.

With regard to water quality, Romania needs to improve its policy according to the intervention logic of the Water Framework Directive. In order to respect the transitory deadlines, set in the Priority Accession Treaty, there are two aspects: (i) investments in urban waste water treatment plants; (ii) reducing pressures in agriculture by better identifying and defining both the compulsory measures that all farmers have to respect and the complementary ones for which they can receive funding.

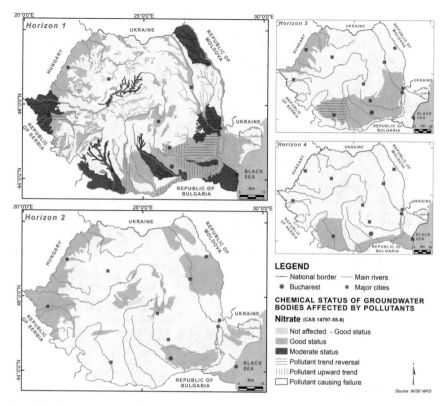

Fig. 3.46 Chemical status of Romanian groundwater bodies affected by nitrate

3.5 Recommendations

- Adopt a holistic approach of water pollution concepts to create effective source control pollution mechanisms rather than end-of-pipe solutions.
- For the protection of water resources, it is essential to continue with the same determination and desire because the improvement of water quality is ongoing.
- Strengthening and raising awareness among local communities and significant decision-makers on the value of aquatic and freshwater ecosystems.
- Restoring the natural functioning of the river by modifying the current infrastructure, ensuring at the same time the safety and proper functioning of the natural and anthropic habitats.
- Restriction, blocking or strict usage of non-deterioration (avoiding the development of a new hydropower plant).

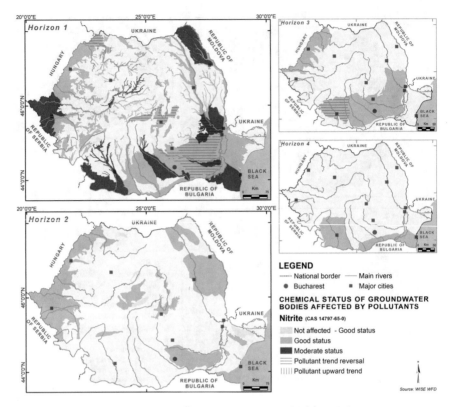

Fig. 3.47 Chemical status of Romanian groundwater bodies affected by nitrite

- Adjusting funding flows for appropriate, useful and efficient actions.
- Participatory administration (to allow effective involvement of all stakeholders).
- Encourage and support more efficient and less rigid public participation.
- The communication on progress so far has not been successful and has not shown interest to the public. Designing a GIS-based communication model for the public of all ages and levels of education that is friendly, attractive, targeted and immediate.
- So far, the WFD has been very useful, fulfilling its purpose, but requires a number of changes to be fruitful and effective.

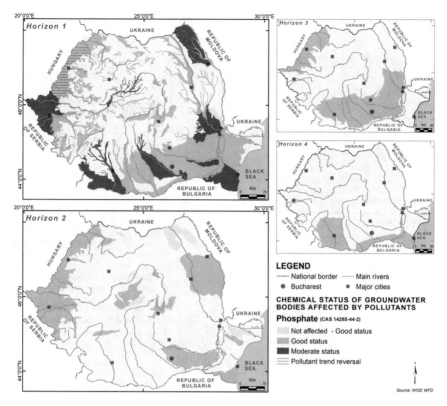

Fig. 3.48 Chemical status of Romanian groundwater bodies affected by phosphate

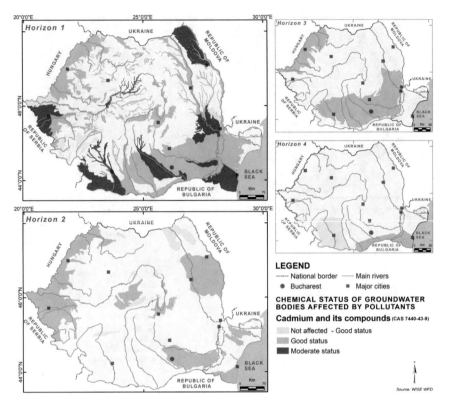

Fig. 3.49 Chemical status of Romanian groundwater bodies affected by cadmium

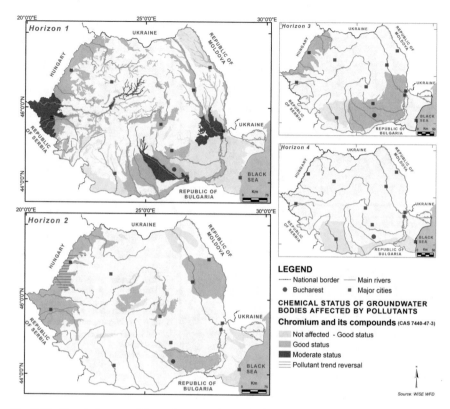

Fig. 3.50 Chemical status of Romanian groundwater bodies affected by chromium

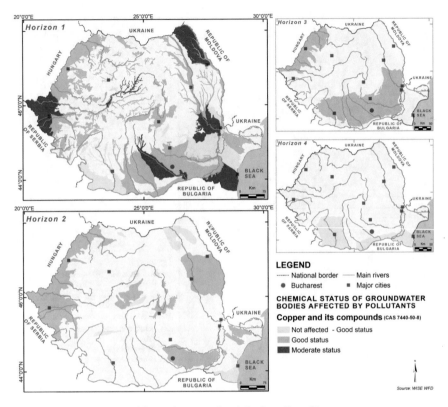

Fig. 3.51 Chemical status of Romanian groundwater bodies affected by copper

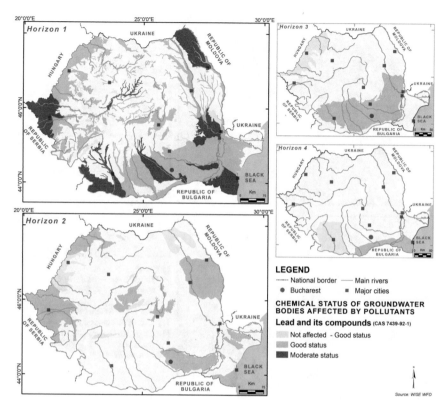

Fig. 3.52 Chemical status of Romanian groundwater bodies affected by lead

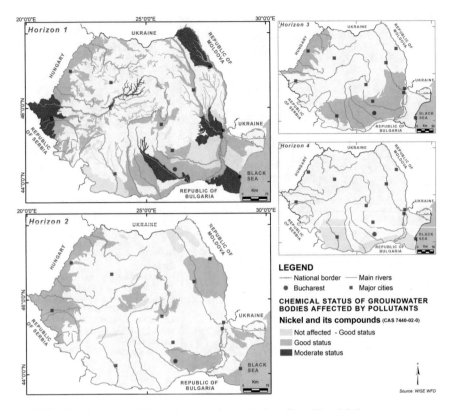

Fig. 3.53 Chemical status of Romanian groundwater bodies affected by nickel

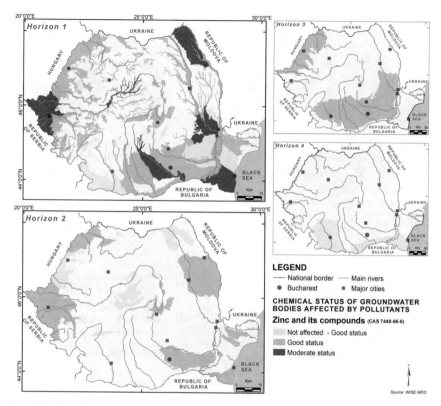

Fig. 3.54 Chemical status of Romanian groundwater bodies affected by zinc

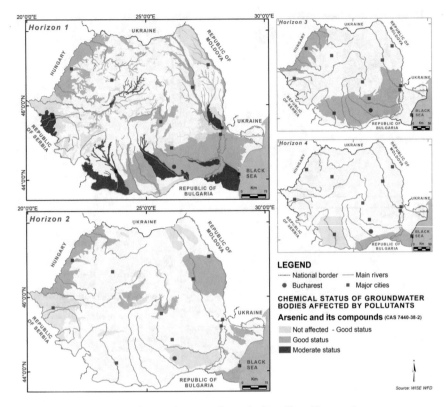

Fig. 3.55 Chemical status of Romanian groundwater bodies affected by arsenic

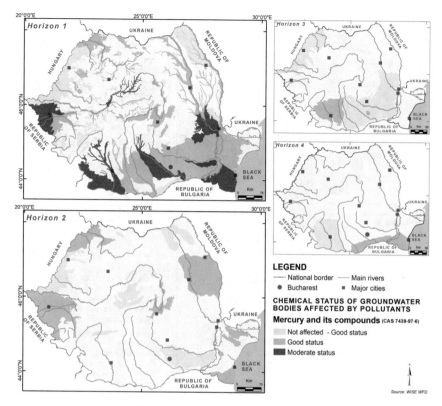

Fig. 3.56 Chemical status of Romanian groundwater bodies affected by mercury

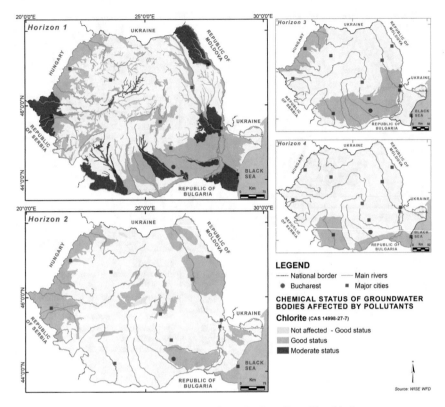

Fig. 3.57 Chemical status of Romanian groundwater bodies affected by chlorite

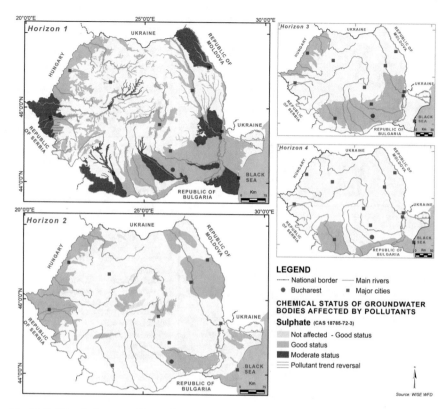

Fig. 3.58 Chemical status of Romanian groundwater bodies affected by sulphate

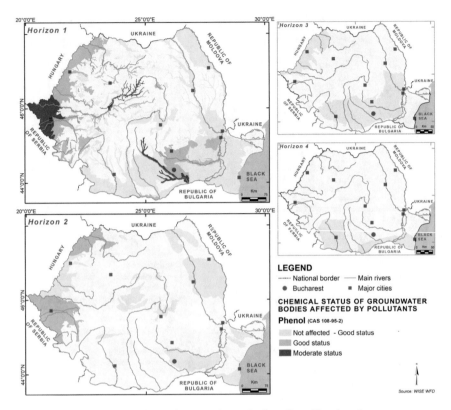

Fig. 3.59 Chemical status of Romanian groundwater bodies affected by phenol

Acknowledgements This work was financially supported by the Department of Geography from the "Alexandru Ioan Cuza" University, of Iasi, and the infrastructure was provided through the POSCCE-O 2.2.1, SMIS-CSNR 13984-901, No. 257/28.09.2010 Project, CERNESIM. Also, the authors would like to dedicate this chapter to the memory of late professor Gheorghe Romanescu, a true scholar, who thrived on learning and helping others learn.

References

1. Adumitroaei MV, Gavriloaiei T, Sandu AV, Iancu OG (2016) Distribution of mineral nitrogen compounds in groundwater in Vaslui County (Romania). Rev Chim 67:2530–2536
2. Anastasiu P, Rozylowicz L, Skolka M, Preda C, Memedemin D, Cogălniceanu D (2016) In: Rat M, Trichkova T, Scalera R, Tomov R, Uludag A (eds) Alien species in Romania in 1st ESENIAS scientific reports: report 1 state of art of aliens species in south eastern Europe, University of Novi Sad, Serbia

3. Anghel AM, Diacu E, Ilie M, Cimpoeru C, Marinescu F, Marcu E, Tociu C (2017) Statistical correlations between physical and chemical indicators in order to assess the water quality of artificial lakes in south Romania, Bucharest-Ilfov area. Biointerface Res Appl Chem 7:2048–2052

4. Babovic N, Markovic D, Dimitrijevic V (2011) Some indicators of water quality of the Tamis river. Chem Ind Chem Eng Q 17:107–115

5. Banica A, Breaban IG, Paiu M (2016) From vulnerability towards resilience approach in water distribution network - the issue of asbestos cement pipes. J Environ Protection Ecol 17(3):885–895

6. Banica A, Breaban IG, Terryn IC, Munteanu A (2016) Vulnerability and resilience of the urban drinking water system in the city of Bacau, Romania. In: SGEM book 1 psychology and psychiatry, sociology and healthcare, education conference proceedings, vol II

7. Barbulescu A, Barbes L (2013) Assessment of Techirghiol lake surface water quality using statistical analysis. Rev Chim 64:868–874

8. Bird G, Brewer PA, Macklin MG, Balteanu D, Serban M, Driga B, Zaharia S (2008) River system recovery following the Novat-Rosu tailings dam failure, Maramures County, Romania. Appl Geochem 23:3498–3518

9. Bird G, Macklin MG, Brewer PA, Zaharia S, Balteanu D, Driga B, Serban M (2009) Heavy metals in potable groundwater of mining-affected river catchments, northwestern Romania. Environ Geochem Health 31:741–758

10. Boariu C, Craciun I, Giurma-Handley CR, Hraniciuc TA (2013) Assessment of the impact of riverbed design on water quality - the case study of Bahlui river, Iasi, Romania. Environ Eng Manage J 12:625–635

11. Botezan C, Ozunu AL, Stefanie H (2015) Vulnerability assessment: the case of the Aries river middle basin. J Environ Protection Ecol 16:1326–1336

12. Bravo AG, Loizeau JL, Ancey L, Ungureanu VG, Dominik J (2009) Historical record of mercury contamination in sediments from the Babeni Reservoir in the Olt River, Romania. Environ Sci Pollut Res 16:66–75

13. Bravo AG, Loizeau JL, Dranguet P, Makri S, Bjorn E, Ungureanu VG, Slaveykova VI, Cosio C (2016) Persistent Hg contamination and occurrence of Hg-methylating transcript (HgcA) downstream of a chlor-alkali plant in the Olt River (Romania). Environ Sci Pollut Res 23:10529–10541

14. Breaban IG, Paiu M, Cojocaru P, Cretescu I (2013) Studies upon the groundwater quality index of the aquifer from Barlad middle basin. In: Geoconference on water resources, forest, marine and ocean ecosystems, pp 317–324

15. Briciu AE, Toader E, Romanescu G, Sandu I (2016) Urban streamwater contamination and self-purification in a central-eastern European city. Part I Rev Chim 67(7):1294–1300

16. Briciu AE, Toader E, Romanescu G, Sandu I (2016) Urban streamwater contamination and self-purification in a central-eastern European city. Part B Rev Chim 67(8):1583–1586

17. Bud I, Duma S, Gusat D, Pasca I, Bud A (2016) Environmental risks of abandoning a mining project already started: Romaltyn Mining Baia Mare. In: IOP conference series: materials science and engineering, vol 144, p 012004. http://iopscience.iop.org/article/10.1088/1757-899X/144/1/012004/pdf

18. Burtea MC, Ciurea A, Bordei M, Romanescu G, Sandu AV (2015) Development of the potential of ecological agriculture in the Village Ciresu, County of Braila. Rev Chim 66(8):1222–1226

19. Burtea MC, Sandu IG, Cioromele GA, Bordei M, Ciurea A, Romanescu G (2015) Sustainable exploitation of ecosystems on the Big island of Braila. Rev Chim 66(5):621–627

20. Calin MR, Radulescu I, Ion AC, Sirbu F (2016) Radiochemical investigations on natural mineral waters from Bucovina region, Romania. Rom J Phys 61:1051–1066

21. Campean RF, Olah NK, Toma C, Dumitru R, Arghir G (2011) In depth variation of water properties for St. Ana lake - Romania related to sediments in suspension. Stud Univ Babes-Bolyai Chem 56:107–117

22. Capatina C, Gamaneci G, Simonescu CM (2012) Impact assessment of the surface mining exploitation on the environment in the district of Gorj, Romania. J Environ Protection Ecol 13:1375–1390
23. Capatina C, Simonescu CM (2008) Management of waste in rural areas of Gorj county, Romania. Environ Eng Manage J 7:717–723
24. Capatina C, Simonescu CM, Lazar G (2013) Preliminary data regarding the content of heavy metals from the soils of Targu-Jiu area. Rev Chim 64:218–223
25. Capsa D, Barsan N, Felegeanu D, Stanila M, Joita I, Rotaru M, Ureche C (2016) Influence of climatic factors on the pollution with nitrogen oxides (NOx) in Bacau city, Romania. Environ Eng Manage J 15:655–663
26. Catianis I, Ungureanu C, Magagnini L, Ulazzi E, Campisi T, Stanica A (2016) Environmental impact of the Midia Port - Black Sea (Romania), on the coastal sediment quality. Open Geosci 8:174–194
27. Caunii A, Butnariu M, Chicioreanu DT, Milosan I (2014) Aspects relating to the organization of the integrated monitoring system in Romania. In: Iacob AI (ed) 2nd world conference on business, economics and management
28. Chicos MM, Damian G, Stumbea D, Buzgar N, Ungureanu T, Nica V, Iepure G (2016) Mineralogy and geochemistry of the tailings pond from Straja Valley (Suceava county, Romania). Factors affecting the mobility of the elements on the surface of the waste deposit. Carpathian J Earth Environ Sci 11:265–280
29. Cirtina D, Capatina C (2017) Quality issues regarding the watercourses from middle basin of Jiu River. Rev Chim 68:72–76
30. Ciubotariu AC, Istrate GG (2016) Physico chemical parameters of water from Galati area (Romania). In: Water, resources, forest, marine and ocean ecosystems conference proceedings, Bk3, vol 1, pp 561–568
31. Cojocariu C, Barjoveanu G, Robu B, Teodosiu C (2012) Integrated environmental impact and risk assessment of the agricultural and related industries in the Prut river basin. Stud Univ Babes-Bolyai Chem 57:151–166
32. Coman A, Chirila E, Popovici Carazeanu I (2006) Investigation of the Constanta surface water's pollution sources. In: Simeonov L, Chirila E (eds) Chemicals as intentional and accidental global environmental threats. NATO security through science series. Springer, Dordrecht
33. Constantin V, Stefanescu L, Kantor CM (2015) Vulnerability assessment methodology: a tool for policy makers in drafting a sustainable development strategy of rural mining settlements in the Apuseni Mountains, Romania. Environ Sci Policy 52:129–139
34. Costescu AI, Nemec NS, Halbac-Cotoara-Zamfir R (2011) Water resource quality modelling in the hydrographic basin of the Bega river. J Environ Protection Ecol 12:2019–2027
35. Cozma AI, Baciu C, Papp D, Rosian G, Pop CI (2017) Isotopic composition of precipitation in western Transylvania (Romania) reflected by two local meteoric water lines. Carpathian J Earth Environ Sci 12:357–364
36. Cozma AI, Baciu C, Moldovan M, Pop IC (2016) Using natural tracers to track the groundwater flow in a mining area. In: Ioja IC, Comanescu L, Dumitrache L, Nedelea A, Nita MR (eds) Ecosmart - environment at crossroads: smart approaches for a sustainable development
37. Cozma DG, Cruceanu A, Cojoc GM, Muntele I, Mihu-Pintilie A (2015) The factorial analysis of physico-chemical indicators in Bistrita's upper hydrographical basin. In: Water resources, forest, marine and ocean ecosystems, SGEM, vol I
38. Craciun I, Giurma I, Giurma-Handley CR, Telisca M (2009) Integrated quantitative and qualitative monitoring system of water resources. In: VIth international conference on the management of technological changes, Alexandroupolis, Greece, vol II, pp 53–56
39. Cristea L (2015) The environmental risks results in Bayer hydrometallurgical technology. In: Ecology, economics, education and legislation, vol III, pp 659–664
40. Cruceanu A, Cojoc GM, Cozma DG, Muntele I, Mihu-Pintilie A (2015) Comparativ study of surface waters quality in the hydrographic upper basin of Bistrita river (Romania). In: Water resources, forest, marine and ocean ecosystems, SGEM, Bk3, vol 1, pp 159–166

41. Csolti A, Botez F, Postolache C (2017) Ecological study on nitrogen biogeochemical cycle after conversion from grassland to cropland in southeastern Europe (Romania). Environ Eng Manage J 16:837–845
42. Cuciureanu A, Kim L, Lehr CB, Ene C (2017) The groundwater quality of the area tailings mining ponds in the north of Romania. Rev Chim 68:1695–1699
43. Curseu D, Sirbu D, Popa M, Ionutas A (2011) The relationship between infant methemoglobinemia and environmental exposure to nitrates. In: Gokcekus H, Turker U, Lamoreaux JW (eds) Survival and sustainability: environmental concerns in the 21st century. Springer, Berlin, Heidelberg
44. Curtean-Banaduc A, Didenko A, Guti G, Banaduc D (2018) Telestes Souffia (Risso, 1827) species conservation at the eastern limit of range - Viseu river basin, Romania. Appl Ecol Environ Res 16:291–303
45. Dabu C (2015) An integrated approach towards the depollution of the Apuseni mountains. In: Scientific papers-series e-land reclamation earth observation, surveying environmental engineering, vol 4, pp 107–115
46. Daescu AI, Holban E, Boboc MG, Raischi MC, Matei M, Ilie M, Deak G, Daescu V (2017) Performant technology to remove organic and inorganic pollutants from wastewaters. J Environ Protection Ecol 18:304–312
47. Despescu DA, Lazaroiu G, Mavrodin ME, Tudor RE (2016) Study regarding the water pollution in Romania. In: Energy and clean technologies conference proceedings, SGEM, Bk4, vol 1, pp 677–682
48. Dinu L, Stefanescu M, Patroescu V, Cosma C, Cristea I, Badescu V, Alexie M (2016) Trials for the improvement of the ettringite process for the mine water treatment. J Environ Protection Ecol 17:83–93
49. Dippong T, Mihali C, Goga F, Cical E (2017) Seasonal evolution and depth variability of heavy metal concentrations in the water of Firiza-Strimtori lake, NW of Romania. Stud Univ Babes-Bolyai Chem 62:213–228
50. Dippong T, Mihali C, Nasu D, Berinde Z, Butean C (2018) Assessment of water physico-chemical parameters in the Strimtori-Firiza reservoir in Northwest Romania. Water Environ Res 90:220–233
51. Dogaru D, Zobrist J, Balteanu D, Popescu C, Sima M, Amini M, Yang H (2009) Community perception of water quality in a mining-affected area: a case study for the Certej catchment in the Apuseni mountains in Romania. Environ Manage 43:1131–1145
52. Dulama ID, Radulescu C, Chelarescu ED, Stihi C, Bucurica IA, Teodorescu S, Stirbescu RM, Gurgu IV, Let DD, Stirbescu NM (2017) Determination of heavy metal contents in surface water by inductively coupled plasma - mass spectrometry: a case study of Ialomita river, Romania. Rom J Phys 62
53. Dumitrel GA, Glevitzky M, Popa M, Vica ML (2015) Studies regarding the heavy metals pollution of streams and rivers in Rosia Montana area, Romania. J Environ Protection Ecol 16:850–860
54. Dunca AM (2018) Water pollution and water quality assessment of major transboundary rivers from Banat (Romania). J Chem
55. Dunea D, Iordache S (2016) Using SWAT model and PROBA-V multispectral data to assess water quality in Calmatui river basin, Romania. In: Water, resources, forest, marine and ocean ecosystems conference proceedings, vol I
56. Eerkes-Medrano D, Thompson RC, Aldridge DC (2015) Microplastics in freshwater systems: a review of the emerging threats, identification of knowledge gaps and prioritisation of research needs. Water Res 75:63–82
57. Enea A, Hapciuc OE, Iosub M, Minea I, Romanescu G (2017) Water quality assessment in three mountainous watersheds from Eastern Romania (Suceava, Ozana and Tazlau Rivers). Environ Eng Manage J 16:605–614
58. European Commission (2000) Directive 2000/60/EC of the European Parliament and of the Council establishing a framework for Community action in the field of water policy. https://eur-lex.europa.eu/resource.html?uri=cellar:5c835afb-2ec6-4577-bdf8-756d3d694eeb.0004.02/DOC_1&format=PDF

59. European Commission (2012) The EU Water Framework Directive - integrated river basin management for Europe. http://ec.europa.eu/environment/water/water-framework/index_en. html

60. European Commission (2017) Special Eurobarometer 468: attitudes of European citizens towards the environment. http://data.europa.eu/euodp/en/data/dataset/S2156_88_1_468_ENG

61. European Environment Agency (2017) Urban Waste Water Treatment. https://www.eea.europa.eu/data-and-maps/indicators/urban-waste-water-treatment/urban-waste-water-treatment-assessment-4

62. European Environment Agency (2018) WISE large rivers and large lakes. https://www.eea.europa.eu/data-and-maps/data/wise-large-rivers-and-large-lakes

63. European Environment Agency (2018) WISE WFD Database. https://www.eea.europa.eu/data-and-maps/data/wise-wfd-2

64. European Environment Agency (2018) EEA Report No 7/2018, European waters assessment of status and pressures 2018

65. Faciu ME, Lazar I, Ifrim I, Ureche C, Lazar G (2014) Exploratory spatial data analysis of heavy metals concentration in two sampling sites on Siret river. Environ Eng Manage J 13:2179–2186

66. Le Faucheur S, Vasiliu D, Catianis I, Zazu M, Dranguet P, Beauvais-Fluck R, Loizeau JL, Cosio C, Ungureanu C, Ungureanu VG, Slaveykova VI (2016) Environmental quality assessment of reservoirs impacted by Hg from chlor-alkali technologies: case study of a recovery. Environ Sci Pollut Res 23:22542–22553

67. Feher IC, Moldovan Z, Oprean I (2016) Spatial and seasonal variation of organic pollutants in surface water using multivariate statistical techniques. Water Sci Technol 74:1726–1735

68. Florescu D, Ionete RE, Iordache A, Horj E, Stefanescu I, Culea M (2011) Optimal measurements of surface water pollution-case study on southern part of Romania. Asian J Chem 23:5209–5212

69. Florescu D, Ionete RE, Sandru C, Iordache A, Culea M (2011) The influence of pollution monitoring parameters in characterizing the surface water quality from Romania Southern area. Rom J Phys 56:1001–1010

70. Galaon T, Petre J, Iancu VI, Stanescu E (2015) LC-MS/MS determination of eight pharmaceuticals and two disinfectants in the Danube river and three major tributaries from Romania. In: Ecology, economics, education and legislation, vol II

71. Ganea IV, Roba C, Gligor D, Farkas A, Balc R, Moldovan M (2017) Assessment of environmental quality in Lacu Sarat area (Braila county, Romania). Carpathian J Earth Environ Sci 12:377–387

72. Gheorghe S, Lucaciu I, Stanescu E, Stoica C (2013) Romanian aquatic toxicity testing strategy under reach. J Environ Protection Ecol 14:601–611

73. Gherman A (2011) Environmental impact of mining industry in the Caliman mountains. In: 17th international conference the knowledge-based organization. Conference proceedings 2: economic, social and administrative approaches to the knowledge-based organization, vol 2, pp 142–149

74. Girbaciu A, Girbaciu C, Petcovici E, Dodocioiu AM (2015) Water quality modelling using Mike 11. Rev Chim 66:1206–1211

75. Hategan RM, Popita GE, Varga I, Popovici A, Frentiu T (2012) The heavy metals impact on surface water and soil in the non-sanitary municipal landfill "Pata Rat"- Cluj-Napoca. Stud Univ Babes-Bolyai Chem 57:119–126

76. Hill MK (2010) Water pollution. In: Understanding environmental pollution, 3rd edn. Cambridge University Press, Cambridge, pp 236–285. Available from: Cambridge Books Online

77. Hoaghia MA, Roman C, Kovacs ED, Tanaselia C, Ristoiu D (2016) The evaluation of the metal contamination of drinking water sources from Medias town, Romania using the metal pollution indices. Stud Univ Babes-Bolyai Chem 61:461–471

78. Horn HA, Bilal E, Ribeiro VE, Trindade MW, Baggio H (2012) The pollution impact on the water, soil and plant generated by foundries in Pirapora and Varzeá da Palma, Minas Gerais state, Brazil. Carpathian J Earth Environ Sci 7(4):211–218

79. https://ec.europa.eu/eurostat/statistics-explained/index.php/Water_statistics
80. https://www.green-report.ro/administatia-nationala-apele-romane-a-inceput-o-operatiune-de-ecologizare-a-lacului-bicaz/
81. https://www.newsinlevels.com/products/plastics-in-the-black-sea-level-3
82. http://oildepol.ro
83. http://www.radiocluj.ro
84. https://www.mediafax.ro
85. Imre K, Morar A, Ili M, Hor F, Bade C, Imre M (2017) Quality of drinking water in rural areas of western Romania with reference to parasites. J Biotechnol 256:S92
86. Ion A, Vladescu L, Badea IA, Comanescu L (2016) Monitoring and evaluation of the water quality of Budeasa Reservoir-Arges River, Romania. Environ Monit Assess 188
87. Ionete RE, Popescu R, Costinel D (2015) An isotopic survey of some mineral water resources in the Carpathian chain (Romania). Environ Eng Manage J 14:2445–2456
88. Ionita A, Toma O (2012) Monitoring the impact of human factors on surface waters to prevent ecological disasters: case study on faecal pollution of Nicolina river, Iasi, Romania. In: Barry DL, Coldewey WG, Reimer DWG, Rudakov DV (eds) Correlation between human factors and the prevention of disasters
89. Iordache M, Branzoi IV, Popescu LR, Iordache I (2016) Evaluation of heavy metal pollution into a complex industrial area from Romania. Environ Eng Manage J 15:389–394
90. Iordache M, Popescu LR, Pascu LF, Iordache I (2015) Environmental risk assessment in sediments from Jiu river, Romania. Rev Chim 66:1247–1252
91. Iordache M, Popescu LR, Pascu LF, Lehr C, Ungureanu EM, Iordache I (2015) Evaluation of the quality of environmental factors, soil and water in the Parang mountains, Romania. Rev Chim 66:1009–1014
92. Iticescu C, Georgescu LP, Topa C, Murariu G (2014) Monitoring the Danube water quality near the Galati city. J Environ Protection Ecol 15:30–38
93. Iticescu C, Murariu G, Georgescu LP, Burada A, Topa CM (2016) Seasonal variation of the physico-chemical parameters and water quality index (WQI) of Danube water in the transborder lower Danube area. Rev Chim 67:1843–1849
94. Jung AV, Le Cann P, Roig B, Thomas O, Baurès E, Thomas MF (2014) Microbial contamination detection in water resources: interest of current optical methods, trends and needs in the context of climate change. Int J Environ Res Public Health 11(4):4292–4310
95. Kim L, Vasile GG, Stanescu B, Calinescu S, Batrinescu G (2015) Distribution and bioavailability of mobile arsenic in sediments from a mining catchment area. J Environ Protection Ecol 16:1307–1315
96. Lacatusu R, Citu G, Aston J, Lungu M, Lacatusu AR (2009) Heavy metals soil pollution state in relation to potential future mining activities in the Rosia Montana area. Carpathian J Earth Environ Sci 4:39–50
97. Lacatusu R, Rauta C, Ghelase I, Carstea S, Kovacsovics B (1994) Local nitrate pollution of soils, waters and vegetables in some areas of Romania. In: Anke M et al (eds) Deficient and excessive levels of macroelements and trace elements in nutrition, pp 491–497
98. Lazar AL, Baciu C, Roba C, Dicu T, Pop C, Rogozan C, Dobrota C (2014) Impact of the past mining activity in Rosia Montana (Romania) on soil and vegetation. Environ Earth Sci 72:4653–4666
99. Levei EA, Senila M, Miclean M, Abraham B, Roman C, Stefanescu L, Moldovan OT (2011) Influence of Rosia Poieni and Rosia Montana mining areas on the water quality of the Aries river. Environ Eng Manage J 10:23–29
100. MAPM (2002) Code of good agricultural practices, 2 vols. The Expert Publishing House, Bucharest, 308 pp
101. Macalet R, Radu E, Radu C, Pandele A (2013) Hydrogeological characterisation of the groundwater bodies from the Southern part of Romania. In: Geoconference on science and technologies in geology, exploration and mining, SGEM, vol 2, pp 211–221
102. Maftei AE, Iancu OG, Buzgar N (2014) Assessment of minor elements contamination in Bistrita River sediments (upstream of Izvorul Muntelui Lake, Romania) with the implication of mining activity. J Geochem Explor 145:25–34

103. Manea C, Podina C, Crutu G, Pordea I, Popescu M, Iliescu M (2011) Radiological risk assessement by determining the additional effective dose received by the population Ciudanovita mining Area (Banat - Romania). Rev Chim 62:986–991

104. Marcu E, Deak G, Ciobotaru IE, Ivanov AA, Ionescu P, Tociu C, Diacu E (2017) The assessment of content and dynamics of nutrients in water and sediments from the Plumbuita Lake, Bucharest I. Study on phosphorous content and distribution. Rev Chim 68:2492–2494

105. Marcu E, Deak G, Ciobotaru IE, Ivanov AA, Ionescu P, Tociu C, Diacu E (2017) The assessment of content and dynamics of nutrients in water and sediments from the Plumbuita lake, Bucharest II. Study on nitrogen content and distribution. Rev Chim 68:2744–2746

106. Marin C, Tudorache A, Moldovan OT, Povara I, Rajka G (2010) Assessing the contents of arsenic and of some heavy metals in surface flows and in the hyporheic zone of the Aries stream catchment area, Romania. Carpathian J Earth Environ Sci 5:13–24

107. Marinov AM, Marinov I, Diminescu MA (2017) Groundwater quality in the proximity of a polluted lake: a joint experimental-modeling study. Environ Eng Manage J 16:1081–1091

108. Martonos IM, Sabo HM (2017) Quality of drinking water supplies in Almasu rural area (Salaj county, Romania). Carpathian J Earth Environ Sci 12:371–376

109. Matache ML, David IG, Dinu C, Radu LG (2018) Trace metals in water and sediments of the Prut river, Romania. Environ Eng Manage J 17:1363–1372

110. Merciu FC, Cercleux AL, Peptenatu D (2015) Rosia Montana, Romania: industrial heritage in situ, between preservation, controversy and cultural recognitio. Ind Archaeol Rev 37:5–19

111. Miclean M, Tanaselia C, Roman M, Cadar O, Roman C (2013) Organochlorine pesticides and metals in Danube water environment, Calafat-Turnu Magurele sector, Romania. In: Geoconference on water resources, forest, marine and ocean ecosystems, pp 261–268

112. Mihu-Pintilie A, Romanescu G, Stoleriu C (2014) The seasonal changes of the temperature, pH and dissolved oxygen in the Cuejdel Lake, Romania. Carpathian J Earth Environ Sci 9(2):113–123

113. Mihu-Pintilie A, Romanescu G, Stoleriu CC, Breaban IG (2015) Physico-chemical parameters in mountain freshwater: Cuejdi River from Eastern Carpathians, Romania. Key Eng Mater 660(257):261

114. Mihu-Pintilie A, Paiu M, Breaban IG, Romanescu G (2014) Status of water quality in Cuejdi hydrographic basin from Eastern Carpathian, Romania. In: 14th SGEM

115. Modoi OC, Roba C, Torok Z, Ozunu A (2014) Environmental risks due to heavy metal pollution of water resulted from mining wastes in NW Romania. Environ Eng Manage J 13:2325–2336

116. Moisii AM, Romanescu G, Breaban IG (2016) State of water quality for Rebricea river by principal component analysis. In: Gâştescu P, Lewis W Jr, Breţcan P (eds) Conference proceedings water resources and wetlands

117. Moldovan OT, Levei EA, Marin C, Banciu M, Banciu HL, Pavelescu C, Brad T, Cimpean M, Meleg I, Iepure S, Povara I (2011) Spatial distribution patterns of the hyporheic invertebrate communities in a polluted river in Romania. Hydrobiologia 669:63–82

118. Moldovan OT, Meleg IN, Levei EA, Terente M (2013) A simple method for assessing biotic indicators and predicting biodiversity in the hyporheic zone of a river polluted with metals. Ecol Ind 24:412–420

119. Morar F (2015) Copper content of surface waters in Mures County, Romania. In: Moldovan L (ed) 8th international conference interdisciplinarity in engineering, Inter-Eng, vol 19, pp 451–455

120. Mosneag SC, Vele D, Neamtu C (2017) Evaluation of drinking water quality Cluj county from Romania regarding nitrate concentration. Carpathian J Earth Environ Sci 12:389–393

121. Muntean LE, Moldovan M, Moldovan D, Bican-Brisan N, Cosma C (2016) Natural radioactivity of water sources related to geology and radiation exposure in Alba county. Carpathian J Earth Environ Sci 11:575–582

122. Murariu M, Ciobanu CI, Bunia I, Surleva A, Butnariu AE (2016) Wheat seeds as environmental markers in heavy metal and arsenic pollution of Tarnita mining area. In: Water resources, forest, marine and ocean ecosystems conference proceedings, SGEM, Bk3, vol 2, pp 685–692

123. NIS (2012) National Institute of Statistics
124. Napper IE, Thompson RC (2016) Release of synthetic microplastic plastic fibres from domestic washing machines: effects of fabric type and washing conditions. Mar Pollut Bull 112(1–2):39–45
125. National Institute of Statistics (2012) Statistical yearbooks of Romania
126. Nesaratnam S (2018) Effects of pollutants on the aquatic environment. https://www.open.edu/openlearn/nature-environment/effects-pollutants-on-the-aquatic-environment/content-section
127. Niacsu L (2012) Geomorphologic and pedologic restrictive parameters for agricultural land in the Pereschiv catchment of Eastern Romania. Carpathian J Earth Environ Sci 7:25–37
128. Nikolaichuk VI, Vakerich MM, Shpontak JM, Karpu'k MK (2015) The current state of water resources of Transcarpathia. Visnyk Dnipropetrovsk Univ Biol Ecol 23:116–123
129. Olaru V, Voiculescu M, Georgescu LP, Caldararu A (2010) Integrated management and control system for water resources. Environ Eng Manage J 9:423–428
130. Osan J, Torok S, Alfoldy B, Alsecz A, Falkenberg G, Baik SY, Van Grieken R (2007) Comparison of sediment pollution in the rivers of the Hungarian Upper Tisza Region using non-destructive analytical techniques. Spectrochim Acta Part B At Spectrosc 62:123–136
131. Ozunu A, Stefanescu L, Costan C, Miclean M, Modoi C, Vlad SN (2009) Surface water pollution generated by mining activities. Case study: Aries river middle catchment basin, Romania. Environ Eng Manage J 8:809–815
132. Paiu M, Breaban IG (2016) Distribution of nitrate concentration in groundwater in some rural settlements from eastern Romania. In: Water, resources, forest, marine and ocean ecosystems conference proceedings, Bk3, vol 1, pp 235–242
133. Papadatu CP, Bordei M, Romanescu G, Sandu I (2016) Researches on heavy metals determination from water and soil in Galati county, Romania. Rev Chim 67:1728–1733
134. Papp DC, Cociuba I, Baciu C, Cozma A (2017) Origin and geochemistry of mine water and its impact on the groundwater and surface running water in post-mining environments: Zlatna Gold Mining Area (Romania). Aquat Geochem 23:247–270
135. Papp DC, Cociuba I, Baciu C, Cozma A (2017) Composition and origin of mine water at Zlatna gold mining area (Apuseni Mountains, Romania). In: Marques JM, Chambel A (eds) Procedia earth and planetary science. 15th water-rock interaction international symposium, Wri-15, vol 17, pp 37–40
136. Paun I, Chiriac FL, Marin NM, Cruceru LV, Pascu LF, Lehr CB, Ene C (2017) Water quality index, a useful tool for evaluation of Danube river raw water. Rev Chim 68:1732–1739
137. Pehoiu G (2009) Waste and environmental impact in urban areas of the county Dambovita (Romania). In: Pardalos P, Mastorakis N, Mladenov V, Bojkovic Z (eds) Proceedings of the 3rd international conference on energy and development - environment - biomedicine, pp 87–90
138. Pehoiu G, Murarescu O (2010) Environment and water resources in Targoviste Plain (Romania). In: Schmitter ED, Mastorakis N (eds) Proceedings of the 5th IASME/WSEAS international conference on water resources, hydraulics & hydrology/Proceedings of the 4th IASME/WSEAS international conference on geology and seismology: water and geoscience, pp 90–95
139. Pivokonsky M, Cermakova L, Novotna K, Peera P, Cajthaml T, Jandac V (2018) Occurrence of microplastics in raw and treated drinking water. Sci Total Environ 643:1644–1651
140. Popa M, Glevitzky M, Popa DM, Dumitrel GA (2014) Study regarding the water contamination and the negative effects on the population from the Blaj area, Romania. J Environ Protection Ecol 15:1543–1554
141. Popa D, Prunisoara C, Carlan C (2016) Aspects concerning the monitoring of the qualitative parameters of groundwaters from Balesti - Tamasesti area, Gorj county, Romania - a nitrate vulnerable zone. Agrolife Sci J 5:119–124
142. Popa M, Dumitrel GA, Mirel G, Popa DV (2015) Anthropogenic contamination of water from Galda River - Alba County, Romania. In: Cimpeanu SM, Fintineru G, Beciu S (eds) Conference agriculture for life, life for agriculture, vol 6, pp 446–452

143. Popa RG (2012) The evaluation of the water environment factor, affected by the oil exploitation within the Ticleni-Romania lease. In: 12th international multidisciplinary scientific geoconference, SGEM, vol 5, pp 875–882
144. Popescu R, Costinel D, Ionete RE, Axente D (2014) Isotopic fingerprint of the middle Olt River basin, Romania. Isot Environ Health Stud 50:461–474
145. Popescu LR, Iordache M, Ungureanu EM, Buica GO (2016) Impact of mercury pollution on soil, surface water and sediment ecosystems in the area of an old mercury mine. Environ Eng Manage J 15:1087–1091
146. Popescu R, Mimmo T, Dinca OR, Capici C, Costinel D, Sandru C, Ionete RE, Stefanescu I, Axente D (2015) Using stable isotopes in tracing contaminant sources in an industrial area: a case study on the hydrological basin of the Olt river, Romania. Sci Total Environ 533:17–23
147. Preda E, Mincea MM, Ionascu C, Botez AV, Ostafe V (2018) Contamination of groundwater with phenol derivatives around a decommissioned chemical factory. Environ Eng Manage J 17:569–577
148. Prioteasa L, Prodana M, Buzoianu M, Demetrescu I (2014) Determination by ICP-MS of heavy metals and other toxic elements in drinking water from several rural areas of Romania. Rev Chim 65:925–928
149. Prunisoara C, Papa SD, Carlan C (2015) The impact assessment of mining activities on surface and ground waters in Gorj county - Romania. In: Water resources, forest, marine and ocean ecosystems, SGEM, Bk3, vol 1, pp 647–654
150. Radu D, Ilinca CM (2009) GIS database for groundwater integrated management. In: Lepadatescu D, Mastorakis NE (eds) Proceedings of the 2nd international conference on environmental and geological science and engineering, pp 96–100
151. Radulescu C, Pohoata A, Bretcan P, Tanislav D, Stihi C, Chelarescu ED (2017) Quantification of major ions in groundwaters using analytical techniques and statistical approaches. Rom Rep Phys 69:2
152. Radulescu C, Stihi C, Dulama ID, Chelarescu ED, Bretcan P, Tanislav D (2015) Assessment of heavy metals content in water and mud of several salt lakes from Romania by atomic absorption spectrometry. Rom J Phys 60(1–2):246–256
153. Radulov I, Lato A, Berbecea A, Lato I, Crista F (2016) Nitrate pollution of water in Romania Serbia cross - border area as a consequence of agricultural practices. In: Water resources, forest, marine and ocean ecosystems conference proceedings, SGEM, Bk 3, vol 3, pp 205–212
154. Roba C, Rosu C, Pistea I, Ozunu A, Mitrofan H (2015) Groundwater quality in a rural area from Buzau county, Romania. Sci Pap Ser Manage Econ Eng Agric Rural Dev 15:305–310
155. Roba C, Cosma C, Burghele BD, Moldovan M, Buterez C (2016) Evaluation of water quality from public distribution systems in the aspiring Buzau Land Geopark, Romania. In: Water, resources, forest, marine and ocean ecosystems conference proceedings, Bk3, vol 1, pp 321–328
156. Romanescu G, Cojoc GM, Sandu IG, Tirnovan A, Dascalita D, Sandu I (2015) Pollution sources and water quality in the Bistrita catchment (Eastern Carpathians). Rev Chim 66:855–863
157. Romanescu G, Cretu MA, Sandu IG, Paun E, Sandu I (2013) Chemism of streams within the Siret and Prut drainage basins: water resources and management. Rev Chim 64(12):1416–1421
158. Romanescu G, Hapciuc OE, Sandu I, Minea I, Dascalita D, Iosub M (2016) Quality indicators for Suceava river. Rev Chim 67:245–249
159. Romanescu G, Iosub M, Sandu I, Minea I, Enea A, Dascalita D, Hapciuc OE (2016) Spatiotemporal analysis of the water quality of the Ozana river. Rev Chim 67:42–47
160. Romanescu G, Miftode D, Mihu-Pintilie A, Stoleriu CC, Sandu I (2016) Water quality analysis in mountain freshwater: Poiana Uzului reservoir in the Eastern Carpathians. Rev Chim 67:2318–2326
161. Romanescu G, Pascal M, Mihu-Pintilie A, Stoleriu CC, Sandu I, Moisii M (2017) Water quality analysis in wetlands freshwater: common floodplain of Jijia-Prut Rivers. Rev Chim 68:553–561

162. Romanescu G, Paun E, Sandu I, Jora I, Panaitescu E, Machidon O, Stoleriu C (2014) Quantitative and qualitative assessments of groundwater into the catchment of Vaslui river. Rev Chim 65:401–410
163. Romanescu G, Sandu I, Stoleriu C, Sandu IG (2014) Water resources in Romania and their quality in the main lacustrine basins. Rev Chim 65:344–349
164. Romanescu G, Stoleriu CC (2014) Seasonal Variation of temperature, pH, and dissolved oxygen concentration in Lake Rosu, Romania. Clean-Soil Air Water 42:236–242
165. Romanescu G, Stoleriu CC, Lupascu A (2012) Biogeochemistry of wetlands in barrage Lacul Rosu catchment (Haghimas - Eastern Carpathian). Environ Eng Manage J 11:1627–1637
166. Romanescu G, Tarnovan A, Sandu IG, Cojoc GM, Dascalita D, Sandu I (2014) The quality of surface waters in the Suha hydrographic basin (Oriental Carpathian Mountains). Rev Chim 65:1168–1171
167. Romanescu G, Tirnovan A, Cojoc GM, Juravle DT, Sandu I (2015) Groundwater quality in Suha basin (Northern Group of Eastern Carpathians). Rev Chim 66:1885–1890
168. Romanescu G, Tirnovan A, Sandu I, Cojoc GM, Breaban IG, Mihu-Pintilie A (2015) Water chemism within the settling pond of valea Straja and the quality of the Suha water body (Eastern Carpathians). Rev Chim 66:1700–1706
169. Romanescu G, Alexianu M, Asandulesei A (2014) The distribution of salt massifs and the exploitation of ancient and current reserves of mineralized waters within the Siret hydrographical basin (Romania) - case study for the eastern area of the Eastern Carpathians. In: Geoconference on water resources, forest, marine and ocean ecosystems, Bk 3, vol 1, pp 731–746
170. Romanescu G, Cojoc GM, Tirnovan A, Dascalita D, Paun E (2014) Surface water quality in Bistrita river basin (Eastern Carpathians). In: Geoconference on water resources, forest, marine and ocean ecosystems, Bk 3, vol 1, pp 679–690
171. Romanescu G, Jora I, Panaitescu E, Alexianu M (2015) Calcium and magnesium in the groundwaters of the Moldavian plateau (Romania) - distribution and managerial and medical significance. In: Water resources, forest, marine and ocean ecosystems, SGEM, Bk3, vol 1, pp 103–112
172. Salihou Djari MM, Stoleriu CC, Saley MB, Mihu-Pintilie A, Romanescu G (2018) Groundwater quality analysis in warm semi-arid climate of Sahel countries: Tillabéri region, Niger. Carpathian J Earth Environ Sci 13(1):277–290
173. Sarmasan C, Draghici S, Daina L (2008) Identification, communication and management of risks relating to drinking water pollution in Bihor county. Environ Eng Manage J 7:769–774
174. Schiopu AM, Ghinea C (2013) Municipal solid waste management and treatment of effluents resulting from their landfilling. Environ Eng Manage J 12:1699–1719
175. Scradeanu M, Tevi G, Mocuta M, Popa I, Scradeanu D (2016) Investigation and preliminary assessment on groundwater oil pollution in an urban area. Case study. Focsani, Romania. J Environ Protection Ecol 17:108–118
176. Sedrati A, Houha B, Romanescu G, Stoleriu CC (2018) Hydro-geochemical and statistical characterization of groundwater in the south of Khenchela, el Meita area (northeastern Algeria). Carpathian J Earth Environ Sci 13(2):333–342
177. Senila M, Levei E, Cadar O, Senila LR, Roman M, Puskas F, Sima M (2017) Assessment of availability and human health risk posed by arsenic contaminated well waters from Timis-Bega area, Romania. J Anal Methods Chem 3037651
178. Sirbu I, Benedek AM (2018) Trends in Unionidae (Mollusca, Bivalvia) communities in Romania: an analysis of environmental gradients and temporal changes. Hydrobiologia 810:295–314
179. Sluser BM, Schiopu AM, Balan C, Prutean M (2017) Postclosure influence of emissions resulted from municipal waste dump sites: a case study of the North-East region of Romania. Environ Eng Manage J 16:1017–1026
180. Spellman FR (2017) The science of environmental pollution, 3rd edn. Taylor and Francis, CRC Press

181. Stanescu B, Batrinescu G, Kim L, Cuciureanu A (2014) Vulnerability of the groundwater to nutrient pollution of a county in the south-eastern of Romania. In: Geoconference on science and technologies in geology, exploration and mining, SGEM, Bk 1, vol 2, pp 1051–1058

182. Stanescu B, Kim L, Lehr C, Stanescu E (2017) Assessment of the environmental aspects in a city area affected by historical pollution. In: 20th international symposium - the environment and the industry, pp 128–133

183. Stoica C, Vasile GG, Banciu A, Niculescu D, Lucaciu I, Lazar MN (2017) Influence of anthropogenic pressures on groundwater quality from a rural area. Rev Chim 68:1744–1748

184. Stoica AI, Florea RM, Baiulescu GE (2009) The influence of gold mining industry on the pollution of Rosia Montana district. In: Corral MD, Earle JL (eds) Gold mining: formation and resource estimation, economics and environmental impact. Nova Science Publishers, Inc., pp 63–89

185. Strungaru SA, Nicoara M, Teodosiu C, Baltag E, Ciobanu C, Plavan G (2018) Patterns of toxic metals bioaccumulation in a cross-border freshwater reservoir. Chemosphere 207:192–202

186. Stumbea D (2013) Preliminaries on pollution risk factors related to mining and ore processing in the Cu-rich pollymetallic belt of Eastern Carpathians, Romania. Environ Sci Pollut Res 20:7643–7655

187. Teodosiu C, Barjoveanu G, Sluser BR, Popa SAE, Trofin O (2016) Environmental assessment of municipal wastewater discharges: a comparative study of evaluation methods. Int J Life Cycle Assess 21:395–411

188. Teodosiu C, Robu B, Cojocariu C, Barjoveanu G (2015) Environmental impact and risk quantification based on selected water quality indicators. Nat Hazards 75:S89–S105

189. Teodosiu C, Gilca AF, Barjoveanu G, Fiore S (2018) Emerging pollutants removal through advanced drinking water treatment: a review on processes and environmental performances assessment. J Cleaner Prod 197:1210–1221

190. Tevi G, Vasilescu M, Grigore F, Rojanschi V (2014) Effects of nutrient pollution on ground water used for drinking purposes. In: Geoconference on water resources, forest, marine and ocean ecosystems, Bk 3, vol 1, pp 223–230

191. Tita M, Tufeanu R, Tita O, Miricescu D (2017) Quality parameters of dairy wastewater in the Mures county, Romania. Sci Study Res Chem Chem Eng Biotechnol Food Ind 18:145–152

192. Trifan A, Breaban IG, Sava D, Bucur L, Toma CC, Miron A (2015) Heavy metal content in macroalgae from Roumanian Black Sea. Rev Roum Chim 60(9):915–920

193. Trifu MC, Ion MB, Daradici V (2013) Different methods for farmer's implication in the nitrate management at river basin scale. In: Geoconference on water resources, forest, marine and ocean ecosystems, vol 1, pp 101–108

194. U.S. Environmental Protection Agency (2017) Chapter 3: water quality criteria. In: Water quality standards handbook. EPA-823-B-17-001. EPA Office of Water, Office of Science and Technology, Washington, DC. https://www.epa.gov/sites/production/files/2014-10/documents/handbook-chapter3.pdf

195. Vasile GG, Petre J, Nicolau M (2013) Metal speciation in sediments from Rosia Montana mining area, Romania. In: Geoconference on ecology, economics, education and legislation, SGEM, vol I, pp 677–684

196. Vica ML, Popa D, Glevitzky M, Siserman C, Matei HV (2017) Quality of drinkable water springs in two Alba county regions - comparative study. J Environ Protection Ecol 18:1389–1397

197. Vosniakos F, Vasile G, Albanis T, Petre J, Stanescu E, Cruceru L, Nicolau M, Kochubovski M, Gjorgjev D, Nikolaou K, Golumbeanu M, Vosniakos K, Vasilikiotis G (2010) Comparison study of physicochemical parameters evaluation between Danube river Delta and Axios-Vardar river valley (2003-2009). In: Simeonov LI, Kochubovski MV, Simeonova BG (eds) Environmental heavy metal pollution and effects on child mental development: risk assessment and prevention strategies. Springer, Dordrecht

198. Vremera R, Costinel D, Ionete RE (2011) Isotopic characterization of the major water source of Rm. Valcea Area, Romania. Asian J Chem 23:5202–5204

199. World Bank Group (2018) Romania water diagnostic report: moving toward EU compliance, inclusion, and water security. World Bank, Washington, DC. © World Bank. https:// openknowledge.worldbank.org/handle/10986/29928
200. Wright RT, Boorse DF (2017) Environmental science. Toward a sustainable future, 13th edn. Pearson Education, Inc.
201. Zobrist J, Sima M, Dogaru D, Senila M, Yang H, Popescu C, Roman C, Bela A, Frei L, Dold B, Balteanu D (2009) Environmental and socioeconomic assessment of impacts by mining activities-a case study in the Certej River catchment, Western Carpathians, Romania. Environ Sci Pollut Res 16:14–26
202. Zoran MA, Savastru RS, Savastru DM, Miclos SI, Tautan MN, Baschir LV (2012) Thermal pollution assessment in nuclear power plant environment by satellite remote sensing data. In: Neale CMU, Maltese A (eds) SPIE proceedings, remote sensing for agriculture, ecosystems, and hydrology XIV, vol 8531

Chapter 4
Management of Surfaces Water Resources—Ecological Status of the Mureş Waterbody (Superior Mureş Sector), Romania

Florica Morar, Dana Rus and Petru-Dragoş Morar

Abstract Water, the environmental factor which makes life on Earth possible, is also the environment which facilitated the emergence of life millions of years ago. The rational use of water at all levels, starting from global consumption to local contexts, in accordance with the principles of sustainable development, is a duty of human society. In order to protect water resources, over 70 regulations and directives were drafted at a European level, which Romania also observes. Romania's water resources consist of surface waters (rivers, lakes, the Danube River) and groundwater. In the present chapter, the authors used statistical data of some indicators which determine the ecological status (ecological potential) of the Mureş waterbody confluence Petrilaca—confluence Arieş, within the Mureş Hydrographic Basin, for a period of three years. Based on the state of the physicochemical and biological elements, the authors assessed the ecological status of the water mentioned above over three sections (Tîrgu Mureş, Ungheni/Mureş and Iernut). The determinations and the interpretations help classify the areas into the following classes: good, maximum and moderate in terms of the analyzed elements.

Keywords Resources · Surface water · Romania · Pollution · Mureş river · Pollution forms · Monitoring · Quality level

F. Morar
Industrial Engineering and Management Department, Faculty of Engineering, "Petru Maior" University of Tîrgu Mureş, Tîrgu Mureş, Romania
e-mail: florica.morar@ing.upm.ro

D. Rus (✉)
Department of Electrical Engineering and Computers, Faculty of Engineering, "Petru Maior" University of Tîrgu Mureş, Tîrgu Mureş, Romania
e-mail: dana.rus@ing.upm.ro

P.-D. Morar
Technical University Bucharest, Bucureşti, Romania
e-mail: dragos.morar@yahoo.com

© Springer Nature Switzerland AG 2020
A. M. Negm et al. (eds.), *Water Resources Management in Romania*, Springer Water,
https://doi.org/10.1007/978-3-030-22320-5_4

4.1   Introduction

Among the global problems currently faced by humankind, water quality degradation and lack of water are issues that are increasingly affecting the quality of life
and the development of human society. That is why the sustainable management of
water resources, which is based on the integrated management of water resources,
must take into account the achievement of the current objectives of the society, but
without compromising the goals of the future generations, while preserving a clean
environment.

4.1.1   Principles and Aspects of Water Resources Integrated Management

For integrated water resource management, the international community has recommended governments to apply the following principles [1]:

- *The basinal principle*—water resources are formed and managed in river basins.
  Sustainable management of water resources integrates water users from a river
  basin;
- *The principle of quantity-quality management*—these two aspects of water management are closely connected. Therefore it seems necessary to have a unitary
  approach leading to optimal technical and economical solutions for both aspects;
- *The principle of solidarity*—the planning and development of water resources
  implies the collaboration of all factors involved in the water sector: the state, local
  communities, users, water households and NGOs;
- *The "polluter pays" principle*—all costs related to the pollution are borne by the
  polluter;
- *The economic principle*—the beneficiary pays—water has an economic value in
  all its forms of use and must be recognized as an economic asset;
- *The principle of access to water*—it is important to recognize that among the
  fundamental rights of being human is the access to clean and sufficient water at
  an adequate price.

These principles underpin the concept of integrated water resource management,
which combines water use issues with the protection of natural ecosystems by integrating water uses [2]. The most important aspects related to the development of
water resource systems are the following:

- Physical sustainability—which means maintaining the natural circuit of water and
  nutrients;
- Environmental sustainability—"zero tolerance" for pollution beyond the self-
  cleaning capacity of the environment. There are no long-term effects or irreversible
  effects on the environment;

- Social sustainability—maintaining water requirements and willingness to pay for water services;
- Economic sustainability—economic support for measures which ensure a high living standard in terms of water consumption for all citizens;
- Institutional sustainability—maintaining the ability to plan, manage and operate the water resource system.

Water resource management requires the involvement of all interested parties—public and private—at all levels and at the right time.

The integrated water resource management is based on the River Basin Management Plan (RBMP), in line with the provisions of the EU Framework Directive 2000/60. Based on knowledge of water body status, RBMP sets the target objectives for a six-year period and proposes measures to achieve good water status for sustainable use [1, 3].

4.1.2 The Importance of Water and Areas of Use

The present stage of economic and social development is characterized by unprecedented dynamics. Given the limited nature of natural resources, water, one of the most important natural resources, must be well managed, in order to prevent the "water crisis", term which is increasingly used in the forecasts of international organizations.

Scientists have repeatedly warned of a possible global water crisis, which has determined the United Nations to set March 22 as "World Water Day".

The care for a rational water management is due to the characteristics of the water as a natural resource with an uneven distribution in time and space, with a limited character, with limited possibilities (from a technical and economic point of view) to be transferred between regions or different river basins [4].

Due to the increase in the world population and other factors, fewer people benefit from drinking water. The water problem can be solved by increasing production, a better distribution, and the wise use of existing resources.

The UNESCO Water Development Report (WWDR 2003) of the World Water Assessment Program notes that in the next 15–20 years, the amount of available drinking water will decrease by 30%, while 40% of the world population do not have sufficient clean water for minimal hygiene. More than 2.2 million people died in 2000 from contaminated drinking water or droughts. In 2004, an English organization, WaterAid, reported that a child dies every 15 s due to water-related diseases that could easily be prevented [5].

At the United Nations Conference on Water Resources, held in Mar del Plata (Argentina, 14–25 March 1977), it was estimated that the total volume of water on Terra was 1400 million km^3. Of this, freshwater held a percentage of 2.7; only 0.46% of the volume of freshwater on Earth is directly used [6].

Relative to relatively limited water resources, human water requirements have seen a steady increase from 1.4 billion cubic meters in 1950 to around 20.40 billion cubic meters in 1989, of which:

- 11% drinking water for the population and the public domain;
- 44% drinking water and industrial water for businesses;
- 45% water for irrigation, animal husbandry and fish farming.

Also, the average water consumption ranges from 575 L/day/person in the US (reaching 100 L/day/person in California, Arizona) to 200–300 L/day/person in the majority of the EU Member States. In poorly developed countries (e.g. Mozambique), the average water consumption is less than 10 L/day/person.

According to the World Health Organization and UNICEF, the minimum water consumption per day and per person is 20 L (from a source situated within one kilometer), a quantity considered sufficient for sanitary purposes and drinking. To ensure daily needs for a decent living, daily water consumption per person is around 50 L [7].

The increase in water consumption has been accompanied by an increase in the amount of waste water without any correlation with the development and vamping of the treatment plants to increase the capacities and the quality of the purification process, while ensuring the need for quality water.

Thus, according to the data from the Romanian Waters Company for 1990, out of the total waste water discharged, it was estimated that only 22% of these waste waters were properly cleaned, according to the legislation, approximately 50% have been cleansed inefficiently, and about 28% have been discharged into the natural environment without treatment, adversely affecting the quality of the environment, especially the aquatic environment [8].

The current trends in water consumption will lead to excessive demand for water resources, disturbing the balance of these resources, which would have adverse long-term effects on the country's economic and social development.

Water uses cover the most diverse forms (drinking water, fish farming, water supply for industry and agriculture, urban and recreational purposes), so its pollution potential is very high.

Avoiding the pollution of water sources and eliminating its effects is currently a major concern for those working in the field of water supply.

4.1.3   Surface Water Resources at a National Level

The water quality in Romania is monitored according to the structure and methodological principles of the Integrated Water Monitoring System in Romania (I.W.M.S.), restructured according to the requirements of European Directives.

The national water monitoring system includes two types of monitoring, as required by Law 310/2004 amending and supplementing the Water Law no. 107/1996, which included the provisions of the Framework Directive 60/2000/EEC in the water domain. Surveillance activities are carried out to assess the status of all water bodies within the river basins, while operational activities aim to monitor bodies of water which risk failing to meet water protection objectives [8, 9].

Water resources in Romania are both at the surface and underground.

In the updated National Management Plan related to the national portion of the Danube river basin which is included in the Romanian territory, the following categories of surface waters are mentioned:

- Rivers (natural, highly modified and artificial)—78.905 km (cadastral rivers);
- Natural lakes—129;
- Transitional waters—781.37 km^2 (619.37 km^2 sea transitional waters and 162 km^2 Sinoe lake);
- Coastal waters—571.8 km^2 (116 km).

In terms of quality, Romanian watercourses are classified into three categories.

Category I—Waters which can become potable for the water supply of populated centers, or which can be used to supply livestock farms and trout farms.
Category II—surface waters which can be used in fish farming (other than salmon culture), for technological needs in industry and recreational activities.
Category III—waters for crop irrigation, power generation in hydropower plants, industrial cooling plants, laundries and other uses for which such quality is acceptable, and the last category includes degraded waters where fish fauna cannot develop.

The assessment of surface water quality after 1999 is based on the processing of primary analytical. This data was obtained monthly in 312 first order surveillance sections located in 14 river basins: Tisa—8, Olt—36; Someş—28, Vedea—8, Crişuri—18; Argeş—34, Mureş—40, Ialomiţa—19, Bega-Timiş—21, Siret—49, Nera-Cerna—5, Prut—19, Jiu—15, Danube—12 [1, 10].

4.2   Water Quality and Surface Water Pollution

The quality of surface water (but also groundwater) is rapidly damaged by pollutants from domestic, agricultural and industrial activities. Surface water pollution and their decontamination is easier to achieve compared to underground, slowly-polluting waters, but the decontamination process is long-lasting and sometimes irreversible.

Water pollution is the alteration of physical, chemical and biological qualities of water, produced directly or indirectly, naturally or anthropically [11, 12].

Pollution can occur:

- Continuously or permanently (e.g. city sewerage, industrial waste discharge);
- Discontinuously (e.g. pollution at certain time intervals);
- Temporarily (e.g. temporary colonies);
- Accidentally (e.g. in case of damage).

Pollution prevention and the recovery of contaminated water resources must be the primary objective of water treatment. The current water resources and those which will be located in the future will be protected by legislation, measurement techniques and land use planning.

The quality of water defined by its physical, chemical and biological characteristics are, in a general way, the condition of the existence of life and, in particular, of human activities.

Water quality is the set of quantified physical, chemical, biological, and bacteriological characteristics which allow the sample to be grouped into one category, thus acquiring the ability to serve a particular purpose.

The United Nations Global Environment Monitoring System (GEMS) stipulates the monitoring of water quality according to three categories of parameters [13]:

• Basic parameters: temperature, pH, conductivity, dissolved oxygen, colibacilli;
• Parameters indicating persistent pollution: cadmium, mercury, organo-halogenated compounds and mineral oils;
• Optional parameters: total organic carbon, biochemical oxygen demand, anionic detergents, heavy metals, arsenic, chlorine, sodium, cyanides, total oils, streptococci [13].

All natural waters contain substances from the environment, natural substances or substances from anthropogenic activities.

The parameters indicating natural water quality and their values can be found in Table 4.1. The units of measure or expression of parameters are: milligrams (mg.), Milligrams/liter (mg/l), millisiemens/centimeters (ms/cm), percentage (%), according to the analyzed parameter [14, 15].

Table 4.1 Values of natural water quality parameters [16]

Parameters	Units of expression	Values
pH	units	6.5–8.5
TDS (dissolved substance)	mg	≈200
EC (electrical behavior)	ms/cm	≈0.3
Alkalinity	mg $CaCO_3$/l	≈100
Hardness	mg $CaCO_3$/l	≈50
Organic carbon	mg C/l	≈25
Organic carbon dissolved	mg C/l	0.1–10
Biochemical consumption, oxygen	mg O_2/l	1.5–3.0
Chemical consumption	mg O_2/l	10–20
Oxygen	%	75–125
Nitrates	mg N/l	1–4
Phosphates	mg P/l	≈0.1

The main source of drinking water, industrial water, but also for agriculture and fish farming is surface water. But this water is constantly subjected to the pressure of the anthropic factors, which causes its pollution, which will bring about the degradation of the environment with all its components.

Water pollution is currently considered a major problem faced by mankind, all the more so as economic agents—in the present case—discharge into the environment (natural environment) more and more quantities of polluting substances and matter [17].

Since pollution is a phenomenon without frontiers, an analysis of the pollution mechanism and of its effects on ecosystems should include a study on the interdependence of the three components of the environment: water, air, and soil, because an affected component also affects the other components.

4.2.1 The Necessity of Water Quality Protection

The treatment of water quality from domestic activity is closely related to taxes, income, but also to the size of the consuming population.

Domestic pollution caused by combustion processes, the location of drainage channels, the location of solid waste disposal lakes, the flow of liquid fuels from tanks and the flow of urban water.

In rural areas, domestic sewage is usually routed directly to surface waters (without pre-treatment) or to groundwater through infiltration. In urban areas, wastewater is usually collected and directed to treatment plants, but these are either not sufficiently equipped or defective or in some cases, they do not exist [18, 19].

Industry affects water and air contamination in many ways. Agriculture, through fertilization and treatment schemes against diseases and pests, is a source of water and soil pollution.

The most important pollutants of surface waters that affect soil and groundwater are nitrates and phosphates, pathogenic elements, bacteria, heavy metals, micro-organic compounds, etc.

The protection of the quantity and quality of existing waters for the future must be done in an integrated international context respecting the laws of the hydrological cycles in hydrographic basins, in aquifer levels or at a regional scale [20, 21].

According to the provisions of the Water Framework Directive, the rivers, coastal areas and lakes of Europe must be classified as having "good ecological status" in terms of quality.

In order to accomplish this, European citizens have an important role, because they need to be informed and involved in the preparation of river basin management plans for the main water courses.

They must also take into account pressures from anthropogenic activities (e.g. domestic and industrial waste water discharges, activities in intensive farming, especially the use of chemical fertilizers and pesticides) [22].

In order to improve the quality of surface waters in Romania, the above mentioned anthropic pressures are taken into consideration and, at the same time, specialists are looking for adequate, real solutions for the implementation of projects aiming at improving the quality of the waters.

Therefore, to increase the efficiency and flexibility of water resources and their consumption, it is necessary to re-examine the technical aspects of water systems, economic issues and water management legislation associated with water consumption priorities [16].

4.2.2 Classification of Water Pollution

Classification of surface water pollution can be done according to several criteria:

I—according to the mode of production of water pollution:

- Natural pollution—for example when water passes through soluble rocks, it is charged with different salts, or when the flowering of water occurs;
- Artificial pollution—due to sources of waste-water from anthropogenic activities;
- Controlled or organized pollution—originated from wastewater transported through the sewerage at certain points;
- Uncontrolled pollution—comes from sources of liquid pollution which emerge naturally, most often through meteoric waters transporting pollutants from the ground into rivers;
- Normal pollution—comes from familiar sources of pollution, collected and transported through the sewerage network to the treatment plant or transported directly to the receiver;
- Accidental pollution—may be the result of industrial process disturbances when large quantities of harmful substances are discharged in the sewage network or accidental uptake of toxic substances directly into emissaries, etc.
- Primary pollution—represented by the deposit of suspended substances in wastewater discharged into the ground, on the bottom or banks of the river;
- Secondary pollution—given by the gases resulting from the fermentation of organic matter from suspended matter in the emissary and which cause the movement of the remaining substances, bringing them to the surface of the water, from where they are then transported downstream by the water stream [23].

II—according to the nature of the polluting agent:

- Biological, bacteriological, viral and parasitological pollution (directly or indirectly linked to human presence);
- Physical pollution—due in particular to radioactive substances as well as to thermal pollution by the discharge of hot fluids into emissaries and also by insoluble elements which are suspended or sedimented;

- Chemical pollution—caused by the discharge of chemical substances in the water [24].
- The factors causing water pollution are grouped as follows:
- Demographic factors—represented by population numbers in a given area, and pollution is generally proportional to population density;
- Urbanistic factors—corresponding to the development of human settlements, which use large quantities of water which are returned into nature in the form of intensely contaminated wastewater;
- Industrial or economic factors—represented by the level of economic and industrial development of a region [25, 26].

4.2.3 Water Pollution Sources

Water pollution sources are classified according to several criteria:

- By their origin: domestic activities; industry; agriculture; transports
- By the origin of pollutants: local sources (e.g. sewerage pipes, discharge platforms); diffuse sources (e.g.: pollutants spread over a large area);
- By position: fixed sources (e.g.: industrial, animal breeding activities); mobile sources (e.g.: cars, houses, moving equipment).

However, the most frequent sources of pollution are domestic, industrial and those caused by livestock breeding activities, causing domestic pollution, industrial pollution, and pollution due to agricultural activities.

Domestic pollution is proportional to the population.

Domestic wastewater mainly contains putrefactive organic matter (carbohydrates, proteins, lipids), most of which decanting, resulting in layers of the organic slurry.

Among the organic materials which are characteristic of domestic pollution are also dissolved salts in the form of calcium, magnesium, bicarbonates, sulphates, etc., which are found in much lower quantities than organic matter.

Industrial pollution—is specific to industrially developed areas. Pollutants are very variable, depending on the field of activity.

They can be represented by raw materials, intermediate products or end products, by-products or co-products.

Pollution caused by agricultural activities—are represented by animal waste, natural or synthetic fertilizers, soil erosion due to inappropriate soil processing, etc.

Another type of pollution is caused by meteoric waters—the meteoric waters cause the movement of various types of wastewater on the surface of the soil but also waste, mineral fertilizers etc., which thus migrate into the soil or reach the surface waters [25].

Table 4.2 Classification of pollutants according to their effects [29]

Compound groups	Effects
Bio-degradable substances	Deoxygenation, unpleasant odors, dead fish
Toxic substances	Animal poisoning, dead fish, and plankton, bioaccumulation
Acids and alkali	Destruction of natural buffering capacity and disturbance of ecological balance
Disinfectants (Cl_2, H_2O_2, formalin etc.)	Selective destruction of microorganisms, unpleasant tastes, and smells in treated waters
Salts	Change of water characteristics (salinity, hardness etc.)
Oxidizing and reducing agents	Consumption of dissolved oxygen, eutrophication, selective development of bacteria, unpleasant smells
Esthetically disagreeable substances	Floating and sedimentary substances, eutrophication, anaerobic benthic deposits, odors, fish mortality
Pathogen agents	Hydric infections and diseases in humans and animals, plant diseases

4.2.4   Water Pollutants

Water pollutants can be grouped according to the following criteria:

- By their nature: organic, inorganic, biological, thermal, radioactive pollutants;
- By the state of aggregation: suspensions (insoluble in water); soluble in water; colloidal dispersions;
- By the degree of natural degradation in water: readily biodegradable; hardly biodegradable (degradation takes place in less than 30 days); non-biodegradable (degradation occurs in 30–60 days); refractory (degradation begins after at least 2 years) [27, 28].

Table 4.2 shows a classification of pollutants according to their disagreeable characteristics and effects.

4.2.5   Legal Aspects in the Field of Water Protection

Humankind has become aware of the hazards posed by pollutants to environmental health, and implicitly to human health. Various conventions and protocols attempt to reduce as far as possible the effects of pollutants. Such examples are the Montreal Protocol (1993), the United Nations Framework Convention on Climate Change (1994), the Kyoto Protocol of the United Nations Framework Convention on Climate Change (2001), the Stockholm Convention on Persistent Organic Pollutants (2001).

Water management in Romania meets both the requirements of the Water Framework Directive 60/2000/EEC and other EU Water Directives:

- Directive 75/440/EEC—surface water intended for drinking;
- Directive 76/464/EEC—gradual elimination of priority substances/priority hazardous substances,
- Directive 91/676/EEC—pollution by nitrates from agricultural sources,
- Directive 78/659/EEC—the quality of freshwaters requiring protection or improvement to support the life of fish,
- Directive 91/271/EEC on Urban Wastewater Treatment etc.

The Water Framework Directive has been transposed into national law by Law no. 310/2004 for amending and completing the Water Law no. 107/1996. This Directive provides the European Commission, the Member States and the candidate countries with the opportunity to cooperate in a new partnership. The objective is to protect inland waters, transitional waters, coastal waters and groundwater, by preventing pollution at source and the establishment of a unitary control mechanism for pollution sources [3].

The success of this Directive is based on close and coherent cooperation within the European Community, among the Member States and at the local level as well as on information, consultation and participation of the public, including users.

The objectives of the Water Framework Directive (2000/60/EC) are to establish a framework for the protection of inland surface waters, transitional waters, coastal waters, and groundwater, aiming at:

- Preventing further deterioration, preserving and improving the status of aquatic ecosystems;
- Promoting sustainable water use on the basis of long-term protection of available water resources;
- Ensuring increased protection and improvement of the aquatic environment, in particular through specific measures to progressively reduce discharges, emissions and losses of priority substances and by gradually stopping discharges, emissions and losses of hazardous and priority hazardous substances;
- Ensuring the gradual reduction of groundwater pollution and the prevention of further pollution;
- Mitigating the effects of floods and droughts;
- Ensuring a sufficient supply of surface water and groundwater of good quality, which is necessary for a sustainable, balanced and equitable use of water;
- Protection of marine and territorial waters;
- Achieving the objectives set out in the relevant international agreements, including agreements aimed at preventing and eliminating pollution of the marine environment;
- Stopping or phasing out discharges, emissions and losses of dangerous substances which imply an unacceptable risk for the aquatic environment.

One must also mention the Council Directive 91/271/EEC on Urban Waste Water Treatment, which has been fully transposed into Romanian legislation by

Government Decision 188/2002 approving certain rules on discharge conditions in the aquatic environment of sewage (MO / Monitorul Oficial Journal 187 / 20.03.2002) [30].

The Water Framework Directive—2000/60EC is a new approach to water management, based on the basin principle and imposing strict deadlines for implementing the measures included in the program. The Directive establishes a number of integrated principles for water management, including inter alia the participation of the public in water management and the integration of economic aspects. Under this directive, the Member States of the European Union were required to ensure the good status of all surface waters by 2015.

The Management Plan of the Hydrographic Basins in Romania, the main implementation tool of the Framework Directive was finalized at the end of 2009. The plan contains the programs of measures (required by Article 11 of the Directive, measures mentioning the progressive achievement of the good status of all surface waters until 2015), the reasons for any delay for the program measures to become operational, and the timetable for the implementation of the measures [1].

Regarding the implementation of the Urban Wastewater Treatment Directive 91/271/EEC, the entire territory of Romania is a sensitive area, requiring a long transition period, namely 12 years [30].

4.2.6 Monitoring Romanian Water Quality

In the complex activity of water quality protection, the main element is the knowledge of its quality at each moment and on the basis of the obtained and analyzed data a prognosis can be made of the evolution trend of the water quality on hydrographic basins or on more restricted areas [22, 31].

The main activities that contribute to the achievement of the water quality objectives are as follows:

- Tracking and measuring activities over a limited period for a specific purpose, such as the establishment of fisheries;
- Continuous monitoring by measurements and observations of water quality, on certain sections or tributaries in the case of waters with certain important uses, such as a source of drinking water;
- Long-term monitoring activity based on standardized measurements for the study of water quality and the evolution of water quality over time and space.

The main purposes of monitoring activities in general and of water, in particular, are:

- Alarm in case of increases in pollution values that can become dangerous;
- Checking the validity of strategies for water quality protection;
- Evaluation and prognosis of the evolution of the water quality;

- A tool for determining water polluters in case of environmental accidents investigations.

The water quality in Romania is monitored according to the structure and methodological principles of the Integrated Water Monitoring System in Romania (I.W.M.S), restructured according to the requirements of the European Directives.

The national water monitoring system comprises two types of monitoring activities, as required by Law 310/2004 amending and supplementing the Water Law 107/1996, which included the provisions of the Water Framework Directive 60/2000/EEC and other EU Directives. In this way, a monitoring activity is carried out to evaluate the status of all water bodies within the river basins, and an operational monitoring activity involves bodies of water at risk of failing to meet water protection objectives [10, 32].

Depending on the qualitative characteristics of the water bodies, various types of monitoring programs were developed in accordance with the requirements of the Water Framework Directive [33].

The following must be mentioned: Surveillance Program (S), the Operational Program (O), the Investigation Program (I) The Reference (R) and Best Available Section Program (BASP), Drinking Water Program (DW), Intercalibration Program (IC), Monitoring Program for Vulnerable Areas for Nitrate Pollution, Ihtiofauna Monitoring Program (IH) for Habitats and Species Protection (HS), the International Conventions Program (CI) and the Highly Modified Waters (HMW) Program.

I.W.M.S contains 6 components (subsystems):

- Flowing surface waters;
- Lakes (natural and of accumulation);
- Transitional waters (river and lake);
- Coastal waters;
- Underground waters;
- Wastewater.

From the point of view of the areas where the monitoring activities are applied, the most important are chemical monitoring, biological monitoring, ecotoxicological monitoring.

In the case of emission monitoring, it is assumed that the limits of the pollutant compounds in the waters discharged into emissaries do not depend on the change in river quality level and that the basic element is to prevent pollution of the emissary. The requirements for reducing the amount of polluting substances in discharged waters into emissaries depend on the toxicity of these substances and their persistence and tendency to accumulate and bio accumulate in the aquatic environment [16].

Emission monitoring should provide an overview of the whole basin with all tributaries that introduce water with varying degrees of pollution. It should also track impacts on the aquatic environment, and the purpose for which water is used, but also study the cumulative effect of pollutants and decomposition products and highlight sources of diffuse pollution.

If it is desired to design a monitoring system for a particular river basin or tributary, it must necessarily take at least two stages:

- Inventory of all possible emissions, in terms of their characteristics: components, concentrations, frequency, variance amplitudes;
- The systematic detection and assessment of the complexity of the chemical composition and of the samples to be performed in the laboratory.

Once these steps have been established, the following variables will be established: monitoring variables, monitoring station location areas, sampling frequency, how laboratory samples are analyzed and how results are interpreted.

In practice, there are a number of quality indicators that have been grouped in 12 classes. The classes are as follows: oxygen regime, water aggression indicators, and salinity indicators, presence of nutrients, general inorganic pollutants, general organic pollutants, microporous—heavy metals, microporous—pesticides, radioactivity, microbiology, biology, and flow.

The allowable values for basic indicators which measure the degree of surface water pollution (river Danube, seawater) are provided in the NTPA-001/2002 [34, 35]. Technical Regulations for Water Protection.

The normative is aimed at establishing the general quality conditions of all categories of wastewater before their discharge into natural receptors, as well as the admissible limit values of the main quality indicators of these waters. (Normativ din 28 februarie 2002).

The norms that aim at establishing the quality of surface water quality are transposed by the Water Framework Directive (Directive 2000/60/EC), depending on the conditions at a national, regional, local level.

Water quality monitoring stations are an integral part of the Romanian Water Quality Surveillance System (WQSS), a system under the responsibility of the Romanian Waters Company, and are integrated into the Integrated Monitoring System for the Quality of Environment in Romania (IMSQER). Romania is making efforts to make compatible the systems and values laid down in the normative with the norms of the European Community, especially because in the case of water, we have to deal with the transboundary transport of the pollutants with the waters of the rivers that leave the territory of Romania (Normativ din 28 februarie 2002).

4.3   Pollutant Impact on Surface Waters

Due to the diversity and complexity of the sources of pollution, the consequences of water pollution are numerous, some of them unforeseen, with long-term effects, others not measurable. Because of this, many diseases originate in the quality of human drinking water, and many species of flora and fauna are affected by the quality of the water they have at their disposal.

Some pollutants that have a remanent character will affect the aquatic environment, and not only in the long run, which has unwanted effects even after decades, even after the source of pollution has ceased.

Here are some of the most important consequences of pollution.

Due to the diversity and complexity of the sources of pollution, the consequences of water pollution are numerous, some of them unforeseen, with long-term effects, others not measurable. Because of this, many diseases originate in the quality of drinking water, and many species of flora and fauna are affected by the quality of the water they have at their disposal [36, 37].

Some pollutants that have a persistent character will affect the aquatic environment in the long run, which has unwanted effects even after decades, even after the source of pollution has ceased.

Here are some of the most important consequences of pollution.

Sanitary consequences:

- Human and animal waste discharged into water causes water sickness (typhoid fever, dysentery, cholera, gastroenterocolitis, intestinal parasites, etc.);
- Toxic substances in industrial wastewater can cause acute and chronical poisoning with unpredictable consequences over time;
- Radioactive substances from the waters discharged from nuclear power plants or from the processing of minerals can cause radiation-related diseases;
- Alteration of the organoleptic characteristics of the waters (smell, taste, and color), etc.

Biological consequences:

- Thermal pollution may lead to a decrease in dissolved oxygen concentration, and temperature rise. These consequences are responsible for a series of unwanted processes (disappearance of fish species, mass development of thermophilic algae, eutrophication, creation of anaerobic conditions favoring the development of anaerobic microorganisms and occurrence of products such as hydrogen sulfide, methane, carbon dioxide etc.
- The unfavorable modification of the structure of aquatic biocenosis in favor of those who have lower claims for good water quality;
- Disturbances of aquatic life through the death of species due to lack of oxygen or destruction of natural habitat; etc.

Economic consequences:

- Increased costs for both treated water and for products requiring a certain quality of water;
- The difficulty of using polluted water sources, which require expensive treatment processes;
- Providing water from other sources, at a great distance;
- Costs of treating people who have become ill due to poor water quality, etc.

Social consequences:

- Compromising the natural areas of recreation and agreement;
- Compromising tourist or urban areas situated in the vicinity of rivers or polluted seaside;

- Population emigration due to deterioration of environmental conditions in general and of water quality in particular;
- Affecting protected areas of scientific or touristic interest.

Hydrological and Hydrogeological Consequences:

- Change in flows due to the pollution of surface water, but especially of underground water, due to changes in superficial tension, viscosity, adsorption power, soil permeability;
- The premature clogging of the river beds and especially of the beds of accumulation lakes.

The effects of pollutants can be assessed both by their behavior over time and by their nature. It is difficult to determine which of the two forms is more dangerous for the aquatic ecosystem because everything depends on the levels at which these effects are manifested.

For example, in the case of accidental pollution (more commonly encountered are those caused by hydrocarbons), the first environmental factor which is affected is surface water.

Due to its immiscibility with water, most of the pollutant rises to the surface of the water, thus preventing atmospheric air diffusion. Pollutants also cause major changes in water quality, making it improper for agriculture, animal husbandry, agreement, feeding the population, etc. [29, 38].

Accidents with hydrocarbons on surface waters, therefore, have a negative influence on abiotic environmental factors by altering their physicochemical and biological properties. At the same time, accidents directly or indirectly cause a number of harmful effects on abiotic factors with which biotic factors are interdependent.

The aquatic environment is much more complex than the land and air in that the interdependence between the environment and the body is much closer in the case of aquatic life than in the case of land. As a result, the effects of the impact of aquatic organisms in case of accidental spills with oil products is much stronger than in terrestrial animals.

The impact on aquatic ecosystems varies due to the degradation of the oil and the long exposure period. In general, acute or chronic exposure levels are toxic and dangerous to aquatic organisms [39, 40].

Numerous laboratory and field studies, as well as studies on discharges, have generated consistent observations about the potential effects of spilled oil products on the marine biotope [41].

Analyzing the data provided by the National Agency "Romanian Waters" (ANAR) (Table 4.3) regarding accidental pollutants it is noticed that more than half of the accidental pollution produced during the period 2000–2011 is represented by oil pollution [42, 7].

As a result of the pollution, a restructuring of aquatic biocenosis takes place in the water stream, the dissolved oxygen concentration and the biological oxygen consumption change. The amount of dissolved oxygen in the watercourse is influenced both actively and passively:

Table 4.3 Statistical situation regarding the number of accidental pollution cases in the Romanian aquatic environment (period 2000–2011)

Year	Number of accidental pollution cases in the period 2000–2011					
	Animal industries	Oil	Slag and suspensions	Chemical products	Others	Total
2000	0	34	0	3	44	81
2001	0	66	6	9	38	119
2002	0	74	10	13	20	117
2003	0	50	5	8	0	63
2004	1	13	2	7	4	27
2005	4	39	5	3	12	63
2006	7	36	3	5	3	54
2007	15	49	8	0	10	82
2008	13	49	0	8	13	83
2009	17	31	0	11	5	64
2010	10	34	0	12	4	60
2011	12	14	0	2	7	35
Total	79	489	39	81	160	848

Source A.N.A.R., quoted by Popescu D, Nistoran D

- Actively—oxygen intake in the mass of water—is achieved by air diffusion through the air-water interface, but also by the photosynthesis process by which oxygen is released in the mass of water;
- Passively—oxygen consumption—which occurs through natural pollution (e.g. erosion), direct pollution by waste-water spillage and aerobic decomposition but, in particular, anaerobic decomposition of organic debris from sediments. It is known that in anaerobic processes there are microorganisms that do not require the presence of dissolved oxygen, the final products of the processes are sulphides, hydrogen sulphide, methane, etc.

4.4 Case Study—Water Quality in the Mureş Hydrographic Basin, Mureş Waterbody Confluence Petrilaca—Confluence Arieş,—Reflected in an Overview of Physico-Chemical and Biological Indicators

We consider it necessary to mention that the study aims to present the values of some indicators characterizing the quality of the Mureş waterbody confluence Petrilaca—confluence Arieş, for the period 2015–2017. We will present and discuss the minimal, average and maximal values of the physicochemical and biological indicators for three sections of the Mureş water body: Tîrgu Mureş, Ungheni/Mureş and Iernut.

We will present some general elements regarding the Mureş River Basin, emphasizing the Mureş Superior Sector which is the object of the present research (Fig. 4.1).

The Mureş River, the second longest river in Romania (after the Danube), is 803 km long, out of which 761 km is on the territory of Romania. It flows from the Oriental Carpathians (Hăşmaş Mountains), it passes through the Giurgeu Depression, it crosses Călimani and Gurghiu Mountains through the Toplita-Deda Gorge. It then crosses the central part of the Transylvanian Plateau through the towns of Reghin, Tîrgu Mureş, Iernut, Aiud, then enters the Plain Arad. It passes through Arad and to the west of Nădlac, defining the border between Romania and Hungary over a distance of 31 km, then it enters Hungary and flows into the Tisa River. It is considered a calmer river, compared to its "stormy" brother, the Olt River. In Antiquity it was known as "Maris" or "Marisia" [42–44].

The course of the Mureş River from the spring until the point where it flows into the Tisa River is divided into four characteristic sectors:

- The upper Mureş, including Giurgeu Depression and the Toplita-Deda Gorge (area studied in the present paper);
- The Middle Mureş situated on the central area of the Transylvanian Plateau, between Deda and Alba Iulia;
- The Lower Mureş, between the Western Carpathian Mountains, the Southern Carpathians and the Banat Mountains, between Alba Iulia and Lipova;
- The Lower Mureş from the Western Plain, between Lipova and the border with Hungary.

Fig. 4.1 The Mureş River—landscape

4.4.1 General Presentation of the Mureş Basin

The Mureş River basin is located in the central and western part of Romania; it borders the Criş and Someş river basins to the north, the Banat hydrographic and the Jiu and Olt river basins area to the south, the Siret hydrographic basin to the west, and the Hungarian border to the east. The Mureş River Basin is located in the area delimited by the Eastern, Southern and Western Carpathians, and its lower sector is located in the center of the Tisa Plain.

The Mureş River Basin covers all relief forms. Relief varies with the diversity of forms, and the altitude has values between 80 m (minimal altitude—where the river exits of the country in the Western Plain) and 2509 m (maximal altitude in the Retezat Mountains). Hills and plateaus occupy 55% of the area of the basin, the mountains are 25% the difference being in plains, valleys, and meadows [45, 46].

The total catchment area (including the Ier channel) is 28,310 km^2 representing 11.7% of the country's surface. The total surface water resources in the Mureş basin area total approx. 5876.3 million m^3/year, out of which usable resources are approx. 1054.07 million m^3/year. They represent approx. 88.9% of the total resources and are formed mainly by the rivers Mureş, Târnave, Arieş, Strei, Cerna and their tributaries [47].

From a climatic point of view, the Mureş basin is characterized by continental-moderate climate with hot summers and long and cold winters, especially in the mountain areas in the northeast of the basin, and in the west the climate is arid, the summers being in generally drier and warmer.

The large variety of the natural environment allowed the designation of a large number of protected areas within the Mureş basin. Thus, there are 2 national parks, 5 natural parks, 71 special conservation areas, 19 avifauna special protection areas and 147 reserves included in the Register of Protected Areas, which is an integral part of the river basin management plan. These protected areas amount to 9833.04 Km^2 and represent about 35% of the surface of the Mureş basin. Most of the areas are located in the mountain and hill areas, on the upper courses of the rivers Mureş, Gurghiu, Arieş, Strei, Mare, and Geoagiu.

From an administrative point of view, the Mureş river basin comprises the Mureş and Alba counties, parts of the counties of Harghita, Sibiu, Cluj, Hunedoara, Arad, Timiş and small areas from the counties of Braşov, Bistriţa-Năsăud, Caraş-Severin. The Mureş River Basin is located in the West Development Region (25.19%), North West (4.69%) and Center (54.03%). At the level of the entire river basin, there are 16 municipalities, 23 towns, 303 communes and 1780 villages [43, 48].

In the following, we refer to elements describing only the Upper Mureş sector, because this is the area covered by this study, namely: the water-course/body Mureş—Petrilaca confluence—confluence Arieş, with sections Tîrgu Mureş, Ungheni/Mureş and Iernut (Fig. 4.2a and b).

The Upper Mureş River, between Izvoare and the Topliţa-Deda Gorge, is situated in a neo-volcanic area (Oaş, Gutâi, Văratec in the north and Călimani, Gurghiu,

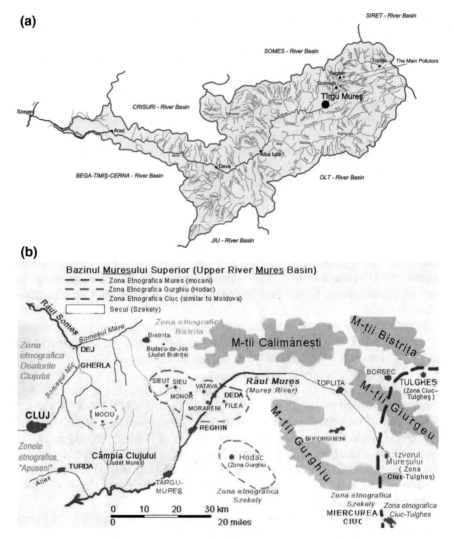

Fig. 4.2 a Mureş River Basin. *Source* http://www.cjmures.ro/Hotariri/Hot2016/anexa_hot135_2016.pdf. **b** Upper Mureş River Basin. *Source* https://www.google.ro/search?q=Mureşul+superior-+harta [48, 9]

Harghita in the south). In the northern area and esites and pyroclastites predominate, while the andesites, riolites, dacites predominate in the south.

In the hydrographic basin of the Upper Mureş river, the average multi-annual temperatures fluctuate between 5 and 6 °C in the depression area and slightly above 0 °C in the mountainous sector, with the coldest month, January (−6 °C … −7 °C) and the warmest month, July (12 °C … 17 °C). Multiannual rainfall averages between 480 and 980 mm, with an average of 610 mm per well [46, 49].

In the upper hydrographic basin of the Mureş River, the water composition is influenced by the anthropic factor through its various activities. Such activities include extraction of the underground resources (ores, salt, mineral waters, etc.), chemical crops farming, animal farms, technological processes resulting in waste water, domestic water evacuation of households, etc.

However, the most significant changes in water quality in the Mureş Superior Basin are produced by the chemical industry, the wood processing industry, the food industry, the construction materials industry, the urban, rural and agricultural localities [50, 51].

4.4.2 Water Contamination Indicators of Water Pollution of the Upper Mureş River Basin (2015–2017)

The Water Quality Laboratories in Tîrgu Mureş, as well as the subordinated laboratories (belonging to Water Management Systems—S.G.A.—Alba, Hunedoara and Arad) carry out quality analyses for the water in the Mureş Hydrological Basin, according to the provisions of the Operations Manual of the Monitoring System. Laboratories collect, preserve, sample, sub-sample and analyze samples from a physico-chemical and biological point of view, according to SR EN ISO/CEI 17025: 2005.

For the three sections mentioned at point 4.4.1, the indicators/parameters are analyzed in Table 4.4 and correspond to Annex 5 of the WFD (Water Framework Directive).

In terms of biological elements, the phytoplankton and the macro-invertebrates were studied.

4.4.2.1 Mineralization and Chemical Composition of River Water in the Upper Basin of Mureş

Mineralization and chemical composition of river water in the Mureş Upper Basin has been addressed in some general studies and studies [43]. The springs and rivers, which originate in meteoric water, have the chemical composition determined by the nature of the rocks with which they come in contact. Mountain springs generally have a low salt content (less than 50 mg/l). However, the variation of water mineralization is closely related to other factors: variation in water flows, type of water supply, waste water spills, precipitation and temperature.

There are authors who, through studies conducted in the Mureş basin, assert that the multi-annual average values of mineralization generally increase from source to sea. Moreover, in the sector belonging to the mountainous area, the waters of the Mureş River have reduced mineralization, slightly exceeding 100 mg/l [45, 52]. The indicators belonging to the Salinity indicator group determined during the three years

Table 4.4 Quality elements taken into consideration for the period 2015–2017

No. cart	Indicators group	Indicator	U/M
1.	Thermal conditions	Air temperature	°C
		Water temperature	°C
2.	Acidification conditions	pH	units
3.	Salinity	Fixed residues	mg/l
		Conductivity	μS/cm
		Chlorides	mg/l
		Sulfates	mg/l
		Dissolved Fe	mg/l
		Total manganese	mg/l
4.	Oxygen status	Dissolved oxygen (concentration)	mgO_2/l
		Dissolved oxygen (saturation)	%
		CBO_5	mgO_2/l
		CCO–Cr	mgO_2/l
5.	Nutrients	$N–NH_4$	mg/l N
		$N–NO_2$	mg/l N
		$N–NO_3$	mg/l N
		N total	mg/l N
		$P–PO_4$	mg/l P
		Total P	mg/l P
6.	Specific pollutants—metals	Total Cadmium	μg/l
		Total Ni	μg/l
		Total Pb	μg/l
		Total Cu	μg/l
		Total Zn	μg/l
		Total Cr	μg/l
		Total As	μg/l

*Source*https://eur-lex.europa.eu/legal-content/RO/TXT/?uri=CELEX:32000L0060

of the study (2015–2017) for the Mureş water body, the Petrilaca confluence—Arieş confluence characterizes the ecological status of the Mureş water body as good.

Table 4.5 contains the average values (according to which the ecological status of the Mureş water body was appreciated) for the 12 determinations carried out in each of the three years considered in the study.

For the Ungheni-Mureş sector, fixed residual values, conductivity, and bicarbonate values were determined (their average value was reported). Thus, the average values of bicarbonates for the Ungheni—Mureş sector are: 985,875 mg/l for the year 2015, 102,525 mg/l for the year 2016; 113,725 mg/l for the year 2017 [53].

Table 4.5 The average values of the general physico-chemical elements that characterize the salinity of the waters for the studied area

Section	Year	Physico-chemical indicators/elements						
		Fixed residues (mg/l)	Conductivity (μS/cm)	Chlorides (mg/l)	Sulfates (mg/l)	Dissolved iron (mg/l)	Total mangan (mg/l)
Tîrgu Mureș sector	2015	169.5	241.166666	26.08	34.041666	0.091241	0.052575
	2016	180.5	237.583333	24.75	21.666666	0.091875	0.052841
	2017	180.0	237.916666	28.18	20.591666	0.083725	0.053441
Ungheni/Mureș	2015	213.375	293.0	–	–	–	–
	2016	216.0	290.0	–	–	–	–
	2017	229.625	307.5	–	–	–	–
Iernut sector	2015	241.5	314.5	33.40	29.17	0.0725	0.04645
	2016	243.0	325.5	32.90	31.70	0.1153	0.105175
	2017	254.75	347.5	37.16	28.96	0.1104	0.122637

4.4.2.2 Physico-Chemical Indicators

In Table 4.6 one may observe that the water temperature in the three sections of the Mureş water body recorded average values between 10.58 °C (2015) and 17.25 °C (2016) in the Iernut section. Higher temperatures were registered in this section compared to the other two sections over the three years of the study. The pH values allow water to be labeled as neutral, slightly alkaline, which does not raise drinking water problems.

The minimal, average and maximal values of the Oxygen Group Indicators presented in Fig. 4.3 (Dissolved Oxygen—Concentration), Fig. 4.4 (Dissolved Oxygen—Saturation), Fig. 4.5 (CBO5), Fig. 4.6 (CCO–Cr), suggest good environmental status/potential for the three sections of the Mureş water body—the Petrilaca confluence—the Aries confluence. Thus, the oxygenation conditions in 2015 were appreciated as having good ecological status and in the next two years these conditions led to the assessment of the ecological status as moderate, due to the CCO–Cr values.

The values of nutrients in the water are presented in Table 4.7. Following the monitoring activities and the calculated determinations, it was appreciated that the ecological status/potential is good for the entire body of water which was assessed. The state of nitrites and nitrates was appreciated as moderate and good.

The specific monitored pollutants and the minimal, average and maximal values are found in Table 4.8.

In 2015, the metals were not monitored for the Ungheni/Mureş section. However, assessments were made, according to the norms in force (and which have been established according to the regulations of the Water Framework Directive) [54].

The quality potential expressed according to the specific pollutants for the year 2015 received the rating "good", and in the following years, the potential was maximum (Table 4.9).

4.4.2.3 Biologic Indicators

The presence of phytoplankton and macro invertebrates is influenced by aeration conditions, by the presence of nutrients, of specific pollutants in the aquatic environment, as well as by the thermal conditions that occur in the area during the year [55].

Average phytoplankton values vary from one year to another; in 2016 there was a decrease in the average value, after which in the following year the value was differentiated by a slight increase compared to the previous years. The presence of macro-vertebrates is more pronounced in 2016 than in 2015, a situation that gives it a maximum status/potential.

We consider it necessary to mention that the investigative environments are water, sediment and biota. The quality elements, the parameters, and minimum monitoring frequencies are in line with the requirements of the Water Framework Directive, depending on the type of program. Monitoring of the water status in Romania based on the monitoring programs established in accordance with Art. 8 (1), (2) of the

Table 4.6 Thermal and acidification conditions

Year	Air temperature (°C)			Water temperature (°C)			pH		
	Min	Med	Max	Vin	Med	Max	Min	Med	Max
Tîrgu Mureș sector									
2015	−6	11.66	30	0	10.58	25	6.8	7.82	8
2016	−13	10.08	23	0	12	23	7.7	8.03	8.6
2017	−5	10.17	28	0	10.58	25	7.1	7.86	8.5
Ungheni/Mureș sector									
2015	−6	12.38	25	0	11.87	20	7.7	7.83	8
2016	6	14.13	30	4	12.75	23	7.7	7.85	7.9
2017	−3	13.63	30	0	13.13	26	7	7.76	8.1
Iernut sector									
2015	7	15.25	22	10	13.75	18	7.8	7.9	8
2016	13	15	18	10	17.25	25	7.8	7.88	8
2017	−1	13.75	28	0	13.25	26	7.4	7.83	8

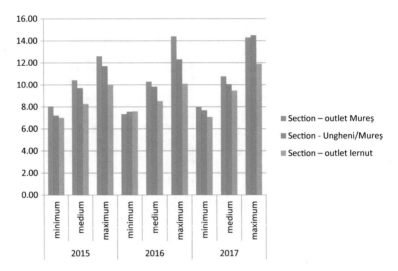

Fig. 4.3 Dissolved oxygen (concentration) (mgO$_2$/l)

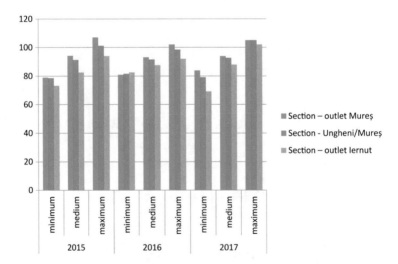

Fig. 4.4 Dissolved oxygen (saturation) (%)

Water Framework Directive is carried out by the Romanian Water Authority through its territorial units (Water Basin Administrations). Surface water monitoring activities include: the surveillance program, the operational program, and the investigation program. By means of these programs, a real assessment of the investigated environments is made.

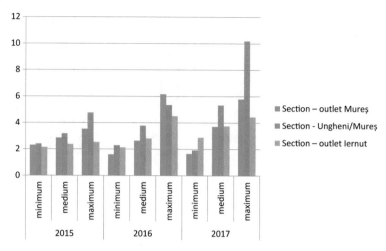

Fig. 4.5 CBO$_5$ (biochemical oxygen consumption over 5 days) (mgO$_2$/l)

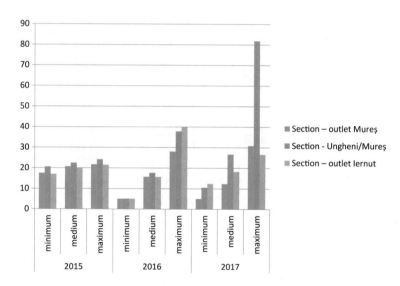

Fig. 4.6 CCO-Cr (oxygen consumption) (mgO$_2$/l)

4.5 Conclusions

Assessment of the ecological status and chemical status of water bodies in accordance with the requirements.

The Water Framework Directive is considered to be a major challenge, as for the first time, it is necessary to apply methods of analysis and evaluation at a European and national level which comply with the principles of Directive 2000/60/EC. At a

Table 4.7 Statistical data regarding the values of the Nutrients

Nutrients	2015			2016			2017		
	Min	Med	Max	Min	Med	Max	Min	Med	Max
Tîrgu Mureş sector									
N–NH$_4$	0.027	0.15575	0.606	0.0125	0.08979	0.311	0.125	0.09604	0.407
N–NO$_2$	0.07	0.02866	0.108	0.01	0.02308	0.05	0.009	0.01875	0.05
N–NO$_3$	0.203	0.707416	0.83	0.095	0.503166	1.06	0.215	0.576166	1.1
Total N	0.4	1.55925	3.6	0.5	0.780833	1.59	0.5	0.935	2.53
P–PO$_4$	0.02	0.04775	0.136	0.009	0.048333	0.083	0.027	0.57333	0.101
Total P	0.0035	0.65541	0.15	0.013	0.0655	0.103	0.031	0.071083	0.11
Ungheni/Mureş sector									
N–NH$_4$	0.168	0.527125	1.16	0.1	0.506	2.35	0.028	0.260625	0.769
N–NO$_2$	0.019	0.047875	0.09	0.013	0.129625	0.65	0.024	0.0575	0.137
N–NO$_3$	0.32	1.311	2.3	0.145	0.751125	1.92	0.602	1.028125	1.82
Total N	1.58	3.2625	4.3	1.02	2.22875	7.5	0.5	1.59375	3.13
P–PO$_4$	0.037	0.059125	0.085	0.028	0.04775	0.097	0.031	0.07025	0.12
Total P	0.065	0.90625	0.15	0.035	0.060875	0.105	0.01	0.074	0.133
Iernut sector									
N–NH$_4$	0.124	0.19025	0.305	0.043	0.35375	1	0.09	0.1805	0.42
N–NO$_2$	0.028	0.079	0.144	0.01	0.06925	0.185	0.023	0.047625	0.085
N–NO$_3$	0.79	1.5625	2.3	0.754	1.13	2.07	0.05	0.8685	1.69
Total N	1.53	3.0075	5	1.27	2.2425	4.45	1.14	1.69125	2.3
P–PO$_4$	0.047	0.10375	0.133	0.01	0.031	0.05	0.019	0.049125	0.131
Total P	0.07	0.12	0.15	0.033	0.053	0.074	0.022	0.060875	0.145

Table 4.8 Statistic data regarding the values of the group-specific pollutants—metals

Metals	2015			2016			2017		
	Min	Med	Max	Min	Med	Max	Min	Med	Max
Tîrgu Mureş sector									
Total Cd	0.0125	0.0615	0.176	0.0125	0.042041	0.081	0.125	0.125	0.125
Total Ni	0.5	1.6825	3.69	0.5	1.605	3.2	2	2	2
Total Pb	0.5	2.5325	4.31	1.15	1.97	3.7	2.5	2.5	2.5
Total Cu	1.26	2.28	4.06	0.598	2.015666	3.75	1.5	1.84666	3.89
Total Zn	5	12.6	39	0.25	14.356666	24	0.25	11.375	25
Total Cr	0.5	0.5	0.5	0.5	0.5	0.5	1	1	1
Total As	1.34	2.3525	4.58	0.799	1.861583	2.78	2	2	2
Ungheni/Mureş sector									
Total Cd	–	–	–	0.0125	0.04325	0.09	0.125	0.125	0.125
Total Ni	–	–	–	1.71	2.225	2.72	2	2	2
Total Pb	–	–	–	1.52	1.9605	2.402	2.5	2.5	2.5
Total Cu	–	–	–	1.86	2.3625	2.71	1.5	1.7375	3.4
Total Zn	–	–	–	6.32	18.13	27.6	0.25	9.1475	25
Total Cr	–	–	–	0.5	0.9825	2.43	1	1	1
Total As	–	–	–	0.811	0.95325	1.22	2	2	2
Iernut sector									
Total Cd	0.041	0.061	0.074	0.032	0.06325	0.086	0.125	0.125	0.125
Total Ni	0.5	2.05	3.99	1.12	2.1625	3.27	2	2	2
Total Pb	1.57	2.47	3.79	2.98	3.445	4.09	2.5	2.5	2.5

(continued)

Table 4.8 (continued)

Metals	2015			2016			2017		
	Min	Med	Max	Min	Med	Max	Min	Med	Max
Total Cu	1.29	1.745	2.57	2.19	3.4125	4.19	1.5	2.725	4.94
Total Zn	5	10.025	19.5	11.7	23.45	32	1.72	12.215	25
Total Cr	0.5	0.5	0.5	0.5	0.5	0.5	1	1	1
Total As	2.55	3.125	3.66	1.52	2.54	3.82	2	2	2

Table 4.9 Average values of phytoplankton and macro invertebrates for the Mureş water body, confluence Petrilaca- confluence Arieş

Indicators	Average values		
	2015	2016	2017
Phytoplankton	0.6776	0.6501	0.7143
Macro invertebrates	0.6353	0.7943	0.7488

national level, efforts have been made to assess the ecological status of the waters. Significant progress has been made, but, the current Management Plan mentions that there are still issues and/or uncertainties which need to be taken into account when interpreting the results of evaluations. The ecological status and the ecological potential have been assessed taking into account some of the biological, hydro morphological and physicochemical elements.

Since some methods complying with the requirements of the Water Framework Directive were developed after the first phase of the European inter-calibration exercise, they were not included in the assessment process, and the confidence in the assessment of ecological status and environmental potential was consequently moderate and low. Also, there was no database of monitoring data over a longer period of time since the monitoring system (which has to comply with the requirements of Directive 2000/60/EC) was implemented starting from 2007, the year of Romania's accession to EU.

Regarding the ecological state of the natural surface water-rivers monitored in the Mureş hydrological basin in the years 2015 and 2016 [53, 47] one can conclude:

- In 2015, 1624.74 km of the Mureş hydrological basin were evaluated, of which 81.04% had good ecological status, 18.49%—moderate ecological status, the difference of 0.47% being evaluated as having a bad ecological status;
- In 2017, 76.39% of the estimated kilometers (1316.12) had good ecological status, and 23.61% were characterized as moderately ecological.

For the year 2017, the documents on the water quality synthesis in the Mureş Hydrological Basin were not finalized, but an improvement of the process of monitoring and assessment of the basin waters is already observed.

It is necessary to continue the process of developing the monitoring system to cover all the elements of quality (biological, hydro morphological, and physicochemical) and all investigation media (water, sediment, and biota) with a frequency that ensures high levels of confidentiality and precision in assessing the condition of water bodies.

Particular attention should be paid to other aspects such as:

- Monitoring of specific pollutants and priority substances (especially organic micro-pollutants), in order to ensure detection/quantification limits which allow compliance with environmental quality standards to improve the assessment of the chemical and ecological status of water bodies;
- Improving quality assurance and quality control systems in analytical laboratories.

The creation, maintenance and development of a database and the use of an instrument (unitary evaluation program) for all elements (biological, physicochemical, hydro morphological) to be integrated with the existing IT systems [1].

4.6 Recommendations

Following the study, we can make the following recommendation:

- The information on the implementation of relevant legislation on the website of competent European and national authorities should be more detailed;
- There should be greater transparency for the public in accessing financial instruments;
- The population should be educated to understand and address information on the quality of the environment and the rational use of natural resources.

Acknowledgements The authors would like to thank the staff from the Department of Water Resource Management, Monitoring and Protection within the Mureş Water Management Administration for their collaboration.

References

1. Planul Naţional de Management actualizat aferent porţiunii naţionale a bazinului hidrografic internaţional al fluviului Dunărea care este cuprinsă în teritoriul României
2. Ghiga C (2004) Infrastructură teritorială şi dezvoltare urbană. Editura Uranus, Bucharest
3. Directiva Cadru privind Apa (Directiva 2000/60/EC)
4. Brezeanu Gh, Simon-Gruiţa A (2002) Limnologie generală. H.G.A, P.H., Bucharest
5. https://ro.wikipedia.org/Utilizarea_apei_de_catre_oameni
6. https://www.google.ro/search=Conferinta+Natiunilor+Unite+asupra+resurselor+de+apa
7. www.mmediu.ro
8. http://www-old.anpm.ro/ResurseledeapaRomnia
9. https://ro.wikipedia.org/wiki/Gospodarirea_apelor
10. https://eur-lex.europa.eu/legal-content/RO/TXT/?uri=CELEX:32000L0060
11. Stoianovici, Ş, Robescu D (1982) Procedee şi echipamente pentru tratarea şi epurarea apei. Ed. Tehnică, Bucharest
12. Trufaş C (2003) Calitatea apei. Editura Agora, Călăraşi
13. Uttomark P, Wall P (1975) Lake classification for water quality management. University of Wisconsin Water Research Center, http://ph.academicdirect.org/CFACI
14. Bucur A (1999) Elemente de chimia apei. H.G.A. P.H, Bucharest
15. Neniţescu CD (1972) Chimie generală. Editura Didactică şi Pedagogică, Bucharest
16. Rusu T (2008) Tehnologii şi echipamente pentru tratarea şi epurarea apelor, vol I. U.T.Press, Cluj-Napoca
17. Varduca A (1997) Hidrochimie şi poluarea chimică a apelor. Editura H.G.A, Bucharest
18. Varduca A (2000) Protecţia calităţii apelor. Editura H.G.A, Bucharest
19. Vişan S et al (2000) Mediul înconjurător—poluare şi protecţie. Ed. Economică, Bucharest
20. Mitrănescu M (1972) Curs de chimie analitică cantitativă, Institutul Politehnic "Traian Vuia" Timişoara. de Chimie Industrială, Fac
21. Negulescu M et al. (1987) Epurarea apelor uzate industriale, vol I şi II. Editura Tehnică, Bucharest
22. Varduca A (1999) Monitoringul integrat al calităţii apelor. Editura H.G.A, Bucharest
23. David I, Sumălan I, Carabeţ A (1996) Transportul poluanţilor prin medii fluide. P.Timişoara, Lito U
24. Popescu D-M, Nistoran DE (2005) Poluarea apelor de suprafaţă cu lichide petroliere. ed. Printech, https://w.w.w.researchgate.met/publication/301292708

25. Negulescu M (1971) Epurarea apelor uzate orăşeneşti. Ed. Tehnică, Bucharest
26. Platon V (1997) Protecţia mediului şi dezvoltarea economică. Ed. Economică, Bucharest
27. Gâştescu P (1998) Limnologie şi oceanografie. H.G.A, Bucharest
28. Ghimicescu G, Hîncu I (1974) Chimia şi controlul poluării apei. Editura Tehnică, Bucharest
29. Chanlett E-T (1973) Environmental protection. Mc Graw-Hill, New-York, (https://www.abebooks.com/book-search/title/environmentaprotection/author/emil-chanlett/)
30. Directiva 91/271/EEC a Tratării Apelor Uzate Orăşeneşti
31. Şerban P, Gălie A (2006) Managementul apelor. Principii şi reglementări europene, Editura Tipored, Bucharest
32. http://www.rowater.ro/daMures/Informarepublica/
33. Părăuşanu V, Ponoran I (2003) Dezvoltarea durabilă şi protecţia mediului. Editura Sylvi, Bucharest
34. Legea Protecţiei Mediului nr. 137/1995, modificată şi completată conform OG 91/2002
35. Normativ din 28 februarie 2002 privind stabilirea limitelor de incarcare cu poluanti a apelor uzate industriale si orasenesti la evacuarea in receptorii naturali, NTPA-001/2002 Publicat in Monitorul Oficial, Partea I nr. 187 din 20 martie 2002
36. Robescu D, Robescu D (1999) Procedee, instalaţii şi echipamente pentru epurarea avansată a apelor uzate. Ed. Bren
37. Sorocovschi V, Serban GH (2008) Hidrogeologie. Ed. Casa Cărţii de Stiinţă, Cluj-Napoca
38. Xenia LA, Gharibeh BA, Aurel A, Dorina A (2008) Fundamente de chimia mediului. Editura Didactica si Pedagogica, R.A, ISBN: 978-973-30-2015
39. Angelescu A, Ponoran I, Ciobotaru V (1999). Mediul ambiantşi dezvoltarea durabilă. Editura ASE, Bucharest
40. Diudea M, Todor Ş, Igna A (1986) Toxicologie acvatică. Dacia, Cluj-Napoca
41. Ornitz BE, Champ MA (2002) Oil spills first principles: prevention and best response. Elsevier Science Ltd, Amsterdam, Netherlands
42. Pora R (1998) Modelarea calităţii râurilor. Editura H.G.A, Bucharest
43. Ujvari I (1972) Geografia apelor României. Ed. Stiinţifică, Bucuresti
44. https://ro.wikipedia.org/wiki/Hidrografia_României
45. Szőcs A (2010) Hidrochimia si poluarea râurilor din Bazinul hidrografic superior si mijlociu al Mureşului. Doctoral thesis summary. Coordonator Prof. Univ.Dr, Sorocovschi Victor, Cluj-Napoca
46. Sorocovschi V (1996) Podisul Târnavelor—studiu hidrogeografic. Ed. CETIB, Cluj—Napoca
47. http://www.rowater.ro/daMure/SintezaprivindcalitateaapelordinBazinulHidrograficMure9F-2015,2016,2017
48. http://www.cjmures.ro/Hotariri/Hot2016/anexa_hot135_2016.pdf
49. Legea nr. 107/1996 (Legea Apelor), M.Of. nr.244/8.10.1996
50. Sorocovschi V (2002) 2004 Hidrologia uscatului. Casa Cărţii de Stiinţă, Cluj-Napoca
51. Sorocovschi V (2005) Câmpia Transilvaniei. Studiu Hidrogeografic, Casa Cărţii de Stiinţă, Cluj - Napoca
52. Sorocovschi V, Szőcs A, Vodă M (2006) Mineralizarea apei râurilor din bazinul hidrografic al Mureşuluisuperior si mijlociu. In: Integrarea Europeană Impact si Consecinţă, Simpozion Stiinţific Internaţional, pag. 387–395, Ed. "Dimitrie Cantemir", Târgu—Mureş
53. Arhiva/ baze de date Administratia Bazinală Mureş
54. Legea 310/2004 pentru modificarea si completarea legii 107/1996, M.Of. nr. 584/30 iunie 2004
55. Luo B, Maqsood I, Yin YY, Huang GH, Cohen SJ (2015) Adaption to climate change through water trading under uncertainty-An inexact two-stage nonlinear programming approach. Environronmental Systems Engineering Program, Faculty of Engineering, University of Regina, Regina, SK S4S 0A2, Canada Adaptation and Impacts Research Group, Environment Canada, Vancouver, BC V6T 1Z2, Canada, Vol 2, Issue 2 (December 2003), https://doi.org/10.3808/jei.200300022

Part III
Water Supply

Chapter 5
Water Supply Challenges and Achievements in Constanta County

A. Constantin and C. St. Nitescu

Abstract Dobrogea is a Romanian region surrounded by waters: Low Danube at West, the Danube Delta at North and the Black Sea East. Although its location is in the proximity of large water bodies, its climate is dry, and it benefits from only a small number of rivers and creeks and some lakes with fresh or brackish water. The groundwater instead is generous, and it represents the main drinking water source. A brief description of the water sources in Dobrogea gives an image of the water supply potential of the region. The first drinkable water supply systems were built long ago, in ancient times, to bring fresh water to Tomis, the city overlapped today by Constanta, the largest Romanian city on the Black Sea coast. Nowadays drinkable water supply systems use mainly groundwater, but also surface water from the Danube. They are organised as local or regional systems which face fundamental issues generated by increased demand, aged infrastructure and a slight decline in quality. Therefore an ample project is in progress, aiming to rehabilitate the existing water facilities and distribution networks and to build new ones. A newly built transmission conduit will interconnect the current systems into a regional one, with more flexibility and resilience. The project also envisages the implementation of surveillance and control methods to diminish the water loss and to increase energy efficiency. The same issues referring to water loss and energy consumption but more acute are faced by the irrigation water systems used in agriculture. The transition from large cultivated areas belonging to the state to small properties, after 1989, increased the water price and made difficult to use the existing huge irrigation systems. A national modernization project is going to improve irrigation which is vital for agriculture in Dobrogea. A particular focus lies on the water pumping stations that are the most critical energy consumers. A numerical simulation carried out with particular software is a beneficial way to analyse the hydraulic parameters and energy coefficients in each configuration or operation possibility of a pumping installation, allowing the engineers to make correct decisions. We present two case studies, one regarding a drinkable water system and the other an irrigation water pumping station. Elements of environmental impact and the challenges of the on-going water supply projects are addressed as well.

A. Constantin (✉) · C. St. Nitescu
Faculty of Civil Engineering, Ovidius University, Constanta, Romania
e-mail: aconstantina@univ-ovidius.ro

© Springer Nature Switzerland AG 2020
A. M. Negm et al. (eds.), *Water Resources Management in Romania*, Springer Water,
https://doi.org/10.1007/978-3-030-22320-5_5

Keywords Water resources · Water supply · Drinkable water distribution network · Irrigation water supply · Water pumping stations

5.1   Overview of Surface and Groundwater Resources in Constanta County

Dobrogea, the land situated between big waters, as the lower Danube with its Delta and the Black Sea, is a region deficient in surface water resources and precipitation.

The Dobrogea-Litoral water basin, Fig. 5.1, is located in the south-east of Romania. It is bordered by the Buzau-Ialomita basin to the West and by the Prut basin to the Northeast. To the East, it is bound by the Black Sea and to the South by the Bulgarian border.

The surface of the Dobrogea-Litoral water basin is 11,809 km², covering Constanta County, Tulcea County and the Big Island of Brăila.

Considering the region composed of the low Danube River, Danube Delta, and Dobrogea Water Basin, 28 river bodies have been identified: 20 bodies are in the natural state, three are heavily modified, and five are artificial. Out of these 28 water bodies, 19 bodies have been monitored [22].

The rivers draining the surface of Dobrogea flow into either the Danube or the Black Sea. Those flowing into the Black Sea are smaller in size than those flowing in the Danube River.

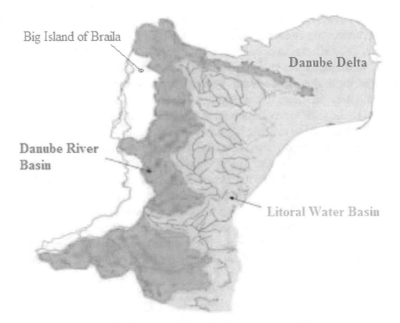

Fig. 5.1 Water bodies in Dobrogea [23]

There have been identified 16 surface rivers with more than 10 km in length.

The 57-km-long Taita River with a basin of 591 km^2 starts as a spring that releases water from a spot at the contact between the Pricopan Summit and the Niculitel Plateau, at an altitude of 240 m. A length of 23 km forms the natural boundary between the two massifs, and then through a sharp turn in the Horia village, it chooses the shortest path to Babadag Lake. In the upper course, Taita has a narrow basin not exceeding 20–50 m [22]. The versants have slopes of 25–30°, and dry valleys fragment them. The valley of the Taita River extends further so that it reaches 150 m in width at Hamcearca and 300 m at Babadag Lake. The most important of its tributaries are the creeks Alba and Taita.

Slava River has a length of 38 km and a basin surface of 356 km^2 [22]. Its lower course and its tributary on the left, Ciucurova, are part of the corridor separating the Babadag Plateau from Casimcea Plateau. Slava joins the Ciucurova River, near the village of Slava Rusa.

Telita River has 48 km in length and covers 287 Km^2. It springs close to the village of Niculitel, from the plateau with the same name, at the altitude of 270 m. Its valley, in the springs area, has a torrential appearance, but downstream the Telita village it already enters the high depression of the Nalbant. The groundwater in the bottom of the valley is close to the surface (1–5 m), which causes Telita to receive a permanent, but very low, underground feed [22].

Casimcea River has a well-developed basin area in the southern part of the Casimcea Plateau with tributaries that head to the Pantelimon Depression. After leaving the springs, its course forms the boundary between North Dobrogea and Tortoman's Plateau and flows into Lake Tasaul. Casimcea springs from a high plateau, at the altitude of 309 m. The length of the river is 69 km, and the surface of the basin is 740 km^2 [22]. After a torrential sector of about 10 km, the river enters the contact area between the Jurassic limestones and the old Caledonian massif of Pantelimon. Here it receives some tributaries, mainly developed to the right, such as the Dereaua Mare, Pantelimon, Valea Seaca and Gura Dobrogei, and to the left Ramnicul and Gradina Mucova. Floods occur suddenly with torrential rains reaching the maximum level in approx. 1–2 h, then drop somewhat slower in 6–12 h. Only 85% of these water bodies are moderately from the ecological viewpoint, and the rest of 15% are in poor ecological condition [22].

The hydrographic network of the Constanta County is divided into two distinct units: the Danubian group and the maritime group.

The rivers in the Danubian group drain the western part of the county, most of them ending with fluvial limes. Among them, we may mention Topologu and Carasu.

The rivers in the maritime group drain the eastern part of the county. The most important of these is Casimcea.

Constanta County has a poor river network whose average density is below 0.1 km/km^2. All rivers have very steep slopes in the springs, after which the slopes swiftly decrease, with the major riverbed becoming very wide.

The specific multiannual average flows are below 1 l/s-km^2, somewhat higher being only in the Casimcea and Topologu River springs. Multiannual average flows are relatively small compared to the size of the basin surface.

The maximum volume of water usually occurs at the end of winter and the beginning of spring (February–April), and the minimum at the end of the autumn and the beginning of winter (November–January) when the average is 33% and 17–18% eastern annual volume of the county [21].

In the Dobrogea hydrographic space, including the Danube Delta, a total of 75 waterbodies of natural lakes have been identified, out of which 17 are monitored.

The principal lakes in the county of Constanta are included in Table 5.1.

The waters of these lakes have no required characteristics for drinking purposes. The primary uses that we can benefit from the water of these lakes are fish farming, irrigation, and nautical sports. Despite the narrow strip of land that separates the Siutghiol Lake from the Black Sea, this lake has fresh water, due to its hydraulic connection to the aquifers beneath. Lake Techirghiol is well known for its curative properties, due to its salty waters and therapeutic sludge on its bottom.

The groundwater resources related to Dobrogea Water Basin (up to the depth of 0–300 m) totalize about 3172 million m^3/year (100.6 m^3/s), out of which 84.8 m^3/s— from deep layers, of outstanding quality and 15.8 m^3/s—drinking water with higher mineralization, coming from the groundwater. Of this total, in South Dobrogea, the exploitable amount is 8.95 m^3/s from the deep layers and 0.2 m^3/s from the phreatic, while in the North and Central Dobrogea, the resources are 2.15 m^3/s from the deep layers and 0.85 m^3/s of the phreatic [8].

There were identified ten bodies of groundwater, Fig. 5.2, out of which four water bodies are free-surface aquifers, and the other six are under pressure aquifers.

In fact, groundwater is kept by three aquifers which overlap each other, having a vast potential. [1] identifies these three water sources beneath Constanta harbour as phreatic water, medium depth groundwater (the Sarmatian aquifer), deep groundwater (the Upper Jurassic—Lower Cretaceous aquifer). The primary source of deep layer groundwater, which lasts on the Southern Dobrogea, is the Prebalkanic Plateau, on the Bulgarian territory. From that Bulgarian region, where the Upper Jurassic deposits crop out groundwater flows to our Upper Jurassic–Lower Cretaceous aquifer [1]. Another source of this aquifer is the downward leakage from the Sarmatian aquifer. In the South-Eastern Dobrogea, the aquifer is also supplied by effective infiltration from precipitation. The piezometric heads of this deep aquifer are higher along the coast than the ones of the Sarmation aquifer; consequently, an upward leakage occurs [1].

Table 5.1 Lakes in Constanta county [23]

Name	Surface area (ha)	Type	Name	Surface area (ha)	Type
Siutghiol	1900	Natural	Techirghiol sarat	1227	Natural
Tabacarie	94	Natural	Techirghiol dulce	240	Natural
Tasaul	2335	Natural	Oltina	2509	Natural
Corbu	520	Natural	Bugeac	1774	Natural
Tatlageac	178	Natural	Vederoasa	150	Natural
Nuntasi	1050	Natural	Tibrin	621	Artificial

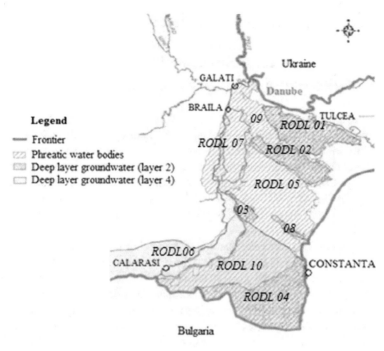

Fig. 5.2 Groundwater bodies in Dobrogea [23]

On the Romanian territory, the aquifer discharges to the Black Sea, and also to the Lake Siutghiol. The water flow direction is from South to North, with West-to-East inflection, close to the city of Constanta and the water quality is appropriate for drinking.

There are references to other sources of the groundwater, namely a component of the Danube, in the Ostrov-Cernavoda area, but also the irrigation works contribute [19].

Nevertheless, water can be found a few tens or a few hundred meters deep in a particular perimeter, in Dobrogea.

The aquifer fed by the Prebalkanic Plateau waters has one of the essential discharges in the Siutghiol Lake, a source of drinking water that has been used since ancient times.

5.2 Drinkable Water Supply Systems

5.2.1 Past and Present of Drinkable Water Supply Infrastructures

The first evidence of potable water supply facilities in Dobrogea dates from the Roman period. In the 1st century AD, when the Romans took the rule in Dobrogea, the region had a few cities that had been founded by the Greek colonists in a Getic space, long before, in the 7th–6th centuries BC. Among these towns, Tomis—overlapped by nowadays Constanta—was the most outstanding in the Roman period. The great Latin poet Publius Ovidius Naso (43 BC–17/18 AD), who spent the last nine years of his life at Tomis, mentions, in his poems Pontice, the brackish water the locals used to drink.

Tomis continued to flourish, becoming the capital of Scythia Minor province. The number of inhabitants increased, and the water demand was higher and higher. Archaeological proves bring data about the drinkable water sources and the aqueducts built to convey water to the city. The archaeologists have different opinions regarding the way Tomis was supplied by water, but they agree the primary sources of fresh water were the aquifers placed on the Western side of Siutghiol Lake. The aquifers were a quality source of fresh water. Besides, the locals used water from lakes and wells.

As Tomis became The Left Pont Metropolis, expensive waterworks were done. The groundwater was taken through a drainage gallery from the aquifer next to the Capidava—Ovidiu fault (from a spot close to Mamaia resort) and was conveyed to the city by two aqueducts. Later on, the continuous increase in freshwater consumption determined the community to build a longer aqueduct, which would have tripled the amount of conveyed water. This high cost construction was possible when the city became the capital of the Scythia Minor province.

The remains of this aqueduct which would have been supplying Tomis during the first centuries AD, helped the archaeologists to restore its way from the source end-Canara-along the Black Sea shore, to the city. The dashed line represents the curved way of this aqueduct of about 20 km in length in Fig. 5.3 [18]. We have to mention that at that time the shoreline was farther into the nowadays sea area, so the aqueduct was not as close to the sea as it seems [2]. In the fortified area, the water, brought by these aqueducts, was kept under the city in a system of galleries, partly dug into the natural rock, partially made of waterproof stone and brick masonry. In several parts of the city, especially in public areas, wells communicating with these galleries allowed people to draw out drinkable water.

In the Middle Age, the city had an extended period of decay. So in the Modern era, at the very beginning of the twentieth century, [12] said that "the inhabitants of Constanta drank the same water the poet Ovidius had drunk" referring to the brackish water in the most of the wells in use.

The first modern drillings, in search of the deep fresh water, were made by a Belgian firm in 1897, close to Constanta, in the Caragea-Dermen area. The water

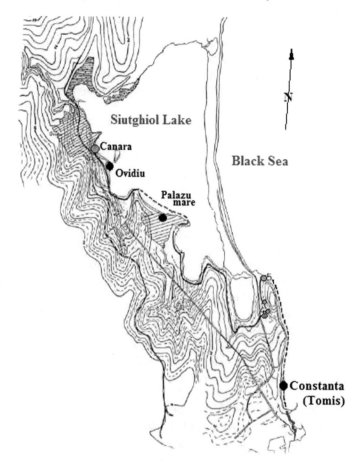

Fig. 5.3 Water supply sources for ancient city Tomis (blue dots) [18]

was found at 35 meters in depth. More other wells were also drilled in 1927 [19]. Despite these efforts, fresh water was currently brought in Constanta only around The First World War facilities [24]. In 1948 there were drilled 27 wells at Caragea-Dermen source, out of which ten were equipped to be put in operation, the rest having inadequate low flows.

During 1950–1957 the water supply system of the seashore was developed by deep drillings carried out at Eforie-Nord, Eforie-Sud, and Techirghiol. A duct of 400 mm in diameter and more than 30 km in length conveyed water from the Caragea-Dermen source to Constanta city and further to the southern resorts on the Romanian Black Sea coast. There were also built 2×1000 m^3 tanks and one 300 m^3 water castle at Eforie-Nord and Eforie-Sud, 2×1000 m^3 tanks at Techirghiol and about 50 km of distribution network.

From 1959 on, Mangalia had its own water supply facilities [24].

The Romanian authorities launched in the 1960s a vast operation to study the underground water phenomenon in Dobrogea. First, there was comprehensive research of Siutghiol Lake, made by local and national authorities. The average depth of the lake was found to be of five to six meters, but there have been discovered places where the depth reaches 17 m, right where the underground springs are [19]. The lake is in contact with the aquifer. This is the reason this lake, which is more than seven kilometers long and two kilometers wide, separated from the Black Sea by a strip of land, still contains fresh water.

Two deep-sea drillings were executed near Neptun. There have been taken samples, and electromagnetic measurements have been carried out. The tests lasted for five years and concluded that there are no sea water infiltrations in the aquifer [1]. Groundwater is the best source of drinking water, being cantoned in a relatively isolated environment from human activities.

At present, Dobrogea uses water from both underground and surface sources.

In Constanta County, there are being into operation 85 underground water sources with 279 equipped wells, offering a total installed capacity of 33520.90 m^3/h and two sources of surface water, respectively the source Galesu and Vifor Hill, whose total installed capacity is 4.57 m^3/h [3].

The water is stored in 176 tanks, with a total storage volume of over 306,000 m^3.

The raw water is treated in two water treatment facilities: Palas—Constanta treatment plant and Vifor Hill - Cernavoda treatment plant, being pumped by 66 drinking water pumping stations, with a total capacity of 157,818 m^3/h [3].

Drinking water is provided to twelve towns and eighty-seven communes and villages in Constanta County. Among these localities, 16 have over 5000 inhabitants, and they are supplied with over 1000 m^3 of water per day.

Particular importance is given to the water quality. Therefore groundwater bodies are monitored through 180 wells spread all over Dobrogea. Moreover, some of them are equipped with level and temperature sensors. Thus, the water chemical state, quantity and temperature are being assessed to maintain a proper and healthy water supply. Lately, six of the ten monitored water bodies have a good chemical status (RODL02, RODL03, RODL05, RODL06, RODL07 and RODL08 in Fig. 5.2). While the remaining four bodies of groundwater have a weak chemical status (due to exceedance in NH_4, NO_3, PO_4, chlorides, Pb) [22].

Both surface water sources, Galesu which is placed on the Danube-Black Sea Canal, and Vifor Hill, placed on the Danube River, are monitored following our legislation, referring to human use. The monitoring activity addresses a complex of issues as physical and chemical parameters of water, identification of water pollution sources, soil erosion degree, the influence of fish farming and so on, aiming to improve the water quality.

5.2.2 Urban and Rural Drinkable Water Supply

The distribution system, in Constanta County, consists of an assembly of local and zonal systems. Due to an increasing demand for water in the Southern part of the region, an old concept of interconnecting the zonal systems is going to be put into operation. Therefore, the groundwater source from Medgidia is meant to supply not only Medgidia but also additional cities including Mangalia and the Southern coast resorts. The regional water supply system will take water from the above mentioned good quality source, located in Medgidia, and convey it to almost all the localities in the Southern coastal area, where the local sources do not fit the water demand or drinkable properties. The proposed system is a resumption of a concept, namely the interconnected seaside system that has been operating for a long time but which was abandoned about 20 years ago. Local water sources were preferred instead, as they were of good quality [21].

The current flow rate abstracted from the five wells in service at Medgidia source is about 417 l/s, less than the current maximum flow rate that can be captured. The test drilling indicates a very high capacity of the aquifer that can be safely exploited.

Constanta, the most important city in Dobrogea, with a population of over 284,000 inhabitants after the 2011 census, is the best example of urban water supply. The city is part of the water supply system Constanta, comprising Constanta, Mamaia Resort, and Palazu Mare. They have five groundwater sources, namely Caragea Dermen, Cismea I, Cismea II, Constanta Nord, and Medgidia, located in the underground cross-border water body RODL06, Fig. 5.2, and a surface source, Galesu Intake.

Caragea Dermen Source consists of 16 wells, with the depth between 35 and 90 m. It has an installed capacity of 2578 m^3/h. The wells are equipped with electric pumps, with flow The specific consumed energy coefficierates between 90 and 504 m^3/h, and pumping heads of 70 mwc.

This is the oldest of the groundwater sources of Constanta, its first wells being drilled in 1915. The abstracted water is chlorinated and conveyed to the storage and pumping facility Calarasi (situated in Constanta city), to supply part of Constanta city. Other ducts take water to the Ovidiu town, close to the source.

The Source Cismea is composed of three abstracting groups, located in the North of Constanta. The source has a total capacity of 8497 m^3/h. Water is pumped from 50 to 120 m deep wells, the head varying between 79 and 82 mwc. The water abstracted from this source is pumped to the storage complex Palas [22].

The Constanta North Source is located in the Northern zone of Constanta city, along the southern shore of Lake Siutghiol. It consists of 5 wells drilled at a depth of 300 m, except for one well, drilled at a depth of only 65 m. The capacity is of 2218 m^3/h [8].

The wells are equipped with submersible electric pumps, with flow rates between 299 and 612 m^3/h and pumping heads ranging between 30 and 90 mwc [23].

The abstracted water is pumped to the Constanta North storage and pumping facility.

The Medgidia Source, located along the left bank of the Black Sea Danube Canal, consists of 11 wells, out of which five are in operation, with depths between 350 and 450 m. The layout of the system, represented in Fig. 5.4 follows the recommendation for water supply systems [15]. The abstracted flow rate is 1500 m³/h. The wells are equipped with submersible pumps, with flow rates between 180 and 300 m³/h, and pumping heads between 75 and 90 mwc. The water is conveyed through a duct of 1200 mm in diameter to the Constanta South storage facility and pumping station, via the Eforie Nord complex. At present, the Dn 1200 duct is no longer functional after a series of damages.

The surface water source Galesu is located on the Poarta Alba—Midia Navodari Canal, branch of the Danube-Black Sea Canal. Its layout might be seen in Fig. 5.5.

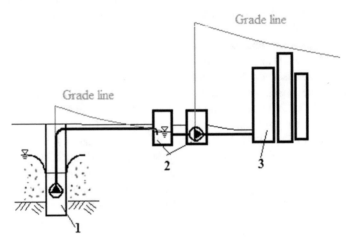

Fig. 5.4 Layout of part of the Constanta water supply system using groundwater source: 1. Well at Cismea source; 2. Calarasi storage-treatment-pumping plant; 3. City

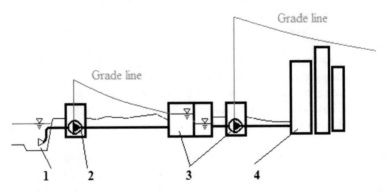

Fig. 5.5 Layout of part of the Constanta water supply system using surface water source: 1. Poarta Alba -Midia Navodari Canal; 2. Galesu pumping station; 3. Palas storage-treatment-pumping plant; 4. City

It currently delivers a flow of 3.75 m^3/s, pumped to the complex storage—treatment facility Palas, Constanta. The pumping station is equipped with five electro-pumps, with pumping heads between 90 and 110 mwc. The pumping station includes a chlorination facility. From the source, the water is taken by the 2 × DN 1000 mm and 2 × DN 1200 mm mains and transported to the storage-pumping facility Palas at a distance of 17.4 km [21].

The surface water source was designed to cope with the increase in drinking water consumption, on the seaside, during the summer, when Constanta receives many tourists.

The management of Constanta water supply system is oriented to water quality and safety, maximal efficiency and minimal environmental impact; therefore a continuous assessment of all its components performance is carried out. As a large water supply system, it faces issues related to the aging of infrastructure, energy efficiency decline and non-revenue water, due to leakage in the distribution system. Generally speaking, the [9] assesses the distribution losses between 5 and 50%, being much larger than production losses that range between 2 and 10%. The 2014 water balance of Constanta water supply system showed considerable water losses; therefore a comprehensive action was launched, focusing on the real losses in the network and unbilled consumption. In fact, this action is part of an extensive regional project for the whole Constanta County water infrastructure development.

In the rural area, medium or small water supply system mainly take water from local underground sources. The most used sources are the wells in Medgidia, in the central part of the county, Costinesti, Albesti, Vartop, Pecineaga, Dulcesti and Biruinta in the Southern part.

We will refer to the Dulcesti water supply system as an example of a small system. The village Dulcesti, situated at approximately 30 km South from Constanta, Fig. 5.6, has 1413 inhabitants. The Dulcesti groundwater source is located in the South-Eastern part of the village and consists of a total of 16 wells with depths ranging from 56 to 80 m. Only 11 wells are in operation and have a total capacity of 1165 m^3/h (324 l/s). At present, three of them are supplying Dulcesti village [23].

The whole source feeds not only Dulcesti but also some resorts on the Southern coast and occasionally Mangalia when extra water demand occurs.

The main that brings water to Dulcesti is 2 km in length and 150 mm in diameter.

There are no treatment and storage facilities, and currently, the village is not provided with the fire water reserve, the volume of compensation and the volume of damage.

The water captured from the Dulcesti source is pumped from the wells directly into the distribution network. The water disinfection is carried out in the pipeline, right at the source and the chlorination plant is rudimentary. The distribution network of Dulcesti is made of HDPE pipes with diameters between 63 and 160 mm. The total unfold length is 15.9 km. The branched water distribution network in Dulcesti is directly supplied by pumps mounted in wells. The network faces large amounts of water loss; therefore, the actual flow rate is 28.11 l/s instead of the demand which is only 15.26 l/s [5].

Fig. 5.6 Dulcesti village
(Google maps)

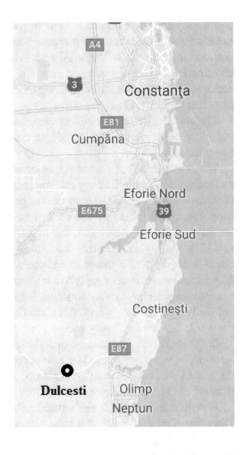

According to [27] it is no longer allowed to pump drinkable water directly from the well into the network. Moreover, the water quality declined, and it no longer meets the requirements for sodium and nitrate indicators. A water facility is about to be built, comprising a storage tank with the capacity of 1200 m³, which contains the necessary volume for the fire reserve, the volume of damage and the volume of compensation for Dulcesti and two other neighbouring villages Mosneni, and Pecineaga. There will be built, as well, an electrochlorination plant, a pumping station which will ensure the minimum necessary pressure in the distribution network of Dulcesti.

In the vision of the newly developed water management project in Dulcesti, the water source is represented by the transport pipeline Medgidia-Eforie-Mangalia, requiring water precloration, according to the standards in force.

Beside Dulcesti, this main will supply with water either Pecineaga or Mosneni villages. To provide the necessary pressure to consumers and fire hydrants, two pumping groups are required in the same station:

- Dulcesti water pumping group, consisting of two identical pumps, equipped with frequency converters.

- Pecineaga and Mosneni water pumping group consisting of three pumps, equipped with frequency converters.

5.2.3 Efficiency of Drinkable Water Supply Systems

The Stockholm Statement [28] mentioned water as the 'bloodstream of the green economy', pointing out its importance for people and environment. In this context, efficiency and environmental impact are two of the most significant goals headed by water management authorities. The efficiency of a water supply system is a comprehensive concept which deals with each component (resources, abstraction, treatment, transport, pumping, and distribution), regarding energy efficiency, water loss, and technical status. Different organisations have developed several performance indicators to conduct a comparative analysis of water supply system, a large number of performance indicators.

Heider et al. [11] makes an extensive review of the literature regarding performance evaluation frameworks including several indicators that cover all the aspects (e.g., physical asset, staffing, operational, customer satisfaction, economical). The most rational and with widespread applicability is considered to be the evaluation system (Heider et al. [11] carried out by International Water Association (IWA). IWA introduced a series of performance indicators to assess water supply systems performance and to compare water utilities in a coherent framework. The difficulty relays in collecting data required to determine the performance indicators.

The Dobrogea-Litoral water utility assesses the most relevant IWA performance indicators for all the zonal and local water systems. As in Constanta the water distribution network has 650 km of pipeline, and half of it is more than 40 years old, the water losses are considerable and moreover, the hidden leakage spots are difficult to locate. In Constanta water supply system, the on-going rehabilitation project was expected to improve the total real water losses, by decreasing this indicator by 6% during 2014–2020. This estimation was based on sonar detection methods of water leakages. As in 2016, the water utility started to use a novel technology based on satellite imagery to detect leaks and non-revenue-water, real losses are expected to be considerably reduced. The hidden leaks are detected in real time, bringing a consistent decrease to background leaks and water loss. The estimation made for the entire network shows potential savings of 695,000 m^3 of water per year [20].

Real water losses include two categories: unavoidable annual real losses and potentially recoverable real losses. Heider et al. [11] identifies four methods to recover part of the second category losses:

- Leakage control;
- Pressure management;
- Speed and quality of repairs;
- Pipeline and asset management.

By decreasing the real water losses, the Infrastructure Leakage Index (ILI) will improve as well. It shows the ratio of Current Annual Real Losses (CARL) to Unavoidable Annual Real Losses (UARL). ILI is a measure of how well the distribution network is being managed and how proper is its maintenance considering the current operating pressure, but it has to be analysed together with other indicators, such as the total system input and the real water losses per service connection and day.

One of EU targets for 2020, included in the package of climate and energy, is a 20% improvement in the EU's energy consumption. Therefore thorough attention has to be paid to the pumping stations—at the source and the distribution network—as they are significant energy consumers.

From the technical point of view, our study focuses on the energy efficiency, an essential indicator with technical, social and economic consequences. The lack of data prevented us from assessing the energy efficiency for large water supply systems, but for small ones, as those in the rural area, we performed a series of operation simulation in EPANET, to assess and compare the specific consumed energy coefficient and all the technical parameters that define the hydraulic system.

The specific consumed energy coefficient, e, expresses the amount of energy consumed to pump 1 m^3 of water and convey it through the distribution pipeline. More on this coefficient is given in [7].

5.3 Irrigation Water Supply Systems

5.3.1 Past and Present in Irrigation Infrastructure

The land improvement works mainly support agriculture in Dobrogea. The region has always been an arid one, but the rapid climate change in the late decades, made agriculture more and more vulnerable and dependant on irrigation. Temperature evolution and water scarcity, and we refer here to surface waters and precipitation, along with the anthropogenic activity led the scientists to the conclusion that a real danger of desertification threatens our land. Droughts, floods, and other climate change issues have a significant impact on the food production, despite the potential of agriculture.

Irrigation, a keystone of food security policies in the face of climatic variability all over the world, is more valuable in Dobrogea. The irrigation infrastructure has been developed since the years 1960. The fast growth recorded in Dobrogea, during the communist era increased the irrigated area from 15,000 ha in 1965 to 520,000 ha in 1989. The irrigated area at that moment represented 80% of agricultural land in Constanta County and 40% in Tulcea County.

The irrigation systems were large. Out of the 100 irrigation systems built in Dobrogea, four systems had more than 100,000 ha. The average area was 28,144 ha [14].

These vast dimensions had as necessary consequences high pumping heads and long distances to convey the water, which resulted in massive energy consumption.

The irrigation systems faced a gradual decline after 1990. The abolition of large-scale exploitation structures, the introduction of differentiated water pricing according to pumping head and delivery point were among the factors that generated the decline. The irrigated area varies every year, according to the precipitation regime, but it can be an indicator for the timeline of this activity in Dobrogea. Thus, if in 2003, a drought-afflicted year, the irrigated area was about 124,000 ha, in 2007, a year with a precipitation regime similar to 2003, there were no more than 26,800 ha (9300 ha in Constanta County and 17,500 ha in Tulcea County) [14].

The infrastructure aged. Therefore, a series of rehabilitation projects were developed. The most recent project, The National Programme for the Rehabilitation of the Main Irrigation Infrastructure in Romania, was launched in 2016 [16]. A rigorous analysis, made as part of this national programme, revealed the main current issues of the irrigation systems:

- low hydraulic efficiency;
- high cost of electricity for the systems still based on pumping (the Danube being the main source of water);
- large water price.

5.3.2 Overview of the Main Types of Irrigation Water Supply Pumping Stations Used in Agriculture

The main irrigation water source is the Danube, water being withdrawn directly from the river or from the canal that connects the Danube to the Black Sea. Water is transported on long distances, by a system of canals and lift pumping stations. The first pumping station, the base one, takes water from the source at a high flow rate and relatively small head and delivers it in a canal. At the opposite end of the canal, there is another one named the first stage pumping station, which pumps the water into another transport canal, operating at a smaller discharge and higher pumping head than the base pumping station. An irrigation system used to have several stage pumping stations so that water was conveyed to agricultural land situated far away from the water source. At present, only the first stage pumping station is still in operation, in most of the irrigation systems. The buried irrigation water distribution networks are supplied by booster pumping stations, which take water from the transport canals and pump it directly into the network. It is obvious that such a large system is a huge energy consumer. It is the only option for irrigation carried out with surface water, as long as gravitational irrigation is not possible in this region. The irrigation systems that operate at present in Constanta County are Galesu, Carasu, Seimeni, Nicolae Balcescu, Mihail Kogalniceanu, Harsova, and part of Sinoe system that continues in Tulcea County.

In the last decade, many efforts were made to rehabilitate and upgrade part of the existing pumping stations, according to the new technological progress.

The engineering design for the rehabilitation of an existing pumping station aims not only the meeting of technical required parameters but also the entire possible energy efficiency during pumping station operation. Thus, every single variant of an old pumping station modernization must be technically and economically analysed before adopt a specific solution and the best method is a numerical simulation by the help of a hydraulic software.

5.3.2.1 Base Irrigation Water Supply Pumping Station Mircea Vodă

Mircea Vodă pumping station is a valuable example for the base irrigation water supply pumping stations. It operates in the irrigation system Carasu, in Dobrogea and it is located on the left side of the Danube- Black Sea Canal. Water is taken from the canal CA_0 by 14 identical electric pumps, of split case type. The impeller of each centrifugal pump is of 950 mm in diameter. The pumps are connected in two groups, and each group, composed of $6 + 1$ pumps operating in parallel, delivers water on its own discharge duct. The individual suction ducts are 1200 mm in diameter. The water is transported by two mains of 1150 m in length and 1900 mm in diameter. A horizontal air chamber of 30 m^3 total volume protects each main from water hammer.

One pump has a discharge of 6480 m^3/h, at the total head of 65 m and operates at the constant rotational speed of 750 rot/min. The driving electric motor of 1600 kW, is powered by a voltage of 6 kV. The pumps did not work properly on all their operation range determined by the static head variation of the system. The pump operation had to be limited within the range 65–61.5 m due to the high value of the net positive suction head, (NPSH). Moreover, the pumping station recorded a low energetic efficiency. Therefore its modernization was needed.

The static head values, determined by the water level variation in the suction and respectively, discharge basin, are $H_{g\,max} = 52.00$ m; $Hg_{med} = 50.00$ m; $H_{g\,min} = 48.50$ m. These data and the total head losses all over the hydraulic system, which were calculated taking into account the fully developed turbulent flow regime [4], led to the following main values for the total pumping head: $H_{max} = 65.00$ m; $H_{med} = 59.80$ m; $H_{min} = 56.00$ m.

By a thorough analysis of the operation of the pumps and the customer's demand, we concluded that the pumps have to be renewed by ones with larger impellers, of 1000 mm in diameter. There were considered similar centrifugal pumps, with the discharge of $Q = 6500$ m^3/h at a total head of $H = 65$ m, with the same geometric features, but with better inner hydraulics, which means efficiency of $\eta = 90\%$. The net positive suction head is NPSH $= 8.5$ m at the above given duty point [4]. The electromotor power, the pump's suction and discharge diameter are the same for both the original and the proposed pumps. The new pumps run in the same range of total head, but at higher discharge values, due to the larger impellers, that means in between $H_{min} = 56$ m (at $Q = 7560$ m^3/h) and $H_{max} = 65$ m (at $Q = 6500$ m^3/h), at a constant speed. The pump's efficiency, at the maximum head $H_{max} = 65$ m, is η_p

Table 5.2 Technical parameters for the main duty points. Old and new Mircea Vodă pumping station [4]

Head	Parameter	Old pump	New pump
$H_{max} = 65$ m	NPSH (m)	8.20	8.50
	Pump's efficiency (%)	83	90
	Absorbed power (kW)	1440	1328
$H_{min} = 56$ m	NPSH (m)	12.50	11.00
	Pump's efficiency (%)	79	83
	Absorbed power (kW)	1520	1447

$= 90\%$ at its overall efficiency is 86%, which results in an absorbed power amount of 1328 kW. The new value of the net positive suction head, for this duty point, is 8.50 m which is convenient for proper operation of these pumps.

Furthermore, at the minimum head $H_{min} = 56$ m, the pump efficiency lowers down and the absorbed power becomes Pa $= 1447$ kW, Table 5.2. The pump's performance curves indicate a smaller net positive suction head, of 8.20 m. This newly proposed pump will operate within its optimal range. Nevertheless, the existing priming installation is recommended to be kept.

Only six pumps in a pumping group were to be replaced. One current pump from each group has to be equipped with a frequency converter, to operate at variable speed. This measure will make possible the pumping station to deliver small discharge values and, consequently, to make considerable energy savings.

The main advantages offered by the new solution consisted of:

• low exploitation costs due to the new possibility to adjust to low discharge values;
• higher efficiency;
• lower consumed power for the same delivered discharge.

5.3.2.2 Lift Irrigation Water Supply Pumping Station

Galesu is a lift irrigation water pumping station, which supplies with water an olericulture area of 3950 ha. The original pumping station had been functioning since 1970.

The pumping station takes water from canal CA_1 and delivers it through two mains of 1000 mm in diameter and 900 m in length. The static head varies between $H_{gmin} = 26$ m and $H_{gmax} = 26.35$ m [6]. As the pumps deteriorated, the operation and maintenance costs became unacceptable. The proposed modernization solution consisted of replacing the internally worn pumps with five new horizontal centrifugal double flux pumps. The piping system (excepting the two concrete made mains) and hydraulic equipment were renewed.

Table 5.3 Technical parameters for main duty points. Galesu pumping station

Duty point	Parameter	Old PS	New PS
$H_{max} = 34$ m, $Q = 5400$ m^3/h (on one single main)	η_p (%)	82	89
	P_a (kW)	2215	1575
$H_{min} = 29$ m, $Q_p = 2,760$ m^3/h (one pump)	η_p (%)	73	87
	P_a (kW)	400	315

where Q discharge flow rate; H total pumping head
η_p pump's efficiency; Pa absorbed electric power

The pumps have individual suction pipelines. On the discharge pipeline system, there is a particular connection for the middle pump that can deliver on each one of the two mains and also on both mains simultaneously. The discharge duct of the pump P3 (the middle one) splits immediately after it exits the building and each of its branches is connected to other individual discharge ducts. Both branches are provided with butterfly valve. Therefore, there are more operation possibilities: one or more pumps, delivering on one or both mains. The installation is symmetrical from a hydraulic viewpoint.

The most relevant parameters for the two main duty points of the pumping station are presented in Table 5.3. The parameters for the old pumping station were determined by neglecting the fact that the efficiency of the pumps has been reduced in time, due to internal wear. By comparing, the theoretical pump's efficiency for similar duty points (for the old and the new station), results in a significant gain of 7–14%.

Along with the renewing of old pumps, pipelines, and fittings, the pumping system was provided with automation and monitoring equipment, for efficient operation and control.

The mains protection from water hammer was enhanced by imposing a special closing law to the check valves. The law was the result of a thorough study of the pressure variation in the most vulnerable sections of the ducts, during water hammer, conducted by numerical simulation.

5.3.3   Towards High-Yield Irrigation Water Pumping Stations

Agriculture in Dobrogea has to be supported by irrigation, and fresh water has to be transported over long distances, as the Danube is the best source compared to the groundwater that can be affected by pumping for large-scale agricultural purposes. This is the reason we cannot compare the energy efficiency of our irrigation systems to similar systems built in other regions, with different topography, climate and water resources. All we can do is to improve, by technical means and personnel skills the energy efficiency of our systems.

For now, the only way to improve efficiency is to rehabilitate the existing infrastructure that means pumping stations, transmission ducts, canals, watering networks and all this action has to be conducted according to a previous total analysis of the hydraulic system. The dimensioning of each element in the hydraulic system depend upon the effective combination of the system's components: water source, intake structure, reservoir, pumps, pipeline, etc. It is more difficult for an engineer to conform to the restrictions imposed by rehabilitation of an old facility when he has to upgrade the installation by replacing only part of it than to plan and design a new one.

Many old pumping stations were oversized from the very beginning when speaking about pumping heads and discharges. Furthermore, pumps operated at constant speed, so the discharge could be adjusted only by combining different pumps to operate in parallel or to partly close the valve on the discharge duct.

The construction and installation costs for an irrigation facility are high. Besides, its operation relies on large amounts of electricity which is expensive. Therefore, it is imperative that the pumps degree of efficiency be high.

The optimal engineering design, construction, and operation of a high-yield irrigation water system depend on:

- careful estimation of the pumps maximal head as a key factor in pump selection;
- expert design, based on hydraulic analysis and numerical simulation;
- use of variable frequency drives for better pressure regulation and for efficiently matching the pump discharge to water demand;
- monitoring and control equipment used in pumping installations;
- water transport canals insulation;
- proper operation and maintenance;
- well-qualified personnel.

All these goals involve large investments. Therefore it is difficult to achieve them simultaneously. However, the on-going national programme envisages bringing the new technology in the existing irrigation infrastructure in Dobrogea, on stages.

The importance of the numerical simulation of a water supply system operation in engineering design is pointed out below in two study cases.

5.4 Numerical Simulation–Fundamental for Water Supply System Engineering Design

The numerical simulation is a useful tool in a decision-making process, concerning technical solutions. It is widely used to assess the hydrological state of the surface water, groundwater, and soil water in the area of interest and to find flood protection solution or a water source for the revitalization of some old dry river bed [26]. Referring to urban and rural water facilities, a reliable and technically sound option may be achieved after a rigorous analysis of different possible configurations and engineering design solutions, made in the first stage of planning. Moreover, the numerical simulation offers a basis on which the engineers can validate their option for a future project or can evaluate the operation parameters of an existing hydraulic system. EPANET is an easy to use software, conceived for water supply systems [25].

5.4.1 Drinkable Water Distribution Network in the Rural Area. Numerical Simulation

The low-pressure water distribution network in Dulcesti is a branched one, directly supplied by pumps mounted in wells. The buried network has the same geometric shape as the streets of the village, as it is represented in Fig. 5.7. The ratio is about 10 m of pipe per inhabitant. The local topography shows a big difference in elevation, up to

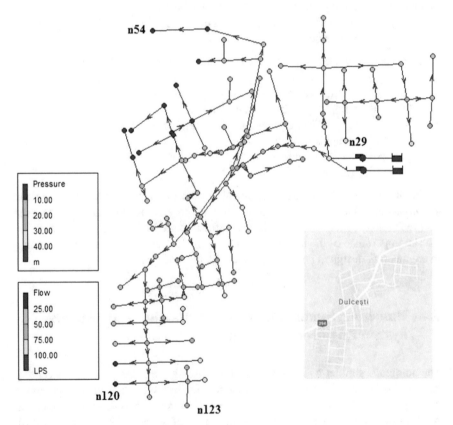

Fig. 5.7 Layout of Dulcesti water supply system, in EPANET, on the left, and the image of the village on the map, on the right (Google maps)

37 m, which is specific for this village. Furthermore, there are essential differences in elevation, albeit the consumers are in proximity. The actual water demand is 15.26 l/s.

The new-proposed pumping station that will supply the villages Dulcesti, Peceneaga, and Mosneni, according to the regulations regarding water supply systems (NP133 2013) will be part of a water facility placed in the South-East of Dulcesti village. Aiming to achieve an optimum energy efficient operation of the pump group that supplies Dulcesti and, consequently, an affordable water price, we've developed a study of the hydraulic parameters and flow control opportunities, in different operation alternatives. The analyses were conducted with the help of EPANET software, taking into account various operation possibilities.

In the proposed configuration, water is pumped from the wells into an open reservoir mounted at the elevation of 8.9 m, and then it is pumped into the network by two identical pumps.

The highest junction, n54 in Fig. 5.7, is in the North-Western part of the village, at an elevation of 43.11 m. On the other hand, the lowest node is n29, at an elevation of only 9 m. In the Southern branch of the network, there are differences up to 17 m between the elevations of nodes n120 and n123, according to the notation in Fig. 5.7. It has to be mentioned that in EPANET, a system is a collection of interconnected links and junctions. A reservoir or a tank are junctions, while a pipeline or a pump are defined as links. All the water demand is concentrated in junctions.

The two identical pumps considered in the proposed solution deliver a flow rate of 7.61 l/s at a total pumping head of 43 m. The study involved two different alternatives constant and variable speed pumps, to get the best variant for the network operation and to compare the results regarding efficiency to the existing network.

The simulation was also carried out for the existing network, with its water losses, that means for a couple of pumps that deliver 28.11 l/s of water, at a total head of 44.5 m, which is considered the reference case for the specific energy consumption.

The standard water demand for the rural area [17] was introduced in the demand pattern and in the pumps operation timetable, which is a control command in EPANET.

All the information on pressure and velocity fields, along with results regarding the specific energy consumption, obtained by simulation, made possible a comparative investigation of the technical parameters for each alternative.

As it was to be expected, we obtained a pressure difference between the South-Eastern and the North-Western part of the village. In all the studied cases, a minimal pressure of at least 7 mwc is provided at the furthermost hydrant. For example, in the node 54, the pressure is 9.26 mwc at the maximal demand value. Pressure increases in a few nodes up to the maximal allowed value of 60 mwc, for the case when constant speed pumps are used, but only in the night time, when the demand is low [5].

The velocity field shows values comprised between 0.3 and 0.8 m/s, in the central pipelines, while in the external ones, where the diameter is large enough for future expansion of the network, the velocity might decrease under 0.3 m/s [5].

The results regarding the energy efficiency are given in the Table 5.4. The specific energy consumption coefficient, e, was calculated as a time-weighted average during 24 h.

Table 5.4 Specific energy consumption coefficient, e. Dulcesti pumping station

Case	e (kWh/m^3)
Constant speed pumps, operating on the existing network	0.199
Constant speed pumps, operating on the rehabilitated network, considering only the unavoidable water losses	0.178
Variable speed pumps, operating on the rehabilitated network, considering only the unavoidable water losses	0.166

The variable speed pumps cope better with the rapid variable water demand during a day. The variant with variable speed pumps led not only to a better coefficient for the specific consumed energy but also to smaller pressure values, in spite the speed could not be lowered more than 78%, due to the high static head of the village network.

We may conclude that the rehabilitation of the distribution network is necessary to eliminate the water losses, not only because water itself is so valuable, but also for the substantial energy saving.

Kanakudis et al. [13] propose that the energy used per system input volume of water (kWh/m^3) to be a performance indicator for the small water supply systems, along with the others proposed by IWA.

5.4.2 Irrigation Water Supply Pumping Installation. Numerical Simulation

The rehabilitation technical solution included the replacement of the old pumps with new ones, more efficient than the original ones, as the pump building technology has progressed in the last few decades. While in the old installation there were valves mounted on each discharge duct of the five pumps, in the new one, the valve of the pump P3 was eliminated and there were introduced two valves on the ducts c34 and respectively c35, Fig. 5.8.

This change allows new possibilities of operation of the pumping installation because the pump P3 can deliver on one or both mains, according to the status of the valves.

It has to be mentioned that in Fig. 5.8, the reservoir R is the same for all the pumps, and the valves are introduced as properties of the pipes (links) they are mounted on.

The numerical simulation was performed in EPANET either for the old or the new pumping installation, for all the operation possibilities. The results referring to the efficiency and power consumption are gathered in Table 5.5. Although we neglected the internal worn of the old pumps, the gain in efficiency is obvious. The pumps were worn, so, after rehabilitation, the gain in efficiency and energy saving is more considerable than these estimations.

Furthermore, by simulation, we have easily got information on the velocity and pressure field, in every single operation possibility.

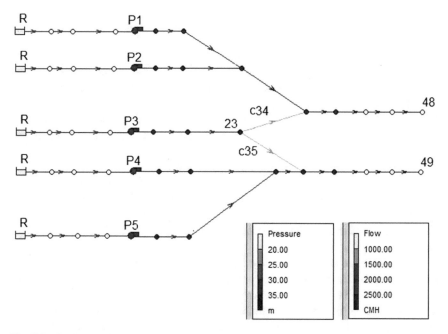

Fig. 5.8 The layout of Galesu re-pumping station, in EPANET

Table 5.5 Results regarding energy obtained by numerical simulation. Galesu pumping station

Pumps in operation	Old pumps		New pumps	
	e (kWh/m³)	P (kW)	e (kWh/m³)	P (kW)
P1	0.10	270	0.09	237
P3 on one main	–	–	0.09	237
P3 on both mains	0.09	264	0.08	235
P1 + P3	0.11 × 2	520	0.09 × 2	474
P1 + P2 + P3 on one main	–	–	0.11 × 3	684
P1 + P2 + P3 on both mains	0.12 × 3	735	0.11 × 3	678
P1 + P2 + P3 + P4 + P5	0.11 × 5	1290	0.11 × 5	1170

5.5 Future Challenges for Reliable and Resilient Water Supply Systems. Environmental Impact

The new regional water supply system will take water from the above mentioned good quality source, located close to Medgidia, and convey it to almost all the localities in the Southern coastal area, where the local sources do not fit the water demand or the drinkable properties. The proposed system is a resumption of a concept, namely the interconnected seaside system, that had been operating for a long time but which was abandoned about 20 years ago. Local water sources were preferred instead, as they were of good quality [21].

The interconnection of the existing systems will bring more flexibility and resilience in operation, allowing a dual way for water flow, in and out the existing systems, according to the demand. The interconnected system offers the possibility of eliminating intermediate pumping stations and supplying by gravity the reservoirs in Constanta Sud from the underground sources Techirghiol or Medgidia. Moreover, there will be no need of dividing the network into pressure zones, even under current operation.

The risk of disruption of the existing systems operation will decrease while the protection of the sources will be strengthened, by avoiding to exceed the intake flow rate from a certain source. For instance, an increased flow rate from the Palas underground source, which is the source of water supply for Constanta and Navodari, could lead to seawater seepage and put in jeopardy the quality of the abstracted water [21].

The assessment studies completed for the implementation of the new projects show no negative impact of the water supply systems under normal operating conditions, on the water bodies. Moreover, by reducing water loss in the distribution networks (quite large in some settlements) will result not only in lower costs for water consumers but also in deviating, from the natural paths, of smaller water volumes, with reduced risks of water depletion, environmental changes, and pollution.

To maintain the drinkable water quality at source, the area of severe regime and sanitary protection are to be fenced. Anti-contamination systems of the source and complete automation of the pumping stations and water treatment facilities will contribute to a good quality of the drinking water.

In the impact assessment studies, the necessary technical measures and waste management plans are conceived to reduce and even eliminate the environmental impact. In addition to the technical and economic criteria, the risk of climate change was considered.

The most critical challenge is the protection of the natural habitats that give unicity to Dobrogea. The new construction works that will intersect Natura 2000 protected natural areas (such as Danube Canarale, Hagieni Forest) will not affect the fauna and the flora, due to the attention paid to the chosen technologies, ensuring the lowest possible impact. Trenchless technologies are going to be used. The site impact assessment corresponds to low negative influences. Strict measures are envisaged, to avoid disruption of some species and habitat alteration during the construction works. The expected long-term result of the project is the reduction of water bodies' pollution and, consequently, the improvement of the natural conditions for the species and habitat of community interest [23].

Nevertheless, during the period of building of new drinking or irrigation water supply systems, we can assume that there will be a moderate impact on the environment, caused by possible inappropriate manipulation of vehicles or accidental leakage of oils or fuels in soil or water.

The rehabilitation of the wastewater collectors and, consequently, the elimination of uncontrolled leakage in the soil are leading factors in avoiding groundwater contamination. New wastewater treatment plants are to be built in isolated areas (such as Castelu-cluster Poarta Alba-cluster) and, furthermore, a new sectorisation of the collecting areas [23] with a more appropriate distribution of waste flows to the plant will be possible. Therefore, a reduced number of wastewater pumping stations will be needed, due to local wastewater treatment, and the risk of long wastewater transport conduits damage will be diminished. Furthermore, will be made a new investment in the wastewater treatment infrastructure for towns with the population between 2000 and 10,000 equivalent inhabitants. Only the two new wastewater treatment plants in the towns of Hârşova and Băneasa will reduce the current volume of untreated wastewater in Constanta County with approximately 78%.

Periodic inspections and sound plans to prevent and combat accidental pollution have an essential role in the surface and groundwater sources protection. In areas where nitrate values have already exceeded the maximum concentration allowed by the standards, water wells will be deeper (Tortomanu 1000 m in depth and Braila Island 500 m). Gleeson and Richter [10] suggest a new metric: the environmental flow response time, that allows water managers to quantify the timescales of the impacts of groundwater pumping. We strongly recommend this metric regarding groundwater abstraction in Dobrogea.

As urban and rural water demand increases, it is of enhanced importance to ensure the water quality and, at the same time, to prevent any possibility to pollute or compromise the underground or surface sources. Industry and agriculture activity may cause pollution, river regulation and hydraulic structures may also disturb natural water circulation. Therefore integrated management for sustainable water sources is the best way to preserve the environment.

New clean energy sources, like solar energy, have to be used in small irrigation systems, taking into account that the insolation in the area is appropriate.

In the close future, other innovative solutions like precision pump control, and irrigation on-demand method, are needed to replace current irrigation with an intelligent one. Desalinisation of brackish water and water reuse are also a must in the context of freshwater scarcity foreseen by the experts.

In our world, where water and energy are more valued than ever, we must get involved in solving efficiency issues and the authorities have to find the means to implement new intelligent technologies.

5.6 Conclusions

The depletion of the groundwater quality is an ongoing phenomenon, which confined us, in Dobrogea, to use deeper aquifers for abstracting drinking water. This phenomenon has to be stopped by better management of the pollution prevention methods.

The poor quality of some water sources in the Southern part of Dobrogea and the considerable increase of the drinkable water consumption during summer (when the tourists come to the continuously extending resorts along the Black Sea coast) require a more flexible supply system and new appropriate water sources for that region.

The proposed interconnected system will provide considerable additional flow rate in the coastal areas where resources are used excessively in the warm season, reducing the variation in groundwater levels of local sources. New sources or the activation of previously preserved boreholes in rural areas (due to reduced consumption), even placed far away from the shore, will compensate seasonal flows through the connection to the regional system.

It is imperative to use surface water for irrigation so that the quality of the groundwater to be preserved. Therefore, the irrigation systems in Dobrogea depend on long distance water transportation as the primary surface water source is the Danube or the canal that connects this river to the Black Sea.

The efficiency and low energy consumption are among the goals of the Romanian authorities, and the designers enrolled in the rehabilitation programmes, as the infrastructure for drinkable and irrigation water supply is aged. The environment protection has to be a priority for all decision-makers regarding the technology used in new construction sites, for maintaining or even restoring the natural habitats that give Dobrogea specific charm and beauty and, above all, assure a healthy living for the inhabitants.

An integrated water resources management and a change in men's philosophy concerning the influence of the more and more variable climate form the best way to the sustainable development of Dobrogea region. This sustainability has to be the responsibility of the nowadays generations.

5.7   Recommendations

According to the conclusions we have drawn, we formulate the following recommendations:

a.   The continuous monitoring of the physical and chemical parameters of the water, especially regarding sodium and nitrate indicators.
b.   The identification and suppression of water pollution sources.
c.   The quantification of the timescales of groundwater pumping impact has to be performed.
d.   New wastewater treatment facilities have to be built mainly in the rural area of Constanta County.
e.   The interconnection of the zonal water supply systems into a regional one has to be planned so that to diminish the number of intermediate pumping stations and to use gravity instead.
f.   The use of numerical simulation in all planning and designing activity.
g.   The extent of sonar methods use for water loss detection across the water distribution networks is needed.
h.   The increase in the number of local wastewater treatment facilities.

i. The use of trenchless technologies and responsible working team, in specially protected areas.

j. The prevention of groundwater pollution by avoiding its use as an irrigation water source, even if for small isolated areas.

k. The use of highly efficient and, if possible, variable speed pumps, to renew the existing irrigation and drainage pumping stations.

l. The assessment of the specific energy consumption coefficient before a pumping station is put into operation and the introduction of a coherent system to compare it.

m. New investments in pressure control and automation equipment are needed to modernise the existing pumping stations.

n. The continuous assessment of the water supply systems performance has to be conducted to keep the yield into a high range all over the region.

o. Investments in new technologies appropriate for our region, such as solar energy driven equipment and seawater desalinisation, have to be considered on long-term in the future.

p. A sound programme for either initial or continuous education of the personnel involved in engineering design, construction and exploitation of the water systems.

References

1. Caraivan G, Dinu I, Fulga C, Radu V (2010) Possibility of extending the drinking water supply for the Constanţa Harbor. Geo-Eco-Marina 16:75–85
2. Ciortan R (2001) Harbour arrangements. Ovidius University Press, Constanta
3. Constanta County Council (2018) Local environmental action plan, Constanta County, www.cjc.ro/dyn_doc/Hotarari/Proiecte/201//Sedinta_03_din27.03.2018/25.pdf
4. Constantin A, Dordescu M, Iordache Gh, Stănescu M, Rosu L, Cusnerenco V (2010) The analysis of technological rehabilitation solutions for the irrigation water supply pumping station Mircea Vodă, from Dobrogea, vol 55, book 2. Scientific Works, Agricultural University Plovdiv Plovdiv, pp 187–192
5. Constantin A, Nitescu CS (2017) Operation optimization of water distribution network in rural area. In: 17th international multidisciplinary scientific geoconference SGEM 2017, vol 17, issue 31, SGEM2017 conference proceedings, 485–492 pp. www.sgem.org
6. Constantin A, Niţescu CS, Stănescu M, Roşu L, Florea M (2008) Discharge pumping ducts protection from cavitation using air. Acta Technica Napocensis, vol 4, Secţ. Instalaţii, pg. 355–363. ISSN 1221-5848, Cluj—Napoca, Romania
7. Constantin A, Stănescu M, Rosu L, Nitescu C (2009) Energy consumption improvement of a drainage pumping station. In: Proceedings of the international scientific conference of the university of architecture, Civil Engineering and Geodesy, Sofia, Bulgaria
8. Dobrogea Litoral Regional Water Branch DLRWB (2013) Plan for prevention, mitigation and protection from flood effects in the Dobrogea-Litoral Water Basin, www.rowater.ro/dadobrogea/list/anunturi1/attachements/62/raport%20de%20mediu_PPPDEI_DL. Accessed 23 April 2018
9. European Environment Agency (2014) Performance of water utilities beyond compliance, EEA Technical report No 5, Luxembourg: Publications Office of the European Union. https://doi.org/10.2800/13253, ISBN 978-92-9213-428-0, ISSN 1725-2237

10. Gleeson T, Richter B (2017) How much groundwater can we pump and protect environmental flows through time? Presumptive standards for conjunctive management of aquifers and rivers, River Res Applic., Wiley, pp 1–10
11. Heider H, Sadiq R, Tesfamariam S (2014) Performance indicators for small- and medium-sized water supply systems: a review. Environ Rev 22: 1–40. www.nrcresearchpress.com/er
12. Ionescu M (1931) Dobrogea-Tomis-Constanţa. Romania, Constanţa, p 76
13. Kanakudis V, Tsitsifli S, Samaras P, Zououlis P, Demetriou G (2011) Developing appropriate performance indicators for urban water distribution systems evaluation at Mediterranean countries, Water Utility J 1:34–40. EW Publications
14. Lup A (2014) Romania's socialist agriculture during 1949–1989 -Myth and reality. Ex Ponto, Constanţa, Romania
15. Manescu Al (1998) Water supply. Ed. HGA, Bucharest
16. Ministry of Agriculture and Rural Development (2016) National Programme for the main irrigation infrastructure in Romania, Ministry of Agriculture and Rural Development, http://www.madr.ro/docs/agricultura/programul-national-reabilitare-irigatii-update.pdf. Accessed 4 May 2018
17. NP 133 (2013) Regulation regarding the design, execution and exploitation of water distribution systems, Bucharest, Romania
18. Papuc Gh (2005) The water supply of Tomis City in Roman and Late Roman Era. Ex Ponto, Constanţa, Romania
19. Pitu N (2012) The legend of the underground river beneath Dobrogea, http://romanialibera.ro/special/reportaje/legenda-fluviului-subteran-de-sub-dobrogea–articol-integral–252520
20. RAJAC (2016) UTILIS case studies, RAJA, Constanta, https://utiliscorp.com/cases/raja-water-utility-romania/
21. RAJAC, Constanta County Water Board (2017a) National Project for water and wastewater infrastructure development within operation area of S.C. RAJA S.A. Constanta, during 2014–2020. Constanta County. Environmental impact assessment report, vol 1. Constanta, http://rajac.ro/wp-content/uploads/2017/02/1_RIM_Constanta_Vol_1.pdf
22. RAJAC, Constanta County Water Board (2017b) National Project for water and wastewater infrastructure development within operation area of S.C. RAJA S.A. Constanta, during 2014–2020. Constanta County. Environmental impact assessment report, vol 2, Constanta, https://rajac.ro/wp-content/uploads/2017/02/2_RIM_Constanta_Vol_2.pdf
23. RAJAC, Constanta County Water Board (2017c) National Project for water and wastewater infrastructure development within operation area of S.C. RAJA S.A. Constanta, during 2014–2020. Constanta County. Environmental impact assessment report, vol 3. Constanta, http://rajac.ro/wp-content/uploads/2017/02/3_RIM_Vol_3_rezumat_NT.pdf
24. RAJAC, Constanta County Water Board (2017) History, http://rajac.ro/istoric/. Accessed 27 April 2018
25. Rossman L (2000) EPANET 2 User Manual Water Supply and Water Resources Division National Risk Management Research Laboratory, US
26. Šoltész A, Čubanová L, Baroková D, Červeňanská M (2018) Hydrological and hydraulic aspects of the revitalization of wetlands: a case study in Slovakia. In: The handbook of environmental chemistry. Springer, Berlin, Heidelberg. https://doi.org/10.1007/698_2017_227
27. STAS 4163-2 (1996) Water supply. Water distributions. Engineering and operation technical directions, Bucharest
28. Stockholm International Water Institute (2011) World Water Week, The Stockholm Statement to the 2012 United Nations Conference on Sustainable Development in Rio de Janeiro, Stockholm

Chapter 6
Drinking Water Supply Systems—Evolution Towards Efficiency

Ciprian Bacotiu, Cristina Iacob and Peter Kapalo

Abstract Throughout humankind history, drinking water supply was crucial for the survival of all communities. In Romania, major cities like Bucharest, Timişoara, Iaşi, Cluj all have a long and rich water history, full of achievements. The limited space of one chapter is allowing us to only discuss the case of Cluj, especially the period after 1989. The first section of this chapter highlights the most critical moments in Cluj-Napoca's water supply history, until the end of the communist regime. A lot of technical data is presented, showing the continuous urban population growth, followed by the investments in the water infrastructure. Between 1945 and 1989, the evolution was primarily focused on the quantitative aspects, in the context of the centralized economy which had its influence on the entire development of the city and also on the activity of the local water operator. The second section discusses the profound transformations that took place after the Revolution (1989) in the field of water supply systems: market economy, inflation, water metering, re-organization of water companies, new materials and equipment, access to European funds, strong measures for efficiency. The following three sections are dedicated to the new technologies and paradigms that had a tremendous positive impact on the activity of the Cluj regional water company. The next section describes the plans for the development of the water supply systems in the Someş-Tisa region. The fight against water losses will be a top priority. Finally, several conclusions and recommendations were presented, with emphasis on the objectives of a modern water operator and on the value of water.

Keywords Water distribution · Cluj · Multicriteria decision · GIS · SCADA

C. Bacotiu (✉) · C. Iacob
Building Services Engineering Faculty, Technical University of Cluj-Napoca, Cluj-Napoca, Romania
e-mail: ciprian.bacotiu@insta.utcluj.ro

C. Iacob
e-mail: cristina.iacob@insta.utcluj.ro

P. Kapalo
Faculty of Civil Engineering, Institute of Building and Environmental Engineering, Technical University of Košice, Košice, Slovak Republic
e-mail: peter.kapalo@tuke.sk

© Springer Nature Switzerland AG 2020
A. M. Negm et al. (eds.), *Water Resources Management in Romania*, Springer Water,
https://doi.org/10.1007/978-3-030-22320-5_6

6.1 Introduction and Short Historical Perspective

Throughout the history of humankind, the degree of civilization of a society was often measured by its achievements in the water supply domain as water means life.

For Romania, the water supply history is very rich and complex. Therefore this chapter will only focus on a particular geographical region, situated in the heart of Transylvania.

An essential economic and academic center, with more than 400,000 inhabitants, Cluj-Napoca is the fourth major city of Romania. It is located in the north-western part of the country and is considered the unofficial capital of the historical province of Transylvania. Surrounded by hills, at about 50 km from the Apuseni Mountains, Cluj-Napoca is situated on both sides of the Small Someş River. It seems that water abundance attracted the first settlers about 80–120 thousand years ago, as the first vestiges of human inhabitance on these territories indicate.

When Dacia was part of the Roman Empire, the town was called Napoca, acquired the municipium status and became the capital of the Dacia Porolissensis province. Regarding water supply and resources, they were apparently of great importance, as demonstrated by a large number of water supply works dating from the Roman period, uncovered by archaeologists.

Not much progress was made in water supply during the next centuries, even though the town's population increased gradually: 6000 inhabitants in 1453, 13,920 inhabitants in 1785, 24,000 inhabitants in 1837 and 26,638 inhabitants in 1869 [17]. The main water source was the Small Someş River, some springs from the Feleacul Hill and a few fountains, from which people used to carry water with buckets and other rudimentary means. As a result of poor sanitary conditions and water infestation, numerous epidemics spread rapidly and decimated the population of the town. Serious measures had to be taken; therefore in 1887 (when the town was part of the Austrian-Hungarian Empire), the authorities started the first modern water supply system of Cluj. It consisted of 5 wells placed in the central part of the city, a pumping station, a cast iron pipe 1677 m long and a small tank, which supplied about 1200 m³ water per day [17].

In 1889, the Town Hall organized a competition aiming at the elaboration of a long-term project regarding the city's water supply and sewerage system [18]. The winner was the Schlieck company from Budapest. His project stipulated that 2275 m³ of water have to be provided daily from deep wells (6–8 m). The proposed network had 26.6 km of pipes with diameters between 40 and 275 mm, 47 valves, 107 hydrants and 65 public drinking fountains [17].

As the problems raised by the water supply and sewerage system became more and more complex and the population of Cluj reached 37,000 inhabitants, the need for a specialized company with the qualified technical staff was evident. As a result, on November 8, 1892, following a Town Hall decision, the Water and Sewerage Factories of Cluj were founded.

In 1893, under this organizational framework, the site for the sources, tanks and networks was opened. In 1900, the Floreşti underground source and a 2000 m³ reservoir were set into operation. Water was transported to town through a 5.5 km pipe. The total length of water ducts was 42.3 km, while the sewerage network had 24.5 km.

By 1910, Cluj had about 60,000 inhabitants, therefore more and more water was needed. In 1918, a new intake front ("Green Cap") was set into operation, 1 km down the river from the Floreşti source [17].

In 1919, after the Unification of Transylvania and Romania, Cluj passed under the Romanian administration. The city continued its development on the northern side of Feleacul Hill, on the Cetăţuia Hill, on Dâmbul Rotund Hill and the left bank of the Someş River [18].

Between 1924 and 1934, third water capturing area was enabled at the Floreşti water source, together with a new adduction pipe (centrifugally spun reinforced concrete pipe—Swiss Vianini license, 700 mm diameter), a re-pumping station (Grigorescu Water Works) and a 3000 m³ water tank [17]. In 1930, Cluj Water and Sewage Plants (UACC) became a public company. By 1940, the supplied water reached 16,000 m³/day for about 110,000 inhabitants and 94 local enterprises, delivered through 160 km of pipes.

After the Second World War, Romania was trapped in the Soviet Union sphere of influence. The communist era began, with severe consequences on urban development and water supply: in order to sustain the rapid industrialization strategy of the government, a large population from rural areas was relocated to the cities. As a consequence, between 1958 and 1968 the population of Cluj-Napoca increased from 155,000 to 185,000 inhabitants and the city was confronted with a severe lack of water, despite the supplied volume of 55,000 m³/day.

In 1968 strong measures had to be taken, so specialists from the Institute for Design of Hydrotechnical Constructions in Bucharest concluded that Cluj needs a new source of water, which should collect and treat the surface water of the Small Someş River. The construction of this new water treatment station began in 1971 in Gilău, up to the Someş River, about 16 km from Cluj. The water was taken from a nearby artificial lake and passed through sieves, settling tanks, quick filters and finally chlorinated in order to complete the treatment process.

The Gilău water treatment plant (WTP) was inaugurated in 1973 [2] and produced cheap and good quality water (about 60,000 m³/day), which was then transported to Cluj by gravity, through a pre-compressed reinforced concrete pipe PREMO, 1000 mm diameter.

In 1978, a new Gilău-Cluj adduction pipe (SENTAB type, 1400 mm diameter) was set into operation [4]. The water distribution network (WDN) was also extended, new pumping stations and new water tanks were built in order to satisfy the continuous development of the city. During the last years of the communist regime, Cluj-Napoca had approximately 330,000 inhabitants [5], the volume of the supplied water reaching 165,000 m³/day.

In 1989, the Revolution ended Ceauşescu's dictatorship and Romania started its first steps in capitalism. This event was also a milestone in the evolution of WDNs.

6.2 Opportunities After 1990

In 1990, the legacy of the totalitarian regime became obvious in the field of drinking water systems. Under the grip of Ceauşescu, Romania's main objective was to repay the external debt as soon as possible. Therefore the decade 1980–1990 was dominated by various "strange" rules and restrictions concerning *quality* investments in the infrastructure. Every equipment or material had to be made by the national industry (as cheap as possible), and imports were forbidden. As an example, cast iron was considered an "energy-intensive" material and was massively substituted by asbestos cement for the new drinking water pipes (which later proved to be a health hazard and banned in many countries).

Therefore, at the end of 1989, the situation was challenging: out-of-date and obsolete equipment (the effect of "savings"), an old drinking water supply network (with some 100 years old pipes), substantial water losses, no general metering.

It is necessary to point out that in the centralized command economy, there was no inflation (this concept was not officially acceptable), all prices were controlled by the government, so water was very cheap. This was not a bad thing per se, but had a dangerous side-effect at that time: almost nobody was seriously concerned about water losses, metering or wasting water. Anyway, the official reports about water losses were carefully "prepared", and the Party leaders said that Romania is going steadfast to the communism...

Then, in December 1989 the Revolution has changed the rules of the game, the old lies were exposed, and the crude reality was sometimes shocking. In the new era of transition from socialism to capitalism, very few people knew how to quickly find the right solutions and to foresee the dangers. However, for the sector of public utilities (drinking water companies included), despite the aforementioned problems, opportunities were identified more easily than in other fields [19]. Maybe because the market for drinking water facilities was solid, clients waiting for quality services at low prices. Nevertheless, local water operators/companies had to adapt themselves in order to survive, following the principles of economic efficiency. Success models from abroad were available as a reference, and the problems of the sector were relatively easy to spot out.

In this context, in 1991 under the authority of Cluj County Council, the Cluj County Water and Sewage Company (RAJAC) was founded. At that moment, the drinking water network had 412 km of pipes, with 18,782 connections to clients, delivering up to 3.1 m^3 of water per second for about 400,000 inhabitants, with a 60,300 m^3 storage capacity in water tanks [17].

The main concern was to identify the imperative investment needs that would lead to an increased quality of the services provided. Thus, with funding from local sources, some important works were completed, such as:

– supplying the drinking water demand by extending the water treatment station capacity by adding a supplementary filtration stage, a new 1000 l/s clarifier and also extending the micro-screening stage, between 1992 and 1996;

- supplying drinking water at high-rise residential buildings especially in some newly built districts of Cluj-Napoca;
- building new water main for pressure balancing, 5 km long and 1400 mm diameter;
- increasing the safety of the regional water supply system by introducing a new 1200 mm pipeline for raw water transportation from Someşul Cald Lake to the Gilău WTP, between 1996 and 2000 (Fig. 6.1).

In 1995, RAJAC began to expand at county scale, taking under its service 22 rural localities located mainly along the route Gilău-Cluj-Gherla [5].

The need to increase the quality of services for consumers, complying with environmental regulations and aligning with the requirements resulted from the transposition of European Union (EU) water directives have imposed the identification of new investment programs, especially with non-refundable funding.

Thus, since 1997, *four major investment projects* have been conducted for the expansion, modernization and rehabilitation of water and sewerage infrastructure in the serviced area (however, the sewerage aspects will not be discussed here).

We should point out a crucial moment: in December 2004, RAJAC was reorganized and became the *Someş Water Company* (Romanian acronym CAS S.A.), a shareholder-owned commercial company. The shareholders were Cluj and Sălaj County Councils, the municipalities of Dej, Gherla and Zalău, the towns of Huedin, Cehu Silvaniei, Şimleu Silvaniei and Jibou.

As of 2006, CAS S.A. becomes the first and largest regional water company in Romania that has crossed the administrative borders of its own county (Fig. 6.2), providing water services to 7 cities and 53 rural localities in Cluj and Sălaj counties [2, 4, 5].

The following *four major investment projects* are currently completed, but for three of them, the external debt is still being repaid, i.e. the repayment of the co-financing credits assumed.

Fig. 6.1 The raw water supply pipe between Someşul Cald Lake and the Gilău WTP (photo by courtesy of CAS S.A.)

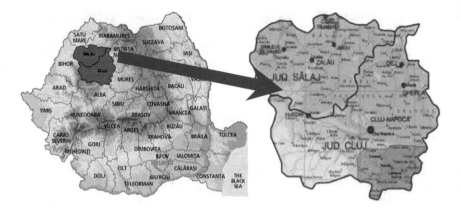

Fig. 6.2 Service area of Someş Water Company (by courtesy of CAS S.A.)

MUDP II—Municipal Utilities Development Program

The works financed through this program were conducted from 1997 to 2002, totaling 38 million US dollars [3]. The funding structure was:

- 45.45% European Bank for Reconstruction and Development (EBRD) credit;
- 24.15% PHARE non-reimbursable fund;
- 30.39% of the public budget and own sources of the water company (Fig. 6.3).

Through this program, which targeted exclusively the public infrastructure in the municipality of Cluj-Napoca, about 117 km of networks were rehabilitated, of which

Fig. 6.3 A water tank built during MUDP II program (photo by courtesy of CAS S.A.)

46 km of water supply pipes on 90 streets and 5.8 km of water pipes for the supply of water tanks. Also, the Gilău WTP has been modernized and rehabilitated. The reimbursement of this credit was completed in 2012.

ISPA—Structural Pre-accession Instrument

The program had a total value of 63.6 million euros, and the work was carried out in the period 2003–2010. The structure of the financing was:

- 55.12% non-refundable EU funds;
- 19.27% state-guaranteed European Investment Bank (EIB) credit;
- 1.92% PHARE non-refundable funds;
- 11.74% of the public budget;
- 11.95% of the company's sources.

The program mainly targeted the municipality of Cluj-Napoca but also had a component of developing sewage networks in the upstream rural area. In total, about 173 km of pipe networks were made or rehabilitated, of which 32 km of water supply pipes on 61 streets and 34 km of water pipes for the supply of water tanks.

Other notable works that were carried out during this program were:

- the raw water transportation pipe from the Tarniţa catchment and the Someşul Cald underwater transportation pipe;
- the micro-hydropower station of 10,000 MWh/year;
- the modernization of the Gilău WTP.

The final term for repayment of the credit is in 2026.

Adopting Lake Tarniţa as the main source of raw water has created the premises for the future expansion of the administered area, including the Sălaj County, and also eliminated the dependence of the municipality of Dej on a private treatment station belonging to a bankrupt company.

The micro-hydropower station developed on the raw water catchment system (Fig. 6.4) allowed the production of green energy and, thus, entering the green certificate market.

In order to ensure sustainable development, to maintain a good water source quality and to prevent pollution (Fig. 6.5) in the entire geographical area of the Tarniţa-Gilău catchment basin, a plan for pollution prevention and source water management was developed [3]. The implementation of this plan was entrusted to a committee in which all responsible authorities and institutions were involved, such as the Cluj County Council, the local authorities, universities, environmental NGOs, public health and environment authorities.

SAMTID—Small and Medium Town Infrastructure Development Program

The program was conducted during the 2005–2009 period in the small and medium-sized towns of the area served by the regional operator in the counties of Cluj and Sălaj (Zalău, Şimleu Silvaniei, Jibou, Cehu Silvaniei—of Sălaj County and Dej, Gherla, Huedin—of Cluj County) and amounted to 14.4 million euros.

Fig. 6.4 The micro-hydropower station downstream of Tarniţa Dam (photo by courtesy of CAS S.A.)

Fig. 6.5 Environmental awareness banner—Tarniţa Lake area (photo by courtesy of CAS S.A.)

The structure of the financing was:

– 50% EBRD credit;
– 37.5% PHARE non-refundable funds;
– 12.5% National Fund contribution.

It represented the first significant program, after 1989, to expand and modernize the water supply infrastructure for this category of urban localities.

The most significant achievements in the water supply infrastructure of Cluj-Sălaj that were attained during this program are:

– replacement of approximately 115.6 km of water distribution network;

– full metering of consumers;
– water supply system monitoring in 62 points;
– rehabilitation of 6 pumping stations;
– rehabilitation of 4 drinking water tanks.

So, old pipes of steel and asbestos cement were replaced by new pipes made of high-density polyethylene (Fig. 6.6), glass fiber or ductile cast iron. Gradually, as a result of network improvements and modernization, water quality increased, pressure fluctuations at the consumer were reduced, and water losses decreased. But beyond any doubt, full metering has had the biggest social and economic impact. Immediately after the full metering, water consumption decreased sharply. This was due mainly to the shift in human behavior, because the bills showed that water is precious…

The payments of the external debt are made annually according to the reimbursement plan established by the Loan Agreement signed by the water company and the EBRD. The payments are made from the royalties paid by the investment beneficiaries—the administrative-territorial units of Dej, Gherla, Huedin, Zalău, Jibou, Cehu Silvaniei and Şimleu Silvaniei. The final term for repayment of the credit is 2019 [3].

SOP ENV—Sectoral Operational Program Environment (POS Mediu)

This project was called "*Extension and Rehabilitation of Water Supply and Wastewater Systems from the Counties Cluj/Sălaj—Improvement of Water Supply and Wastewater Systems from the Cluj-Sălaj area*".

The project was conducted in the 2007–2016 period, amounted to 196.9 million euros and was financed under the following financing structure [3]:

Fig. 6.6 Works at the Bologa-Huedin water supply pipe (photo by courtesy of CAS S.A.)

- 74%-Cohesion funds (EU grant);
- 11.3%-Romanian government;
- 1.9%-local authorities;
- 12.8%-credit to the EIB guaranteed by the administrative-territorial units (the shareholders).

The project comprises 23 contracts of which: 15 contracts of works, 2 product supply contracts, 2 supply contracts with service provision, 3 service contracts for technical assistance, 1 service contract for the project audit. All 23 contracts of the Cluj/Sălaj project were completed within the deadline, i.e. 30.06.2016. In total, 303 km of water and sewerage networks were rehabilitated or constructed of which 92 km of water supply networks, 91 km of extensions for water supply networks. The program also included the modernization of Vârşolţ WTP (Fig. 6.7) and the modernization of 3 water catchment systems. The payments of the external debt are made from the company's own sources (net profit) and the final term for repayment of credit is the year 2028 [3].

The second part of this program included a *complementary project* called "Improvement of water supply and sewerage systems in the Cluj-Sălaj area within the allocations available from the Cohesion Fund—Axis 1 SOP ENV 2007–2013". This complementary project was financed from the savings achieved within the first project, and it included 7 contracts, of which: 4 contracts of various site works, 1 technical assistance service contract, 1 advertising service contract, 1 service contract for the project audit. All 7 contracts of the Cluj-Sălaj project were completed within the deadline, i.e. 30.06.2016.

The works were conducted between 2015 and 2016 and were aimed at:

- rehabilitation of water supply pipelines (from the catchment area to the Grigorescu pumping station) in Cluj-Napoca municipality;

Fig. 6.7 The modernized Vârşolţ WTP (photo by courtesy of CAS S.A.)

– extension and rehabilitation of water networks in the municipality of Dej.

The SOP ENV program also included an additional acquisition project, financed from savings from the EIB credit [3]. The total value of the acquisitions was around 10 million euros.

6.3 GIS Technology

Geographic Information System (GIS) is one of the most revolutionary and exciting technologies of the last decade in our field of activity [16]. In short, it is a system designed for storing, analyzing and displaying spatial data.

Any GIS integrates five main components: hardware, software, methods, data and people. People are by far the most important. Servers, desktop PCs, software, networks, routers, storage or wireless devices play a big role in the effective deployment of GIS. However, the physical system must be coordinated with procedures and workflows. Utility companies have to decide how their people are deployed and when and where data needs to be accessed [8].

According to a classic definition formulated in 1990 by the National Center for Geographic Information and Analysis (NCGIA), GIS purpose is "to facilitate the management, manipulation, analysis, modeling, representation and display of geo-referenced data to solve complex problems regarding planning and management of resources".

In general, modern GIS technology should be capable of answering questions such as:

- Where is …?
- How long is …?
- When did it …?
- Which … is near …?
- What would happen if …?
- Can we combine … with data from …?
- What has changed since …?

In the past, the utility industry has always relied on hardcopy maps to manage facilities, so it was normal that electricity, gas or water companies should be among the first users of digital mapping software [12]. The first computer-driven achievements during the 1970s and early 1980s were focused around producing digital maps, those systems being called Automated Mapping/Facilities Management (AM/FM). Promising, but very expensive at that time. Pretty maps worth nothing without an active involvement in the operational activities. GIS applications should make the routine job functions easier to perform, which will improve productivity and reduce costs [16].

Nowadays, FM applications may be grouped into the following categories [12]:

- query and display;
- design/work order processing;
- equipment maintenance;
- network analysis;
- customer service/service call analysis.

As an example, GIS applications can help identify trends in water main breaks to prioritize pipe replacement and rehabilitation projects [10]. The database is structured using the desired criteria such as pipe material, diameter, age, surrounding soil conditions, proximity, fault/break history, water quality, coordination with other public works projects, etc. These criteria can be represented spatially in a GIS and associated with the pipe inventory.

The coupling of a GIS system with hydraulic modeling software will allow managers to diagnose the water network, to study solutions to various problems and to predict future behavior of the water system. Additional benefits can be achieved by maintaining connectivity between the model and GIS [10]: fire flow analysis, drinking water source analysis, water usage demand allocation, establishing facility elevations, etc.

In Cluj-Napoca, the local water company RAJAC started in 1994 the first steps in implementing a GIS application for managing its water and sewerage networks. It is evident that GIS technology brings several advantages (saves time and money, offers a decision-aid tool, has the potential for integration with other utilities and gives active communication abilities), but prices were still very high at that moment.

Therefore, RAJAC chooses to implement an ad hoc GIS application, in collaboration with a local GIS developer, SC EGH SRL Cluj-Napoca. It was a cost-effective solution, following many abroad examples: instead of a full-size professional GIS, a simplified ad hoc mapping tool can offer almost the same functionality, with lower hardware requirements [12]. Another reason for such an approach was that owning a GIS will increase the value of the company, thus facilitating the access at tenders for governmental or EU financed projects.

The name of this ad hoc GIS software platform was CADMOS-AC (Fig. 6.8), and it provided:

- a graphical database structured on layers (including topographical maps and water network maps overlaid);
- a tabular database structured as a set of alphanumerical attributes, where data were obtained from archives or directly from terrain measurements;
- GIS specialized analysis: query reports, thematic maps and charts that further support the decisional process and control of the network.

As input data, CADMOS-AC used:

- topographic maps scale 1:500 coming from Cluj-Napoca municipality (digital format);
- water network maps (analog format), tabular descriptive data (analog format).

Fig. 6.8 CADMOS-AC early implementation (by courtesy of CAS S.A.)

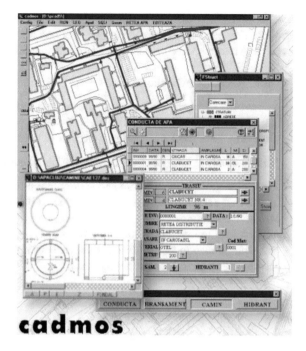

The main phases during the GIS implementation process were:

– acquisition of the basic topographic map, scale 1:500, in Stereo 70 coordinates;
– gathering the input information from different sources;
– conversion of pipes and other network elements, using a data model established by the application designer together with RAJAC.

There was a plan to integrate CADMOS with some hydraulic calculation and modeling applications such as AutoCad-HYDRA-EPANET, in order to facilitate the design, rehabilitation and extension of the water distribution network. With a Romanian interface, CADMOS had the advantage of at-hand customization to new requirements, specific for RAJAC, subsequent to the implementation phase.

However, though useful, over time CADMOS has shown its limitations:

– does not allow connections to relational databases such as Microsoft SQL Server and Oracle;
– does not perform spatial interrogations;
– does not perform geocoding (i.e. geographic positioning of objects, based on coordinates stored in alphanumeric fields);
– does not perform raster georeferencing;
– does not operate in a client-server architecture in a computer network.

In the present, the GIS system used by our local water operator CAS S.A. is GeoMedia Professional, a software product made by INTERGRAPH (the world second larger GIS developer, after ESRI). GeoMedia Professional tools are preferred by a large majority of users during the phase of spatial data production and primary exploitation, because of the ease of use (compared with the ESRI family products, such as ArcInfo, ArcView and ArcGIS) and including an impressive number of specific GIS tools.

Some examples of applications:

– prepare work orders for the network inspection and management;
– identify the valves that should be shut off when a pipe break occurs;
– surveying the activities regarding water losses detection in pipes.

The main advantages of GeoMedia Professional are the following:

– easy to integrate with AutoCad and MicroStation maps specific to CAS S.A.;
– improves the security and lifespan of CAD and GIS data through the consolidation of CAS S.A. services;
– full potential to integrate the existing CAS S.A. paper maps (scanner A3 → A0);
– potential connection to other information systems within CAS S.A. (SCADA, accounting programs, client billing, etc.).

6.4 SCADA Technology

Water supply is vital for people. Therefore continuous supervision of water treatment and water distribution processes is needed in order to maintain the normal functioning parameters or to quickly solve any problem that could appear. SCADA is the right solution. This acronym stands for Supervisory Control And Data Acquisition. Basically, it means that various computer-based control systems allow operators to monitor and control a facility's equipment either locally or remotely. SCADA may turn devices and equipment on or off, display real-time operational data, provide details and views of various processes, display large amounts of data and set alarms.

Proper solutions imply automation and monitoring architectures which contain [6]: a supervision and control system for the real-time installation, programmable logic controllers with basic functions (communication, adjusting, measuring), communication systems, interfaces with sensors, electrical drive elements, measuring devices, etc. (Fig. 6.9).

In Cluj-Napoca, the regional water operator CAS S.A. has started the implementation of its SCADA systems since 2003.

Now, the central dispatcher of Cluj has a complex structure. The core hardware components are 2 redundant rackable Siemens industrial servers IPC IPC547D (e.g. i7, 8 GB RAM) with RT licenses, server extensions and redundant extensions. Each important application/license is associated with a dedicated Siemens industrial PC. For example, there is one PC for the archive server, one for the WinCC RC license, one for the Step 7 license, one for the WebNavigator license, one for each OPC server and one for each WinCC RT 128 client. The entire SCADA platform is made by Siemens (SIMATIC WinCC 7.0). The communication rack is situated in a climatized

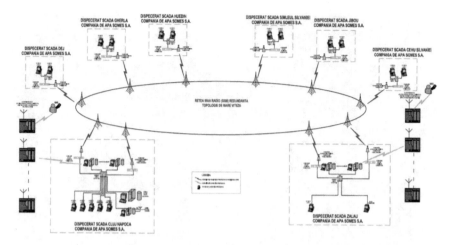

Fig. 6.9 SCADA architecture of the Someș Water Company (by courtesy of CAS S.A.)

room and has multiple SCALANCE switches from Siemens. The communication interfaces are OPC, namely OPC DA. The software application OPC Sinaut Micro ensures the communication with RTU points equipped with S7 200 automata and GPRS MD720-3 modems. The software application OPC KEPServerEX 5 ensures the communication with RTU points equipped with other types of automata, which have Modbus RTU or Modbus Ethernet communication protocols. The software application OPC Inventia ensures the communication with RTU points equipped with Inventia datalogger.

A big VideoWall structure is used for the detailed monitoring of the technological processes (Fig. 6.10). As controller for the VideoWall, a Transform AX6 computer is used (Windows XP, WinCC 7.0 SP3 client). The entire system is reliable and works well.

There are other local dispatchers that have their client workstations (e.g. in Cluj county we have Huedin, Gherla and Dej dispatchers). The communication system is using the GSM network, each monitoring point of the entire network being watched by the client workstations of the dispatchers.

Local SCADA systems are running at the Gilău WTP, the Govora pumping station, the Grigorescu pumping station and the Zorilor pumping station. This approach is modular and ensures flexibility, with the purpose of an easier configuration and maintenance of the system.

In conclusion, SCADA is a powerful tool which allows [6]:

- a significant decrease in all operating costs, while improving system performance and reliability;
- more efficient treatment processes, leading to water quality improvement;
- short intervention time in case of equipment/network failure, targeting a continuous water supply to the population;

Fig. 6.10 SCADA in the Cluj-Napoca central dispatcher (photo by courtesy of CAS S.A.)

- permanent real-time monitoring of various technological parameters and energy consumption;
- increased productivity of the pumping stations;
- user assistance in elaborating technical analyses;
- offering valuable information to the decision makers for taking optimal decisions;
- continuous recording of data, e.g. no need for operators to manually take meter readings;
- remote accessed through an internet connection on the desktop computer, laptop, smartphone or tablet.

6.5 Multiple Criteria Decision-Making Paradigm

In the urban water services, the most important objective is to have a good and reliable infrastructure. As many water pipe systems are approaching the end of their lifespan, municipalities and water companies are facing a complex task: choosing a proper material for replacement. The same difficult choice also appears when designing entire new water systems.

Choosing the right water pipes is an extremely important decision, which in Romania early after 1990 was sometimes a game of chance, in the absence of specific regulations [14], combined with the interests of both material suppliers and (quite often) senior officials of local government or water system operators.

However, during the last decade, in order to fight corruption and to ensure a clean and transparent decision-making process, new approaches were adopted. For example, multi-criteria decision making (MCDM) or multi-criteria decision aid (MCDA) is a discipline aimed at supporting decision makers when faced with numerous and sometimes conflicting evaluations. MCDA manages to highlight the complexity of those irreconcilable variables and provide a way to come to a "best" compromise, in a transparent process.

Application of these methods generally require the following steps [7]:

- choice of a set of possible options (alternatives);
- creation or acquisition of attributes/criteria that are applied to all identified options;
- allocation of scores that reflect the impact or "weight" of each attribute/criterion.

There are many MCDA/MCDM methods in use today. They are not "miracle" tools, in the sense that sometimes, each decision maker has to feed them with subjective information. Therefore, the results are not always unique because they are dependent on decision maker's preferences. If the results are arguable, new scenarios may be taken into account, by negotiating on those preferences.

The American school uses the *full aggregation* approach: "a score is evaluated for each criterion, and these are then synthesized into a global score. This approach assumes compensable scores, i.e. a bad score on one criterion is compensated for by a good score on another" [11].

The French school uses the *outranking* approach: a bad score may not be compensated for by a better score. The order/rank of the alternatives may be partial because the notion of incomparability is allowed. Two alternatives may have the same score, but their behavior may be different, and therefore they remain incomparable [11].

From the multitude of MCDA methods available, the *Electre 1* method (French school) was preferred for our choice problem: the selection of the water pipe material. The ELECTRE acronym comes from the French "Elimination Et Choix Traduisant la Realité", which means *elimination and choice expressing the reality* [9].

Electre 1 follows a partial aggregation procedure, using the outranking relationship as a basis, together with concordance and discordance indexes and finally with sensitivity analysis. The method is based on a pairwise comparison of alternatives, done systematically, criterion by criterion, following the Condorcet principle (seen as a milestone for the outranking theory):

An action X outranks an action Y when X is *at least as good* as Y with respect to the majority of criteria, without being *substantially less good* than Y with respect to the other criteria. [15]

Therefore, the hypothesis "X outranks Y" involves simultaneously a *concordance condition* and a *non-discordance condition*.

We applied the Electre 1 method for a typical situation when the water operator had to choose "the best" type of pipe material among five competitors:

- Polyvinyl chloride pipe (PVC);
- Steel pipe, spirally welded (STSW);

Table 6.1 Criteria and weights used for the choice of water pipe material

No.	Criterion name	Abbreviation	Weight
1	Hydraulic smoothness	HYDRA	1
2	Mechanical strength	MECH	2
3	Chemical stability	CHEM	1
4	Diameter range	DIAM	1
5	Handling and installation	INST	2
6	Operational problems and service life	LIFE	3
7	Environmental aspects	ENV	1
8	Costs/financial effort	COSTS	3

– Ductile cast iron pipe (DCI);
– Fiberglass pipe (FG);
– Pre-stressed concrete pipe (PC).

They are the so-called "actions" (or *alternatives* or *options*).

In the next step, the decision maker has established what are the criteria that should be used in order to compare the potential actions (competitors). As a general rule, those criteria should be coherent, comprehensive and non-redundant. For our case, Table 6.1 summarizes the eight criteria taken into account and their weights.

It is important to point out that each criterion may include other sub-criteria, so the discussion is far more complicated than it seems. An aggregation procedure may take place. Regarding the weights, they represent the importance of a certain criterion for the decision maker, in this case, the privileged ones are "Life" and "Costs".

Apparently, the number of criteria, their exact meaning and their weight come from a negotiation process with all the actors involved in the decision process. Each decision maker has some degree of subjectivity when it comes to choosing the weights.

Every MCDA procedure needs as input data, the evaluation of the alternatives against the criteria. Therefore, the next step was to evaluate each option with respect to each criterion. Depending on the criterion, the assessment may be objective with respect to some scale of measurement (e.g. money) or, it can be subjective, reflecting the personal opinion of the evaluator. In the end, a so-called *performance matrix* was obtained.

The performance of each alternative with respect to each criterion is usually expressed by a grade, e.g. very good, good, medium, satisfying, unsatisfying. The entire procedure for converting the grades into numbers which are then used for computing the concordance and discordance indexes is detailed in Schärlig [15].

Based on Schärlig [15], we developed in our department a software tool called AMEL1 [1], which was used to make all the needed indexes calculations and to perform sensitivity analysis in order to identify the "best" water pipe material (Fig. 6.11).

The ELECTRE 1 method can be used to construct a partial ranking and finally choose one of the most promising alternatives. Using the concordance and discordance indexes followed by a sensitivity procedure, this method is based on "common

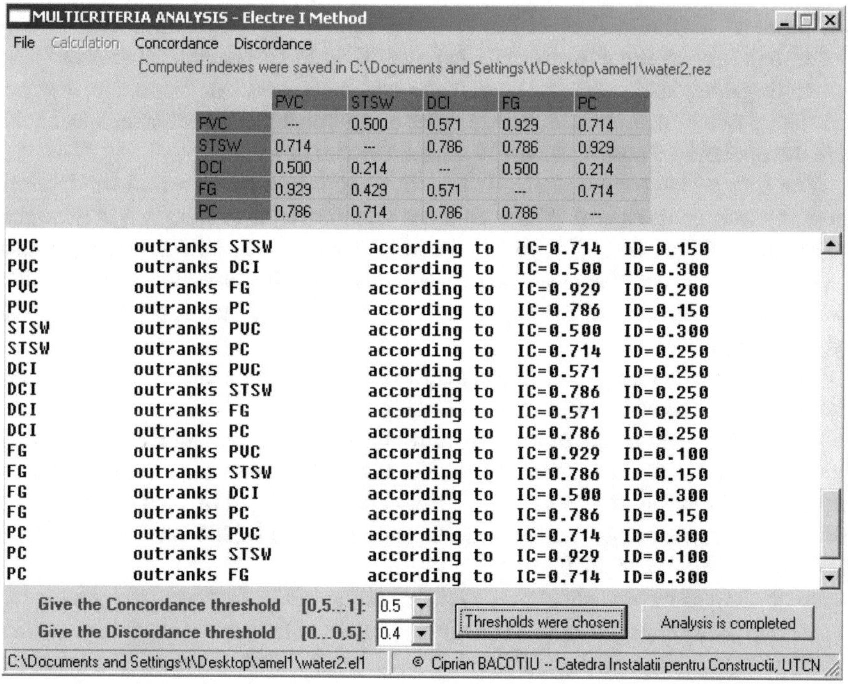

```
MULTICRITERIA ANALYSIS - Electre I Method                                    _ □ ×
File  Calculation  Concordance  Discordance
             Computed indexes were saved in C:\Documents and Settings\t\Desktop\amel1\water2.rez

                         PVC     STSW    DCI     FG      PC
                  PVC    ---     0.500   0.571   0.929   0.714
                  STSW   0.714   ---     0.786   0.786   0.929
                  DCI    0.500   0.214   ---     0.500   0.214
                  FG     0.929   0.429   0.571   ---     0.714
                  PC     0.786   0.714   0.786   0.786   ---

PVC      outranks STSW        according to   IC=0.714   ID=0.150      ▲
PVC      outranks DCI         according to   IC=0.500   ID=0.300
PVC      outranks FG          according to   IC=0.929   ID=0.200
PVC      outranks PC          according to   IC=0.786   ID=0.150
STSW     outranks PVC         according to   IC=0.500   ID=0.300
STSW     outranks PC          according to   IC=0.714   ID=0.250
DCI      outranks PVC         according to   IC=0.571   ID=0.250
DCI      outranks STSW        according to   IC=0.786   ID=0.250
DCI      outranks FG          according to   IC=0.571   ID=0.250
DCI      outranks PC          according to   IC=0.786   ID=0.250
FG       outranks PVC         according to   IC=0.929   ID=0.100
FG       outranks STSW        according to   IC=0.786   ID=0.150
FG       outranks DCI         according to   IC=0.500   ID=0.300
FG       outranks PC          according to   IC=0.786   ID=0.150
PC       outranks PVC         according to   IC=0.714   ID=0.300
PC       outranks STSW        according to   IC=0.929   ID=0.100
PC       outranks FG          according to   IC=0.714   ID=0.300      ▼

    Give the Concordance threshold   [0,5...1]:  0.5 ▼
    Give the Discordance threshold   [0...0,5]:  0.4 ▼    Thresholds were chosen    Analysis is completed

C:\Documents and Settings\t\Desktop\amel1\water2.el1    © Ciprian BACOTIU -- Catedra Instalatii pentru Constructii, UTCN
```

Fig. 6.11 Searching for the "best" water pipe material with AMEL1 software

sense" and may be a valuable tool for any decision maker. Several scenarios can be run, allowing the decider to better understand his alternatives.

It could be stated that in this particular case there is no miracle material, i.e. there is no absolute winner for any situation. Ductile cast iron won by a very small margin. Several future scenarios may be taken into account, by slightly reconsidering some grades or some weights in the performance matrix.

6.6 Steps to the Future

For the further development of the water sector in the Cluj-Sălaj region, long-term investment programs have been promoted, in order to provide the population with adequate water and wastewater services, at the required quality and at affordable rates.

The Operational Program for Large Infrastructure (POIM) was elaborated to meet Romania's needs for development identified in the 2014–2020 Partnership Agreement and in full compliance with the Common Strategic Framework and the Position Document of the European Commission's services. Within the list of predefined projects to be financed by POIM (Priority Axis 3, Specific Objective 3.2) we found the present "**Regional Project for the Development of Water and Wastewater Infrastructure in the counties of Cluj and Sălaj in the period 2014–2020**". By promoting integrated water and wastewater systems in a regional approach, the cost-efficiency maximization is aimed, in order to optimize the global investment costs and the operational costs induced by such investments.

The first package of measures is targeting **the rehabilitation and modernization of water sources and WTPs**. In order to improve the existing water treatment processes at Gilău WTP (Fig. 6.12), a series of works will be conducted to eliminate the current deficiencies. The capacity of the treatment facilities at the Gilău WTP depends directly on the maximum daily flow rate, which is equal to 2650 l/s or 228,960 m^3/day.

The water source for the treatment plant is Lake Tarniţa, having as backup sources the reservoirs at Someşul Cald and Gilău (Fig. 6.13).

The quality parameters of the treated water must comply with the requirements laid down by the Council Directive 98/83/EC on the quality of water intended for human consumption, transposed in the Romanian legislation by the Law 458/2002, with subsequent modifications and additions (the Law 311/2004 and Ordinances no. 11/2010, no. 1/2011 and 22/2017).

According to the statistical data available at the WTP, the treatment process has been maintained constant, and the quality indicators of the treated water have permanently complied with the legal provisions mentioned above. However, the improved treatment technology will be able to guarantee both the long-term compliance of quality indicators with even more severe requirements of the European and Romanian drinking water legislation and increased operational efficiency.

Reducing the pressure head of raw water at the entrance to the WTP

This was a technologically necessary measure, but at the same time, it allowed to use the water-energy potential to cover part of the station's electricity consumption.

Fig. 6.12 The Gilău WTP (photo by courtesy of CAS S.A.)

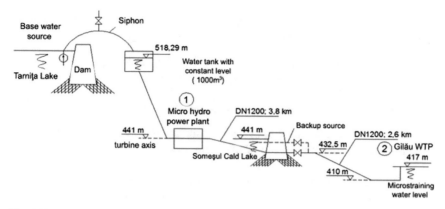

Fig. 6.13 General schematics of the raw water circuit before the WTP (by courtesy of CAS S.A.)

A feasibility study determined the optimal solution, from both technical and financial point of view, to harness the hydrodynamic energy of water before entering the Gilău WTP. The optimal scenario is the installation of a hydro-mechanical device composed of:

– two Francis turbines, with an average annual flow of 1.65 m per second and a net fall of 17 m.
– an asynchronous generator with the nominal power of 290 kW and a nominal voltage of 0.4 kV;
– five valves—for the inlet and outlet of the turbines and the bypass pipe;
– a power room and a control cabinet.

Adjusting the water flow through the turbines can be achieved in two ways:

– through the steering device of the turbine, controlled by a water-level sensor placed at the microstraining station;
– with the help of the floating valve, placed at the microstraining station.

Rehabilitation of the microstraining station

The existing rotating microstrainers are used for the retention of floating material, micro-flora and aquatic micro-fauna and are equipped with a continuous water jet wash system. The microstraining station consists of four sieves with a capacity of 800 l/s per unit.

The microstrainers operate with different rotational speeds, depending on the raw water loading and on the washing process intensity. The drum washing is carried out continuously by three pumps. The water level sensors are mounted on the raw water inlet and in the microstraining chamber.

The proposed improvement works for this treatment stage are:

• replacement of existing microstrainer drums with ones with variable speed motors and stainless steel mesh fabric;

- replacement of the microstrainer's drive system with a system capable of support-ing the weight of the stainless steel fabric, which includes:
  - replacement of the backwashing system, including the pumps;
  - installing a flow meter on the discharge pipe of the microstrainer's washing station;
  - provision of electrical sluice gates to the microstrainers.

- installing an automatic control system for the microstrainers (for automatically controlling the rotational speed of the drum drive depending on the turbidity of raw water);
- installing four turbidity sensors on the microstrainer's supply pipes;
- replacing the valves at the entrance of the microstrainers with electric valves, which can be controlled according to the water level in the distribution chamber located before the microstraining station;
- installing an automatic dehumidification system and restoration of the micros-training station's building.

Rehabilitation of the existing clarifiers

The clarifiers have the role of retaining the suspensions in the raw water, after the coagulation-flocculation process, and concentrating and discharging the resulting sludge. The station is composed of:

- 3 clarifiers with mechanical recirculation of the sludge, each having a 30 m diam-eter and a nominal capacity of 800 l/s, equipped with variable speed mixers and lamellar modules.
- 1 solids contact clarifier with mechanical recirculation, having a 48 m diameter and a nominal capacity of 1000 l/s, equipped with a rotating scraper bridge. From all the three clarifiers, water is gravitationally transported to the filtration lines.

The proposed improvement works for this stage of treatment are:

- the rehabilitation of the structure for the four clarifiers;
- the rehabilitation or replacement of the scraper bridge;
- the rehabilitation of the electrical and automation system for the radial clarifiers, replacing the ordinary electrical control panel with an electrical and automation panel for the control of the scraper bridge;
- replacing the vertical mixers in the reaction chambers (one pair in each reaction chamber) of the three solids contact clarifiers. It is also necessary to replace the electrical panels for the supply and control of the mixers;
- replacing the lamellar modules, as well as their fastening and support structure;
- replacing the valves on the sludge discharge pipes, with electric gate valves, which can be programmed to open at a certain time or depending on the turbidity measured in the sludge pockets;
- replacing the existing flow control valves on the inlet pipe with electrically operated control valves.
- installing turbidity sensors in the clarifier's sludge pockets;

- replacing the sludge recirculation equipment;
- replacing the metallic parts, including the weir and the valves;
- providing electrical lifting devices for the mixers in the reaction chambers and a pivoting crane at 360° for the lamellar modules.

Rehabilitation of the filtration station

Water filtration is a physical process of separating suspended and colloidal particles from water by passing through a granular material (sand). The filters have the role of lowering the settled water turbidity up to the limit of 1 NTU.

The filtration station at Gilău WTP has 15 rapid double flow filter units each having a surface of 60 m^2, grouped into 3 filter lines. The 15 filters are located in three different buildings.

The existing air system for the backwashing of filters (pipelines, fittings, air distributors, etc.) is physically and morally worn-out, with air leaks, which implies the overworking of air compressors and, implicitly, their premature wear.

From the filtration station, the treated water is gravitationally transported to Cluj area by two supply pipes of 1000 mm, respectively 1400 mm diameter.

The proposed improvement works for this stage of treatment are:

- replacing the hydraulic system in the pipe gallery up to the connection with the washing pumps and the air blowers, with stainless steel pipes;
- replacing the existing valves in the pipe gallery, with electro-pneumatic actuators;
- replacing the float valves from both the inlet and outlet of the filters, as well as the hydraulic pipe assembly in the filter unit;
- replacing the filter distributors with stainless steel distributors;
- providing an automatic level control system in the filters by installing ultrasonic level sensors at each filter and automating the inlet and outlet valves in the filters;
- replacing the air system, including the air compressor;
- providing an automatic system for filter operation and backwashing and rehabilitation of the filter control panels;
- installing a dehumidification system in the filter room;
- providing a sand transportation system for the replacing or discharging of sand from the filters, consisting of a configurable conveyor belt, a giraffe crane and a sand pump.
- rehabilitation of the building.

Building a pre-oxidation station using chlorine dioxide

The provision of the pre-chlorination stage using chlorine dioxide will eliminate large quantities of organic substances that may occur at certain times, avoiding the formation of trihalomethanes.

For pre-chlorination at the Gilău WTP, chlorine dioxide solution has to be prepared on site from chlorine gas and sodium chlorite ($NaClO_2$). Gaseous chlorine can be delivered by vacuum from the chlorine tanks storage room, which also supplies chlorine gas for post-chlorination. The preparation of chlorine dioxide will be done with two generators, having a production capacity of 5 kg ClO_2/h each. The pre-oxidation process can be operated and monitored by the SCADA system.

Water recovery from filter backwashing and thickening of sludge

The water resulting from the backwashing of the filters will be collected in a radial clarifier provided with a scraper bridge. The settled water will be collected and pumped, through a new pumping station, back into the treatment chain, before the microstraining station.

The sludge from the existing clarifiers and recovered from the radial settler of the filter backwashing water will be gravitationally collected in a storage and thickening tank, equipped with a scraper bridge and a weir for discharging the supernatant. The supernatant, from both gravitational thickening and mechanical dewatering, will be transported to a supernatant pumping station, from where it will be pumped upstream of the microstraining station.

Sludge dewatering

The sludge resulted from the mechanical thickening tank will be collected in a storage and homogenization buffer, from where it will be transferred to a sludge dewatering equipment—a press filter. The minimum dry matter content of mechanically dehydrated sludge shall be at least 35%.

Laboratory and laboratory equipment

A new laboratory building will be developed for the new operating conditions of the treatment plant. The building will have three levels and will include a chemistry laboratory and a microbiology laboratory.

Building a pilot treatment station

At certain times the aluminum concentration in the raw water entering the treatment plant increases significantly, sometimes above the accepted limit. In order to find an efficient solution for the removal of all pollutants, including aluminium, special treatability studies are required, which must be carried out for at least one year, in order to analyze the studied parameters in all characteristic periods that could influence the quality of the water. For this purpose, a pilot station is proposed with a flow rate of 1 m^3/h.

Flow monitoring at the outlet of the WTP

In order to facilitate production balances and calculate the technological water consumption, the installing of flow meters at the outlet of the treatment plant is proposed, on both supply pipes. The flowmeters shall be ultrasonic, with an accuracy of $\pm 1\%$ of the flow and shall be located where access can be made easily, at a distance of approximately 1 km downstream of the WTP. The flowmeters will be mounted according to the manufacturer's specifications and integrated into the SCADA system.

Upgrading the SCADA system at Gilău WTP

Although at the Gilău WTP there is a functional SCADA system, it is obsolete and inadequate for such critical infrastructure. From a hardware point of view, a completely new SCADA structure is required. The optimum dosage of chemicals to be administered will be permanently calculated and adjusted by the station's laboratory, guided by a SCADA system. The doses of chlorine for the final disinfection process will be adjusted automatically, depending on the free chlorine input value, the plant being equipped with an on-line analyzer of residual chlorine in the drinking water.

Another package of measures is dedicated to the **extending of the existing water network**:

In 2016, the water supply connection rate in the area Cluj-Sălaj allocated to the Someş Water Company (Table 6.2) is about 79% (of which 65% comply in terms of continuous supply of water meeting the required quality parameters).

After the investments made through this project, the connection and compliance rate will increase to 95% in 2022 (Table 6.3), which represents a supplement of 242,429 inhabitants that will be supplied with drinking water from compliant sources at the end of the project.

Other **future measures** envisaged in the POIM project:

- Taking into account the extension of the operating area—both of the water supply system and of the wastewater system, part of the existing operational buildings at the branch level do not meet the future needs, so it is necessary to supplement the equipment used for interventions and maintenance;
- At Someş Water Company there is no Integrated Information Management System similar to those used by other companies in other countries. In the context of the current evolution of information and innovation technology, it is imperative to optimize the operation and to achieve the performance indicators for both

Table 6.2 Compliant water supply services in the area of the project—before the project

	Before the project (2016)						
	Population 2016 (according to National Statistics Institute)	Water supply services complying with the Directive 98/83/EC		Population connected to water supply services not complying with the Directive 98/83/EC		Population not connected to water supply services	
	No.	No.	Percentage (%)	No.	Percentage (%)	No.	Percentage (%)
Total	807,413	528,212	65	115,727	14	163,474	21
Urban	*487,048*	*399,273*	*82*	*79,906*	*16*	*7869*	*2*
Rural	*320,365*	*128,939*	*40*	*35,821*	*11*	*155,605*	*49*

Source CAS S.A.

Table 6.3 Compliant water supply services in the area of the project—after the project

	After the project (2022)			Water supply services from other sources of funding	
	Estimated population in 2022	Water supply services complying with the Directive 98/83/EC			
	No.	No.	Percentage (%)	No.	Percentage (%)
Total	808,608	770,641	95	37,967	5
Urban	484,137	483,197	100	940	0
Rural	324,471	287,444	89	37,027	11

Source CAS S.A.

the existing and the future infrastructure implemented through SOP ENV, ISPA, MUDP and other sources of financing;

- The current SCADA system implemented within the SOP ENV investments 2007–2013 can no longer integrate other objectives. Thus, in order to ensure efficient operation and the integration of the new objectives, it is necessary to extend the existing SCADA system within the works financed by the POIM— Large Infrastructure Operational Program (2014–2020);
- In 2016, the number of system breakdowns on water networks was equal to 5380, due to the high wear of the asbestos cement, cast iron and steel pipes with exceeded lifetime. It is necessary to rehabilitate 7.5 km of water transportation pipes and to replace 132.2 km of water mains and networks, which represent 4.2% of the total length of water supply networks. It is also necessary to replace the High-Density Polyethylene (HDPE) and PVC pipes that do not provide the necessary hydraulic capacity;
- The water balance at the operator level before and after the project is presented in Table 6.4. Loss of drinking water due to system breakdown is more than 40% of the intake water (unfortunately). Thus rehabilitation measures are needed, especially for the networks where the water loss is high, as well as measures to improve the operator's capacity for the identification of losses.

6.7 Conclusions

According to the National Statistics Institute [13], in 2016, in Romania the population connected to the public water supply system was 12,853,110 persons (65.2% of the total population). The majority were living in urban areas, i.e. 10,040,392 persons (94.9% of the urban population), and the remainder of 2,812,718 persons were living in rural areas (30.8% of the rural population). These numbers show that Romania still has to solve a lot of problems in this domain, even though during the last decade some achievements have been made. Unfortunately, there is a huge gap between

Table 6.4 Water balance before and after the project

Description	Units of measure	Before the project (2016)	After the project (2022)
System input raw water volume	m³/year	62,687,524	71,553,700
Domestic water consumption	m³/year	23,377,484	29,496,834
Non-domestic water consumption	m³/year	9,016,248	9,602,128
Unbilled authorized consumption	m³/year	5,114,911	5,084,661
Water losses	m³/year	25,178,881	27,370,077
	%	40.17%	38.25%
Non-revenue water	%	48.33%	45.36%
Connected population	Inhabitants	643,939	770,641
Domestic specific water consumption	l/person/day	99	105
Total specific consumption (including water losses)	l/person/day	267	254

Source CAS S.A.

urban areas and rural areas concerning the access to public water supply systems. Slowly, things are changing, because urban water operators tend to expand their activity in the nearby rural areas, but this will only be possible with big investments, based mainly on EU structural funds.

In Cluj-Sălaj region, we have described the extraordinary evolution of a robust water enterprise, Someș Water Company (CAS S.A.), evolution oriented towards efficiency. As a solid proof, in 2018 CAS S.A. has obtained the *second place* in the national ranking of the lowest price per cubic meter of drinking water delivered to the consumers. This is a remarkable performance, taking into account the fact that Cluj water quality was maintained to exceptional levels, in compliance with all standards and regulations.

6.8   Recommendations

Like CAS S.A., every modern local or regional water operator should be permanently concerned to fulfil its mission [4], by taking specific sectoral measures:

– providing appropriate water supply to localities and citizens;
– increasing the number of customers connected to the centralized drinking water supply system;

- extending the existing centralized drinking water supply system;
- permanently increasing the quality of water and services;
- finding new investment financing sources;
- integrating environmental and quality management systems;
- improving operational and financial performances;
- avoiding an uncontrolled increase in the drinking water price;
- reducing consumptions and network water losses;
- increasing the use of information technology/automation systems;
- permanently improving customer relations;
- decreasing the duration of repairs;
- increasing the reaction speed to emergencies and complaints;
- permanently improving employees' skills;
- complying with the European standards concerning the environment.

The future will bring new specialized technologies such as the Water Management Suite (WMS), by integrating and connecting to complex databases several applications like GIS, SCADA, Maintenance Management Systems, Laboratory Information Management Systems, etc. The Smart City concept with "intelligent" pipes and valves is also very promising.

However, despite all the progress in technology, the danger of a drinking water crisis is lurking before us. Therefore the most important message to pass to the future generations is *"Please cherish water, don't waste it!"*

Acknowledgements The authors would like to express their gratitude to the CAS S.A. (Someş Water Company) for supporting this work and providing all the necessary information.

References

1. Bacotiu C, Vitan E, Hotupan A, Kapalo P (2017) Pipe material selection for urban sewerage systems. In: No Conference Proceedings of Scientific Papers, CASSOTHERM 2017, Civil Engineering Faculty, Institute of Building and Environmental Engineering, Kosice, Slovak Republic
2. CAS S.A., Someş Water Company (2018) History. Retrieved from http://www.casomes.ro/en/?page_id=1769. Accessed on 21 May 2018
3. CAS S.A., Someş Water Company (2018) Raport de activitate al Companiei de Apă Someş S.A. pentru anul 2017 (Activity Report of Someş Water Company for 2017). Retrieved from http://www.casomes.ro/wp-content/uploads/2014/12/Raport-activitate-CA-SOMES-pt.-2017.pdf (in Romanian). Accessed on 22 May 2018
4. Croitoru V-L (2018) Someş water company. Presentation brochure, Cluj-Napoca, Romania
5. Croitoru V-L, Ciatarâş D, Neamţu C (2010) Cluj water supply between past and future. In: Air and water components of the environment. Presa Universitară Clujeană, Cluj-Napoca, pp 232–239. Retrieved from http://aerapa.conference.ubbcluj.ro/2010/Croitoru_Ciataras_Neamtu.htm. Accessed on 22 May 2018
6. Dobriceanu M, Bitoleanu A, Popescu M, Enache S, Subtirelu E (2008) SCADA system for monitoring water supply networks. WSEAS Trans Syst 10(7):1070–1079

7. Dodgson JS, Spackman M, Pearman A, Phillips LD (2009) Multi-criteria analysis: a manual. Department for Communities and Local Government, London. Retrieved from http://www.communities.gov.ukcommunity. Accessed on 18 May 2018
8. ESRI (2007) Enterprise GIS for utilities—transforming insights into results, An ESRI ® White Paper. Retrieved from https://www.esri.com/library/whitepapers/pdfs/enterprise-gis-for-utilities.pdf. Accessed on 18 May 2018
9. Figueira J, Mousseau V, Roy B (2016) Electre methods. In: Figueira J, Greco S, Ehrgott M (eds) Multiple-criteria decision analysis. State of the art surveys, International series in operations research and management science. Springer, New York, pp 133–162
10. Ginther P (2007) Use of GIS growing in the municipal water, wastewater business. Retrieved from https://www.waterworld.com/articles/print/volume-23/issue-4/editorial-feature/use-of-gis-growing-in-the-municipal-water-wastewater-business.html. Accessed on 21 May 2018
11. Ishizaka A, Nemery P (2013) Multi-criteria decision analysis: methods and software. Wiley, Chichester
12. Meyers JR (2005) GIS in the utilities. In: Longley PA, Goodchild MF, Maguire DJ, Rhind DW (eds) Geographical information systems: principles, techniques, management and applications. Wiley, Chichester, pp 801–818
13. National Statistics Institute (2017) Press Communique no. 244 from 29.09.2017, regarding water supply in 2016, (in Romanian). Retrieved from http://www.insse.ro/cms/sites/default/files/com_presa/com_pdf/distributia_apei16r_0.pdf. Accessed on 18 May 2018
14. Popescu A, Onofrei M, Kelley C (2016) An overview of European good practices in public procurement. East J Eur Stud 7(1):81–91. Retrieved from http://ejes.uaic.ro/articles/EJES2016_0701_POP.pdf. Accessed on 20 May 2018
15. Schärlig A (1985) Décider sur plusieurs critères: panorama de l'aide à la décision multicritère. Presses polytechniques et universitaires romandes, Lausanne (in French)
16. Shamsi UM (2005) GIS applications for water, wastewater, and stormwater systems. CRC Press, Boca Raton
17. Truță L (2005) Apa Clujului (The waters of Cluj). Studia, Cluj-Napoca (in Romanian)
18. Truță L (2007) The water museum. Studia, Cluj-Napoca
19. Vlaicu I, Hațegan I (2012) Alimentarea cu apă a Timișoarei: istorie, prezent și perspective (Water supply of Timișoara: history, present and perspectives). Ed. Brumar, Timișoara (in Romanian)

Part IV
Antropic Influence to Water Resources

Chapter 7
The Vulnerability of Water Resources from Eastern Romania to Anthropic Impact and Climate Change

Ionuț Minea

Abstract The regional and local strategies of the past decade have attempted to introduce the principles of sustainable development, which would allow Romania to both meet economic demands and face increasingly pronounced climate change. With a surface of 20,569 km^2 and a population of 2.2 million, the eastern part of the country, spread between the rivers Siret and Prut and sharing borders with the Ukraine (to the north) and the Republic of Moldova (to the east). All the region is classified as vulnerable from an economic and social point of view, but also as far as the impact of climate change upon water resources is concerned. Anthropic activities and climate change are triggering the modification of the hydrological regime (in both quantity and quality) and an increase in the severity of issues associated with water bodies shared with other countries (such as the river Prut, which acts as the border between Romania and the Republic of Moldova), thus rendering an already overexploited resource even more vulnerable. The conclusions of scientific analyses aimed at assessing the effects of climate change upon water resources allow an evaluation of their degree of vulnerability in the face of irreversible challenges. The trend toward phenomena such as global warming, evaluated for the region at 0.2–0.3 °C for the last 50 years, together with an increase in the frequency of extreme temperature values and precipitation volumes emphasize the degree of vulnerability of water resources to current climate change.

Keywords Anthropic impact · Climate change · Eastern Romania · Vulnerability · Water resources

7.1 Introduction

Water sources may be rendered vulnerable through both natural and anthropic mechanisms, which interfere in the hydrological cycle, modifying the natural regime of the water volumes of aquatic bodies. The anthropic mechanisms which lead to water

I. Minea (✉)
Department of Geography, Faculty of Geography and Geology, "Alexandru Ioan Cuza" University of Iasi, Iasi, Romania
e-mail: ionutminea1979@yahoo.com

© Springer Nature Switzerland AG 2020
A. M. Negm et al. (eds.), *Water Resources Management in Romania*, Springer Water,
https://doi.org/10.1007/978-3-030-22320-5_7

resources becoming vulnerable involve changes in the natural flow regime and evolution of hydrographic networks and groundwater bodies through reservoirs, derivations from one hydrographic basin to another, desiccation, aqueducts or the collection of large volumes of water for economic and social purposes [36]. The changes that such anthropic actions bring to water bodies are both quantitative and qualitative. In areas of water scarcity, large volumes of water are translated across seasons or from neighboring areas, the anthropic impact becoming exponential [16, 43, 48]. Eastern Romania is such an area, its water deficit during the summer months being counteracted with water volumes either translated from one season to another, or from neighboring basins, through various catchments [24]. The role of the latter is, firstly, to ensure a certain volume of water during periods of pluviometric deficit and, secondly, to minimize the effects of extreme hydrologic phenomena (flash floods, floods, drought) [44, 46, 49].

Onto this historical background of water management, current climate change adds itself in an increasingly pronounced manner. Through the cumulation of natural mechanisms, climate change gradually leads to increase of natural extreme phenomena. In this region in the category of the most important extreme phenomena we can include aridization, as a result of an increase in the frequency of extreme temperature values and precipitation volumes [7, 9, 20, 52]. Also the concentration of precipitation within limited time intervals lead to an increase in the frequency of catastrophic floods, in maximum annual flow for the hydrographic network and changes in the groundwater level [3, 4, 6, 8, 60]. Such processes are, as expected, ever more noticeable in areas where resource-related vulnerability is added to economic and social vulnerability.

Through both regional and local strategies, general and coherent sustainable development principles are currently being introduced, so that areas with increased vulnerability of water sources to the impact of anthropic activities and climate change are able to meet economic demands and face increasingly significant climate change [10–12, 61].

The issue of water resources and the anthropic impact upon them is of great scientific interest at international level by assessing the pollution of groundwater [19, 28, 50, 53, 54] or evaluation on water balance in different hdyrological extreme condition [31, 32]. In Romania anthropic impact on water resources was focused on wetlands [33, 34], to capitalize on the hydropower potential [35], or evaluation of surface water pollution in different natural condition [2, 14, 37–41, 47, 59]. For Eastern Romania, given the current issue of drought, the emphasis is laid on the exploitation of groundwater sources, particularly within the Moldavian Plateau [5, 17, 23, 25, 29]. To minimize the effects of drought, large-scale ponds or lakes have been created within the Moldavian Plain and on the most significant streams of the area, namely Siret, Prut, Jijia etc. [42, 45].

As a result, the present paper seeks, primarily, to identify the main issues associated with the anthropic impact and climate change upon the water resources of Eastern Romania, taking into account the social and natural characteristics of the region.

7.2   General Features of the Region

The area under study occupies a surface of 20,569 km^2, is inhabited by approximately 2.2 million, and covers the eastern portion of Romania, between the rivers Siret and Prut, bordering Ukraine (to the north) and the Republic of Moldova (to the east). Economic and social studies classify it as vulnerable to an increase in the severity of issues associated with water bodies that act as borders or are shared with other countries (such as the river Prut, which represents the border between Romania and the Republic of Moldova) or certain groundwater bodies.

Tableland shapes dominate the topography, with altitudes between 5 and 450 m. Geologically speaking, sands prevail in the southern part, where the high infiltration rate of the water from precipitation diminishes the runoff. In the center, there is a mixture of sands and clays, which favor landslides and gully erosion [21, 27]. The presence of clay deposits implies a more rapid reaction of the runoff. The changes in land use of the last two centuries, with continuous deforestation, have increased the runoff in this central area, as well [56, 57]. In the north, clay and marl deposits are dominant, while in the northwest clay and sandstone deposits are more frequent [29].

Regarding the climate of the region, semi-dry continental features are dominant, with maximum temperatures and precipitations in the summer (June or July) and minimum values being recorded during the winter (January or February). The mean annual temperature increases from 7 to 8 °C in the north to 9–10 °C in the south. The annual amounts of precipitation range from 620 to 480 mm and decrease from north to south. In the winter, precipitation is mainly solid, and most of it remains in the form of a snow layer until spring because of frequent negative temperatures, which vary from −4 to −6 °C in the north, and −2 to −3 °C in the south [52]. Seasonal variation indicates that the precipitation amounts recorded in the summer are twice to three times higher than those falling during the winter [15, 55].

In Eastern Romania, the discharge values of small rivers (under 50 km in length) are quite low, less than 1 m^3/s annually, and their regime is characterized by a spring maximum, as well as by late summer and early autumn minimum. The spring maximum runoff is based on precipitation and the melting of snow (see Fig. 7.1).

7.3   Methodologies Implemented in the Assessment of Water Resource Vulnerability

The anthropic impact upon water sources can be evaluated through a series of methodologies. Given, however, the specific natural conditions and the social and demographic evolution of the area under analysis, we consider that the water exploitation index (WEI) is the most suitable in this respect. This index refers to the degree of strain that anthropogenic activities exercise upon natural water sources within a certain space (hydrographic sub-basin, hydrographic basin, national territory or region),

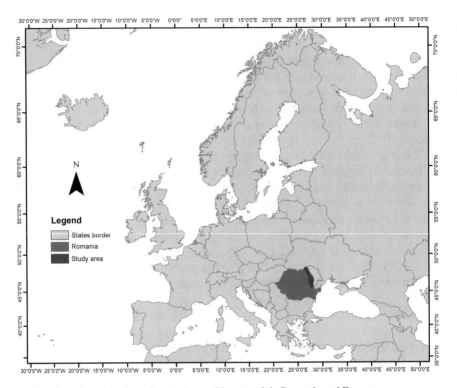

Fig. 7.1 Location of the Siret-Prut hydrographic network in Romania and Europe

and is used to identify areas prone to water deficit. Mathematically speaking, the former is calculated as the ratio between the mean water demand and the long-term water resources of a region. The index is also employed to highlight the anthropic impact upon water sources through their use in various sectors. Developed by the European Environmental Agency, it was then translated from the national level and to that of river basin or sub-basin (WEI+) [13].

If the index is below 10%, then the water sources are not subjected to anthropic strain. If it is between 10 and 20%, then the water sources are exposed to reduced anthropic strain, while values between 20 and 40% suggest increased anthropic strain. Values of the WEI over 40% indicate extreme strain upon water sources, with devastating medium- and long-term effects. The shortest time span taken into account for the calculation of the long-term annual mean of the WEI+ is 20 years. The data sent by Romania to Eurostat between 1990 and 2016 has allowed the calculation of a national water exploitation index and, by applying the WEI+, the indicator was then calculated for the eastern part of the country.

To highlight the impact of climate change upon water sources, 5 series of data were used, namely the seasonal and annual mean values for air temperature, precipitation, river discharge and phreatic level. The climate-related data were acquired at six meteorological stations: Botosani, Iasi, Vaslui, Bârlad, Galati and Roman (for the

1961–2016 interval, under the management of the National Meteorological Administration). The data on river flow were obtained at 17 hydrometric stations located on rivers without hydrotechnical intervention (for the 1961–2016 interval). The values regarding groundwater level were provided by 56 hydrogeological wells (for the 1983–2016 time span, under the management of the Prut-Bârlad Water Basin Administration).

The determination of trends and their slopes was carried out using the nonparametric Mann-Kendal test and the Sen method, developed by researchers from the Finnish Meteorological Institute [51]. The methods in question have been described and widely used to identify the trends of certain climatic and hydrological parameters [7, 58]. In the present study, in order to ensure statistical significance, a value of 0.5 was chosen for the α coefficient.

The correlations between climatic parameters, river flow and variations in the phreatic level were established using the Bravais-Pearson linear correlation coefficient. This index was applied both to the seasonal data series and the annual data series. The pairs of data series were devised based on the minimum distance between meteorological, hydrological and hydrogeological stations, and the same time span (1983–2016) was used for the correlation between precipitation and phreatic level. It was, thus, considered that values of the Bravais-Pearson coefficient of at least 0.5 indicate an acceptable degree of correlation between the data series analyzed, given the spatial variation of natural conditions (topography, geology etc.).

7.4 The Hydrographic Network, Natural and Artificial Lakes and Groundwater Bodies

The total surface of the Siret-Prut hydrographic area represents 8.63% of the territory of Romania. The hydrographic network encompasses 392 streams (according to the official data published by the National Water Administration on www.rowater.ro) [1], with a total length of 7.696 km and a mean density of 0.38 km/km^2. On Romanian territory, the Siret-Prut hydrographic area comprises the middle and lower basin of the river Prut, the hydrographic basin of the river Bârlad, and all the left-side tributaries of the river Siret across the counties of Botoşani, Iași, Vaslui, and Galați (see Fig. 7.2).

The Siret-Prut hydrographic area comprises the following categories of surface water bodies: rivers (natural and highly modified by humans)—with a length of 7.696 km (rivers recorded in official documents). Permanent rivers, with a length of 2.269 km, represent 29.48% of the total of streams, and non-permanent rivers, with a length of 5.427 km, represent 70.52% of the total of streams. To all this are added: 7 natural lakes with surfaces greater than 0.5 km^2, and a natural lake that has been significantly altered anthropically, as well as 72 water accumulations with surfaces that exceed 0.5 km^2, and 262 ponds.

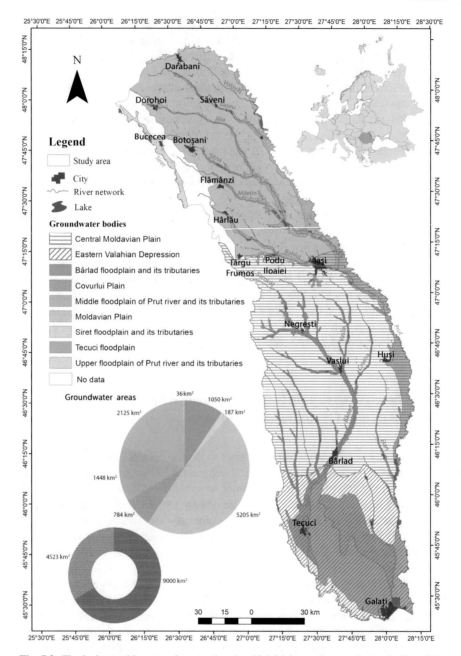

Fig. 7.2 The hydrographic network, natural and artificial lakes and groundwater bodies of the Siret-Prut hydrographic area

The implementation of the principles of the Water Framework Directive EC 2000 has allowed the grouping of surface water bodies (rivers and lakes) based on typology. Thus, 324 surface water bodies have been identified and classified as follows: 230 natural water bodies (223 rivers and 7 natural lakes). Of these, 45 water bodies, corresponding to rivers, have suffered significant anthropic intervention, 1 water body in the form of a lake that has been significantly modified by humans, to which it is added 45 artificial lakes and 3 other artificial water bodies (canals and derivations). The total volume of surface water of the Prut-Bârlad area is around 3.661 M m^3/year, of which roughly 960 M m^3/year is usable. This represents 94% of the total volume and is provided, mainly, by the rivers Prut and Bârlad and their tributaries. The Siret-Prut area also includes 72 significant artificial lakes (with surfaces exceeding 0.5 km^2), of which 49 serve complex purposes and encompass a volume of usable water of 614.85 M m^3.

When calculated in relation to basin population, the specific usable resource is 437.16 m^3/inh./year, while the specific resource, calculated in relation to the volume that is theoretically available (mean and multiannual), is 1.667.12 m^3/inh./year. The water resources of this hydrographic space can, therefore, be considered limited and unevenly distributed across time and space. The mean multiannual discharge values of the main rivers of the area are the following: river Prut 105 m^3/s (3.314 M m^3/year) at its confluence with the Danube, river Jijia 10 m^3/s (316 M m^3/year), river Bârlad 11 m^3/s (347 M m^3/year) at its confluence with the Siret, river Vaslui 1 m^3/s (31.56 m^3/year), river Tutova 1 m^3/s (31.56 M m^3/year). Of the total length of the water bodies of the Siret-Prut basin, the non-permanent streams account for 80% (River Basin Management Plan for the Prut-Barlad hydrographic basin, 2016–2021).

Across the Siret-Prut hydrographic area, 7 groundwater bodies have been identified, delineated and described, of which one is shared with the Republic of Moldova [5]. All 7 belong to the porous type, the water being stored in deposits of Quaternary or Sarmatian-Pontian age. The greatest number of groundwater bodies (6-ROPR01, ROPR02, ROPR03, ROPR04, ROPR06, and ROPR07) has been delineated in the floodplains or on the terraces of the rivers Prut, Bârlad and Siret, being stored in porous and permeable alluvial-fluvial deposits of Quaternary age. Being located close to the surface, they are unconfined, therefore subjected to anthropic strain both quantitatively and qualitatively. The shared groundwater body ROPR05 (Central Moldavian Plateau), stored in Sarmatian-Pontian deposits, is under anthropic pressure, yet has limited economic importance. The groundwater resources are estimated at 251.4 M m^3 (7.97 m^3/s), of which 34.7 M m^3 (1.1 m^3/s) come from phreatic sources, and 216.7 M m^3 (6.87 m^3/s) from sources located at depth.

7.5 Demographic Backgrounds

From an administrative point of view, the Siret-Prut hydrographic area covers almost entirely the counties of Botoşani (90%), Iaşi (83%) and Vaslui (100%), and partially the counties of Neamţ, Bacău, Vrancea, and Galaţi. Its total population is approxi-

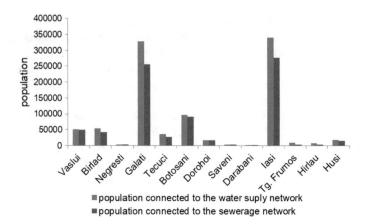

Fig. 7.3 The cities of the Siret-Prut hydrographic area with centralized water distribution and sanitation systems

mately 2.2 million, of which 50.5% live in urban areas. 13 of the 15 cities and towns benefit from a centralized water distribution and sanitation system, and 14 of them also have water treatment stations (see Fig. 7.3). Nevertheless, numerous issues of surface water and groundwater pollution are faced [22].

In the rural areas, only 35% of the communes benefit from a centralized water distribution system, and only 10% have a wastewater treatment plant. The increase in population, coupled with economic activities related to agriculture (pisciculture, livestock farming and the cultivation of arable land occupying approximately 1 million ha) and industry (automotive, industrial equipment, textiles, glass and ceramics, siderurgy), have led to a growing demand for water. As a result, water adductions from neighboring basins have been carried out along with the catchment of significant water volumes from the main hydrographic network (the rivers Siret and Prut) (e.g., the water adduction from the Moldavian basin, located in the vicinity of a mountainous area, to the largest city of the region, Iași, carried out at the beginning of the 20th century, and given new dimensions in the mid-60s),

7.6   The Anthropic Impact upon Water Resources

Such an impact upon the water sources of a densely populated region such as the Siret-Prut hydrographic area manifests itself at various levels (see Fig. 7.4). The first one is related to the changes that occur within the hydrographic network through hydrotechnical constructions, which influence the hydromorphological features of surface water bodies and impact their ecosystems.

Hydrotechnical constructions such as dams, weirs, and bottom sills block the flow of rivers, affecting the hydrological regime, sediment transport and, particularly, the migration of wildlife. Constructions along the banks (levees, regularization and

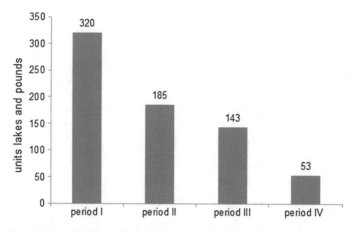

Fig. 7.4 The evolution of accumulations with the Siret-Prut hydrographic area

consolidation works) disrupt the lateral connection between water bodies and their floodplains and reproduction areas, thus altering the natural state of the environment (Figs. 7.5 and 7.6). Consequently, hydro-morphological alterations upon water bodies hinder the migration of fishes, leading to a decline in natural spawning, a reduction in biodiversity and species abundance, as well as changes in the composition of animal populations. There is insufficient knowledge, both at a regional and at a European

Fig. 7.5 Series of lakes on the Bahlueț River

Fig. 7.6 The River Jijia during the floods of the summer of 2010

level, of the relationship between hydro-morphological pressures and their impact, the synergic manner in which the former often act making it difficult to establish a proper connection between the type of pressure and effect.

Another form of anthropic impact is related to the collection and retrieval of water volumes from surface bodies and groundwater bodies, with consequences both on the hydrological regime and the biota of the area.

The volumes of water collected vary according to the socio-economic demands and availability of each year, the amount of water available depending on atmospheric contribution.

The analysis of water collection and retrieval has revealed, in 80% of cases, a drop in the natural discharge values registered at the main hydrometric stations of the region (Fig. 7.7).

The difference between the discharge values measured and those reconstituted for the natural flow regime, calculated taking into account the water lost through collection for various purposes, is 6.8 m³/s, obviously with higher values for the main rivers. For example to the river Prut, a value of 4.3 m³/s was calculated as different from the natural flow, which indicates that over 60% of the total discharge is lost within the hydrographic network of the area. At the same time, the water from sewage systems or from other sources leads to an increase in discharge at certain hydrometric stations. The values summed up for the entire region exceed 4.5 m³/s, substantially modifying the hydrological regime, particularly in the case of rivers that flow through urban agglomerations.

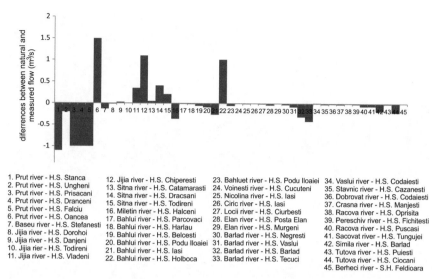

1. Prut river - H.S. Stanca
2. Prut river - H.S. Ungheni
3. Prut river - H.S. Prisacani
4. Prut river - H.S. Dranceni
5. Prut river - H.S. Falciu
6. Prut river - H.S. Oancea
7. Baseu river - H.S. Stefanesti
8. Jijia river - H.S. Dorohoi
9. Jijia river - H.S. Danjeni
10. Jijia rier - H.S. Todireni
11. Jijia river - H.S. Vladeni

12. Jijia river - H.S. Chiperesti
13. Sitna river - H.S. Catamarasti
14. Sitna river - H.S. Dracsani
15. Sitna river - H.S. Todireni
16. Miletin river - H.S. Halceni
17. Bahlui river - H.S. Parcovaci
18. Bahlui river - H.S. Harlau
19. Bahlui river - H.S. Belcesti
20. Bahlui river - H.S. Podu Iloaiei
21. Bahlui river - H.S. Iasi
22. Bahlui river - H.S. Holboca

23. Bahluet river - H.S. Podu Iloaiei
24. Voinesti river - H.S. Cucuteni
25. Nicolina river - H.S. Iasi
26. Ciric river - H.S. Iasi
27. Locii river - H.S. Ciurbesti
28. Elan river - H.S. Posta Elan
29. Elan river - H.S. Murgeni
30. Barlad river - H.S. Negresti
31. Barlad river - H.S. Vaslui
32. Barlad river - H.S. Barlad
33. Barlad river - H.S. Tecuci

34. Vaslui river - H.S. Codaiesti
35. Stavnic river - H.S. Cazanesti
36. Dobrovat river - H.S. Codaiesti
37. Crasna river - H.S. Manjesti
38. Racova river - H.S. Oprisita
39. Pereschiv river - H.S. Fichitesti
40. Racova river - H.S. Puscasi
41. Sacovat river - H.S. Tungujei
42. Simila river - H.S. Barlad
43. Tutova river - H.S. Puiesti
44. Tutova river - H.S. Ciocani
45. Berheci river - S.H. Feldioara

Fig. 7.7 Difference between natural and measured flow in the Siret-Prut area

The collection of water for economic activities is divided into various sectors (Figs. 7.8 and 7.9). The smallest quantities are used in livestock farming, with mean annual values of 2.1 M m^3, and the irrigation of agricultural land, with mean annual values of 90 M m^3. The various forms of industry require the largest amounts of water in the region, with mean annual values of over 140 M m^3. In the Siret-Prut area, the volume of water collected annually for economic purposes reaches a mean of 450 M m^3, with variations based on natural conditions and the dynamics of the activities requiring it.

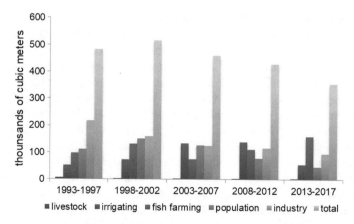

Fig. 7.8 Water collection in the Siret-Prut area, on sectors of economic activity over the period 1993–2017

Fig. 7.9 The Soleşti accumulation, meant to supply water to the city of Vaslui and to minimize flooding

7.7 The Evaluation of the Anthropic Impact

The official data published by the European Environmental Agency has revealed that Romania suffers from relatively low water stress/deficit, the mean annual value of the WEI+ being around 19.6%, with a minimum of 15.2% recorded in 2013, and a maximum of 41.4% recorded in 1990.

The same type of methodology was applied to Eastern Romania. The region was divided into hydrographic basins corresponding to the typology suggested by the Water Framework Directive EC 2000. The lake units were taken from the vectorial states of Corine Land Cover (Fig. 7.10), while the morphometric data regarding depth and volume were provided by the Prut-Bârlad Water Basin Administration (River Basin Management Plan for the Prut-Bârlad hydrographic basin 2009–2015).

The results obtained are different from the national ones, given the natural conditions and demographic background of the region. The mean annual water resources for the entire region reach values of up to 5185 M m^3. During years with drought, these resources are considerably diminished, down to under 2000 M m^3. As illustrated (in the previous chapter, see Fig. 7.8), the amount of water collected varies between 350 and 525 M m^3. As a result, for the entire region, the values of the WEI+ from 1990 to 2016 are between 5.4 (1994) and 32.5 (2000) (Fig. 7.11). The greatest pressure upon water resources is exercised during the dry years (2000, 2007, 2012), when the values of the WEI+ index exceeded 30. The mean value of the WEI+ is 15.7,

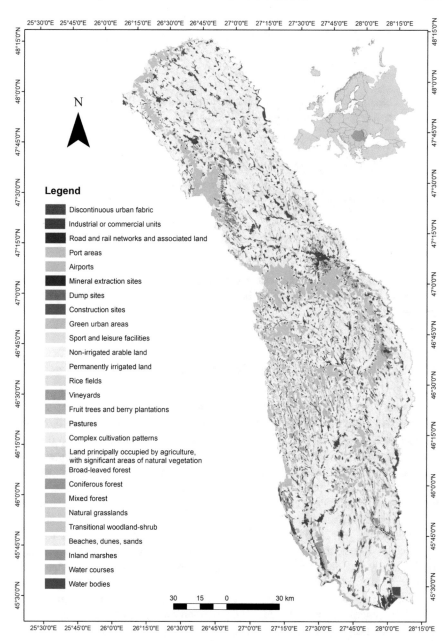

Fig. 7.10 Land use in Eastern Romania (from Corine Land Cover 2012, EEA)

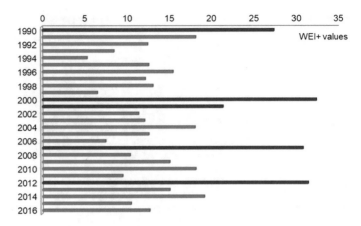

Fig. 7.11 Variation of WEI+ values in Eastern Romania between 1990 and 2016

indicated limited pressure upon water resources, but with the potential of increasing in severity over the following years because of growing demand and the effects of climate change.

7.8 The Impact of Climate Change upon Water Resources

The rapid economic changes caused by increases in the global population and the rate at which resources are being depleted by humanity, by the constant technological advance and socio-political shifts, also imply a series of modifications brought to natural components. The most important one is related to climate change due to the greenhouse effect, which will have a significant impact on the environment and socio-economic activities. Global warming processes have led to an increase in the frequency and amplitude of extreme events that impact the evolution and management of water resources. According to "The annual report on the state of the environment in Romania for the year 2016" [26], climate change will have direct effects upon sectors such as agriculture, forestry, water management. These will lead to an increase in the frequency and intensity of extreme meteorological and hydrological phenomena (storms, floods, drought).

Climate change in Romania follows the global pattern, with certain regional particularities regarding air temperature and precipitation [9]. These modifications have also been highlighted within the *ADER—A system of geo-referential indicators at different spatial and temporal scales for the evaluation of the vulnerability and coping mechanisms of agroecosystems in the face of climate change (2011–2014)* project, devised by the National Weather Administration and financed through the Sectorial Plan for Research and Development in Agriculture and Rural Development 2011–2014 and ADER 2020, coordinated by the Ministry for Agriculture and Rural Development. In this project some climatic scenarios were devised for 2021–2050, compared to 1961–2010, along with the quantifiable effects upon the mean multi-annual temperature and precipitation in Romania. Thus, for 2020–2029, a series of modifications to climatic parameters, similar to those occurring at European level, are expected nation-wide, with increases in mean annual temperature between 0.5 and 1.5 °C. As far as the precipitation regime is concerned, the analyses for the 1901–2010 interval indicate a nation-wide trend, particularly after 1961, towards a decrease in the annual quantity of precipitation, along with a pronounced increase in precipitation deficit for southern and eastern Romania.

Croitoru and Minea [8] and Minea and Croitoru [24, 25] have revealed the effects of climate change upon river flow and phreatic level in the eastern part of the country. The nonparametric Mann-Kendal test, used to determine the trends of the main climatic parameters (temperature and mean seasonal and annual precipitation), hydrological parameters (mean seasonal and annual flow) and hydrogeological ones (mean monthly and annual values of the piezometric level). The following aspects have been identified for the region under study:

– A warming trend is obvious, given the rise recorded in mean annual air temperature values at most of the meteorological stations of the region. The rise rate varies between 0.039 and 0.181 °C/decade. It is noteworthy that the air temperature values recorded during the winter and spring months are statistically significant, in agreement with the observations regarding the rise in air temperature in the Northern Hemisphere made in the 2014 IPCC report.
– There is an increase in the annual values of atmospheric precipitation, yet this increase does not have statistical significance. When analyzing data recorded during different seasons, one can remark that the highest slopes are specific for summer, the season with the highest amounts. Almost half of the slopes recorded during the summer season are also statistically significant, varying from 10.255 to 14.115 mm/decade (Table 7.1). Positive trends are also specific for spring and

Table 7.1 Slopes of temperature (°C/decade) and precipitation (mm/decade) trends in Eastern Romania over the period 1960–2010

Series	Botosani	Iasi	Roman	Vaslui	Barlad	Galati
Annual (°C/decade)	**0.147**[a]	**0.105**	**−0.234**	**0.114**	0.039	**0.181**
Annual (mm/decade)	13.499	14.655	**13.923**	15.225	−3.315	6.750
Winter (°C/decade)	0.238	0.088	**0.653**	0.043	0.048	0.308
Winter (mm/decade)	**−3.562**	**−3.956**	**−4.163**	−3.738	**−6.218**	−4.779
Spring (°C/decade)	**0.282**	**0.306**	**0.477**	**0.258**	0.128	0.230
Spring (mm/decade)	2.408	3.433	2.750	**8.099**	1.460	2.804
Summer (°C/decade)	0.043	0.111	**−0.333**	0.101	0.044	**0.333**
Summer (mm/decade)	**13.513**	6.310	**14.115**	**10.255**	2.381	7.139
Autumn (°C/decade)	0.023	−0.020	−0.103	−0.152	0.059	−0.027
Autumn (mm/decade)	3.891	4.286	3.313	2.265	0.387	4.652

[a]Values in bold are statistically significant

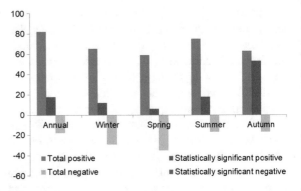

Fig. 7.12 The frequency of positive and negative slopes of river discharge between 1960 and 2015

autumn while decreasing amounts of precipitation characterize winter for all locations considered. The majority have significant slopes, which range from −3.562 to −6.218 mm/decade.

- More than half of the rivers that drain the eastern part of Romania display a trend toward a rise in mean annual and seasonal flow, in agreement with the regional trend also identified in the southern portion of the country [6, 30]. The highest frequency of statistically significant positive trends is recorded in autumn (over 50%), while the statistically significant positive trends of summer and winter are between 10 and 20% (Fig. 7.12). Negative trends, of a decrease in mean seasonal discharge values, with a high percentage during spring (over 30%), have also been identified. They, however, lack statistical significance.
- Given the underground storage conditions, the variations in phreatic level display a delay trend, compared to the water input from precipitation or the hydrographic network. The greater the depth, the less significant is the influence of surface hydric conditions. As a result, starting at depths of 5–6 m, there is no more direct influence from them [24]. The analysis of hydrostatic level values has revealed a slight increasing trend in most of the hydrogeological wells, with a mean value of that 0.3 cm/decade (Fig. 7.13). The explanation lies in the rise in the amounts of precipitation and river discharge values [3], which lead to an increase in the volume of surface water seeping into the underground. The most significant increase occurs at the 200–300 cm depth interval, where the annual increase trend values of the hydrostatic level exceed 20 cm [25]. The trends fade (values under 5 cm) towards the topographic surface and at depths greater than 300 cm, even becoming negative for annual values and those recorded during spring.

The Bravais-Pearson linear correlation coefficient used in the identification of correlations between climatic parameters and river flow indicates a positive connection between the annual data series (in over 50% of cases). The highest values of the coefficient have been established for the central and northern portions of the region (between 0.4 and 0.6). At seasonal level, the best correlations were identified for summer and autumn, when the exclusively liquid precipitation reaches the hydrographic network with ease. The highest values of the coefficient were established for the summer season (0.8), when the rich convective precipitation, which

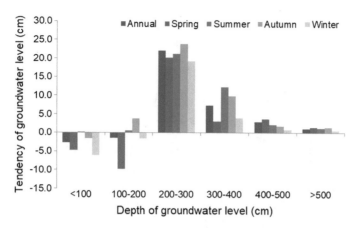

Fig. 7.13 Trends of groundwater level by the depth of drilling in Eastern Romania over the period 1983–2015

sometimes exceeds 100 mm over a few hours, triggers the rapid flow of water downslope, towards river channels. High values of the coefficient were also identified between precipitation and liquid discharge in autumn, explainable through the reduction in vegetative cover, which favors the rapid flow of water towards the hydrographic network. Low values of the correlation coefficient were recorded for winter and spring, when, due to low temperatures, the precipitation is kept in the form of a snow and ice layer, with no influence on river flow.

Given that variations in hydrostatic level are nearly entirely dependent upon the geological features of host rocks and the water input from the surface, the results regarding the correlation between precipitation and groundwater level can vary greatly, even in areas that are hydrogeologically homogeneous. However, by applying the Bravais-Pearson linear correlation coefficient to the seasonal and annual data series, relatively high values (often over 0.3) were obtained for the correlation between precipitation and hydrostatic level for 70% of the datasets analyzed. The geological deposits of the area play an undeniable role, being composed predominantly of sands and silty clays, which favor the infiltration of precipitation, allowing it to reach the water table quickly. As in the case of the correlation between precipitation and river flow, the highest values (with a maximum of 0.6) were obtained for summer and autumn (Fig. 7.14).

The low values calculated for winter and spring (generally under 0.3) are a direct consequence of the reduction in the amount of precipitation fallen throughout the past decades during the two seasons, as well as the water from precipitation being locked in the form of snow and ice. This allows the redistribution of the volume of atmospheric water in the underground across seasons.

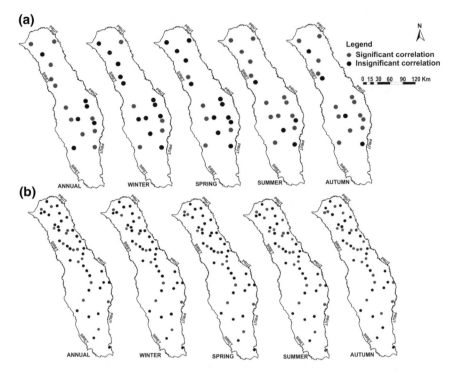

Fig. 7.14 Correlation between precipitation, river flow (**a**) and hydrostatic level (**b**) in Eastern Romania

7.9 Conclusions

Anthropic activities have led to radical changes in the natural conditions under which the water bodies of Eastern Romania have been evolving. Onto them, the regional climatic changes of recent decades have been added, manifesting themselves through modifications in the seasonal thermal and rainfall regime, as well as a secondary trend toward the increase in river discharge and in the underground hydrostatic level. This overlaps with the increasingly negative anthropic impact of the last decades, in the form of significant water collection for the needs of the population, but also for various economic activities (particularly in the industry and the irrigation of agricultural land). These changes require the rethinking of water management systems, both at the surface and in the underground. When devising these systems, one must take into account various climatic scenarios and formulate strategies for the adaptation to climate change both at a regional and at a European level, through the introduction within water management plans of measures regarding the long-term effects of anthropic impact and climate change.

7.10 Recommendations

On the basis of the conclusions of the study on the water resources vulnerability from the eastern part of Romania to anthropic impact and climate change we can make a series of theoretical and practical recommendations. Firstly, it is necessary to include in the local and regional management plans of water resources the scenarios regarding the changes of climatic parameters (both temperature and precipitation) in order to identify the vulnerable areas.

Secondly, in the context of increasing the demand for water for economic and social purposes, it is necessary either to make new large-scale hydro-technical works (such as reservoir lakes) or additions from adjacent water-surplus basins. Last but not least, there is a need for massive investment in expanding water supply systems and waste water collection (especially in rural areas) to reduce river and groundwater pollution.

Acknowledgements The author acknowledges the climatic, hydrological and hydrogeological data provided Moldova Regional Meteorological Center, and Prut-Birlad Basins Branch of Romanian Waters Administration.

References

1. Administrația Națională Apele Române (2018) River Basin Management Plan for hydrographic basin Prut-Barlad 2016–2021. http://www.rowater.ro/daprut/Plan%20management%20bazinal/Forms/AllItems.aspx
2. Banaduc D, Panzar C, Bogorin P, Hoza O, Curtean-Banaduc A (2016) Human impact on Tarnava Mare river and its effects on aquatic biodiversity. Acta Oecologica Carpatica IX:187–197
3. Birsan MV, Zaharia L, Chendes V, Branescu E (2014) Seasonal trends in Romanian streamflow. Hydrol Process 28:4496–4505
4. Blöschl G, Hall J, Parajka J, Perdigão RAP, Merz B, Arheimer B, Aronica GT, Bilibashi A, Bonacci O, Borga M, Čanjevac I, Castellarin A, Chirico GB, Claps P, Fiala K, Frolova N, Gorbachova L, Gül A, Hannaford J, Harrigan S, Kireeva M, Kiss A, Kjeldsen TR, Kohnová S, Koskela JJ, Ledvinka O, Macdonald N, Mavrova-Guirguinova M, Mediero L, Merz R, Molnar P, Montanari A, Murphy C, Osuch M, Ovcharuk V, Radevski I, Rogger M, Salinas JL, Sauquet E, Šraj M, Szolgay J, Viglione A, Volpi E, Wilson D, Zaimi K, Živković N (2017) Changing climate shifts timing of European floods. Science 357(6351):588–590
5. Bretotean M, Macalet R, Tenu A, Tomescu G, Munteanu MT, Radu E, Radu C, Dragusin D (2006) Romania's cross-border underground water bodies. Revista Hidrogeologia, Bucharest 7(1):16–21
6. Croitoru AE, Toma FM (2011) Climatic changes and their influence on stream flow in Central Romanian plain. In: 5th atmospheric science symposium, 27–29 Apr 2011, Istanbul-Turkey, pp 183–194
7. Croitoru AE, Piticar A, Dragotă CS, Burada DC (2013) Recent changes in reference evapotranspiration in Romania. Glob Planet Change 111(December):127–137. https://doi.org/10.1016/j.gloplacha.2013.09.004
8. Croitoru AE, Minea I (2015) The impact of climate changes on rivers discharge in Eastern Romania. Theoret Appl Climatol 20(3–4):563–573. https://doi.org/10.1007/s00704-014-1194-z

9. Croitoru AE, Piticar A, Burada DC (2016) Changes in precipitation extremes in Romania. Quatern Int 415:325–335. https://doi.org/10.1016/j.quaint.2015.07.028
10. Directive 2000/60/EC of the European Parliament and of the council establishing a framework for the Community action in the field of water policy (2000), EU Water Framework Directive 2000/60/EC (2000)
11. Directive 2007/60/EC of the European Parliament and of the council on the assessment and management of flood risks (2007) EU Water Framework Directive 2007/60/EC (2007)
12. Edenhofer O, Pichs-Madruga R, Sokona Y, Minx JC, Farahani E, Kadner S, Seyboth K, Adler A, Baum I, Brunner S, Eickemeier P, Kriemman B, Savolainen J, Schlomer S, von Stechow C, Zwickel T (2015) Contribution of Working Groups III to the Fifth Assessment Report of the Intergovernmental Panel on Climate Change (2014) Climate Change 2014: Mitigation of climate change. Summary for Policymakers. Cambridge University Press, Geneva
13. European Environmental Agency (2017) Water Exploitation Index Plus WEI+ for river basins districts. http://www.eea.europa.eu/data-and-maps/explore-interactive-maps/water-exploitation-index-for-river-1
14. Iosub M, Iordache I, Enea A, Hapciuc OE, Romanescu G, Minea I (2016) Drought analysis in Ozana drainage basin. Aerul şi apa componente ale mediului, pp 392–399
15. Irimia LM, Patriche CV, Quenol H, Sfica L, Foss C (2018) Shifts in climate suitability for wine production as a result of climate change in a temperate climate wine region of Romania. Theoret Appl Climatol 131(3–4):1069–1081. https://doi.org/10.1007/s00704-017-2033-9
16. Jora I, Romanescu G (2010) Hydrograph of the flows of the most important high floods in Vaslui river basin. Aerul şi Apa. Compenente ale Mediului, pp 91–102
17. Jora I, Romanescu G (2011) Groundwater in the hydrographical basin of the Vaslui river. Lucrările Seminarului Geografic "Dimitrie Cantemir" 31:21–28
18. Klein Tank AMG, Wijngaard JB, Konnen GP, Böhm R, Demarée G, Gocheva A, Mileta M, Pashiardis S, Hejkrlik L, Kern-Hansen C, Heino R, Bessemoulin P, Müller-Westermeier G, Tzanakou M, Szalai S, Pálsdóttir T, Fitzgerald D, Rubin S, Capaldo M, Maugeri M, Leitass A, Bukantis A, Aberfeld R, van Engelen AFV, Forland E, Mietus M, Coelho F, Mares C, Razuvaev V, Nieplova E, Cegnar T, Antonio López J, Dahlström B, Moberg A, Kirchhofer W, Ceylan A, Pachaliuk O, Alexander LV, Petrovic P (2002) Daily data set of 20th century surface air temperature and precipitation series for the European climate assessment. Int J Climatol 22:1441–1453. https://doi.org/10.1002/joc.773
19. Kouamé I, Kouassi L, Dibi B, Adou K, Rascanu I, Romanescu G, Savané I, Sandu I (2013) Potential groundwater pollution risks by heavy metals from agricultural soil in Songon area (Abidjan, Côte d'Ivoire). J Environ Prot 4(12):1441–1448
20. Maftei C (2015) Extreme weather and impacts of climate change on water resources in the Dobrogea region. Information Science Reference (an imprint of IGI Global), United States of America
21. Margarint MC, Niculita M (2017) Landslide type and pattern in Moldavian Plateau, NE Romania. In: Rădoane M, Vespremeanu-Stroe A (eds) Landform dynamics and evolution in Romania. Springer, Berlin, pp 271–304. http://dx.doi.org/10.1007/978-3-319-32589-7_12
22. Mihai FC (2017) Waste collection in rural communities: challenges under EU regulations. A case study of Neamt County, Romania. J Mater Cycles Waste Manag. https://doi.org/10.1007/s10163-017-0637-x
23. Minea I (2017) Streamflow-base flow ratio in a Lowland area of North-Eastern Romania. Water Resour 44(4):579–585. https://doi.org/10.1134/S0097807817040121
24. Minea I, Croitoru AE (2015) Climate changes and their impact on the variation of groundwater level in the Moldavian Plateau (Eastern Romania). In: International multidisciplinary scientific geoconferences, SGEM 2015, 15th geoconference on water resources, forest, marine and ocean ecosystems, conference proceedings, vol I, Hydrology and water resources, pp 137–145
25. Minea I, Croitoru AE (2017) Groundwater response to changes in precipitations in north-eastern Romania. Environ Eng Manag J 16(3):643–651
26. Ministry of the Environment (Romania) (2018) The annual report on the state of the environment in Romania, year 2016-ANPM. http://www.anpm.ro/ro/raport-de-mediu

27. Niculiță M, Mărgărint MC, Santangelo M (2016) Archaeological evidence for Holocene landslide activity in the Eastern Carpathian lowland. Quatern Int 415(10):175–189. https://doi.org/10.1016/j.quaint.2015.12.048
28. Ouattara I, Kamagate B, Dao A, Noufe D, Savane I (2016) Groundwaters mineralisation process and transfers of flow within fissured aquifers: Case of transboundary basin of Comoe (Cote d'Ivoire, Burkina Faso, Ghana, Mali). Int J Innov Appl Stud 17(1):57–69
29. Panaitescu E (2007) Acviferul freatic si de adancime din bazinul hidrografic Barlad. Casa Editoriala Demiurg, Iasi
30. Plesoianu D, Olariu P (2010) Câteva observatii privind inundatiile produse in anul 2008 in bazinul Siretului. Analele Universitatii "Stefan cel Mare" Suceava, Sectiunea Geografie XIX:69–80
31. Radevski I, Gorin S, Milevski I, Dimitrovska O, Zlatanoski V (2015) Estimating annual simple water balance on Kriva Reka catchment using multiple linear regression. Geogr Rev 48:35–42
32. Radevski I, Gorin S, Dimitrovska O, Milevski I, Apostolovska-Toshevska B, Talevska M, Zlatanoski V (2016) Estimation of maximum annual discharges by frequency analysis with four probability distributions in case of non-homogeneous time series (Kazani karst spring in Republic of Macedonia). Acta Carsologica 45(3):253–262
33. Romanescu G (2003) Hidrologie generală. Editura TERRA NOSTRA, Iaşi
34. Romanescu G (2006) Hidrologia uscatului. Editura Terra Nostra, Iaşi
35. Romanescu G, Lasserre F (2006) Le potentiel hydraulique et sa mise en valeur en moldavie roumaine. In: Brun A, Lasserre F (eds) Politiques de l'eau. Grands principes et réalités locales. Presses de l'Université du Québec, Canada, pp 325–346
36. Romanescu G (2009) Evaluarea riscurilor hidrologice. Editura Terra Nostra, Iaşi
37. Romanescu G, Stoleriu C, Romanescu AM (2011) Water reservoirs and the risk of accidental flood occurrence. Case study: Stanca–Costesti reservoir and the historical floods of the Prut River in the period, July–August 2008, Romania. Hydrol Process 25(13):2056–2070
38. Romanescu G, Zaharia C, Stoleriu C (2012) Long-term changes in average annual liquid flow river Miletin (Moldavian Plain). Carpathian J Earth Environ Sci 7(1):161–170
39. Romanescu G, Cretu MA, Sandu IG, Paun E, Sandu I (2013) Chemism of streams within the Siret and Prut Drainage basins: water resources and management. Rev Chim (Bucharest) 64(12):1416–1421
40. Romanescu G, Sandu I, Stoleriu C, Sandu IG (2014) Water resources in Romania and their quality in the main lacustrine basins. Rev Chim (Bucharest) 65(3):344–349
41. Romanescu G, Paun E, Sandu I, Jora I, Panaitescu E, Machidon O, Stoleriu C (2014) Quantitative and qualitative assessments of groundwater into the catchment of Vaslui River. Rev Chim (Bucharest) 65(4):401–410
42. Romanescu G, Zaharia C, Paun E, Machidon O, Paraschiv V (2014) Depletion of watercourses in north-eastern Romania. Case study: the Miletin River. Carpathian J Earth Environ Sci 9(1):209–220
43. Romanescu G (2015) Managementul apelor. Amenajarea hidrotehnică a bazinelor hidrografice și a zonelor umede. Editura Terra Nostra, Iaşi
44. Romanescu AM, Romanescu G (2015) Modificări antropice în arealul cuvetei lacustre Stânca-Costeşti. Studiu de caz asupra riscului hidrologic. Editura Terra Nostra, Iaşi
45. Romanescu G, Zaharia C, Sandu AV, Juravle DT (2015) The annual and multi-annual variation of the minimum discharge in the Miletin catchment (Romania). An important issue of water conservation. Int J Conserv Sci 6(4):729–746
46. Romanescu G, Stoleriu C (2017) Exceptional floods in the Prut basin, Romania, in the context of heavy rains in the summer of 2010. Nat Hazards Earth Syst Sci 17:381–396
47. Romanescu G, Pascal M, Pintilie Mihu A, Stoleriu CC, Sandu I, Moisii M (2017) Water quality analysis in wetlands freshwater: common floodplain of Jijia-Prut Rivers. Rev Chim (Bucharest) 68(3):553–561
48. Romanescu G (2018) Inundațiile: calamitate sau normalitate? Studii de caz: bazinele hidrografice Prut și Siret (România). Editura Transversal, Târgoviște

49. Romanescu G, Mihu-Pintilie A, Stoleriu CC, Carboni D, Paveluc LE, Cimpianu CI (2018) A comparative analysis of exceptional flood events in the context of heavy rains in the summer of 2010: Siret Basin (NE Romania) Case Study. Water 216, 10(2):216:1–17
50. Salihou Djari MM, Stoleriu CC, Saley MB, Mihu-Pintilie A, Romanescu G (2018) Groundwater quality analysis in warm semi-arid climate of ahel countries: Tillabéri region, Niger. Carpathian J Earth Environ Sci 13(1):277–290
51. Salmi T, Maata A, Antilla P, Rouho-Airola T, Amnell T (2002) Detecting trends of annual values of atmospheric pollutants by the Mann-Kendall Test and Sen's Solpe estimates the excel template application MAKESENS. Finnish Meteorological Institute, Helsinki
52. Sandu I, Pescaru VI, Poiana I (2008) Clima României. Editura Academiei Române, Bucharest
53. Sedrati A, Houha B, Romanescu G, Chenaker H (2017) Determination of the contamination level in groundwater in the Sebkha of Elmahmel area, north eastern of Algeria. Air Water Compon Environ 367–376
54. Sedrati A, Houha B, Romanescu G, Stoleriu CC (2018) Hydro-geochemical and statistical characterization of groundwater in the south of Khenchela, el Meita area (northeastern Algeria). Carpathian J Earth Environ Sci 13(2):333–342
55. Sfica L, Croitoru AE, Iordache I, Ciupertea AF (2017) Synoptic conditions generating heat waves and warm spells in Romania. Athmosfere 8(3):50. https://doi.org/10.3390/atmos8030050
56. Stanga IC, Niacsu L, Iacob AM (2016) Environmental approach of land cover at local level: Studinet catchment (Eastern Romania). Environ Eng Manag J 15(6):1–12. WOS:000374795100001
57. Stanga IC, Niacsu L (2016) Using old maps and soil properties to reconstruct the forest spatial pattern in the late 18th century. Environ Eng Manag J 15(6):1369–1378. WOS:000383100900019
58. Tabari H, Hosseinzadeh Talaee P (2011) Analysis of trends in temperature data in arid and semi-arid regions of Iran. Glob Planet Change 79(1–2):1–10. https://doi.org/10.1016/j.gloplacha.2011.07.008
59. Teodosiu C, Cojocariu C, Mustent CP, Dascalescu IG, Caraene I (2009) Assessment of human and neutral impacts aver water quality in the Prut river basin, Romania. Environ Eng Manag J 8(6):1439–1450
60. Wrzesiński D, Choiński A, Ptak M, Skowron R (2015) Effect of the North Atlantic Oscillation on the pattern of lake ice phenology in Poland. Acta Geophys 63(6):1664–1684
61. Zelenáková M, Fijko R, Diaconu DC, Remenáková I (2018) Environmental impact of small hydro power plant—a case study. Environments 5(12):1–10

Chapter 8
Romanian Danube River Floodplain Functionality Assessment

Cristian Trifanov, Alin Mihu-Pintilie, Marian Tudor, Marian Mierlă, Mihai Doroftei and Silviu Covaliov

Abstract The floodplain of the Lower Danube was formed by the complex action of erosion and river accumulation, under the influence of the ascension trend of the river riverbed during the Holocene and the oscillation of the water levels and discharges. Almost full embankment and the construction of non-submersible dams have affected both the hydro-geomorphological system and the local and regional topo-climates, which is a pronounced phenomenon in the conditions of global climate changes. All these alterations have led to a significant transformation of the ecosystems. Transforming these ecosystems of the Lower Danube floodplain into dry land ecosystems has reduced their ecological, recreational, aesthetic and educational functions to only one function: the economic one. The natural capital of the Lower Danube floodplain has a productive capacity that must be known by its functional cells in order to avoid degradation, destruction under anthropogenic impact and to favor the sustainable use of its support capacity. Ensuring sustainable socio-economic development in the Lower Danube floodplain area is also based on knowledge of ecological sustainability, ecosystem integrity, environmental sustainability, ecological, and regional ecosystem balance. The ecological and economic resizing program of the managed areas of the Lower Danube floodplain was designed

C. Trifanov (✉) · M. Mierlă
Informational System and Geomatics Department, "Danube Delta" National Institute for Research and Development, 165 Babadag Str., 820112 Tulcea, Romania
e-mail: cristian.trifanov@ddni.ro

A. Mihu-Pintilie
Alexandru Ioan Cuza University of Iaşi, Interdisciplinary Research Department—Field Science, 54 Lascăr Catargi Str., 700107 Iaşi, Romania

M. Tudor
Management Department of the "Danube Delta" National Institute for Research and Development, 165 Babadag Str., 820112 Tulcea, Romania

M. Doroftei
Biodiversity Conservation and Sustainable Use of Natural Resources Department, "Danube Delta" National Institute for Research and Development, 165 Babadag Str., 820112 Tulcea, Romania

S. Covaliov
Ecological Restoration and Species Recovery Department, "Danube Delta" National Institute for Research and Development, 165 Babadag Str., 820112 Tulcea, Romania

© Springer Nature Switzerland AG 2020
A. M. Negm et al. (eds.), *Water Resources Management in Romania*, Springer Water,
https://doi.org/10.1007/978-3-030-22320-5_8

and launched to assist the Romanian Government in the long-term strategic planning process, to achieve the objectives of the Water Framework Directive and to effectively implement prevention, protection and mitigation of floods, mentioned in the National Strategy for Flood Risk Management. In this sense, a thorough and complex study, based on LiDAR measurements, hydraulic modeling and economic assessment, was conducted to evaluate the functionality of the floodplain as a whole and within agricultural units to determine the equipotential areas for flood-free future and sustainable development of the region. Thus, there were identified three types of areas: the first type represents the areas with only agricultural potential, the second type are, the areas with potential to be ecologically restored (to establish the natural flow of energy and circulation of matter) and the third one is a combination of the first two.

Keywords Economic activities · Ecosystem balance · Equipotential areas · Floodplain · Lower danube

8.1   Introduction

The Danube River is, after River Volga, the second biggest in Europe with an area of 817,000 km^2 and a length of 2778 km from which 1075 km borders Romanian territory in the southern part (Fig. 8.1). The catchment area of River Danube covers at present territories of 18 riparian states. Out of which 13 states hold territories in the Danube Basin bigger than 2000 km^2 [1–5].

The lower floodplain of the Danube River is formed by the complex action of lateral erosion and alluvial sedimentation process under the influence of the general upward trend of the riverbed and the seasonal oscillations of the water levels and flows. According to the existing bioindicators data, it has been established that the Danube river water quality may be framed in the beta-mesosaprobic limits. In Romania 7 river basins are connected to the Danube, 97.8% of the Romanian surface are included in the Danube River Basin, 30% of the Danube River Basin is in Romania [6, 7]. Biocenoses are total or partial impaired in the polluted sections such as downstream confluence with Dambovita River, where plant and animal organisms were not identified. Dambovita River intercepts domestic and industrial wastewaters from Bucharest. The wastewaters disturb the ecological balance, with biological devastation phenomenon. The flow of the Arges in the Danube determines alteration of the water river quality in the confluence area and adverse changes of planktonic and benthic biocenoses downstream [8–20].

The main water management issues to be solved along the Danube are flood effects mitigation, drought effects mitigation, control of torrents, soil erosion and land degradation, use of water power potential level and protection of the Black Sea coast against erosion and beach rehabilitation [21–27]. Romania aims to increase the water treatment and connectivity of citizens to centralized water systems by 70%. During 1960 during 1960–1989, the water quality of inland rivers and Danube

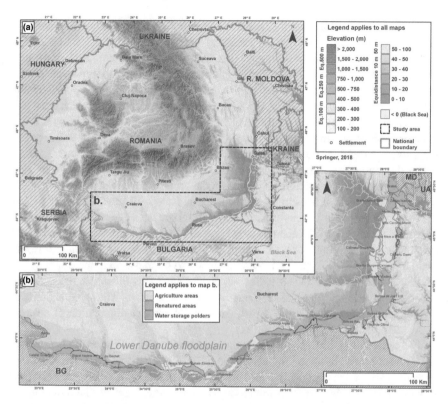

Fig. 8.1 The position of the lower floodplain of the Danube River: **a** within the territory of Romania, and **b** between Iron Gates II (upstream) and Isaccea (downstream)

River 1989, the water quality of inland rivers and Danube River has significantly worsened, but since 1989, the water quality has significantly worsened, but since 1989, the water quality has been improved [28]. In relation to population, Romania's water supplies are quite modest. They are provided by the rivers, which meet 89% of the present demand (48% from the Danube and 41% from the interior rivers), and by the underground sources which account for 11% [29].

In Romania, there is a unitary economic mechanism for the Danube water management products and services, which consists of prices, tariff, penalties, and allowances in order to manage rational the water resources. In this way, the users may follow the quality limits admitted for water discharges and can prevent their quality deterioration. The cost of water are the same all over country but differ according to the source of water: groundwater, inland rivers, Danube River, and the category of users such as: agriculture, industry, households, power plants, fisheries. During the communist period (until 1989), the water demand in the agricultural and industrial sectors increased continuously. After 1989, the water consumption for these sectors decreased, but increased water demand for households [30–32].

The water quality has slightly improved due to both the diminution of agricultural and industrial pollution and to the measures imposed on economic agents by the local EPA. The watercourses length from first-class quality increased from 35% (year 1985) to 66% (year 2002). Nevertheless, 6.6% of the monitored rivers do not meet the quality standards which are an important component in the sustainable development of the aquatic fauna. At the same time, there are no statistics related to the efficiency of water use in industry, although there is a water recycling system that indicates values ranging from 10% to 95% depending on the type of industry and factories. In agricultural activities, a large amount of water is allocated to irrigations, where the efficiency of the water use ranges from 60 to 80%. The average specific water consumption per inhabitant per day in the urban area at country level is about 513 L out of which: domestic use—294 L; public use—70 L; economic activities use—122 L; network losses—134 L. In the rural areas, where the specific water consumption is over 150 L, there is a real concern for the water losses reduction. At national level, the water losses will decrease from 34%, value recorded in the present time, to 15% in 2020. If the real water costs would be introduced and applied, this would represent a good incentive for the economic agents to reduce the losses in water supply and increase the recirculation and water reuse. The most significant matters include the low level of investments and ineffective application of reforms in the national economy [31].

The agricultural industry is the main source of water pollution with methane (CH_4) from animal farms, nitrogen dioxide (NO_2) and ammonia (NH_3), the last one exceeding 95% from total. This is the reason why the water pollution in the upstream area of the Danube floodplain is increasing. The structure, function and dynamics of animal communities have always been strictly dependent on the Danube's hydrology and hydrochemistry, as well as on its interaction with human activities. The floodplain of the Lower Danube it is acknowledged as the whole phenomena the Danube assembled through alluvial processes to the direct action of its waters developed upstream from Calarasi to Balta Borcea and Balta Braila [33–37].

The marshes of the Danube are generated by the action of two or more arms, in which the meaning of sedimentation processes, whether done directly or through flooding is directed from the periphery to the interior. During the last century, the floodplain of the Danube, the Romanian side, was mostly embanked, the ecosystems of this wetland being altered and largely desiccated for agricultural purpose. The river dynamics can no longer maintain or enlarge the wetland ecosystems, the embanked sections of the floodplain tend towards ruderal areas understood as a phenomenon of changing the wetland ecosystem into terrestrial ecosystem under anthropogenic impact [38–50].

The general features of the floodplain relief are a direct result of the intensity of floods (seasonal or peaks), which is understood both in terms of frequency, amplitude and duration. The complexity of the geological base, its mobility and the particularities of the geomorphological evolution of the region, explain the very complicated course of the Danube with many changes of direction, large loops and meanders. The erosion, transport and sedimentation occur simultaneously on the entire riverbed by the continuous tendency of formation or enlarging of the alluvial islands. The intensity of river processes varies a great deal over a year, depending on hydrological changes in the riverbed, from the lowest levels, when the processes are carried out only within the limited range of the riverbed to the maximum, when the processes extend over a surface much larger, including the floodplain [51–60].

The almost complete embankment of the Danube floodplain with in-submersible dams affected both the hydro-geomorphological system and the local and regional topo-climates, a phenomenon enhanced by the global climate changes. Due to the elimination of the floodplain reedbeds, the nutrient retention capacity has been exceeded since the 1970s, with Danube waters being affected by a strong eutrophication which has led to the reduction or loss of submerged macrophytes, the change of the specific algal spectrum and the proliferation of competitive species in the conditions of excess of nutrients (green-blue algae). These phenomena affect trophic cycles, leading to the decline of biodiversity through the disappearance of some species, many of which are of high conservation or economic value. Also, the hydromorphological processes with implications in the size of the riverbed on which the flow depends in good conditions have been affected by displacements of the Danube stream [38–44, 61–66].

Another action, whose effects were not considered, consisted of the construction of Iron Gates and the development of build-up lakes for electricity purpose. Their presence led to the alteration of the flood regime and to the decrease of the quantity of sediments transported by the Danube, due to the decanting of the waters, resulting in major changes in the dynamics of the Black Sea Romanian coastline. Another effect of the Iron Gates is the disruption of the migration routes for sturgeon and shad species. The transformation of the downstream natural floodplain ecosystems into anthropogenic ones, for economic purposes, has reduced their functions in terms of ecological, economic, recreational, aesthetic and educational perspectives [1–5].

Embankment works carried out in the 1960s, along the 800 km of the Romanian bank, in order to obtain arable land, led to the disappearance of the floodplain reedbeds as water filters and shallow waters, favorable breading, spawning, nesting and feeding habitats for species, in general. At a larger scale, much later, the increase of the nutrients amount, from the intensive agriculture and wastewater from riverine settlements transported along the river are the effects of eutrophication of the downstream lakes and changes or decrease of the fish species composition and populations [43, 44, 56, 67].

In the 40's, the scientist Grigore Antipa has promoted the naturalistic concept of extensive landscaping, predominantly fisheries on higher elevation terrain—which impose the interrupted embankment of the Danube floodplain thus restricting the embankment area to only 130,000 ha. The proposed hydrotechnical constructions was submersible dams, so that once every 10 years (average), the flood-defended lands can be flooded, and their usage to change in fishing and forestry [38–42].

This concept can influence the incapacity of the in-submersible dams to ensure a definite defense of the land and the inevitable disasters that would result, exemplified by the dyke breaches observed over time. Furthermore, changing the hydrological equilibrium of the river, which, without the floodplain, would raise its level to floods to such an extent that ports, riverine settlements would be flooded, the dikes would overflow, and the riverbed would lose its stability. Nowadays, more than ever, knowing these undeniable truths, scientists everywhere has focused their attention on the study of watercourse ecosystems [2, 3].

The Danube River must be considered to be much more than a subject of hydraulic modeling. Together with its floodplain, as natural catchment area, it is a very complex ecosystem that provides habitat for a particularly rich flora, and fauna and also supports for socio-economic activities. It has been assessed the current status of the Danube floodplain and developing management plans to ensure integrated management. Moreover, there were established protected areas along the Danube floodplain, which have retained their original characteristics of ponds, lakes, reservoirs, in order to create a network of natural sites such as Natura 2000, and for designation of wetlands that meets the conditions for a green corridor as refuge areas for species. The further steps are to evaluate the ecological and economic resizing of some drained areas for agricultural purposes, which have now been flooded or abandoned and no longer correspond to the purpose for which they were arranged [19, 49, 50, 68].

8.1.1 Reconsidering the Flooding Lines of Defense of the Settlements

The ecological and economic resizing program of the Danube floodplain and delta was designed and launched to assist the Romanian Government in the long-term strategic planning process, to achieve the objectives of the Water Framework Directive and to effectively implement prevention, protection and mitigation of floods, stipulated by the National Strategy for Flood Risk Management. Their complexity determines the acute need of an overview of the principles and methods of representing natural phenomena. To establish and reconsider the lines of defense of the localities located in the Danube floodplain, it is necessary to elaborate a complex study of integrated system analysis of the hydro-morphological conditions.

Mapping of the hydro-geo-morphological units of the Danube floodplain on the Romanian sector has been declining over the last 25 years (until 2006). This situation required the creation of a digital terrain model (DTM) on which the flood defense strategy and territorial planning could be conceived and grounded. The content of the digital terrain model defined the design support of the local defense lines but also created a useful tool for analyzing the dynamics of hydro-geo-morphological phenomena. The correlation between these phenomena and land use could only be made on such a model. Also, the characterization of landscape indicators and indices could only be done on the DTM. The elaboration of the digital terrain model was the first mandatory stage in defining the defense strategy of the localities in the Danube Floodplain. The DTM involved mapping of the hydrological and geomorphological units using the LiDAR technology. Given that the hydro-morphological dynamics of the Danube riverbed is significant, the DTM supports the modeling and simulation of such complex phenomena. The DTM has dual functionality: on the one hand, it gathers information on the nature of the terrain, giving concrete details of the morphological features of the relief; on the other hand, it sees the changes processes due to natural and anthropogenic phenomena. The adopted hydraulic model had designed considering the flow and water level, various flood scenarios.

8.1.2 Assessing the Suitability of Economic Activities Towards a Multipurpose Polder

The almost complete embankment of the Danube Floodplain with non-submersible dykes affected both the hydro geomorphological system and the local and regional topo-climates, a phenomenon accentuated by the global changes. In this context, a socio-economic study was carried out for the assessment of the economic activities within Danube's floodplain. Along with the activities defined in first objective, a program of ecological and economic assessment was required in the polders of the Danube's floodplain.

The objective of this study was the deep and coherent scientific substantiation of the sustainable development, which offered the idea of alternative technical solutions for the sustainable use of the polders within the Danube's floodplain. The transformation of the Danube floodplain ecosystems into terrestrial ecosystems has reduced their functions (ecological, economic, recreational aesthetic and educational), which has had a major impact on the natural capital. To restore the balance of this socio-ecological complex, renaturing activities are required in the degraded polders. Some criteria for carrying out the pre-feasibility study on renaturing the polders were necessary. Therefore, thematic studies were done and their outputs have deliniated the renatured areas that were to be included in the national network of protected areas. The national ecological network is created in order to preserve the natural genetic diversity of the species present in ecosystems and natural complexes, as well as to ensure productive living conditions and to achieve sustainable development on the respective territories [69–73].

8.2 Materials and Methods

The expected results are based on the materials and methods used for the "Ecological and Economical Resizing of Danube's Floodplain in Lower sector" project. These are briefly detailed by the flowchart described in Fig. 8.2.

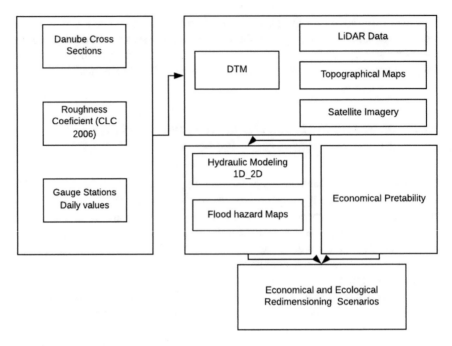

Fig. 8.2 The project workflow

8.2.1 Developing the Digital Elevation Model (DEM)

The elaboration of digital elevation model represents the first compulsory phase for grounding the strategy of defense flooding lines for Lower Danube Floodplain localities. To achieve this goal, it is mandatory to map the hydro-geo-morphological units using LIDAR (Light Detection and Ranging) method.

Technical characteristics

- Mapping the floodplain on the almost whole course of the Danube River in Romania—1075 km;
- The width of the band is variable from 1 km till approximately 80 km;
- Considering that the area has a large network of canals and dykes, topographic cross sections were mandatory.
- Z precision is about 5 cm.

The main activity for reaching the objective of the project was the execution of the terrain numerical model by scanning flight with a LIDAR embankment system in the Danube floodplain. Between 9/05/07 and 19/07/07 the French company SINTEGRA SRL (in its capacity as a subcontractor) managed to carry out DEM observation flights using a LIDAR type system. The laser scanner used in the LIDAR system was Riegl LMS-Q560 mounted on a PartenAvia plane (Fig. 8.3).

Fig. 8.3 Parten Avia plane for LiDAR (DEM- LIDAR: PARTENAVIA P3; Owner: APEI sa; Engine: 2×; LYCOMING 180 CV; Max Altitude: 25 000 Ft; Speed: 145 kts; Equipment: complete IFR/BR nav; LiDAR scanning hatch; Autonomy: 5 h)

The LIDAR (Lidar Detection and Ranging) model is based on echo/retro fusion detection of a laser pulse by measuring its response time (e.g., terrestrial surface). The laser scanner used in the LIDAR system that carried out the work was: Riegl LMS-Q560. Detection allows recording all the echoes of each pulse emitted. For this study, it was proposed to record the first and last echoes to obtain the land numeric model and thus allow segmentation of the vegetation-covered areas. Flight parameters were chosen according to the specifications of the project to achieve the required precision and the required and sufficient point density (i.e. 4–5 pts/m^2) to properly detect the dams, considering the random cloud distribution (Table 8.1).

Table 8.1 LiDAR flight parameters

LIDAR parameters	Value
Flight altitude	450–500 m
Flight speed	45 m/s, 90 kts
Band width	520 m
Lateral coverage	20%
Distance between bands	415 m
Laser frequency	65–130 kHz
Scanning angle	60° (±10°)
Scan frequency	75 Hz
Altimetric precision (standard deviation)	5 cm
Planimetric precision (standard deviation)	20 cm
Average point density	2.8 pts/m^2
Average distance between points (flight direction and perpendicular)	0.6 m

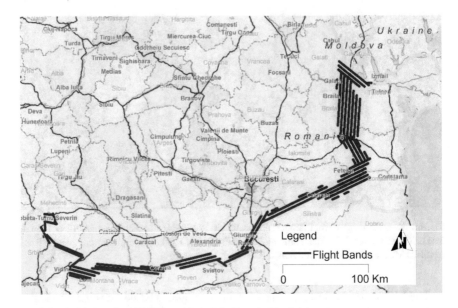

Fig. 8.4 Flight plans for LiDAR scanning of the Danube floodplain

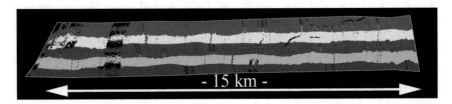

Fig. 8.5 Different LiDAR flights of approximately 500 m width and 15,000 m long with a lateral coverage of 20% in Galati area

To start building the LiDAR database, there were lots of papers and approvals in respect to operate a plane especially cross the border with other countries, and one first step was developing the flight pan (requiring crossing Romania's borders with Bulgaria) of the PartenAvia LiDAR aircraft (Fig. 8.4).

By processing the raw data of the LiDAR flight not only the Digital Terrain Model (DTM) can be extracted and viewed in different shades but also lots of by-products (Figs. 8.5, 8.6, 8.7, 8.8, and 8.9).

8.2.2 Building the Hydraulic Model

Considering the high hydro-morphological dynamics of Danube riverbed, the DTM represents the support for modeling and simulation of complex phenomena. The

Fig. 8.6 The echo intensity of LiDAR corresponding to the same surface (the Danube River is identified with black color to the left which means the LiDAR lack of signal)

Fig. 8.7 The altitude color coded view. The Danube River is in the left with black and the hills in the bottom right of the image

Fig. 8.8 Elevation classes. Low vegetation extract example

Fig. 8.9 Hillshade view of the digital terrain model in which one can easily distinguish the positive relief from the canals, the river embankment and other land management planning

DTM holds a double significance: on the one side collecting information about land characteristics showing the real morphological characteristics of the relief and on the other side shows any changes due to anthropic and natural phenomena that can be very hard to distinguish with bare eyes.

The adopted hydraulic model was capable of designing, in term of water discharge and water level, different flooding scenarios. Based on this hydraulic model new and updated attention water levels alarms should have been implemented. This model was done using Sobek Model from Delft Hydraulics, and it required all the terrain data that could get beside the digital terrain model. Data regarding the river cross-sections were collected through a collective effort of different teams from different institutes. The river cross-sections were made using single beam sonars, and the distance between the profiles was 250 m along the whole lower Danube. Additional topographic determination on several dykes along the Danube where the situation was critical, historical data regarding high and low water regime of the Danube and so on. All this data served as calibration variables to get the most precise model at that time.

Using the hydro-ameliorative schemas of the Danube Floodplain Polders, which exist only in National Institute of Research and Development for Land Reclamation Bucharest (ISPIF) archive, the consortium experts designed the new defense lines and the economic objectives for this area—the flood and menace levels will correspond to the Romanian National Institute for Hydrology and Water management (INHGA) objectives. The development strategies and programs for long-term land use along with the macro-regional and globalization integrative processes have been concentrated on the political program objectives to assure a durable development.

8.3 Results and Discussions

8.3.1 Hierarchy of Spatial Distribution of the Categories of Ecological Equipotential Areas

The natural capital of the lower Danube floodplain has a productive capacity that must be known by its functional cells to avoid degradation, destruction under anthropogenic impact and to favor the sustainable use of its support capacity. Ensuring sustainable socio-economic development in the Danube floodplain area is also based on knowledge of ecological sustainability, ecosystem integrity, environmental sustainability, ecological, and regional ecosystem balance. The biological diversity, functionality and naturalness of the ecosystems in the floodplain of the Lower Danube is a result of their evolution over time and the succession of the various "civilizations" that have disturbed the balance of the components of the initial environments.

The pan-European strategy for the conservation of biodiversity and ecosystem function has clear objectives, among which for Romania and especially for the Danube Delta, the awareness and participation of local communities must occupy a very important place. The equipotential ecological area corresponds to the geotope in the taxonomic scale of natural environments elaborated by J. Richard in 1975. The use of the overlaid analytical map method implies a first assessment of the hierarchization of ecological equipotential areas categories. The landscape individuality of the lower Danube floodplain is based on the interactions established between three main components: ecological potential (ecological support), biological exploitation and anthropic action. They provide the common dynamics of the geosystem expressed through a certain type of landscape. Often, the dynamics of a component may be different from the dynamics of the assembly, and then the change in the relationships between the components imposes a new dynamic trend expressed by landscape modification. Geosystems can evolve between three defining states. The first one represents the relationships of imbalances between components due to natural or anthropogenic causes. The second one represents the disequilibrium between the constituent elements and the relations between them being artificialized by the effect of anthropic activity. And the third one represents the equilibrium relations between ecological support and biological exploitation (morpho-structural stability of the components). These three "states" cause the degradation of the ecological support and/or biological exploitation. The effects then being transmitted, mutually, between all the components [18, 50, 61, 73–75].

The analyzes allowed for the identification of the stability phase of the relief due to the absence of erosion under the conditions of a permanent vegetation layer present on the entire unspoiled surface of the Danube meadows. The areas that include zone of urban/rural localities as the phase of instability in the evolution of the relief due to erosion, in the absence of a permanent vegetation layer, are characteristic of the areas set for agriculture. From the map developed by the European Environment Agency—CLC2000, regarding the land cover, there were extracted the five types of areas as follows: 1—artificial territories; 2—areas with agricultural use; 3—land with forests and semi-natural areas; 4—wetlands and 5 are water surfaces. In this way, the map of equipotential areas was created from which the variation ranges were statistically analyzed. In this respect, one can deduce that in the entire Danube Floodplain the highest percentage is held by 73% (Fig. 8.10).

The artificial areas occur in the form of arable land, permanent crops, pastures and heterogeneous agricultural areas and the occupied spaces, summed up and expressed

Fig. 8.10 Danube floodplain equipotential areas statistics (percentages)

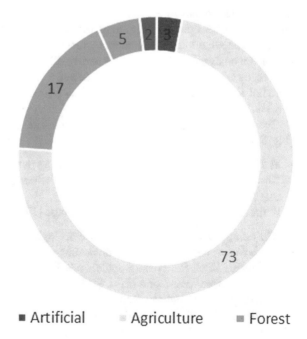

in hectares, representing 418,542,788 ha. The lowest percentage area (3%) is characterized by urban areas, industrial or commercial areas and communications networks, mining areas, landfills and yards, and non-agricultural artificial land, and the occupied space expressed in hectares is of 17,651,021 Ha. A 24% share and is represented by forests and semi-natural areas 17%, wetlands 5% and water areas 2%, and the occupied area in hectares is 143,160,973 ha. The ratio between the different areas is 3.05 and represents the imbalance between natural and man-made areas. Researches on the situation in the Danube floodplain reveal a natural imbalance. Among many anthropogenic interventions in the last 20–30 years and their adverse effects on the environment, these are the most damaging actions: reduction of the Danube hydro-dynamic area, draining of the ponds, cutting the channels that fragment the landscape both longitudinally and transversely, poaching of any nature, extermination of specific forests from the floodplain and on the terraces (especially the high valuable species—oak, black poplar). As a consequence of all these actions, the flora, fauna and natural diversity have catastrophically diminished. The erosions and landslides have grown more and more, especially at the edges of the Danube terraces, with catastrophic consequences for the neighboring localities, which require an improvement of the equipotential relationship through rehabilitation, restoration and ecological rehabilitation works.

8.3.2 Hydrological Scenarios in Danube's Floodplain

Based on the analysis of the efficiency indicators and the cost-benefit ratio carried out in the previous phase of the study, resulted in the map of the economic activity in the Danube floodplain (Romanian sector) (Fig. 8.11).

For the analyzed hydrological scenarios, all the polders suitable for agriculture and renaturation were taken into consideration, and of those suitable for water storage were not selected the small ones and the fish farms, the decrease of the storage area is 6.1%.

The hydrological scenario took into consideration the following polders (Fig. 8.12):

- The inclusion of suitable polders for renaturing purposes (75.439 ha);
- "Inclusion" (does not apply for the hydraulic schematization) of agricultural polders (205.006 ha);
- Without the polders suitable for water storage (193.111 ha).

The comparative results between the hydrological scenario with the reforestation of the above-mentioned agricultural enclosures and the hydrological scenario without any intervention on the existing polders are displayed for the hydrometric stations where differences were found (Figs. 8.13, 8.14 and 8.15). The maximum values of the decrease of the Danube water level, in the case of renaturing the analyzed polders, are consistent in certain areas along the Danube river but in other areas, the level decrease is negligible or less than 3 cm (Fig. 8.16).

Fig. 8.11 The share suitability (percentages) of polders from Danube's floodplain

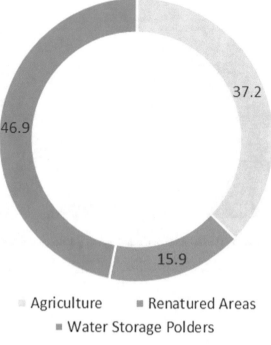

Fig. 8.12 The share suitability (percentages) of polders from Danube's floodplain for hydrological scenarios

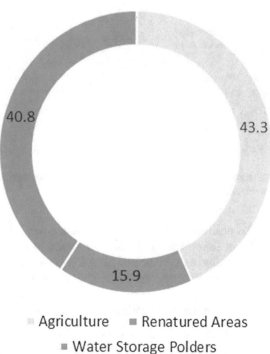

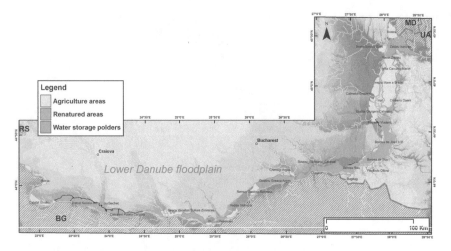

Fig. 8.13 The functions of the agriculture polders considered for hydrological scenarios

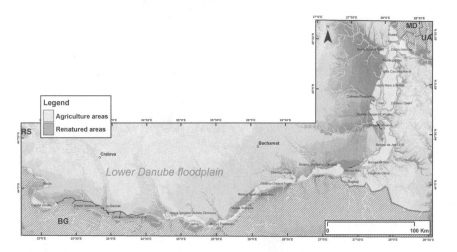

Fig. 8.14 Hydrological scenario corresponding to the renaturation of agriculture polders suitable for this action

The positioning of the flood thresholds was set upstream for each agricultural polder in front of the old lakes connected to the Danube or the irrigation channels. Running of the combined hydrological scenario (renaturing and water storage), was done only during the "active" flood period, namely 10.03.2006–25.05.2006 (Fig. 8.16).

To facilitate the presentation of the results, the following abbreviations were made:

- **S_N**: reference scenario without any intervention on agricultural polders;
- **S_R**: the inclusion of only the areas suitable for renaturation;

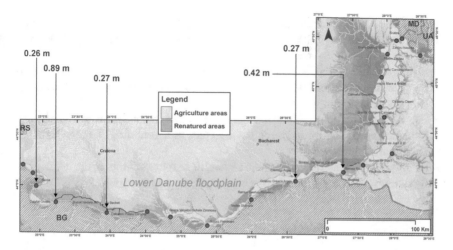

Fig. 8.15 Maximum values of the Danube water levels decrease in case of renaturing the analyzed polders

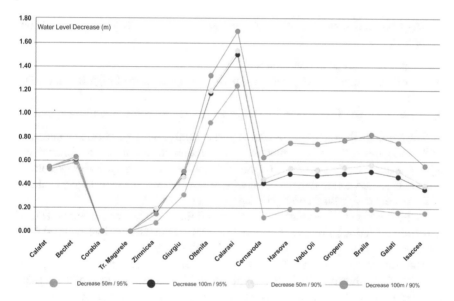

Fig. 8.16 The decrease of the Danube's water level for the analyzed scenarios at the hydrometric stations

- **SR_100 m_95%**: the inclusion of the area suitable for renaturation, the width of the food threshold of 100 m, and its height to a value representing 95% of the Danube's maximum water level;
- **SR_100 m_90%**: the inclusion of the area suitable for renaturation, the width of the food threshold of 100 m, and its height to a value representing 90% of the Danube's maximum water level;
- **SR_50m_95%**: the inclusion of the area suitable for renaturation, the width of the food threshold of 50 m, and its height to a value representing 95% of the Danube's maximum water level;
- **SR_50m_90%**: the inclusion of the area suitable for renaturation, the width of the food threshold of 50 m, and its height to a value representing 90% of the Danube's maximum water level.

8.3.3 The Hydrographical Danube River Plan

The hydrological scenarios were designed to quantify the decrease of the Danube water level:

- Renaturing of some agricultural polders;
- Use of some agricultural polders for water storage in case of high Danube water level;
- Combined scenarios (renaturing and water storage for some agricultural polders).

The hydraulic layout of the renatured agricultural polders was made by including in the cross-section of the Danube the cross section corresponding to the analyzed agricultural polder (Fig. 8.17). To be able to sketch the water storage in the agricultural polders was used a global methodology for removing from the hydrological balance of the water storage flow corresponding to each polder (Fig. 8.18).

In the present study, was used the "Overland flow" module of the Sobek 1D, 2D hydraulic modeling software. This module allows 2D modeling of hydrological flood scenarios to be integrated in 1D hydraulic modeling (Fig. 8.19).

The connection channel between the Danube and the agricultural polder can be considered as a threshold in the magistral dam, which allows the flood of the agricultural polder when the Danube level exceeds its quota. Hydrological scenarios were made for several variants of the threshold width, 50 and 100 m, and the threshold height was calculated for a value representing 95 and 90% of the maximum value of the Danube level in 2006. The positioning of the flood thresholds was set upstream for each agricultural polder in front of the old lakes connected to the Danube or to make a connection to the irrigation channels (Figs. 8.19 and 8.20). Hydrological scenarios were run for representative hydrological regimes of the Danube, namely: 2003 (minimum levels), 2004 (average levels) and 2006 (maximum levels).

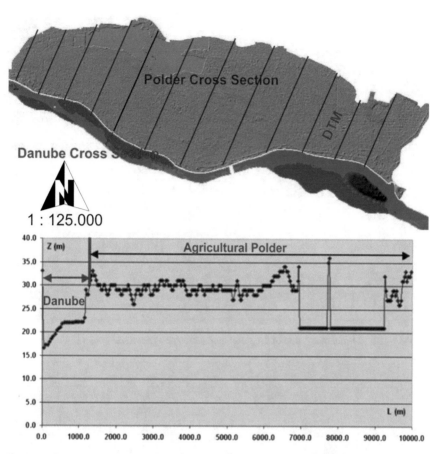

Fig. 8.17 Cross sections used for flooding scenario for the agricultural polders

Fig. 8.18 Schematization of the storage flow for each agricultural polder

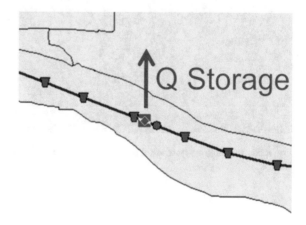

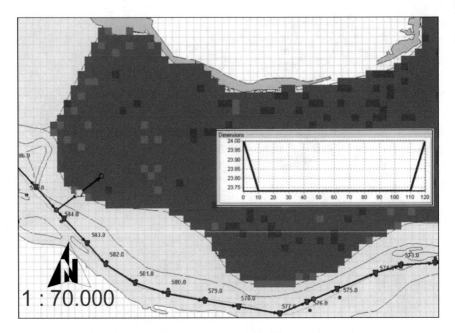

Fig. 8.19 1D_2D schematization of the water storage inside agricultural polder

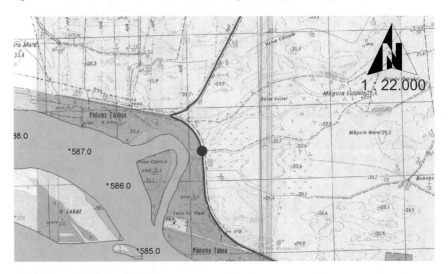

Fig. 8.20 Example of flooding threshold positioning for agricultural polders needed for water storage

8.4   The Consequences of Changing the Hydrological Conditions

Objectives pursued:

- Reconsideration of economic activities within the embanked polders according to the cost/benefit ratio for investments for the maintenance of the defense dykes and the other existing hydro-technical constructions;
- Establishment of a controlled flood regime for storing water at maximum levels of the Danube which endangers the defense systems;
- The establishment of the flood regime and the retention of water in the embanked polders obtained through drainage procedure of the former lakes Bistret, Potelu, Suhaia, Greaca, Calarasi, etc., to renature them.

The recent reduction of wetlands is mainly due to the development of agriculture through the embankment. This reduction also occurs by draining some surfaces, or by regulating some watercourses.

There are many mechanisms through which human activity influences the stability of a wetland, such as deliberate intervention by improving drainage to expand agricultural land or through peat exploitation. This type of intervention has attracted the attention of public opinion, especially through the mass media of the danger of loss of habitats for many species that depend on the existence of these areas.

In the case of the Danube floodplain, the change of the hydrological conditions had the following consequences:

- unproductive lands and the increase of soil salinity;
- reduction of water exchange with the surrounding areas;
- habitat loss for birds;
- major changes in the structure and composition of the vegetation;
- discontinuation of the fish movement from nearby areas towards the polder where they had optimal conditions for spawning;
- loss of organic matter through mineralization;
- blocking the filtration role of the sediments and nutrients gained through floods.

Thus, under these circumstances, the best option is the use of strict wetland policies in the Lower Danube floodplain, followed by a well-developed alert monitoring system as well as many advanced tools to explore strategies and policies to address to the detected threats. With the increased effort on the system and the complexity of the problems, the need for planning tools is changing rapidly.

8.5   Conclusions

Assessment of the socio-economic systems way of function within the socio-ecological complexes of the lower Danube floodplain as well as their performance in relation to the impact on the structure, the quality, the productive and supporting capacity of the foundations that support and feed them, can be achieved in a coherent and useful form for adaptive management using the ecological balance quantification process.

At the beginning of the century and the millennium, the industrialized civilization reached the level of globalization, having at their disposal the huge powers of science and technology, but still faces very serious problems. Adapted to the situation, but preserving its dramatic character, Hamlet's words fit the moment in the form of "to be or not to be". Salvation can still only be possible through sustainable development. This is unanimously considered as the only "safety exit" to avoid a near collapse by interrupting the processes of self-destruction and bypassing the sad destiny of all previous civilizations after long periods of amazing progress. Of course, mankind is threatened by many dangers (pandemics, meteorites, earthquakes, volcano eruptions, nuclear accidents etc.), but those that result from environmental degradation and depletion of resources are the most important, being extended to the entire planet and virtually irreversible.

The concept of sustainable development forsakes the struggle with nature in exchange for its friendship, seeking to halt any process susceptible to prejudice the future generations and to repair everything that is possible. It is a solidarity manifestation beyond the limits of life, proof of the maturity of our civilization in the era of globalization and extraterrestrial searches. Rapid profound changes are needed in the spheres of industrial, economic, educational, cultural, administrative, political, science and technology activities, with a decisive role. On this road, above all, there is a need for another way of thinking and other attitudes in the spirit of "thinking globally, acting locally". Everything needs to be done quickly, on the go. They must learn both the children and mature people not to be exposed and become, from one day to another, incompatible with the requirements of the place where they work or, worst of all, suffer from "outperformed manager syndrome" that blocks and sickens an entire sector, leading him to failure.

From the sphere of principles, sustainable development must materialize in actions with effects in many areas, and regarding reducing energy consumption. For this, engineers have great responsibilities. Although scientific research can still bring important results in reconsideration of economic activities, although the technological solutions that can be applied immediately, exist and should be introduced without delay, with government support.

In summary, the following conclusions can be retained:

- This is the beginning of the transition towards sustainable development, which is urgently needed, and the necessary transformations must go into all sectors of activity, starting with the energy and education of the population;
- The necessary changes must be made during the current generation. Otherwise it will be too late!
- Often, benefits do not only occur where investments are required. That is why the state and international structures need to be substantially involved through legislation, economical mechanism and financial support.

The shift to sustainable development is a process of a large-scale, achievable at the same time by the cooperation of all countries because the effects on the environment do not know limits imposed by the borders.

8.6 Recommendations

The following three aspects represent a great deal of importance for the spatial planning problems:

- The systems must be considered as a whole. Therefore, while a manager intervenes directly only in a limited part of the system, the links will convey the consequences of these policies to other parts of the system. It may be that the problems faced by the administrator have originated in actions that have taken place in other parts of the system to solve simplistically other problems.
- Secondly, human systems and natural systems are dynamic and evolving but never in balance. That is why administrators intervene in changing the system and, at a critical point; the consequences of a small intervention may be of great importance.
- The consequences of planning policies depend on the spatial context in which they are implemented, and how they modify this context.

The Ecological and Economic Resizing Program of the Romanian Danube Floodplain presented a spatial planning tool (SPT) developed in accordance with the three characteristics and built to design, analyze and evaluate long-term policies in a social, economic and ecological context.

The main purpose of SPT is to explore the effects of alternative policies on the quality of the socio-economic and natural environment and to stimulate and facilitate conscious actions and discussions before public debates are taken place. SPT should not seek to optimize economic, environmental and social dimensions, but rather to maximize the whole. Although this involves the loss of details, the advantage of this approach is the strong integrative side of the resulting system, in which autonomous processes play an important role. The actions of the current and proposed policies act against the stakeholders and can be introduced into the SPT system by means of zonal maps and control parameters that behave like constraints on the autonomous dynamics of the system.

The main component of the SPT should be a dynamic land use model applied to the entire territory of the Danube floodplain. To represent the processes that make and change the configuration of the Danube floodplain requires an overlayered model to represent processes on three geographical levels: national, regional and local.

Acknowledgements This chapter was compiled within the project "Ecological and Economic Resizing on Romanian Sector of Danube's Floodplain" implemented between 2006 and 2008 for Ministry of Environment as beneficiary. The study was carried out by Danube Delta National Institute for Research and Development (D.D.N.I.) along with National Institute of Research and Development for Land Reclamation (I.S.P.I.F.), National Institute of Hydrology and Water Management, (I.N.H.G.A.), and EnviroScopy, from Romania and W.W.F.—Karlsruhe University from

Germany. A project would not work without the support of a functioning partnership. Thus, we thank all our project partners and their respective teams, the whole lead partner team with special thanks to Dr. Eng. Iulian Nichersu—project leader, Dr. Eng. Ion Grigoras—LiDAR data processing and other GIS related products and Eng. Adrian Constantinescu—hydraulic modeling.

References

1. Romanescu G (1996) Delta Dunarii. Studiu morfohidrografic. Editura Corson, Iasi
2. Romanescu G (2003) Hidrologie generala. Editura Terra Nostra, Iasi
3. Romanescu G (2003) Morpho-hydrographical evolution of the Danube Delta, I, aquatic surfaces and inner lands. Editura PIM, Iasi
4. Romanescu G (2005) Morpho-hydrographical evolution of the Danube Delta, II, management of water resources and coastline evolution. Land use and the ecological consequences. Editura Terra Nostra, Iasi
5. Romanescu G (2006) Complexul lagunar Razim-Sinoie. Studiu morfohidrografic. Editura Universitatii „Alexandru Ioan Cuza", Iasi
6. Romana Academia (1967) Limnologia sectorului romanesc al Dunarii. Studiu Monografic. Editura Academiei Republicii Socialiste Ramania, Bucuresti
7. Diaconu C, Nichiforov ID (1963) Zona de varsare a Dunarii. Monografia hidrologica. Editura Institutului de Studii si Cercetari Hidrotehnice, Bucuresti
8. Apostu M, Tantaru G, Vieriu M, Bibire N, Panainte AD (2018) Study of the presence of lead in a series of foods of plant origin. Rev Chim (Bucharest) 69(5):1223–1225
9. Banaduc D, Rey S, Trichkova T, Lenhardt M, Curtean-Banaduc A (2016) The Lower Danube River-Danube Delta–North West Black Sea: a pivotal area of major interest for the past, present and future of its fish fauna—a short review. Sci Total Environ 545–546:137–151
10. Buzea E (2011) Flooded areas and their importance in maintaining biodiversity. Meadows lower Danube. J Wetl Biodivers 1:23–46
11. Čech M, Čech P (2013) The role of floods in the lives of fish-eating birds: predators loss or benefit? Hydrobiologia 717(1):203–211
12. Cical E, Mihali C, Mecea M, Dumuta A, Dippong T (2016) Considerations on the relative efficacy of aluminium sulphates versus polyaluminium chloride for improving drinking water quality. Stud Univ Babes-Bolyai, Chem 61(2):225–238
13. Ciomos A, Imre L, Mihaiescu T, Mihaiescu R (2015) Perspectives on the responsible consumption of the water resources in Romania. Pro Environ 8:606–615
14. Danu M, Messager E, Carozza JM, Carozza L, Bouby L, Philibert S, Anderson P, Burens A, Micu C (2018) Phytolith evidence of cereal processing in the Danube Delta during the Chalcolithic period. Quat Int
15. Kahit FZ, Zaoui L, Danu MA, Romanescu G, Benslama M (2017) A new vegetation history documented by pollen analysis and C14 dating in the alder of Ain Khiar—El Kala wet complex, Algeria. Int J Biosci 11(6):192–199
16. Merecki N, Agič R, Šunić L, Milenković L, Ilić ZS (2015) Transfer factor as indicator of heavy metals content in plants. Fresenius Environ Bull 24(11c):4212–4219
17. Omer I (2016) Water quality assessment of the groundwater body RODL01 from North Dobrogea. Rev Chim (Bucharest) 67(12):2405–2408
18. Raischi MC, Oprea L, Deak G, Badilita A, Tudor M (2016) Comparative study on the use of new sturgeon migration monitoring systems on the lower Danube. Environ Eng Manag J 15(5):1081–1085
19. Trifanov C, Romanescu G, Tudor M, Grigoras I, Doroftei M, Covaliov S, Mierla M (2018) Anthropisation degree of coastal vegetation areas in Danube Delta biosphere reserve. J Environ Prot Ecol 19(2):539–546

20. Wu S, Wiessner A, Braeckevelt M, Kappekmeyer U, Ding R, Müller J, Kuschk P (2013) Influence of nitrate load on sulfur transformations in the rhizosphere of *Juncus effusus* in laboratory-scale constructed wetlands treating artificial domestic wasewater. Environ Eng Manag J 12(3):565–573

21. Adopo KL, Romanescu G, N'Guessan AI, Stoleriu C (2014) Relations between man and nature and environmental dynamics at the mouth of the Komoé river, Grand-Bassam (Ivory Coast). Carpathian J Earth Environ Sci 9(4):137–148

22. Adopo LK, Romanescu G, N'Guessan AY, Stoleriu C (2014) Nature and dynamic of sediments at the mouth of Komoé river (Ivory Coast). Lakes, Reserv Ponds 8(1):28–41

23. Adopo KL, N'Guessan MY, Sandu AV, Romanescu G, Sandu IG (2016) The spatial distribution and characterization of sediments and the bottom morphology of the hydroelectric lake in Ayamé 2 (Ivory Coast). Int J Conserv Sci 7(2):567–578

24. Romanescu G (2013) Alluvial Transport Processes and the Impact of Anthropogenic Intervention on the Romanian Littoral of the Danube delta. Ocean Coast Manag 73:31–43

25. Romanescu G (2016) Tourist exploitation of archaeological sites in the Danube Delta biosphere reserve area (Romania). Int J Conserv Sci 7(3):683–690

26. Romanescu G, Mihu-Pintilie A, Carboni D, Stoleriu CC, Cimpianu CI, Trifanov C, Pascal ME, Ghindaoanu BV, Ciurte DL, Moisii M (2018) The tendencies of hydraulic energy during XXI century between preservation and economic development. Case study: Fagaras Mountains, Romania. Carpathian J Earth Environ Sci 13(2):489–504

27. Romanescu G (2013) Geoarchaeology of the ancient and medieval Danube Delta: Modeling environmental and historical changes. A review. Quat Int 293:231–244

28. Galatchi LD (2008) Urban water resources management in Romania - perspectives for the sustainable development in order to supply water to human settlements. In: Hlavinek P, Bonacci O, Marsalek J, Mahrikova I (eds). Dangerous pollutants (xenobiotics) in urban water cycle, Springer, pp 23–34

29. Zavoianu I (1993) Romania's water resources and their use. GeoJournal 29(1):19–30

30. Romanescu G, Gabriela Romanescu, Stoleriu C, Ursu A (2008) Inventarierea si tipologia zonelor umede si apelor adanci din Podisul Moldovei. Editura Terra Nostra, Iasi

31. Romanescu G (2015) Managementul apelor. Amenajarea hidrotehnica a bazinelor hidrografice si a zonelor umede. Editura Terra Nostra, Iasi

32. Romanescu G (2018) Inundatiile: calamitate sau normalitate? Studii de caz: bazinele hidrografice Prut si Siret (Romania). Editura Transversal, Targoviste

33. Burtea MC, Sandu IG, Cioromele GA, Bordei M, Ciurea A, Romanescu G (2015) Sustainable exploitation of ecosystems on the Big Island of Braila. Rev Chim (Bucharest) 66(5):621–627

34. Burtea MC, Ciurea A, Bordei M, Romanescu G, Sandu AV (2015) Development of the Potential of Ecological Agriculture in the Village Ciresu, County of Braila. Rev Chim (Bucharest) 66(8):1222–1226

35. Romanescu G, Cojocaru I (2010) Hydrogeological considerations on the western sector of the Danube Delta—a case study for the Caraorman and Saraturile fluvial-marine levees (with similarities for the Letea levee). Environ Eng Manag J 9(6):795–806

36. Romanescu G, Purice C (2013) The sector of fluvial limans in the southwest of Dobruja—genesis, morphographic and morphometric features. Air and water. Components of the environment, Presa Universitara Clujeana, pp 47–54

37. Romanescu G, Mihu-Pintilie A, Trifanov C, Stoleriu CC (2018) The variations of physico-chemical parameters during summer in Lake Erenciuc from the Danube Delta (Romania). Limnol Rev 18(1):21–29

38. Antipa G (1910) Regiunea inundabila a Dunarii. Starea ei actuala si mijloacele de a o pune in valoare. Instit. de Arte Grafice Carol Gobl, Bucuresti

39. Antipa G (1912) Regiunea inundabila a Dunarii. An. Inst. Geol. Rom., II

40. Antipa G (1913) Trei memorii privitoare la ameliorarea Regiunei inundabile a Dunarei. Bucuresti

41. Antipa G (1916) Pescaria si pescuitul in Romania. Publicatiile Fundatiei Vasile Adamachi. Academia Romana, Bucuresti

42. Antipa G (1922) Dunarea si problemele ei stiintifice si economice. Academia Romana, Bucuresti
43. Botzan M (1991) Valorificarea hidroameliorativa a Luncii Dunarii romanesti si a Deltei. Redactia de Propaganda Tehnica Agricola, Bucuresti
44. Botzan M (2004) Calauza pentru Dunarea romaneasca. Editura Academiei Romane, Bucuresti
45. Ionescu-Sisesti G (1933) Lunca Dunarii si punerea ei in valoare. Imprimeria Centrala, Bucuresti
46. Looy KV, Honnay O, Bossuyt B, Hermy M (2003) The effects of river embankment and forest fragmentation on the plant species richness and composition of floodplain forests in the Meuse Valley, Belgium. Belg J Botany 136(2):97–108
47. Vidrascu IG (1911) Istoricul indiguirilor fluviale si maritime. Tipografia Curtii Regale F. Göbl F. II, Bucuresti
48. Vidrascu I (1921) Valorificarea regiunii inundabile a Dunarii. Tipografia Urbana, Bucuresti
49. Visinescu I, Bularda M (2007) Valorificarea luncii indiguite a Dunarii, nu reinundarea terenurilor agricole. Revista «Profitul agricol» 51–52:35–36
50. Visinescu I, Bularda M (2007) Exploatarea ameliorativa complexa a incintelor indiguite – solutii antiseceta de productii agricole. Cereale si plante tehnice 12:11–13
51. Ionus O, Licurici M, Patroescu M, Boengiu S (2013) Romania assessment of flood-prone stripes within the Danube drainage area in the South-West Oltenia Development Region. Nat Hazards 75:S69–S88
52. Manfreda S, Iacobellis V, Gioia A, Fiorentino M, Kochanek K (2018) The impact of climate on hydrological extremes. Water 10:802. https://doi.org/10.3390/w10060802
53. Mierla M, Romanescu G (2013) Hydrological flood risk assessment for Ceatalchioi locality, Danube Delta. Seminarul Geografic "Dimitrie Cantemir. Iasi, Romania 36:11–22
54. Mierla M, Nichersu I, Trifanov C, Nichersu Iuliana, Marin E, Sela F (2014) Links between selected environmental components and flood risk in the Danube Delta. Acta Zoologica Bulgarica Suppl. 7: 203–207
55. Mierla M, Romanescu G, Nichersu I, Grigoras I (2015) Hydrological risk map for the Danube delta—a case study of floods within the fluvial delta. IEEE J Sel Top Appl Earth Obs Remote Sens 8(1):98–104
56. Romanescu G, Stoleriu C (2014) Anthropogenic interventions and hydrological-risk phenomena in the fluvial-maritime delta of the Danube (Romania). Ocean Coast Manag 102:123–130
57. Ten Brinke WBM, Knoop J, Muilwijk H, Ligtvoet W (2017) Social disruption by flooding, a European perspective. Int J Disaster Risk Reduct 21:312–322
58. Van Leeuwen B, Pravetz T, Liptay ZA, Tobak Z (2016) Physically based hydrological modelling of inland excess water. Carpathian J Earth Environ Sci 11(2):497–510
59. Zeleňáková M (2009) Preliminary flood risk assessment in the Hornád watershed. River Basin Manag 5:15–24
60. Zeleňáková M (2015) Methodology of flood risk assessment from flash floods based on hazard and vulnerability of the river basin. Nat Hazards 79(3):2055–2071
61. Donita N, Popescu A, Pauca-Comanescu M, Mihailescu S, Biris IA (2006) Habitatele din Romania. Modificari conform amendamentelor propuse de Romania si Bulgaria la Directiva Habitate (92/43/EEC). Editura Tehnica Silvica, Bucuresti
62. Donita N, Biris LA, Filat M, Rosu C, Petrila M (2008) Ghid de bune practici pentru managementul padurilor din Lunca Dunarii. Editura Silvica, Bucuresti
63. Mihailovici JM, Gabor O, Petru S, Randasu S (2006) Solutii propuse pentru amenajarea fluviului Dunarea pe sectorul romanesc. Hidrotehnica 51:9–20
64. Moraru N, Ioanitoaia H, Minazzi M (2007) Principii privind realizarea lucrarilor de aparare impotriva inundatiilor a localitatilor situate in lunca Dunarii. Hidrotehnica 52(4–5):24–30
65. Romanescu G, Stoleriu C, Dinu C (2010) The determination of the degree of trophicity of the lacustrine wetlands in the eastern carpathians (Romania). Forum geografic. Studii si cercetari de geografie si protectia mediului 9(9):65–74
66. Zeleňáková M, Fijko R, Diaconu DC, Remeňáková I (2018) Environmental impact of small hydro power plant—a case study. Environments 5(1):1–12

67. Ioanitoaia H, Dobre V, Moraru N (2007) Un secol (1906–2006) de lucrari de indiguiri si amenajari hidroameliorative in lunca Dunarii. Hidrotehnica 52(1–2):41–45
68. Romanescu G, Pascal M, Pintilie-Mihu A, Stoleriu CC, Sandu I, Moisii M (2017) Water quality analysis in wetlands freshwater: common floodplain of Jijia-Prut Rivers. Rev Chim (Bucharest) 68(3):553–561
69. Hohensinner S, Sonnlechner C, Schmid M, Winiwarter V (2013) Two steps back, one step forward: reconstructing the dynamic Danube riverscape under human influence in Vienna. Water History 5:121–143
70. Hzami A, Amrouni O, Romanescu G, Stoleriu C, Mihu-Pintilie A, Saâdi A (2017) Satellite images survey for the identification of the coastal sedimentary system changes and associated vulnerability along the Western Bay of the Gulf of Tunis (Northern Africa). In: Kallel A, Ksibi M, Dhia HB, Khélifi N (eds) Recent advances in environmental science from the Euro-Mediterranean and surrounding regions. Advances in Science, Technology & Innovation, pp 1627–1631
71. Ebert S, Hulea O, Strobel D (2009) Floodplain restoration along the lower Danube: a climate change adaptation case study. Climate Dev 1(3):212–219
72. Sanders LM, Taffs K, Stokes D, Sanders CJ, Enrich-Prast A, Amora-Nogueira L, Marotta H (2018) Historic carbon burial spike in an Amazon floodplain lake linked to riparian deforestation near Santarém, Brazil. Biogeosciences 15:447–455
73. Schindler S, O'Neill FH, Biro M, Damm C, Gasso V, Kanka R, van der Sluis T, Krug A, Lauwaars SG, Sebesvari Z, Pusch M, Baranovsky B, Ehlert T, Neukirchen B, Martin JR, Euller K, Mauerhofer V, Wrbka T (2016) Multifunctional floodplain management and biodiversity effects: a knowledge synthesis for six European countries. Biodivers Conserv 25:1349–1382
74. Romanescu G (2014) The catchment area of the Milesian colony of Histria, within the Razim-Sinoie lagoon complex (Romania): hydro-geomorphologic, economic and geopolitical impli-cations. Area 46(3):320–327
75. Romanescu G, Dinu C, Stoleriu C, Romanescu AM (2010) Present state of trophic parameters of the main wetlands and deep waters from Romania. Present Environ Sustain Dev 4:159–174

Chapter 9
Deforestation and Frequency of Floods in Romania

Daniel Peptenatu, Alexandra Grecu, Adrian Gabriel Simion, Karina Andreea Gruia, Ion Andronache, Cristian Constantin Draghici and Daniel Constantin Diaconu

Abstract The extent of forest cuts is one of the greatest challenges of contemporary society, principally through the effects on ecosystems, air quality, hydrological regime and society. Changes in water leakage regime from the slopes are immediate responses of the ecosystem to short time cutting of forests from large areas. In this study, an assessment was made of deforested surfaces in the period 2000–2016, as well as the frequency of floods at the level of each administrative unit from Romania for the same time. Detailed analyzes focused on the main mountain groups and on a Danube meadow sector where the extent of deforestation is very great. The results highlight the obvious influence of forest cuts on large areas on the frequency of floods.

Keywords Deforestation · Hydrological regime · Floods · Mountain · Assessment · Frequency · Ecosystem

9.1 Introduction

The forest is one of the most important terrestrial ecosystems, the reasons being the area occupied by it (estimated by the United Nations at about 1/3 of the land area), the fact that it hosts over 80% of the terrestrial species (UNEP) and the important role in the chemical circuit of nature [1]. The existence of this ecosystem brings enormous benefits to human society, both from the point of view of protecting the environment and mitigating climate change, as well as economic and social. From this perspective, forest areas might play a key role in preserving biodiversity and mitigating the negative effects of climate change (increasing the proportion of greenhouse gases and increasing CO_2 due to the expansion of the deforested areas) [2–9].

The forest stands for an important part of the fight against pollution, climate change and in slowing the increase in greenhouse gas emissions by its atmospheric carbon

D. Peptenatu · A. Grecu · A. G. Simion · K. A. Gruia · I. Andronache · C. C. Draghici · D. C. Diaconu (✉)
Faculty of Geography, Research Center for Integrated Analysis and Territorial Management, University of Bucharest, Nicolae Bălcescu Blv. 1, District 1, 010041 Bucharest, Romania
e-mail: ddcwater@yahoo.com

© Springer Nature Switzerland AG 2020
A. M. Negm et al. (eds.), *Water Resources Management in Romania*, Springer Water,
https://doi.org/10.1007/978-3-030-22320-5_9

dioxide capture and storage function [10]. Also, the reduction of biodiversity and the occurrence of hydrological hazards (erosion enhancement, floods, landslides), represent other negative effects that can be mitigated by protecting and conserving the forest areas [11–18]. Thus, the growth of the deforested areas produces a negative impact on all climatic parameters, monitoring the forested areas is an action that has become very important, because any intervention can have negative consequences for both, the forest ecosystem and the territorial systems that are based on their existence and exploitation.

One way to positively influence climate change is to reduce the deforested areas [19–25], of this action resulting a number of significant benefits such as: reducing carbon emissions, protecting river basins, conserving biodiversity and soil quality [25–30].

The multiple benefits provided by the forest fund in economic, social and environmental terms make the forest multifunctional, where the production of wood can be supplemented by the ecological and recreational protection [31–33]. Multifunctionality gives it a very important global role in the fight against more topical issues in the social and environmental spheres: poverty, pollution and environmental protection. This role is recognized by a large number of official documents, both at regional level [34], and globally [35–37], an example is the Kyoto Protocol where the beneficial role played by the forest in regulating the concentration of carbon dioxide (CO_2) in the atmosphere [38]. According to these documents, the growth of deforested areas has become a matter of general interest over time, which has led to the need for vigorous action to mitigate the negative effects. Thus, the relationship between forested and deforested areas is a continuous concern at the global level, with sustained efforts being made to increase or at least to keep relatively constant the forested area [39–41].

Relevant, with regard to the benefits of the existence of forest areas, is also the crucial role played by the sustainability of water supply as well as in the alluvial transit of rivers [42]. Recent findings have provided correlations between forest relief and floods [12]. The study concluded that a 10% drop in natural forest area led to an increase in the flood frequency from 4 to 28%. In addition, the same 10% reduction in forests in the interviewed countries, led to an increase of 4–8% in total long-term floods. It is thus obvious that the forest has the natural capacity to absorb water when it rains and to release it gradually, later, in the rivers. Thus, deforestation in the area of major river basins is the cause of strong floods, as the water flow rate increases, causing erosion to the slopes. The hydrographical basin is a complex system where its morphometric and morphological elements generate, under the influence of water and energy inputs, the particularities of the hydrological regime. The river basin includes several areas, some closer to the natural state (forests, pastures, rocks), which similarly retain some precipitation, instead, other anthropogenic areas exhibit completely different manifestations. This is particularly the case for agricultural crops, which, depending on the type of crops, retain different amounts of water, of waterproofed land (access ways, platforms, civil and industrial buildings) which generally retain small amounts of water in the soil and have a high and rapid leakage [43, 44].

In a hydrographic catchment, upstream actions can influence the downstream sides. Therefore a better understanding and assessment of the effects of land use changes on the hydrological regime is of great importance for prediction and mitigation of flood risk as well as for planning and the sustainable development of space [45]. An example is the impact of the forest on maximum flows that depend on the different stages of growth, types of species, climate zones, soil types, morphology and general land management. Changing the precipitation leakage regime at the level of the natural or anthropogenic catchment area is well detailed and known by specialists in the field. The surface leakage, for example, ranges from about 10% in the natural mode to 55% in the anthropogenic regime. This increase in surface leakage occurs in the detriment of the infiltration of water (decreasing from 25% to only 5% in the anthropic environment) as well as the reintroduction in the hydrological circuit by evapotranspiration (from 40 to 30%) [46]. A consequence of the change in rainfall is the increase in the frequency of floods.

This represents a hydrological phenomenon that affects increasingly large areas of land generating economic, cultural and human losses. In recent decades, strong flood events have been recorded, with the entire European continent and the economies of many countries suffering significant losses [47–49]. In Europe, floods have become increasingly worrying for citizens, authorities, insurance companies, and policymakers, given that in the last 15 years the number of people affected by floods in European river basins has increased from 11 to 64 per decade [50, 51]. European countries with large floods registered since the 1970s include Romania, the Czech Republic, the Slovak Republic, the United Kingdom, Germany, Italy and Austria, Romania being affected by the highest frequency [52].

The most catastrophic floods are generated by high-intensity heavy rains. Due to deforestation, the occurrence of high-intensity precipitations in mountain areas often generates a liquid runoff of more than 100–200 mm/24 h, exceeding the soil absorption capacity and river transport capacity [53]. Between 1950 and 2006, there were 12 significant flood events in Europe (flash floods and overflows) with a number of deaths exceeding 100 in each case [54]. The severe floods in Europe, recorded in the first part of this century, were caused mostly by heavy rainfall. The years 2002 and 2006 have proven to be record years, with major flood events, recorded in six EU Member States (Austria, Czech Republic, France Germany, Hungary and Romania). Part of the upward trend in flood damage can be attributed to socio-economic factors such as population growth, extensive urbanization in flood-prone areas, as well as changes in land use, such as increasing deforestation and loss of wet or natural areas, for example through dyke construction, clogging of whales, etc. [55].

Europe is one of the most urbanized continents, with about 75% of its population settled in urban areas. The report by the European Environment Agency (EEA) in 2012 [56], shows that about one fifth of the European cities with more than 100,000 inhabitants are very vulnerable to floods produced by overflowing rivers and heavy rain such as "flash floods". In fact, for decades, urban drainage systems have been optimized to evacuate quantities of water of a certain size. Taking into account the trends of climate change and urbanization that have an ascendant course, this "transport capacity" has already proved to be inadequate in a large number of cities

[56]. Therefore, reducing the preventive risk through the spatial planning process of land use (the way forest exploitation is used for example) is an important but interdependent process with the management of flood risk [57, 58].

Besides the roles that play in social and environmental terms, the forest has also a very important impact globally and economically, role given mainly by the products it provides for many human communities, whose economic functionality is closely related to it. In this context, it is imperative to monitor and study the evolution of deforested areas and the causes of this evolution, especially since it is economically important for local communities whose economy is based on its existence [31, 59–62].

With the expansion of deforested areas, major imbalances in the territorial systems can occur, imbalances that may have negative consequences for the local community in terms of development. Therefore, identifying the key factors that lead to the amplification or favoring of what deforestation means and their analysis, gain an important role for the scientific environment, but especially for the authorities, due to the complexity of existing relationships between certain systemic components where the forest plays an important role [25, 27, 29, 63–66]. Although from a socio-economic and ecological point of view, the benefits of forest areas are well known, the deforested areas extend from year to year, so restraining forest areas has become one of the greatest challenges for decision makers. These factors are facing a great deal of pressure on the existing forestry background on the economic avenue and, in particular, on the economic sectors based on the exploitation and processing of wood.

In many of the specialty studies, there are several factors identified by the growth of deforested areas around the globe: agriculture, by increasing demand for agricultural land (demand for food), corruption, poverty, and the price of wood materials [67–72]. These factors determine the excessive exploitation of the forest fund with negative consequences on the functionality of the ecosystem, as well as the difficulties in the supply of wood [73]. Despite the fact that the protection of forest areas and the reduction of deforested areas are central concerns worldwide [5], the deforested area has increased year-on-year, with the deforestation rate for the period 2000–2010 of 13 million hectares per year, according to UNEP [74].

The amplification of the deforestation phenomenon is based on favorable causes such as the over-development of the economic sectors that are based on the exploitation and processing of wood, over the local support capacity, the smaller incomes of the local communities and, of course, the legislative gaps. On this line, clarifying are the experiences of states with large forest areas presented in some studies that talk about, for example, the pressure exerted on the forest ecosystem in Maine Province of the United States of America, where the deforestation of 1970–1990 with a devastating impact, was based on the intense activities of papermaking companies [75]. Another example may be Portugal, where in the times when the economy was on the rise, the forest management system was forced to adapt to the needs of development [76].

In Europe, the demand for wood for construction and fuel has become increasingly important and is a determining factor in transforming the natural environment by increasing the deforested areas [77, 78]. In the case of Romania, the forested area represents 27.5% of the total area, with 13.5% less than the European Union average (41%), the forest mainly occupying the mountainous area—59.7%. Romania is the country where the deforested area has accelerated in recent years, the most affected are mountain areas. Over the past period, deforested areas, both legally and illegally, have expanded very much, reaching alarming proportions, the main cause being the demand on an upward path, internally being the supply of raw materials for industry and fuel for heating homes, and externally, the export of lumber and logs. In a report of the Court of Accounts of Romania, it is shown that during the period 2005–2011, was illegally harvested 633,500 m^3 of timber, which represents an area of approximately 292,000 ha, and in the 2003 Greenpeace report, the volume of illegally taken wood in 2012 was about 120,000 m^3, which represents 4 million euros [79, 80]. The direct consequence of these economic activities is the reduction of forest areas and the multiplication of negative effects on both the natural and the socio-economic environment.

The prospects of the future global economy indicate an increase in the pressure of the economic component on forest areas. Thus, the creation of an effective environmental risk management systems is a key objective for decision-makers, especially in affected communities [10]. The forest ecosystem is subject to an increasing anthropogenic pressure [81–84], which makes diminishing or even stopping the phenomenon of deforestation becoming one of the greatest challenges that current decision makers are forced to face. Any intervention on this ecosystem may have negative consequences both at the level of the forest ecosystem and at the level of the territorial systems based on the exploitation of the forest. Consequently, the management of the forest fund becomes a matter of public interest of the highest importance [85]. If this management were carried out at a high level, it could have a positive impact on the environment and local economies [61, 86], so that the protection of forest areas would become a priority action that should be implemented as soon as possible [21].

A solution for sustainable wood exploitation is the development of efficient territorial management strategies that will have the role of reducing the imbalances generated by destabilizing forest ecosystems. These strategies should target the regulation of property rights, reducing deforestation by limiting industrial exploitation [87], reforestation of deforested or non-agricultural areas [88, 89], tightening legislation for illegal deforestation, supporting local sustainable economic activities by central and local public authorities, the involvement of the authorities in the better protection and management of the private forestry fund and facilitating access to technology, all of these measures representing priority action lines [90]. All these strategies represent a great need for the territorial systems whose functional profile is based on the resources offered by the forest areas. Thus, forest protection and management are priority actions that need to be implemented as soon as possible [21, 91].

9.2 Study Area

The researches focused on the pressure exerted on the forest fund by deforestation and the impact on the frequency of floods in Romania. To highlight the relationship, detailed analyzes were carried out on the main mountain groups, where deforestation has generated a clear increase on the frequency of floods.

9.3 Methods

The information on the territorial administrative units with the largest deforested forest areas was obtained by creating a statistical database. This was based on the extraction of time-series of 654178 Landsat 7 ETM+ satellite images, characterizing areas covered with forests and changes that took place from 2000 to 2016 [92]. The original raster was converted to Stereo 70 projection, cutting the image corresponding to the administrative boundaries of Romania. After redesigning the original file, using the ArcGIS 9.3 platform, the Raster image has been converted into a point vectorial file, the area represented by it is 741,941 m, with all 12 classes corresponding to the period 2000–2016.

To quantify areas at the local level, a spatial join was made for each administrative-territorial unit, the newly created file containing the attribute table, including the column that sums the pixels in each administrative-territorial unit. By multiplying the number of pixels, existing within each of the latter, with the surface of each pixel, the deforested areas were obtained for each administrative-territorial unit.

Considering the concentration of forests in the mountain area (Fig. 9.1), detailed analyzes were carried out at mountain group level, except the Poiana Ruscă, where the deforested areas are smaller.

In hydrology, the correct assessment of the hazard to which an area is subjected, is called as flood frequency analysis (FFA). A rather abundant literature [93], was published on the development and comparison of different methods of frequency analysis, to estimate the highest intensity and frequency of floods. The presence or absence of extreme events during the observation period makes it difficult to assess the occurrence of this extreme phenomenon in some river basins.

For this reason, alternative approaches have emerged. They are based on enriching the data set with another source of information. These approaches remain purely statistical. On the other hand, it is possible to use a more accurate approach based on the hydrological process. In this case, information on rainfall recorded on the basin is used. The main idea of the study is to use the statistical properties of precipitation and information on the precipitation flow in liquid leakage on the slopes. The rationalization of deforested areas and localities where floods occur, can provide information for decision-making in integrated water resource management. Indeed, many countries have developed their methodology to unify the techniques used by various flood risk practitioners (United Kingdom [94, 95], Spain [96], USA [97, 98], Australia [99, 100], and a number of European countries [101] (Fig. 9.2).

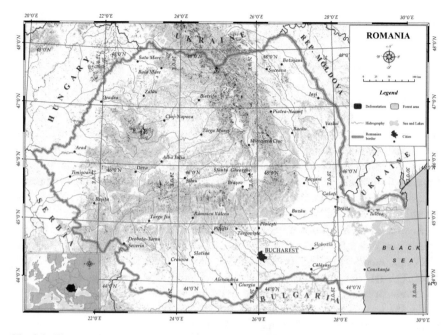

Fig. 9.1 The repartition of forests and deforested areas (2000–2016) in Romania

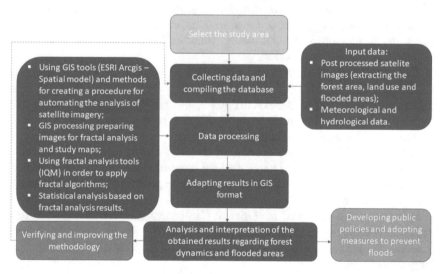

Fig. 9.2 Flowchart of the methodology for analyzing the interaction of the forest area and the floods

9.4 Results

9.4.1 The Evolution of Deforested Areas in Romania

Today, Romania has more than 6.5 million hectares of forests, which had continuously decreased since 1800 when the forested area of the Romanian historical provinces was 8,500,000 ha of forest, or 36% of the territory. The significant reduction was due to the agrarian reform of 1864 and the law on the establishment of communal sheds in 1920. Thus, more than one million hectares of forests have been removed from the forest fund, many of them being deforested, for conversion into the pasture. In 1948, after the adoption of the Constitution and the nationalization of forests, the area of the forest fund was 6,486,000 ha.

According to the National Institute of Statistics, in 1990, on December 31st, the area of the forest fund was 6,367,660 ha, of which the forest area—6.248.990 ha and other lands—118,670 ha. The surface of the forest fund on 31.12.2010 was 6,515,173 ha, out of which 4,363,000 ha, in public property and 2,152,173 ha, in private property. The public property of the state was 3.338.898 ha, of which 3.224.951 ha was covered with forests and 113,947 ha of other lands. The increase is due to the introduction of degraded areas in this category.

After 2000, there has been increasing pressure on forest resources, because of increased demand for export raw materials or processing capacities from Romania. At the same time, illegal cuts of forest have taken a major role in contributing to increasing the pressure on forest resources. Between 2000 and 2016, legal and illegal exploitation, approach 400.000 ha.

The largest deforested areas are in the North Group and the Central Group of the Oriental Carpathians (Figs. 9.3 and 9.4), where 64,812 and 100,613 ha were deforested in the period 2000–2017. In Fig. 9.3, the trend of compaction of deforested surfaces is observed, which leads to an amplification of the consequences, especially with regard to leakage on the slopes.

The Central Group of the Oriental Carpathians is distinguished by the extent of deforested forest areas, over 25% of the land deforested nationwide being concentrated here. After the year 2006, there is a spectacular growth of the restituted forests that have been cut.

In the Southern Group of the Oriental Carpathians, deforested areas are very fragmented at the beginning of the analyzed period (Fig. 9.5), but after 2007 there is a tendency for concentration, which anticipates an amplification of medium and long-term effects.

Făgăraș Mountains are distinguished throughout the analyzed period, by a compaction of the deforested areas, determined by the morphological characteristics of this group, which makes accessibility to forest resources much lower (Fig. 9.6).

Under these conditions, the surfaces that allow easy access to the operating equipment is exploited.

Parâng Group is in the category of mountain units where the pressure on forest resources is increasing after the year 2010 (Fig. 9.7). Even if in the period analyzed,

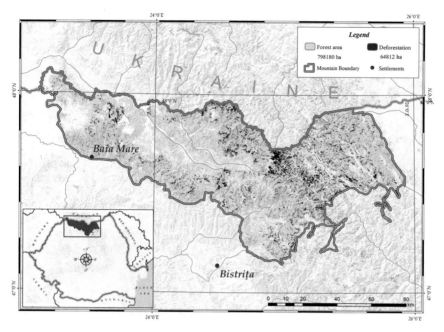

Fig. 9.3 The distribution of forests and deforested areas (2000–2016) of the North Carpathian Group

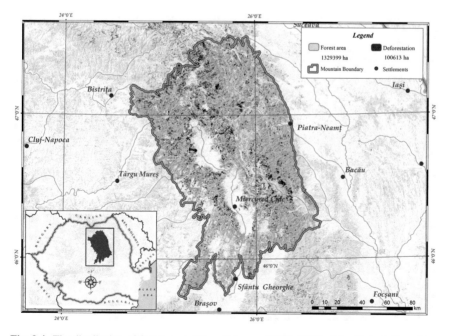

Fig. 9.4 The distribution of forests and deforested areas (2000–2016) of the Central Carpathian Group

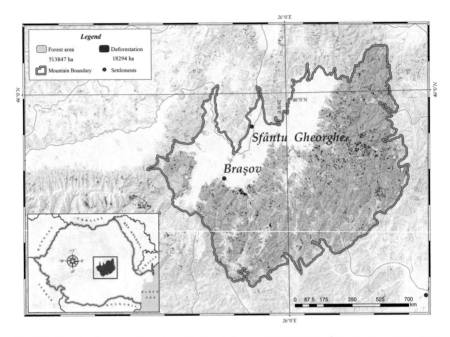

Fig. 9.5 The distribution of forests and deforested areas (2000–2016) of the Southern Carpathian Group

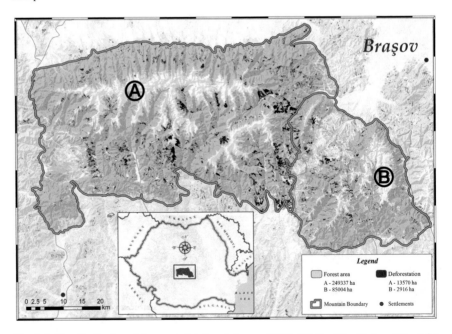

Fig. 9.6 The distribution of forests and deforested areas (2000–2016) of the Făgăraş Mountains (A) and the Bucegi Mountains (B)

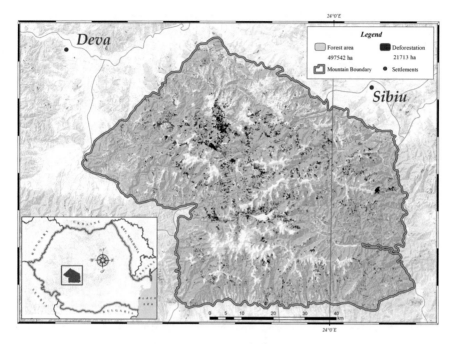

Fig. 9.7 The distribution of forests and deforested areas (2000–2016) of the Parâng Mountains

was deforested 21,713 ha, forecasts are increasing for the areas cut each year. From this rule, the Retezat-Godeanu Mountains (Fig. 9.8) deviates, where only 6079 ha have been cut due to the restrictions of the protected areas.

The economic pressure on forest resources in the Western Carpathians manifests strongly in the Apuseni Mountains (Fig. 9.9), where nearly 40,000 ha have been deforested. The compaction trend is increasingly visible in the northern area where the amplified manifestations of water dynamics on slopes uncovered by vegetation are evident. In the Poiana Ruscă Mountains and the Banat Mountains (Fig. 9.10), deforestation is a very fragmented process, with punctual compactation trends after the year 2010.

The economic pressure on forest resources is manifested in all areas occupied by the forest, the floodplains of the great rivers have been subjected in recent years to very high pressure. In the Danube meadow sector under analysis, over 3500 ha have been deforested (Fig. 9.11), to an important extent due to illegal cuts. Uncontrolled cuttings will lead to significant changes in the morphology of the meadows, as well at the level of drainage.

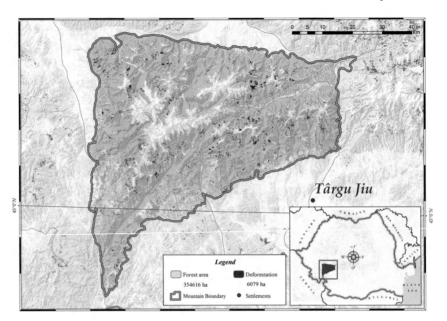

Fig. 9.8 The distribution of forests and grazed areas (2000–2016) of the Retezat-Godeanu Mountains

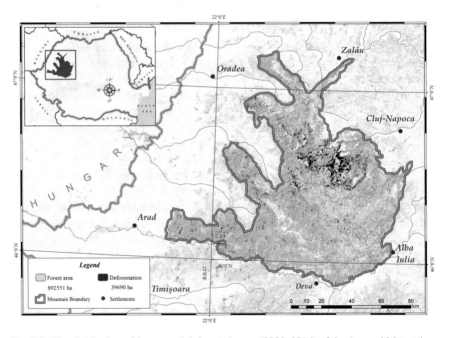

Fig. 9.9 The distribution of forests and deforested areas (2000–2016) of the Apuseni Mountains

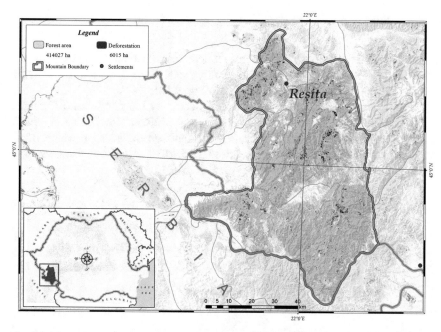

Fig. 9.10 The distribution of forests and deforested areas (2000–2016) of Banat Mountains

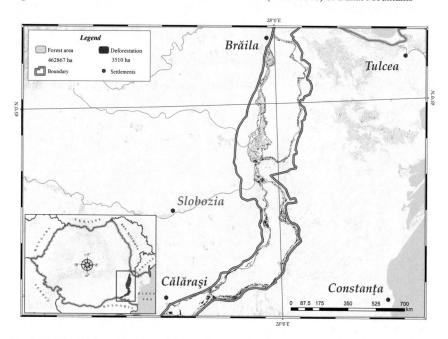

Fig. 9.11 Distribution of forests and deforested areas (2000–2016) in the Danube riverside between Călăraşi and Brăila

9.4.2 Frequency of Floods in Romania

Romania is administratively divided into 41 counties, which totals 3181 communes. Of this total of territorial-administrative units, a total of 2664, were affected by the floods at least once between 2005 and 2016 (Fig. 9.12).

In the period 2005–2016, the localities affected by the floods from Romania, has varied numerically but also as a geographical position. This is due to climate manifestations and land use. It is clearly distinguished the years in which floods occurred in the largest number of localities (2005, 2010, 2014), but also those with the lowest number (2009, 2011, 2012, 2015) (Fig. 9.13).

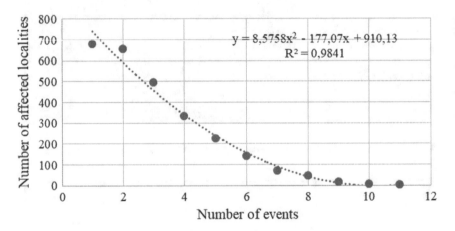

Fig. 9.12 The numerical frequency of the localities affected by floods in the period 2005–2016

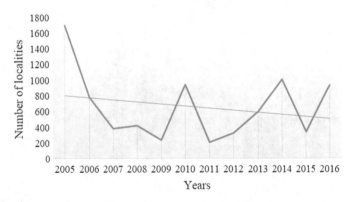

Fig. 9.13 The numerical evolution of the localities in which the floods occurred between 2005 and 2016, from Romania

Economic damages and human life losses caused by floods, are reported at the level of the territorial administrative unit, and not at hydrographic basin level. On the hydrographic basins, is being monitored the total affected area, the length of flooded sectors of the riverbed, the duration of the manifestation and the amount of water flow out.

Performing a territorial analysis of the territorial administrative units that were affected by the floods in the period 2005–2016, it is noted the existence of high-frequency localities but also some that are rarely affected or not at all.

There are situations when the floods affect almost the whole country (Fig. 9.13), as it did in 2005, or only regions within it—2007, 2009, 2011, 2015 (Fig. 9.14).

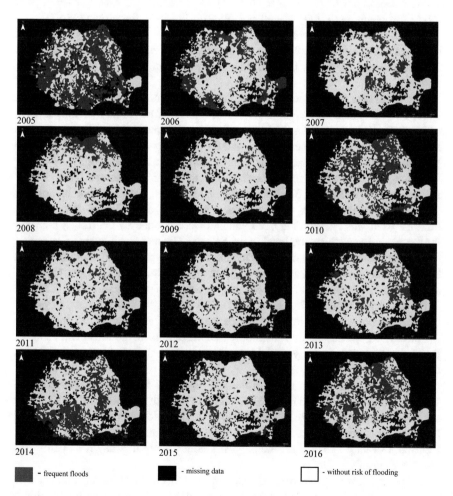

■ - frequent floods   ■ - missing data   □ - without risk of flooding

Fig. 9.14 The localities affected by the floods from 2005 to 2016

Between 2005 and 2016, in Alba County, out of 78 communes, 75 were affected by floods. Half of these have recorded events in at least 3–5 years of this time, and only one locality in 10 years out of 17.

Arad registers 51 localities with events, out of 78, where only four communes were affected by floods in 6 years.

Argeş recorded floods in all 102 localities, three localities being affected the most (in 7 years out of 17). In about 60 localities, there have been floods in at least 2–4 years. Also, other counties record floods at the level of all administrative units, as is the case with Neamţ County (83/83), Suceava (114/114), Iaşi (98/98), Vâlcea (89/89) or Caraş-Severin (77/77). Very close to flooding the entire surface are the counties Botoşani (76/78), Vrancea (70/73), Covasna (41/45), Galaţi (64/65), Sălaj (60/61), Hunedoara (67/69) as well as Dolj (100/111) and Teleorman (93/97).

The county with the fewest localities affected by the flood is Timiş with only 3 localities out of a total of 99. There are also localities that almost annually register at least one extreme event. Localities along the Danube, Olt, Jiu, Siret or Prut are affected.

Historical storms recorded on the Danube are those from April 2006 and June–July 2010. The first had a maximum flow at the entrance to the country (Baziaş section) of 15,800 m^3/s, representing the highest flow rate produced during the observation and hydrometric measurement period, and the second one, 13,350 m^3/s. Significant exceedances of flood rates across the Danube have been recorded for 27–69 days.

Damage to rivers has led to the flooding of 88.841 ha of 9 enclosures: Ghidici-Rast-Bistreţ, Bistreţ-Nedeia-Jiu, Jiu-Bechet, Bechet-Dăbuleni, Dăbuleni-Potelu-Corabia, Olteniţa-Surlari-Dorobanţu, Călărasi-Răul, Făcăeni-Vlădeni, Ostrov-Pecinega. The Calarasi Raul and Făcăeni Vlădeni incidents were flooded by controlled breaches for the reduction of the water level in the Feteşti - Cernavodă highway.

The June–July 2010 flood on the Danube was a two-peak flood, maximum flows being of 13,200 m^3/s in mid-June and 13,350 m^3/s in early July. The causes that generated these exceptional flood waves were large amounts of precipitation (100–230 l/m^2 in the upper basin of the Danube and 100–200 l/m^2 in the Savei basin), as well as the waterproofing of the sloping surfaces and the embankment of the minor bed. All these, conjugated, led to catastrophic manifestations on the lower course of the Danube, although, here, only partial tributaries rivers contributed to the formation of these floodplains.

In the Dobrogea region, the hydrological regime of the rivers has a number of features. High precipitation rates are recorded in a relatively short time (2 h) as was the case from 28/08/2004 in Constanţa County. The recorded rainfall has exceeded the known historical values in certain areas (the average annual rainfall value for Constanţa County is 377 l/m^2, the biggest amount fallen in a day is 111 l/m^2 and registered on July 1, 1992).

Large amounts of precipitations fallen in a relatively short time combined with leakage on the slopes and undersized sewer networks have caused severe floods affecting 14 localities, being recorded important damage. Among the most well-known floods recorded in this area are the ones from 2001, 2002, 2004, 2005.

Analyzing the effects of the 1975 and 1979 floods in the Argeș river basin, it is clear that the flood of 1975 (generated by a uniform rain on the basin) is the most important, totalizing on the Arges and Sabar rivers a flooded area of 50.850 ha. This hydrographic area is now highly regulated (having numerous dams, protective dikes, adduction channels, etc.) being hydrologically controlled for the most part. The problems generated here by the deforested surfaces are the generation of alluviums that contribute to the clogging of the accumulation lakes, thereby diminishing the flood-mitigation ability.

The main floods recorded on the Buzau River in the past years are 1969, 1971, 1975, 1980, 1984, 1991, 2005, and occurred either in May or in July. The flood of May 2005 affected the lower part of the basin (downstream of Măgura), respectively the upper course, the area of Întorsura Buzăului. The upstream area of the basin records some periodical local floods at the end of winter, especially when the melting of the ice causes the separation of some pieces of ice, that cause a blockage. The largest floods were recorded at the Măgura hydropower station (1948, 1975—the maximum flow of 2100 m^3/s, 1971, 1969, 1980, 1984), that took place before the construction of the Siriu dam (1994). However, the dam does not protect the downstream floods caused by the tributaries of the Buzau River, as happened in July 2004 at Nehoiu, where the tributaries of Buzău (Bâsca Rozilei, Nehoiu) flooded the lower part of the locality. In the Ialomița basin, the strongest floods (2001, 2005), were recorded on the Ialomița, Prahova and Teleajen rivers, but also on the tributaries of these rivers. Here, there were abundant precipitations, with increases in flows on the watercourses, reaching and exceeding the defenses on some rivers.

On the Siret river, in the county of Brăila, precipitation ranging from 22 to 130.7 l/m^2, was recorded in 2005 and the flood was caused by the destruction of the defense infrastructure. Among the most known floods produced in this hydrographic area are the ones from 1994, 1999, 2001, 2005.

In the last 100 years, the Olt river basin was affected by significant floods in the years 1923, 1924, 1930, 1932, 1948, 1955, 1970, 1972, 1975, 1991, 1998, 2000, 2007, 2014. Between 1930 and 1970, the largest floods in the Olt river basin occurred in 1932 (April) and 1948 (June). After 1970, the main floods occurred in: 1970 (May), 1971 (July), 1972 (October), 1973 (March), 1975 (July), 1991 (May-June-July) 2005 (May–July–August). The maximum flows recorded during the flood in 1975 were on the Olt River of 2.134 m^3/s in Râmnicu Vâlcea and 2570 m^3/s in Stoenești.

By comparison, in 1991, during the periods 26 May–5 June and 2–7 July, on the Lower Olt, at all hydrometric stations, maximum flows were recorded with the probability of exceeding of 5% (Q max at Izbiceni = 2542 m^3/s), smoothly transiting, as the lower Olt storage lakes, which function as dams, are rated at maximum insurance rates of 1 and 0.1%. In the valleys of the lower Olt, in the counties of Vâlcea and Olt, about 36.000 ha, 1700 households, 800 km of roads, 15 industrial facilities and public institutions, 60 bridges were flooded. The duration of the flood waves was between 2 and 15 days.

Because of the arrangement of the Jiu basin in the north-south direction and the maximum width development of the basin in the upper third, the floods that occur in the basin are concentrated on the middle course and attenuated in the lower course.

The statistical analysis of floods for a period starting from 1914, the year of the establishment of Podari, shows that most of the floods (over 90%) are of pluvial origin. On small tributaries, flood level elevations are given by rains that exceed 10 mm/24 h. On higher watercourses, rains that produce level increases and can cause floods are those that exceed 30 mm/24 h, and over the main course of the Jiu River, those that exceeds 40–50 mm/24 h. Among the most famous floods, there are mentioned the ones from 1961, 1965, 1966, 1969, 1970, 1972, 1975, 1978, 1989, 1991, 1998, 1999, 2000, 2004, 2005, 2006.

The most significant damages occurred in 1999 on the Jiu rivers (downstream of the Motru confluence) and Motru (downstream of Paduş) as a result of the floods that affected an area of 53.52 km^2, respectively 41.68 km^2. The floods that affected the Jiu basin in 1999, were born as a result of discharging of a very large amounts of water generated by the displaced air masses from southwest Europe that were stationed in the western part of Oltenia. Cumulative rainfall for July 11–15, 1999 measured up to 212.8 mm at Tarniţa pluviometric station and 243.9 mm at Tarmigani station, Mehedinţi County; 268.2 mm at Siseşti station, 123.8 mm at Câmpul lui Neag, respectively 125.7 mm at Valea de Peşti station from Hunedoara county; 191.5 mm at Godineşti station from Gorj county, 42 mm at Filiaşi station and 53 mm at Răcari pluviometric station from Dolj county.

In 1966 floods were flooded throughout Banat and culminated with the breakage of the defensive dam on the Timiş river, left shore in Gad locality area. The known cause of these floods was a large amount of rainfall falling on the surface of the basin. This exceptional flood generated peak flows of 1100 m^3/s in Lugoj and 1416 m^3/s at the Şag. The floods from February 1999, were caused by floods of pluvio nival origin and occurred in the Bega Veche, Bega (middle and lower basin), Timiş (lower basin), Pogăniş, Moraviţa, Bârzava (middle and lower basin).

In 2000, due to the existence in the mountain and hill area of a consistent snow cover, over which there were recorded large quantities of liquid precipitations (120 mm in 24 h), the rivers came out of the riverbed. In April 2005, a pluviometric surplus was recorded in Banat. In 2006, in April, rainfall exceeded the multiannual average (10.4 mm at Cebza to 28.6 mm at Gataia). The nature of the precipitations and the recorded water quantities led to the formation of a single flood on most of the watercourses, a flood that had the effect of overtaking defense.

The flood produced in the upper basin of the Arieş river from 10 to 15 March 1981 was produced because of intense and hot rains and the release of water from the snow layer due to the sudden warming of the weather. Maximum-recorded debits: 35 m^3/s at Câmpeni, 860 m^3/s at Baia de Arieş, 820 m^3/s at Buru and 780 m^3/s at Turda. Between December 1995 and January 1996, the floods produced in the Mureş river basin were due to precipitation in the form of rain and melting of snow in mountain areas due to high temperatures (13–14 °C). The large rainfalls falling in the period 11.06.1998–20.06.1998 represent the cause of flood formation in 1998. The highest recorded flows were 1660 m^3/s in Alba Iulia, 1716 m^3/s in Acmariu and 1660 m^3/s at Gelmar on Mureş. The 2005 and 2010 floods were local, with torrent character, generated by precipitation in the form of local avers.

In the Crişuri river basin, floods are formed in all seasons of the year, depending on the moisture intake brought by the air masses. The floods of 1970 had as a triggering factor a heavy rainfall regime. The whole period from January 1 to May 10 was rich in rainfall, with double quantities of water in relation to the normal multiannual values, which caused the infiltration of at most 20–25% of the amount of water lost, the rest entering the drainage process. The value of damage to the Crişuri catchment area amounted to more than 128 million euro at present (390 million lei in 1970). Between May 7 and 9, 1989, in the river basins of the Barcău, Crişul Repede and Crişul Negru rivers, large amounts of precipitations fell, which overlapped over a period when the soil was soaked with water, which caused most of the rainfall to contribute to surface leakage. The floods formed between December 1995 and January 1996 had the following causes: warming and rapid melting of the snow layer, especially in mountainous and hilly areas, the disappearance of ice on rivers, liquid precipitations marked quantitatively in the conditions of a frozen soil, unable to allow infiltration, and water leakage on the slopes to the riverbeds.

Among the most famous floods in the Someş—Tisa hydrographic area, there are mentioned 1970, 1974, 1975, 1978, 1979, 1980, 1981, 1989, 1993, 1995, 1998, 2000, 2001, 2006 and 2008.

In the 1970s, floods were due to long-lasting heavy rainfall, which exceeded the critical thresholds, causing the overflow of most rivers. In 1974, during the June-July period, the floods occurred due to precipitations that far exceeded the average annual precipitation in the watersheds Tur, Someş, Crasna and Vişeu.

In 1980, the floods were due to a repeated fall on a saturated soil, which exceeded the critical thresholds, exceeding the defense rates on most of the watercourses in this hydrographic area. On the Almaş, Crasna, Agrij, Sălaj, Budac, Zalău, Olpret, Iza and Batarci rivers, have overcome 2–3 times the hazard. On the Tur river, the hazard was exceeded 4 times. The most important floods occurred in July when at the Sighet Weather Station were registered 178.6 l/m², at Ocna Şugatag 158.6 l/m², 146.5 l/m² at the Negreşti Oaş pluviometric station, 137.5 l/m² at Vama and 116.9 l/m² at Turulung. On the lower courses, due to the small slope, the floods lasted up to 7 days.

In 1981, the floods were due to precipitation, melting of snow and leakage on the slopes, causing on most of the watercourses in the basin, rises in levels that have exceeded the defenses. In 1995 and 2001, the floods occurred in December due to precipitation in the form of rain and rapid melting of snow, favoring the emergence of flood waves on Şomesul Mare and tributaries, Tisa and tributaries, Lăpuş, Cavnic, Tur and tributaries. In 1998, the floods from June were due to particularly abundant precipitation that caused rich leaks on the slopes and in 2006, on the Ilişua river, a tributary of the Somesul Mare river, was a rapid flood, in June, as well. The total duration of the flood was 34 h and the maximum flow about 212 m³/s (Cristeştii Ciceului hydrometric station). This has caused 13 human lives and a lot of material damage.

In 2008, major damage occurred due to the floods from April, May, July (Maramureş County) and May (Satu Mare County). Watercourses on which damage has been reported were: Tisa, Vişeu, Iza and Mara from Maramureş County, plus the Barlogelor Valley and Lechincioara in Satu Mare County.

The maximum flows recorded on the Siret river and its tributaries were: 865 m³/s on Siret at Şerbaneşti Huţani, 1140 m³/s at Lespezi and 1.920 m³/s in Drageşti, in July 1969. Exceptional flow of 1700 m³/s was recorded on the Trotuş river at Radeana—Vrânceni in May 1975, of 3270 m³/s on the Siret river at Cosmeşti in May 1991 and 1550 m³/s on the Tazlau river at Helegiu, in July 1991.

In 2005, in the Siret and Trotuş hydrographic basins, floods had occurred due to exceptional floods. The flood produced in the summer of 2005 measured the maximum flow at the Vrânceni hydrometric station on the Trotuş river (2800 m³/s—the probability of 0.5%). On the inferior sector of Siret river, at the Lungoci hydrometric station, the peak of the flood had an estimated flow rate of 4500 m³/s, respectively a probability of exceedance of 0.5%, Siret becoming the inner river with the highest flow recorded on the territory of Romania.

In 2005, the largest floods ever recorded occurred on the Putna and Râmnicu-Sărat rivers. On the Putna river, at Boţărlău station, the peak flow value was 1.323 m³/s, being considered the first of the chronological series of maximum flows with the probability of exceeding 2.5%.

The main historical floods produced in Prut-Bârlad, are those recorded in 1965, 1969, 1985, 1988, 1991, 1998, 2005, 2008 and 2010. As a result of the maximum precipitations, there were floods that led to increases in debts on the main rivers Prut, Jijia, Bahlui and Bârlad, as well as their tributaries. The maximum flows recorded in different sections on these water courses, mentioned above during floods, are as follows: the year 1965-2240 m³/s on the Prut river in Rădăuţi—Prut and approx. 200 m³/s on Bârlad at Negreşti, Vaslui and Bârlad, year 1969-731 m³/s on the Prut river at Rădăuţi—Prut, 394 m³/s on Jijia at Todireni, approx. 300 m³/s on Bârlad at Negreşti and Vaslui, and 380 m³/s at Tecuci. In 1985 there were recorded 1250 m³/s on the Prut river in Rădăuţi—Prut, 212 m³/s on the Jijia river at Todireni, approx. 390 m³/s on Bârlad in Negreşti and 250 m³/s in Vaslui. In 1988, the maximum flows were 1780 m³/s on Prut to Rădăuţi—Prut and 104 m³/s on Jijia to Victoria. The flood produced on the Prut River in June–July 2010 reached a maximum flow rate of 2137 m³/s in Rădăuţi—Prut, which was propagated downstream.

By performing a comparative analysis, at the level of the territorial administrative unit, in terms of deforested areas and respectively the number of years in which floods occurred, there is a direct relationship between them (Figs. 9.15 and 9.16).

The Neamţ and Suceava counties are the most affected during the last 15–20 years of deforestation and implicitly floods. Neamţ County records a very high frequency of floods, localities being often affected by rapid floods. Localities as Agapia, Alexandru cel Bun, Damuc and Piatra Neamţ, are very often flooded. The locality from the country, which recorded the highest frequency during this time frame is Tarcău, from Neamţ county.

A second county affected, as surface and frequency, is Suceava, where localities such as Cornu Luncii, Fălticeni, Liteni and Ulma record the most frequent floods. The region is complemented by Iaşi County, where there are also frequently flooded localities, such as Tibana. Another county that is highlighted by an unusually high frequency is Alba County, where Zlatna locality is affected in the analyzed period in 10 of the 17 years by floods.

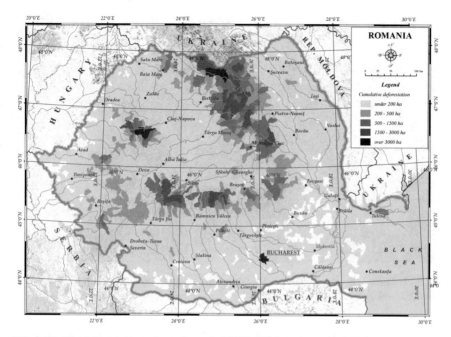

Fig. 9.15 The deforested area from the period 2000–2016, at the territorial administrative unit level, Romania

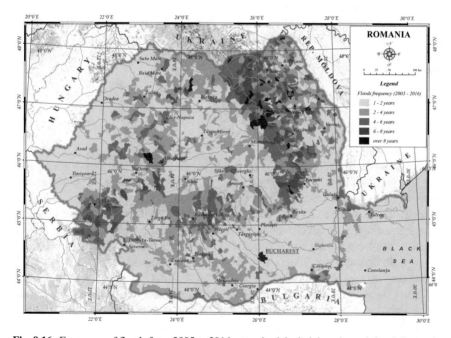

Fig. 9.16 Frequency of floods from 2005 to 2016, at territorial administrative unit level, Romania

Fig. 9.17 The torrent character of the leakage, the alluvial transport and the equipment used in the wood processing, Neamț County, July 2018

By maintaining and developing forest areas, is obtained the increasing of water retention in slopes, the decreasing of the flow rate and implicitly the increasing of the infiltration into the soil [102].

Decreasing soil erosion, which leads to degradation of the slopes, clogging the riverbeds and the storage basin of the lakes, it is another important role of the forest that contributes to the decrease of the number of exceptional events as well as their size (Fig. 9.17).

9.5 Conclusions

The pressure on forest resources is one of Romania's biggest problems after 1990. The low incomes of the population have made that illegal cut, for supplement revenue, to have a great scale. The lack of the general cadastral survey has made that the evidence of forest areas to be uncertain, favoring illegalities in resource management. Encouraging export of woody raw materials, the uncontrolled development of primary wood processing capacities and the tacit support from the authorities, for the exploitation of wood, have led to the cutting of compact forest areas in a very short time, without replanting on the deforested surfaces.

Under these circumstances, cutting the forest to nearly 400,000 ha has caused major imbalances in ecosystems, changing the drainage of the water on the slopes being the immediate consequence.

The zoning and the analysis of hydrological or physical-geographic parameters can provide the premises for a detailed analysis of the processes taking place in a river basin. This method of analysis provides the possibility of taking information from various researchers, which in turn, by their own methods, to deepen and to present the link from the parameters in different forms (analytical, graphical, etc.). The combination of the two parameters is real and obvious, the deforestation action having significant implications for floods.

With the ever-increasing global climate change, with the recording of extreme climatic and hydrological phenomena, more and more frequently, environmental initiatives and policies must take into account the impacts of forests on water resources at local, regional and continental level.

The role of forests on water resources and of the hydrological regime as well as the climate provides a powerful mitigation tool for the effects of global climate change.

9.6   Recommendations

It is necessary to carry out a spatial analysis of the deforested areas and the scale of the phenomenon, detailing the impact on changes on climatic factors, the degradation of the slopes, and last but not least on the local and regional economy.

A call to action on forests, water and climate is emerging on many fronts. Taking into account the effects of forests on water and climate, suggests that this call is urgent. Stimulating regional and continental approaches can contribute to the development of more appropriate governance, thus increasing the chances of success.

Acknowledgements This work was supported by the Research Center for Integrated Analysis and Territorial Management.

References

1. Kolström M, Lindner M, Vilén T, Maroschek M, Seidl R, Lexer MJ, Netherer S, Kremer A, Delzon S, Barbati A, Marchetti M, Corona P (2011) Reviewing the science and implementation of climate change adaptation measures in European Forestry. Forests 2:961–982
2. Pachauri RK, Reisinger A (2007) Climate change 2007: synthesis report, report of the intergovernmental panel on climate change. IPCC, Geneva, Switzerland
3. Pachauri RK, Meyer LA (2014) Climate change 2014: synthesis report, report of the intergovernmental panel on climate change. IPCC, Geneva, Switzerland
4. Thomas CD, Cameron A, Green RE, Bakkenes M, Beaumont LJ, Collingham YC, Erasmus BFN, De Siqueira MF, Grainger A, Hannah L, Hughes L, Huntley B, van Jaarsveld AS, Midgley GF, Miles L, Ortega-Huerta MA, Peterson AT, Phillips OL, Williams SE (2004) Extinction risk from climate change. Nature 427(6970):145–148
5. Foley JA, De Fries R, Asner GP, Barford C, Bonan G, Carpenter SR, Chapin FS, Coe MT, Daily GC, Gibbs HK, Helkowski JH, Holloway T, Howard EA, Kucharik CJ, Monfreda C, Patz JA, Prentice IC, Ramankutty N, Snyder PK (2005) Global consequences of land use. Science 309(5734):570–574

6. Spittlehouse DL (2005) Integrating climate change adaptation into forest management. The Forestry Chron 81:691–695
7. Streck C, Scholz SM (2006) The role of forests in global climate change: Whence we come and where we go. Int Aff 82:861–879
8. Betts RA, Falloon PD, Goldewijk KK, Ramankutty N (2007) Biogeophysical effects of land use on climate: Model simulations of radiative forcing and large-scale temperature change. Agric For Meteorol 142(2–4):216–233
9. McKinley DC, Ryan MG, Birdsey RA, Giardina CP, Harmon ME, Heath LS, Houghton RA, Jackson RB, Morrison JF, Murray BC, Pataki DE, Skog KE (2011) A synthesis of current knowledge on forests and carbon storage in the United States. Ecol Appl 21(6):1902–1924
10. Asante P, Armstrong GW, Adamowicz WL (2011) Carbon sequestration and the optimal forest harvest decision: a dynamic programming approach considering biomass and dead organic matter. J Forest Econ 17(1):3–17
11. Dymond JR, Ausseil AG, Shepherd JD, Buettner L (2006) Validation of a region-wide model of landslide susceptibility in the Manawatu-Wanganui region of New Zealand. Geomorphology 74:70–79
12. Bradshaw CJA, Sodhi NS, Peh KSH, Brook BW (2007) Global evidence that deforestation amplifies flood risk and severity in the developing world. Glob Change Biol 13:2379–2395
13. Whitehead D (2011) Forests as carbon sinks–benefits and consequences. Tree Physiol 31(9):893–902
14. Zanini KJ, Bergamin RS, Machado RE, Pillar VD, Müller SC (2014) Atlantic rain forest recovery: successional drivers of floristic and structural patterns of secondary forest in Southern Brazil. J Veg Sci 25:1056–1068
15. Rudel TK, Sloan S, Chazdon R, Grau R (2016) The drivers of tree cover expansion: global, temperate, and tropical zone analyses. Land Use Policy 58:502–513
16. Blistanova M, Zeleňáková M, Blistan P, Ferencz V (2016) Assessment of flood vulnerability in Bodva river basin, Slovakia. Acta Montanistica Slovaca 21(1):19–28
17. Benchimol M, Talora DC, Mariano-Neto E, Oliveira TLS, Leal A, Mielke MS, Faria D (2017) Losing our palms: the influence of landscape-scale deforestation on Arecaceae diversity in the Atlantic forest. For Ecol Manage 384:314–322
18. Borrelli P, Panagos P, Märker M, Modugno S, Schütt B (2017) Assessment of the impacts of clear-cutting on soil loss by water erosion in Italian forests: first comprehensive monitoring and modelling approach. CATENA 149(3):770–781
19. Watson RT, Noble IR, Bolin B, Ravindranath NH, Verardo DJ, Dokken DJ (eds) (2000) IPCC. Cambridge University Press, UK, p 375
20. Houghton RA (2005) Tropical deforestation as a source of green house gas emissions. In: Mountinho P, Schwartzman S (eds) Tropical deforestation and climate change, IPAM: Belem, Brazil and Environmental Defense, Washington, DC, pp 13–21
21. Stern N (2006) Part VI: international collective action. The economics of climate change. The Stern Review Cambridge University Press, Cambridge
22. Nabuurs GJ, Masera O, Andrasko K, Benitez-Ponce P, Boer R, DutschkeM, Elsiddig E, Ford-Robertson J, Frumhoff P, Karjalainen T, Krankina O, Kurz WA, Matsumoto M, Oyhantcabal W, Ravindranath NH, Sanz Sanchez MJ, Zhang X (2007) Forestry. In climate change (2007): Mitigation. Contribution of Working Group III to the fourth assessment report of the intergovernmental panel on climate change. In Metz B, Davidson OR, Bosch PR, Dave R, Meyer LA (eds). Cambridge University Press, Cambridge, United Kingdom and New York, NY, USA
23. Eliasch J (2008) Part I: the challenge of deforestation. Climate change: financing global forests the eliasch review, Earthscan, London. https://www.gov.uk/government/uploads/system/uploads/attachment_data/file/228833/9780108507632.pdf. Accessed 06 March 2015
24. Strassburg BBN, Rodrigues ASL, Gusti M, Balmford A, Fritz S, Obersteiner M, Kerry Turner R, Brooks TM (2012) Impacts of incentives to reduce emissions from deforestation on global species extinctions. Nat Clim Change 2(5):350–355
25. Thu-Ha DP, Brouwer R, Davidson M (2014) The economic costs of avoided deforestation in the developing world: a meta-analysis. J Forest Econ 20(1):1–16

26. Sedjo RA, Wisniewski J, Sample AV, Kinsman JD (1995) The economics of managing carbon via forestry: assessment of existing studies. Environ Resour Econ 6(2):139–165
27. Chomitz KM, Kumari K (1998) The domestic benefits of tropical forests: a critical review. The World Bank Res Obs 13(1):13–35
28. Stickler CM, Nepstad DC, Coe MT, McGrath DG, Rodrigues HO, Walker WS, Soares-Filho BS, Davidson EA (2009) The potential ecological costs and cobenefits of REDD: a critical review and case study from the Amazon region. Glob Change Biol 15(12):2803–2824
29. World Bank (2011) Estimating the Opportunity Costs of REDD+ : a training manual. http://www.theredddesk.org/sites/default/files/resources/pdf/2011/oppcostsreddmanual.pdf. Accessed 06 March 2015
30. Strassburg BBN, Rodrigues ASL, Gusti M, Balmford A, Fritz S, Obersteiner M, Kerry Turner R, Brooks TM (2012) Impacts of incentives to reduce emissions from deforestation on global species extinctions. Nat Clim Change 2(5):350–355
31. Turner MG (1989) Landscape ecology: the effect of pattern on process. Annu Rev Ecol Syst 20:171–197
32. Price C, Rametsteiner E, Guldin R (2003) Substantive element "economic aspects of forests", including "trade". Science and technology—building the Future of the World's Forests. Contribution to the third Session of the United Nation Forum on Forests in Geneva, 26 May–6 June 2003, IUFRO Occasional Paper, 15, pp 5–8
33. Tempesta T, Marangon F (2008) The total economic value of Italian Forest Landscapes. In: Cesaro L, Gatto P, Pettenella D (eds) The multifunctional role of forest—policies, methods and case studies. EFI Proceedings, 55, pp 319–326
34. European Commission (2008) The UE forest action plan 2007–2011. http://ec.europa.eu/agriculture/fore/publi/2007_2011/brochure_en.pdf. Accessed 21 March 2015
35. UNCED (1992). United Nations Conference on Environment & Development Rio de Janerio, Brazil, 3 to 14 June 1992, chapter 11. https://sustainabledevelopment.un.org/content/documents/Agenda21.pdf. Accessed 26 March 2015
36. UNCSD (2012) Report of the United Nations Conference on Sustainable Development Rio de Janeiro, Brazil 20–22 June 2012, 37. http://www.uncsd2012.org/content/documents/814UNCSD%20REPORT%20final%20revs.pdf. Accessed 03 April 2015
37. FAO (2014) State of the world's forests enhancing the socioeconomic benefits from forests. http://www.fao.org/3/cf470fab-cc3c-4a50-b124-16a306ee11a6/i3710e.pdf. Accessed 07 March 2015
38. Brown AHD, Young AG, Burdon JJ, Christidis L, Clarke G, Coates D, Sherwin W (2000) Genetic indicators for State of the Environment Reporting Department of Environment, Sports and Territories Technical Report, Canberra, Australia
39. Chakravarty S, Ghosh SK, Suresh CP, Dey AN, Shukla G (2012) Deforestation: causes, effects and control strategies. In: Okia CA (eds) Global Perspectives on Sustainable Forest Management, pp 1–28
40. Gao Z, Cao X, Gao W (2013) The spatio-temporal responses of the carbon cycle to climate and land use/land cover changes between 1981 and 2000 in China. Front Earth Sci 7(1):92–102
41. Boucher D (2014) How Brazil Has Dramatically Reduced Tropical Deforestation. Solutions 5(2):66–75
42. Haigh MJ, Jansky L, Hellin J (2004) Headwater deforestation: a challenge for environmental management. Glob Environ Change 14:51–61
43. Chester LA, Gibbons J (1996) Impervious surface coverage: the emergence of a key environmental indicator. J Am Plan Assoc 62(2):255
44. Konrad CP (2003) Effects of urban development on flood. U.S. Geological Survey
45. Chen Y, Xu Y, Yin Y (2009) Impacts of land use change scenarios on storm runoff generation in Xitiaoxi basin, China. Quat Int 208(1–2):121–128
46. How Urbanization Affects the Water Cycle. http://www.swrcb.ca.gov/rwqcb2/water_issues/programs/stormwater/ISDC/Nemo_Fact_Sheet.pdf. Accessed 7 April 2017
47. Balica S, Dinh Q, Popescu I, Vo TQ, Pham DQ (2014) Flood impact in the Mekong Delta, Vietnam. J Maps 10(12):257–268

48. Miklin J, Hradecky J (2016) Confluence of the Morava and Dyje Rivers: a century of landscape changes in maps. J Maps 12(4):630–638
49. Olariu P, Obreja F, Obreja I (2009) Unele aspecte privind tranzitul de aluviuni din bazinul hidrografic Trotus si de pe sectorul inferior al raului Siret in timpul viiturilor exceptionale din anii 1991 si 2005 (in romanian). Analele Universitatii Stefan cel Mare Suceava, Geografie, XVIII, pp 93–104
50. Brinke WBM, Knoop J, Muilwijk H, Ligtvoet W (2017) Social disruption by flooding, a European perspective. Int J Disaster Risk Reduct 21:312–322
51. Van Alphen J, Martini F, Loat R, Slomp R, Passchier R (2009) Flood risk mapping in, experiences and best practices. J Flood Risk Manag 2(4):285–292
52. Faccini F, Luino F, Sacchini A, Turconi L, De Graff J (2015) Geohydrological hazards and urban development in the Mediterranean area: an example from Genoa (Liguria, Italy). Nat Hazard Earth Syst Sci 15:2631–2652
53. Romanescu G, Stoleriu CC (2013) Causes and effects of the catastrophic flooding on the Siret River (Romania) in July–August 2008. Nat Hazards 69:1351–1367
54. Barredo JI (2006) Major flood disasters in Europe: 1950–2005. Nat Hazards 42(1):125–148
55. Feyen L, Barredo JI, Dankers R (2009) Implications of global warming and urban land use change on flooding in Europe. Institute for Environment and Sustainability, DG JRC, European Commission
56. EEA Report (2012) Climate change, impacts and vulnerability in Europe 2012. ISSN 1725-9177
57. Friesecke F (2004) Precautionary and sustainable flood protection in Germany—strategies and instruments of spatial planning. In: 3rd FIG regional conference Jakarta, Indonesia, October 3–7
58. Montz BE (2000) The generation of flood hazards and disasters by urban development of floodplains. In: Parker DJ (ed) Floods, vol I. Routledge, London & New York, pp 116–127
59. Zhang J, Alavalapati JRR, Shrestha RK, Hodges AW (2005) Economic impacts of closing national forests for commercial timber production in Florida and Liberty County. J Forest Econ 10(4):207–223
60. Gios G (2008) Multifunctionality and the Management of Alpine Forest. In: Cesaro L, Gatto P, Pettenella D (eds) The multifunctional role of forest—policies, methods and case studies. EFI proceedings, 55, pp 47–54
61. Goio I, Gios G, Pollini C (2008) The development of forest accounting in the province of Trento (Italy). J Forest Econ 14(3):177–196
62. Paletto A, Ferretti F, Cantiani P, De Meo I (2012) Multi-functional approach in forest landscape management planning: an application in Southern Italy. Forest Syst 21(1):68–80
63. Sedjo RA, Wisniewski J, Sample AV, Kinsman JD (1995) The economics of managing carbon via forestry: assessment of existing studies. Environ Resour Econ 6(2):139–165
64. Peptenatu D, Sîrdoev I, Pravalie R (2013) Quantification of the aridity process in south-western Romania. Iranian J Environ Health Sci Eng 11:5
65. Zelenáková M, Dobos E, Kovácová L, Vágo J, Abu-Hashim M, Fijko R, Purcz P (2018) Flood vulnerability assessment of Bodva cross-border river basin. Acta Montanistica Slovaca 23(1):53–61
66. Pravalie R, Sîrdoev I, Peptenatu D (2014) Changes in the forest ecosystems in areas impacted by aridization in south-western Romania. Iranian J Environ Health Sci Eng 11(1):5
67. Angelsen A, Kaimowitz D (1999) Rethinking the causes of deforestation: lessons from economic models. The World Bank Res Obs 14:73–98
68. Meyfroidt P, Lambin EF (2011) Global forest transition: prospects for an end to deforestation. Annu Rev Environ Resour 36:343–371
69. Thongmanivong S, Fujita Y, Fox J (2005) Resource use dynamics and land-cover change in Ang Nhai village and Phou Phanang national reserve forest, Lao PDR. Research 36(3):382–393
70. Gibbs HK, Ruesch AS, Achard MK, Clayton MK, Holmgren P, Ramankutty N, Foley A (2010) Tropical forests were the primary sources of new agricultural land in the 1980s and 1990s. Proc Natl Acad Sci 107(38):16732–16737

71. Phalan B, Onial M, Balmford A, Green RE (2011) Reconciling food production and biodiversity conservation: land sharing and land sparing compared. Science 333(6047):1289–1291
72. Salam MA, Noguchi T (1998) Factors influencing the loss of forest cover in Bangladesh: an analysis from socioeconomic and demographic perspectives. J Forest Res 3(3):145–150
73. Sasaki N (2006) Carbon emissions due to land-use change and logging in Cambodia: a modeling approach. J Forest Res 11(6):397–403
74. UNEP (2011) Forests in a green economy-a synthesis. http://www.un-ngls.org/spip.php?page=article_s&id_article=3455. Accessed 21 April 2018
75. Acheson JM, McCloskey J (2008) Causes of deforestation: the maine case. Hum Ecol 36(6):909–922
76. Reboredo F, Pais J (2014) Evolution of forest cover in Portugal: A review of the 12th–20th centuries. J Forestry Res 25(2)
77. Kaplan JO, Krumhardt KM, Zimmermann N (2009) The prehistoric and preindustrial deforestation of Europe. Quatern Sci Rev 28(27–28):3016–3034
78. Stanga IC, Niacsu L (2016) Using old maps and soil properties to reconstruct the forest spatial pattern in the late 18th century. Environ Eng Manag J 15(6):1369–1378
79. Court of Accounts of Romania (2013) Synthesis of the audit report on "The Heritage Situation of the Forestry Fund in Romania, 1990–2012" (in romanian). București, 99–102, http://www.curteadeconturi.ro/Publicatii/economie7.pdf. Accessed 15 May 2018
80. Greenpeace România (2013) Illegal deforestations in the forests in Romania-2012. http://www.greenpeace.org/romania/ro/. Accessed 15 May 2018
81. Mertens B, Sunderlin WD, Ndoye O, Lambin EF (2000) Impact of macroeconomic change on deforestation in South Cameroon: integration of household survey and remotely-sensed data. World Dev 28(6):983–999
82. Fearnside PM (2008) The roles and movements of actors in the deforestation of Brazilian Amazonia. Ecol Soc 13(1):23
83. Pattanayak SK, Wunder S, Ferraro PJ (2010) Show me the money: do payments supply environmental services in developing countries? Rev Environ Econ Policy 4(2):254–274
84. Meyfroidt P, Lambin EF, Erb KH, Hertel ThW (2013) Globalization of land use: distant drivers of land change and geographic displacement of land use. Curr Opin Environ Sustain 5(5):438–444
85. Juutinen A, Kosenius AK, Ovaskainen V (2014) Estimating the benefits of recreation-oriented management in state-owned commercial forests in Finland: a choice experiment. J Forest Econ 20(4):396–412
86. Daniels SE, Hyde WF, Wear DN (1991) Distributive effects of forest service attempt to maintain community stability. For Sci 37:245–260
87. DeFries RS, Rudel TK, Uriarte M, Hansen MC (2010) Deforestation driven by urban population growth and agricultural trade in the twenty-first century. Nat Geosci 3:178–181
88. Zhang P, Shao G, Zhao G, Le Master DC, Parker GR, Dunning Jr JB, Li Q (2000) China's Forest Policy for the 21st century. Science 23, 288(5474):2135–2136
89. Sanchez-Cuervo AM, Aide TM (2013) Identifying hotspots of deforestation and reforestation in Colombia (2001–2010): implications for protected areas. Ecosphere 4(11), art143
90. Dolisca F, McDaniel JM, Teeter LD, Jolly CM (2007) Land tenure, population pressure, and deforestation in Haiti: the case of Forêt des Pins Reserve. J Forest Econ 13(4):277–289
91. Juutinen A, Kosenius AK, Ovaskainen V (2014) Estimating the benefits of recreation-oriented management in state-owned commercial forests in Finland: a choice experiment. J Forest Econ 20(4):396–412
92. Hansen MC, Potapov PV, Moore R, Hancher M, Turubanova SA, Tyukavina A, Thau D, Stehman SV, Goetz SJ, Loveland TR, Kommareddy A, Egorov A, Chini L, Justice CO, Townshend JRG (2013) High-resolution global maps of 21st-century forest cover change. Science 342(6160):850–853
93. Castellarin A, Kohnová S, Gaál L, Fleig A, Salinas JL, Toumazis A, Kjeldsen TR, Macdonald N (2012) Review of applied-statistical methods for flood-frequency analysis in Europe. NERC/Centre for Ecology & Hydrology, 122

94. Kjeldsen TR, Jones DA, Bayliss AC (2008) Improving the flood estimation handbook (FEH) statistical procedures for flood frequency estimation. Environment Agency, Bristol, UK
95. Robson AJ, Reed DW (1999) Statistical procedures for flood frequency estimation. Institute of Hydrology, Wallingford, UK, p 338
96. Lázaro MJ, Sánchez Navarro JÁ, García GA, Romero EV (2016) Flood frequency analysis (FFA) in spanish catchments. J Hydrol 538:598–608
97. United States Geological Survey (USGS) (1982) Guidelines for determining flood flow frequency: bulletin 17B of the hydrology subcomittee. United States Geological Survey, Reston, VA, USA
98. Jennings ME, Thomas WO, Riggs HC (1993) Nationwide Summary of U.S. Geological Survey Regional regression equations for estimating magnitude and frequency of floods for ungaged sites. United States Geological Survey: Reston, VA, USA
99. Ball J, Weinmann E, Kuczera G (2016) Peak flow estimation. In: Ball J, Babister M, Nathan RJ, Weeks W, Weinmann E, Retallick M, Testoni I (eds) Australian rainfall and runoff: a guide to flood estimation, Commonwealth of Australia: Canberra, Australia, 3
100. Ball J, Weinmann E (2016) Flood hydrograph estimation. In: Ball J, Babister M, Nathan RJ, Weeks W, Weinmann E, Retallick M, Testoni I (eds) Australian rainfall and runoff: a guide to flood estimation, Commonwealth of Australia: Canberra, Australia, 5
101. Patric JH, Reinhart KG (1971) Hydrologic effects of deforesting two mountain watersheds. Water Resour Res 7:1182–1188
102. Ellison D, Morris CE, Locatelli B, Sheil D, Cohen J, Murdiyarso D, Gutierrez V, van Noordwijk M, Creed IF, Pokorny J, Gaveau DLA, Spracklen DV, Tobella AB, Ilstedt U, Teuling AJ, Gebrehiwot SG, Sands DC, Muys B, Verbist B, Springgay E, Sugandi Y, Sullivan CA (2017) Trees, forests and water: Cool insights for a hot world. Glob Environ Change 43:51–61

Part V
Hydrology

Chapter 10
Hydrological Impacts of Climate Changes in Romania

Liliana Zaharia, Gabriela Ioana-Toroimac and Elena-Ruth Perju

Abstract This chapter provides a comprehensive synthesis of researches on hydro-climatic changes in Romania and presents some original results on hydrological responses to climate changes in Valea Cerbului River basin (area of 26 km^2) located in the Carpathian Mountains, based on the analysis of historical data and hydrological simulations. Although there are spatial differences, on the whole of Romania, after 1960, a general decreasing trend of the mean annual streamflow was detected. More or less significant changes in the annual flow regime were also noticed: upward trends in winter (related to the increase of the air temperature and liquid precipitations to the detriment of snowfall), downward trends in summer (induced by the general warming and increase of evaporation), and upward trends in the autumn flow. By 2050, the simulations under climate scenarios indicate a general decline of the mean multiannual discharges, significant increases of the discharges during winter and pronounced decreases in late summer and autumn. In Valea Cerbului River basin we investigated the changes in the magnitude and frequency of floods and low flows based on observational data, as well as the expected streamflow changes, as projected by simulations with WaSiM-Eth model (under B1, A2 and A1B climatic scenario, for the period 2001–2065 relative to 1961–2000 period). The simulations indicate a slight decrease in mean annual discharge (of 2% up to 6%) by 2065, an increase of mean monthly discharges from January to April, and a decline during May–December.

Keywords Hydrological impacts · Climate changes · Streamflow · Romania

L. Zaharia (✉) · G. Ioana-Toroimac
Faculty of Geography, University of Bucharest, 1 Nicolae Bălcescu, 010041 Bucharest, Romania
e-mail: zaharialili@hotmail.com

G. Ioana-Toroimac
e-mail: gabriela.toroimac@geo.unibuc.ro

E.-R. Perju
National Institute of Hydrology and Water Management, Şos. Bucureşti-Ploieşti 97E, Bucharest 013686, Romania
e-mail: ruth_barbu@yahoo.com

© Springer Nature Switzerland AG 2020   309
A. M. Negm et al. (eds.), *Water Resources Management in Romania*, Springer Water,
https://doi.org/10.1007/978-3-030-22320-5_10

10.1 Introduction

In the last decades, significant climatic variations and changes were detected from global to regional and local spatial scales, causing multiple and various impacts on both natural and human systems [1, 2]. One of the most sensitive natural systems to the climate variability is the hydrological one. Thus, the alteration of the climatic parameters has direct and/or indirect effects on the hydrological cycle, causing changes in streamflow regimes and water resources, with negative societal and environmental consequences. Such changes have already been identified in the variability of the observational data and are estimated for the future through model-based simulations under different climate scenarios. Therefore, the water availability and quality could become major issues for societies and the environment under climate change [3]. To address these issues, societies are concerned about the development of appropriate adaptation strategies to the hydrological effects of the climate change [4].

This chapter focuses on Romania, aiming (i) to provide an overview of changes observed and projected for the main climatic and hydrological parameters and (ii) to investigate the hydrological responses to climate changes in a catchment located in the Romanian Carpathians (based on trend detection analysis of recorded data and hydrological simulations). The hydrological changes in Romania are integrated into the context of the changes signaled at global and European levels which are synthesized in the first part of the chapter.

The paper presents a comprehensive synthesis of the main research findings on the hydrological impact of climate changes in Romania and provides new and original information on this topic, derived from our research.

10.2 Synthesis of Studies on Streamflow Change Detection at Global and European Scale

10.2.1 Observed Hydrological Changes

Due to its scientific and practical interest, the analysis of the rivers flow variability and trends, in the context of climate change, based on recorded data, has been a concern for researchers worldwide. Thus, studies with such issue have been conducted in many countries, such as Canada [5–7], China [8–18], Malaysia [19], Russia [20, 21], U.S.A. [22, 23], Turkey [24–26], as well as in countries/regions from the South America [27–29] and from Africa [30]. Some studies investigated the streamflow variation on a global scale [4, 31, 32].

In Europe, a large number of researches on rivers' flow trends have been carried out in the last decades, in many countries or regions: Austria [33], Czech Republic [34], France [35, 36], Germany [37–40], Iceland [41], Poland [42], Slovakia [43, 44], Spain [45–47], Switzerland [48–51], UK [52–57], and the Nordic and Baltic countries [58–62]. Some studies have referred to several countries and rivers within Europe [4, 31, 63].

The results of the studies mentioned above show heterogeneous spatial patterns of the identified streamflow trends, both at regional and national scales. The magnitude and statistical significance of the identified trends are sensitive to the period examined. On a global scale, according to [4], 70% of the total of 195 stations analyzed do not exhibit any statistical trend at the 10% significance level, and the remaining stations had either positive or negative trends in the annual maximum flow (for observation periods of at least 40 years). During 1948–2004, only about one-third of the 200 investigated rivers have registered statistically significant trends, 45 with downward trends and 19 with upward trends [32].

Regionally, some spatial differences can be distinguished. Thus, a streamflow decrease was generally noticed in regions affected by lower rainfall (e.g., vast areas in Africa, eastern and southeastern Asia and eastern Australia) [4, 10, 16, 32 etc.]. In other regions, where rainfall and temperatures increased, higher streamflow was identified, as in some areas of northern Eurasia, Central and Far-East Asia, northern North America, southern South America [4, 21, 22, 27, 28, 31].

In Europe, a regionally coherent picture of mean annual streamflow tendency was identified (during 1962–2004), with negative trends in southern and eastern regions, and generally positive trends elsewhere [63].

Concerning the changes in observed extreme streamflow, the general conclusion is that in recent decades there are no clear trends in flood discharge at a regional or a national scale in Europe [64]. In terms of floods, geographically organized patterns of climate-driven changes in their magnitude and frequency could not be detected, but an increase in the number of large floods in Europe during 1985–2010 has been identified [65]. For some small regions, increases in flood peak discharges were detected, as in some alpine basins [66], in Switzerland [51], northern part of Austria [33], northeast France [35, 36], as well as in some regions in Germany [37–40], and northern and western UK [56]. Decreases in peak discharges were noticed in some areas or at hydrometric stations in the Baltic region [59, 60], in Poland [42], etc.

10.2.2 Future Hydrological Changes

Assessing the potential impacts of climate changes on the hydrological cycle and water resources in the future has become, together with change detections studies based on observational data, one of the most important concerns in contemporary hydrology. This issue has a major interest in developing and implementing appropriate measures to adapt the society to the climate changes related effects. Consequently, worldwide, a large number of researches has focused on the future projections of the hydrological impacts of climate changes. The assessments are made by using model-based simulations, considering different climate scenarios. Depending on the specific climate scenario and model used, the projections may show different results, with more or less severe hydrological impacts. Generally, each climate and hydrological projection is subject to limitations and uncertainties in its ability to model the climate and the water-related system. Most of the hydrological projection are based on climate change/emission scenarios, without considering other influences such as water uses and political or socio-economic impacts [1]. Therefore, the results projected by the models should be considered with caution, given the associated limitations and uncertainties.

On a global scale, there are many studies on possible future hydrological responses to climate changes [e.g., 67–77]. At continental and regional/national level, numerous studies have been carried out, mainly in Europe [64, 78–88], but also in US [89] and Asia [90–92]. Several papers [93–96] investigated the potential effect of climate change on river flow regimes in some large watersheds on different continents (e.g., Amazon, Rhine, Tagus, Niger, Blue Nil, Mississippi, Mackenzie, Yellow, Yangtze, Lena, Darling).

At the global scale, according to [67], by 2100, the models project a general increase of mean discharges with more than 10%. Some global patterns of change in discharge regimes for 2100 may be distinguished, such as significant decreases in streamflow for southern Europe, southern Australia, south and north of Africa and southwestern South America. Consistent decreases for most African rivers, the Murray and the Danube rivers, as well as slightly increases of discharges for monsoon-influenced rivers are expected. In the sub-Arctic and Arctic regions runoff increases and a phase-shift towards earlier peaks are projected by 2100. An increase in the seasonality of river discharges was identified (both an increase of high flow and a decrease of low flow) for about one-third of the global land surface area for 2071–2100 relative to the reference period 1971–2000 [68]. Most of the 21st climate models used in [69] projected increases in average annual runoff by 2050 in Canada and high latitudes of Eastern Europe and Siberia, and decreases in runoff in central Europe, around the Mediterranean Sea, the Mashriq, central America and Brazil.

A study on impacts of climate change on European hydrology at 1.5, 2 and 3 °C mean global warming above preindustrial level showed that there are clear changes in local impacts on mean, low and high runoff. Important increases in streamflow will affect the Scandinavian countries and northern Poland. Decreases in mean annual runoff were projected only in Portugal at 1.5 °C warming, but at 3 °C warming, the runoff will decrease on the entire Iberian coast, the Balkan Coast and parts of the French coast [86].

Of high interest is the assessment of future changes in the characteristics of the hydrological extreme events (severity and frequency). Therefore, many studies were devoted to these phenomena. Thus, several papers focused on the assessment of future floods and high flow [e.g., 64, 81, 88, 97], while other studies investigated the projections of future low flow and hydrological droughts [70, 71, 73, 74, 83, 87, etc.]. Some authors investigated both hydrological extremes [e.g., 68, 84, 86, 93].

At a global scale, a study on impacts of climate change on river flood risk shows that in 2050 the current 100-year flood would occur at least twice as frequently across 40% of the globe [69]. At 4 °C global warming, countries representing more than 70% of the total population will face increases in flood risk more than 500% [77]. The largest increase in the flood risk will be in Asia, Europe and U.S.A.

In Europe, it was found that, on average, flood peaks with return periods above 100 years are projected to double in frequency within 3 decades [97]. According to [64], hydrological projections of peak flows illustrate changes in many areas of Europe, both positive and negative, with a general decrease in flood magnitude and earlier spring floods for catchments with snowmelt-dominated peak flows. Similar results were found by Rojas et al. [81], showing that in eastern Germany, Poland, southern Sweden and, to a lesser extent, the Baltic countries, the signal is dominated by the significant reduction in snowmelt induced floods and decrease in 100-year discharge. In other areas, as in western Europe and northern Italy, a strong increase in future flood hazard was projected, mainly due to a pronounced increase in extreme rainfall [81]. The flood magnitudes are expected to increase significantly south of 60 N latitude (mainly in large parts of Romania, Ukraine, Germany, France and north of Spain) and to decrease in most of Finland, northwestern Russia and north of Sweden [84]. In a recent study [88], the climate impact simulations show significant decreasing in high flows for the Mediterranean region, from -11% at 1.5 °C up to -30% at 3 °C global warming, mainly due to reduced precipitation. Small changes ($<\pm10\%$) are expected for river basins in Central Europe and the British Isles, under different levels of warming. In northern regions, high flows are projected to increase due to rising precipitation, but floods are projected to decrease due to less accumulated snow and thus less snowmelt [88].

The impact of climate change on the low flow and hydrological drought charac-teristics is projected to be severe and large areas are expected to be affected by water scarcity in the Mediterranean region, Middle East, southeast U.S., Chile, southwest Australia and large parts of Asia [70, 74]. According to [71], the river basins in arid regions will become even drier, while in the cold regions, a shift of the snow melt peak and an increase of low discharges were identified. For most rivers in humid and temperate climates, the results of the projections are considered uncertain.

In Europe, throughout the 21st century, the models indicate that streamflow droughts will become more severe and persistent in many parts of the continent, except for northern and northeastern regions where the drought hazard will decrease. The most affected by the low flows reduction will be the southern regions [83]. The projections showed that the hydrological drought magnitude and duration might increase in Spain, France, Italy, Greece, the Balkans, south of the UK and Ireland [84]. The low flows are expected to increase in the alpine and northern regions [87].

10.3 Overview of Changes in Main Climate Parameters Controlling the River Flow in Romania

Located in the central part of Europe, Romania (area of 238,391 km^2) overlaps almost entirely (97% of its territory) the Danube River Basin, occupying approximately one-third of it [98]. The rest of 3% of the national territory is drained by small rivers flowing into lakes located on the Black Sea coast. In the south and south-east of Romania flows the Lower Danube River on a length of 1075 km (i.e., 38% of its entire length of 2780 km) before reaching the Black Sea through the Danube

Delta (Fig. 10.1). The Romanian territory is almost equally distributed between mountainous (Carpathians), hilly and plain territories, with altitudes varying between zero (on the Black Sea coast) and 2544 m amsl.

Due to its location, Romania has a transitional climate between temperate oceanic and continental, with regional differences induced by the orography and external influences. The main climatic influences affecting the Romanian territory are oceanic (in the western part), Mediterranean (in the southwest), Baltic (in the north), excessive continental (in the east), and Pontic (in the southeast, induced by the proximity to the Black Sea). The mean multiannual temperature in Romania varies with latitude (from 8 °C in the North to 11 °C in the South) and altitude (from 11 to 12 °C in the plains to less than 0 °C on the highest mountains). The average multiannual amounts of precipitation increase from less than 300–400 mm (in the southeast) to over 1200 mm on the Carpathians' heights [99].

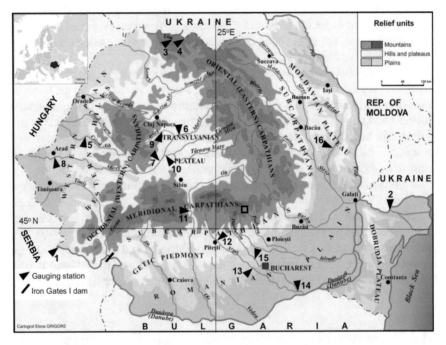

Fig. 10.1 Physical map of Romania with location of the Valea Cerbului River catchment (the black square) and of the gauging stations mentioned in the text: 1 and 2: Baziaş and Ceatal Izmail/Chilia on Danube River; 3: Vadu Izei on Iza River; 4: Bistra on Vişeu River; 5: Gurahonţ on Crişul Alb River; 6, 7, 8: Luduş, Alba Iulia and Arad on Mureş River; 9: Turda on Arieş River; 10: Mihalţ on Târnava River; 11: Cornet on Olt River; 12, 13, 14: Piteşti Ştrand, Malu Spart and Budeşti on Argeş River; 15: Lunguleţu on Dâmboviţa River; 16: Bârlad on Bârlad River

10.3.1 Observed Climate Changes

Many previous studies have dealt with climatic variability in Romania based on observational data, at a national or local scale. Several works investigated the variability of thermal parameters [99–106]. Many researches focused on the variability of parameters related to precipitation [e.g., 107–111]. The temporal dynamics of the snow pack was analyzed in [112–114]. Some studies have dealt with changes in reference evapotranspiration [115] and analyzed the dynamic of the aridity, by using different climate specific indices [116, 117]. The spatio-temporal variability of dryness/wetness in the Danube River Basin (from 1901 to 2013) was analyzed in [118] based on the Standardized Precipitation Index (SPI) for an accumulation period of 6 months. Some studies analyzed the climatic variability and changes on the whole Romanian territory by integrating several parameters [119–123] or at the regional scale, such as the Carpathians [124–126].

Regarding the air temperature, an evident warming trend was noticed on the whole country after 1900, more pronounced after 1980 (Fig. 10.2a). Thus, during 1981–2013, the mean air temperature in Romania reached 10.1 °C, by 0.5 °C higher than the mean air temperature of the period 1901–1980 (9.6 °C) and by 0.4 °C higher than the one of the period 1961–1990 (climatic reference period in Romania) [127] (Fig. 10.2a). During the period 1901–2012, the mean annual air temperature increased by 0.8 °C, with regional differences: higher warming in the southern and eastern parts of the country (by up to 0.8 °C in the Bucharest, Constanța and Roman areas) and insignificant thermal variation in Carpathian areas [128, 129].

Studies on recent climatic changes (1961–2013) in Romania [120, 122, 127], based on observational data recorded at more than 150 weather stations (w.s.), showed statistically significant upward seasonal trends in air temperature, especially in spring and summer and partially in winter. The greatest increases of mean monthly air

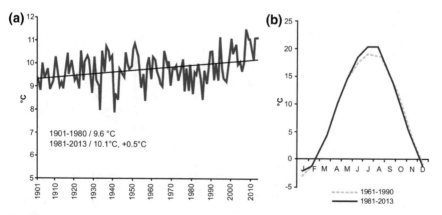

Fig. 10.2 **a** Mean annual air temperature variation in Romania and the linear trend (1901–2013); **b** changes observed in the monthly average air temperature during 1981–2013 relative to the climate reference period 1961–1990 (based on data from 160 weather stations; adapted from [127])

temperature during 1981–2013 when compared to the reference period 1961–1990, were noticed in January–February (1 °C) and in June–August (0.9–1.2 °C), while in autumn months and in December, the modifications were statistically not significant [127] (Fig. 10.2b). The warming tendencies were confirmed by the trends in the annual thermal extremes, in the number of summer days, as well as by other thermal indices, a warming signal being consistent throughout the country. These results are consistent with those found by Busuioc et al. [123] for the period 1961–2010, which detected significant increasing trends for the temperature extremes in all seasons, except for autumn, with the highest increasing rate in summer and the lowest in spring.

In the Carpathian area, general warming was detected, proved by positive trends in warm-related indices and negative trends in cold-related indices. Significant positive trends of air temperature were detected in all seasons, except for autumn which seems to be thermally stable [124–126, 130].

Regarding the precipitation, on the entire Romania, between 1901 and 2013, the annual amount is rather stable (Fig. 10.3a). During 1961–2013, significant upward trends were identified for several stations in autumn (especially in the central and western parts of Romania), while for the other seasons, precipitation did not show significant trends, except for a few isolated stations which recorded negative trends [120, 122]. In the annual regime of precipitation over Romania, for the period 1981–2013, relative to 1961–1990, it was found (based on the analysis of over 150 w.s.), a decline of monthly precipitation especially in spring and summer (by 5–8 mm/month in May and June). A rise was also identified in the first part of autumn, by 8 mm in September and by 5 mm in October [127] (Fig. 10.3b). The annual precipitation extremes indices showed mixed signals, but the majority of the stations had non-significant trends [120]. Upward trends were identified at several weather stations for daily maximum rainfall (mainly in summer and autumn), which increase

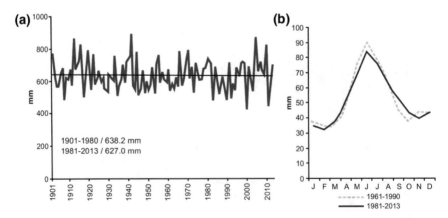

Fig. 10.3 **a** Annual precipitation variation in Romania and the linear trend (1901–2013); **b** changes observed in the monthly precipitation during 1981–2013 relative to the climate reference period 1961–1990 (based on data from 160 weather station; adapted from [127])

the flood risk in these seasons [120, 122]. Regarding precipitation extremes, during 1961–2010, significant increasing trends were detected over large areas in both the frequency of very wet days and maximum daily amount during autumn, and in the maximum duration of dry spells, during summer [123].

The snow-related parameters recorded substantial changes in Romania in the last decades. During the period 1961–2010, general decreasing trends were found in: mean snow depths, number of days with snow cover, number of days with snowfall and continuous snow cover duration. These changes are related mainly to the recent warming and to the slight decrease in winter amounts of precipitation [125]. In the previously mentioned period, the trend analysis of the data recorded at 104 w.s. in Romania showed significant downward trends for the number of days with snow pack at 40% of the stations, and for the mean snow depth, at 20% of the analyzed stations [114]. The most dramatic change concerns the number of snowfall days, which tends to decrease at 82% of the locations. The intra-Carpathian, western and northeastern regions are the most affected by the changes in snow-related parameters.

During 1961–2007, upward trends in the annual reference evapotranspiration were detected in almost 3/4 (72%) of the total of 57 w.s. investigated in Romania, of which 30% is statistically significant. The highest frequency of the positive trends, as well as their absolute maximum magnitude were found during summer in almost the entire country, while in autumn negative trends of the reference evapotranspiration were identified at more than 80% of the locations (but only 20% are significant, mostly in the southern half of the country) [115].

10.3.2   Climate Changes Projections

In order to estimate the future climate in Romania, several studies were made, based on different climate models (both global and regional) and scenarios. Most of the studies were conducted by researchers from the National Meteorological Administration, within European research programs/projects (e.g., ENSEMBLE, SEERISK, ORIENTGATE, etc.).

It is worth noting that all scenarios used for Romania indicate an increase of mean annual air temperature, with different intensities, depending on the time interval and type of scenario considered. The magnitude of the warming varies all over the year, with higher increases projected for summer and winter, and lower increases for autumn [119, 122]. In summer, on the whole country, according to the RCP8.5 scenario, the temperature could rise during 2021–2050 by up to 3 °C (Fig. 10.4a) and by up to 5 °C for 2061–2090 when compared to 1961–1990 (Fig. 10.4b).

Using Global Climate Models—GCMs (available in CMIP 3 and CMIP programs), as well as regional models (available in the EuroCORDEX program) and based on simulations with 6 regional models of EuroCORDEX program (considering RCP 8.5 scenario), it was estimated that, for the 2021–2050 period relative to 1971–2000, the average air temperature will rise by up to 2.5 °C in winter, and by up to 4.6–4.9 °C in summer. In winter, the warming is higher in extra-Carpathians

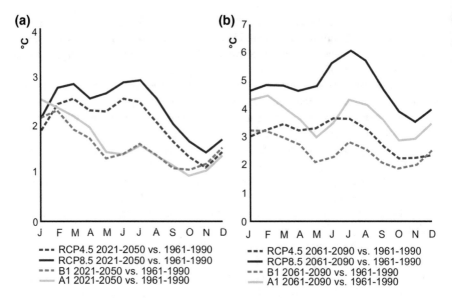

Fig. 10.4 Differences estimated in the multiannual monthly mean air temperatures (averaged for Romania), between the periods 2021–2050 and 1961–1990 (**a**) and between the periods 2061–2090 and 1961–1990 (**b**) (adapted from [122])

regions from eastern and southern Romania, as well as in the depression area from the central part of the country, while in summer, the highest heating is projected for the southern part (in Romanian Plain) [122].

Concerning the precipitation, the above-mentioned models and scenario show regional differences in projected changes: decrease of up to −10% of mean summer precipitation amount for the 2021–2050 period relative to 1961–1990, in south and southeast of Romania, and increase of up to 10% in east, west and the Carpathian area. In the annual regime of precipitation, all scenarios indicate a decline over the 21st century, more intense towards the end of the time horizon (2061–2090), when a reduction of −20 to −30% relative to the period 1961–1990 could be registered (Fig. 10.5a, b). In winter and spring months, the trends indicated by various scenarios are incoherent, with both increase and decline phases. A possible explanation is related to the variability of precipitation in winter and partially in spring due to other factors such as the North Atlantic Oscillation [122].

As a direct effect of projected warming, changes in the characteristics of the snowpack are expected, with impact on river flow. The models and scenarios above-mentioned projected the decrease by more than −60% of the snowpack across the country. In some areas of the Carpathian Mountains, the decrease will reach even −80 to −90% [122].

Information on climate projections in Romania can also be found in studies at continental level [131] or regional scale, in southeastern Europe [132, 133].

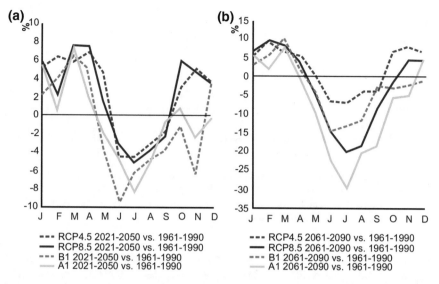

Fig. 10.5 Differences estimated in the multiannual monthly mean precipitation (averaged for Romania), between the periods 2021–2050 and 1961–1990 (**a**) and between the periods 2061–2090 and 1961–1990 (**b**) (adapted from [122])

10.4   Review of Studies on Streamflow Trends and Hydrological Impacts of Climate Changes in Romania

The issue of variability and changes in river flow in Romania has been addressed in several studies focused on streamflow trends detection, based both on historical data and under climate scenarios. In this part, a synthesis of such studies is presented, in order to offer an overview of the Romanian researches in this field.

10.4.1   Observed Changes in River Flow at the Country and Regional Scale

The first studies in Romania investigating the changes in streamflow variability, based on the analysis of trends in observational data, have been published in the 1990s. After 2000 the number of studies on this topic has increased significantly [e.g. 134–143, etc.]. They were conducted mostly at regional scale or watershed level, and have analyzed parameters related to the mean, maximum and minimum streamflow.

At the Romanian spatial scale, a first study investigated the rivers' flow trend during 1976–2005, based on the analysis of mean monthly streamflow from 51 gauging stations (g.s.) corresponding to undisturbed watersheds [140]. The results

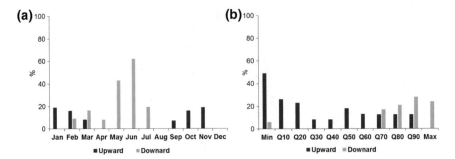

Fig. 10.6 a The percentage of gauging stations with monthly trends in mean streamflow at 10% significance level (1976–2005; 51 stations) (adapted from [140]); **b** relative frequency of statistically significant trends in annual streamflow quantiles Q10–Q90, minima (Min) and maxima (Max) (1961–2009; 25 stations) (adapted from [141])

revealed substantial changes in seasonality, more than half of the stations having negative trends in streamflow at the end of spring and in the early summer. The most important changes occurred in June, where over 60% of the stations presented decreasing trends. From April to July the trends were exclusively negative, while during autumn the streamflow increased (Fig. 10.6a). In February and March, the trends were mixed, with a low percentage (under 20%), and no change has been found for August and December. At the annual temporal scale, only the downward trends were statistically significant [140].

A second study on streamflow trends at national spatial scale is based on the mean daily discharges data series from 25 catchments for the 1961–2009 period, and from 44 catchments for 1975–2009 period [141]. It revealed increasing trends in winter streamflow (related to the increase in air temperature and more liquid precipitation than snow) and rising of minimum spring streamflow (because of the climate warming, leading to an earlier snowmelt and a decrease of the snowpack). Upward trends were also found in autumn flow (explained by the increase of precipitation amounts), while in summer were identified negative trends, which can be related to the general warming and, consequently, to the increase of evaporation [141]. During 1961–2009, statistically significant upward trends for small and medium annual discharges (corresponding to the minima and quantiles Q10–Q60) were detected, and downward trends for high discharges (corresponding to the streamflow quantiles Q70–Q90 and maximum discharge) (Fig. 10.6b). Significant increasing trends in winter streamflow (from minimum discharge to quantile Q80), as well as the decreasing trends in spring median quantiles (Q50 and Q60) and summer low flow (Q10–Q40) were identified. The increase in autumn streamflow since 1961 was found significant for all quantiles. For the minimum flow, significant positive trends in winter, spring and autumn were found, while the maximum flow recorded significant negative trends in winter and summer and positive trends in autumn [141].

More recently, in a study conducted also for the entire country, a trend detection analysis of monthly, seasonal and annual mean streamflow was performed, considering 46 g.s., with data recorded during 1935–2010 [144]. The results show increasing trends (at 95% confidence level) in winter, spring, autumn and at annual time scale over the northwestern part of the country and decreasing trends (at 95% confidence level) in spring over the southern part of the country.

As for the Lower Danube River (on the Romanian territory), it was found by Mikhailova et al. [145] that during 1840–2000, at the entrance to its delta, there were no significant changes in mean annual discharges, but a slightly decreasing trend was identified after 1970. A decreasing trend in mean annual discharges was also noticed after 1975 in [146]. A similar negative trend in the annual mean discharges of the Danube River at the entrance to the delta (at Ceatal Izmail or Ceatal Chilia) was identified during 1948–2004 in [32] and was explained by the global warming, as well as by changes due to human activities (e.g., withdrawal of stream water for different uses, dams and reservoirs).

Between 1931 and 2010, according to [147], the annual average discharge of the Danube River had a slightly decreasing trend at its entry in Romania (at Bazias g.s.). Meanwhile, at the entrance to the delta, at Ceatal Izmail/Chilia, a slight increasing trend was found, in contradiction to previous studies mentioned above (the divergence could be due to differences in the analyzed periods; additionally, the linear trends with a low slope were visually established without being statistically tested). During the analyzed period (1931–2010), the minimum annual discharges of the Danube River had upward trends at the entry in Romania, as well as at the entrance to the delta, more obvious after 1985. Concerning the maximum annual discharges, at Bazias, no trend was detected, while at Ceatal Izmail/Chilia, a slight upward trend was found especially after 1990 [147].

During 1930–2008, shifts in seasonal discharge of the Danube River at the entrance to its delta (at Ceatal Izmail/Chilia) were identified. Thereby, after 1991 the present-day spring high flows and floods tend to occur earlier, the summer runoff is diminished and the low flows in autumn are slightly higher (in October–November) compared to those of the early 20th century [85].

The alterations in annual flow and regime of the Danube River on Romanian territory can be attributed to climate variability/changes, water management activities, as well as land use change. Since 1970, the reservoir dam of Iron Gates I, located at the entrance of the river in Romania (Fig. 10.1), had a major influence on the streamflow and sediment load of the Lower Danube River. The dam caused alterations of the natural annual flow regime due to its regularization role. Thus, during low water periods, the discharges increase (in winter and autumn), while during high waters (in spring and early summer) they are reduced in order to mitigate the flood magnitude [148].

At the regional/watershed scale, the identified trends in streamflow are heterogeneous, being influenced by local climate changes (in the lack of major anthropic pressures). In the Carpathian Curvature region, during 1962–2006, no significant trends were found in the annual flood peak variability and the number of floods exceeding the attention stage (corresponding to the yellow warning code) [137]. Similar results were found in the eastern Carpathians, in the Trotuş River catchment [138]. In the Southern Carpathians (more precisely, in the Bucegi Mountains), a significant upward trend in the magnitude of annual floods was identified between 1961 and 2010. After 1990, an increase in the frequency of floods with high magnitude was detected, as well as an increase in the frequency and magnitude of floods occurred in March–June and September–October [142]. In the Lower Buzău River basin (located in plain area, in the eastern part of Romania), between 1960–2009, a general downward trend in average (annual and monthly) flow series was detected, except for the autumn, when upward trends were found, in accordance with increasing precipitation during this season (especially in October) [143].

In watersheds from eastern Romania, during 1950–2006, a rise of the mean monthly discharges of rivers was found, correlated with the precipitation increase, except for winter when declines of the streamflow were identified [149]. In southwestern Romania (in Oltenia region), between 1961 and 2009, the runoff decline was noticed, as an effect of the increase in climatic water deficit. Statistically significant negative trends in streamflow were identified especially in spring [150].

10.4.2 Projected Hydrological Changes

Among the first studies in Romania on the hydrological impact of climate changes can be named the ones in the 1990s [151, 152]. The number of such studies increased after 2000, but especially after 2010 [144, 153–166]. This issue has been also addressed in several European research projects, with the participation of Romanian institutions (e.g. National Meteorological Administration, National Institute of Hydrology and Water Management, Institute of Geography of the Romanian Academy etc.). In this regard we mention the project CLAVIER (*Climate Change and Variability: Impact on Central and Eastern Europe*) in which ongoing and future climate changes were investigated (based on observational data and on climate projections) and regional impacts were assessed [167]. The project CECILIA (*Central and Eastern Europe Climate Change Impact and Vulnerability Assessment*) aimed to provide climate change impacts and vulnerability assessment in targeted areas of Central and Eastern Europe (including Romania [168]). We also mention the projects SEERISK (*Changing Risks in Changing Climate*), focused on changes in floods and droughts under future climate in the Danube macro-region, including Romania [169] and the project CLIMHYDEX (*Changes in Climate Extremes and Associated Impact in Hydrological Events in Romania*) [170]. In the Carpathian region (including the whole European Carpathian chain with the adjacent areas of the Carpathian Basin, where Romania is located), researches on the climate changes and impacts on water resources were performed within some projects such as CarpathCC (*Climate Change in the Carpathian Region*) and CARPIVIA (*Climate Change Vulnerability and Ecosystem-based Adaptation Measures in the Carpathian region*) [171].

The studies in Romania on the hydrological impact of climate change have been carried out, generally, at regional or river basin scale, with various areas. The first papers on this topic were focused on estimating the possible alterations in river flow regime as a consequence of changes in air temperature and precipitation—data simulated by the Canadian Climate Center Model (CCCM), supposing that the CO_2 concentration doubles by 2075. The results showed that in the Argeş River basin (situated in the southern part of the country), at Pitești Ștrand g.s. (Fig. 10.1), when compared to the reference period 1950–1992, the average annual discharge will decrease by almost 22%. Significant decreases (over 40–50%) are expected in spring (mainly in April and May) and autumn (September–October) while in the winter months, the average discharges will be much higher (with up to 86% in January) [152]. In the Târnava River basin (located in the central part of the country, in the Transylvanian Plateau) at Mihalț g.s. (Fig. 10.1), projections for 2075 with the same climatic model (CCCM) showed a high increase in the average monthly flow in winter (over 70% in January) and decreases of −20 to −30% in the spring months (especially in March and April) and in September, relative to the reference period of 1961–2000 [153].

In the Buzău and Ialomița watersheds (covering External Carpathian Curvature region and the neighboring lowlands—hills and plains), the climate changes impacts on streamflow were estimated in [154, 159], by using three GCMs for three future time horizons (2025, 2050, and 2100), relative to the reference period 1971–2000 (the researches were performed within the European project CECILIA). The hydrological simulation (made with the water balance model WatBal) indicated notable changes in mean annual and monthly streamflow. Thus, the mean annual discharges decrease as the time horizon is larger and could reach 30%. During the winter and early spring periods (from November/December to February/March), an increase in mean monthly flow was detected on long-term, more pronounced for the rivers draining the mountainous area. The simulations also showed the seasonal variability amplification [159].

Also in the Ialomița River basin, it was investigated the potential impact of climate change on maximum flow regime. The discharges for the future period 2011–2050 and the reference period 1951–2010 were simulated with the CONSUL hydrological model, by considering the A1B scenario [157]. The projections show an earlier occurrence of the floods in spring in the future period relative to the reference period. The monthly maximum discharges will be higher in winter and July, and lower in April–June as well as in September–October. While in the reference period the highest number of annual flood peaks was recorded in October, in the future (2011–2050) the most annual floods will occur in March and April [157].

In the Siret River basin (the largest catchment in Romania, draining the eastern part of the country; Fig. 10.1) the climate change impact on maximum flow was investigated based on simulations made for the period 2011–2050, relative to the reference period 1951–2010 [156]. The projections show that in the future, most annual flood peaks will appear earlier in the year, in March–July, compared to the reference period, when they occur during April–August. In the same watershed, according to [172] for the period 2031–2060 (by simulations based on ENSEMBLES data under A1B SRES scenarios) a decline in average runoff from April–September (up to −25%) and tendencies towards an increase in winter (January and February) are expected. The cited reference indicates that during the period 2031–2060, the majority of the climate scenarios applied so far show a clear reduction in summer runoff for the main tributaries of the Lower Danube Basin (e.g., the Olt River at Cornet g.s.), while in the rest of the year, the scenario trends are partially unclear, mainly for the winter months.

In the Olt River basin, one of the largest watersheds in Romania, draining the Carpathians and the central part of the Romanian plain (Fig. 10.1), the potential impact of climate changes on the maximum streamflow was investigated by hydrological simulations (using the CONSUL model) for two periods: 1951–2010 (as reference period) and 2011–2050 [161]. The results indicated a slight increase in the future (up to 5%) in the annual maximum discharges. Significant increases of maximum monthly discharges are expected in winter, July and August, while during the spring months, in June and autumn, the projections indicate a decrease of the maximum streamflow. In the future period, it is likely that floods will become more frequent in winter and summer compared to the reference period, and rarer in spring and autumn (except October). The high discharges (with return periods of 1000, 100, 50, 20 and 10 years) tend to rise in the upper part of the Olt River basin and to slightly decrease in the lower part [161].

Within the European project CLAVIER (conducted between 2006 and 2009), in which several Romanian institutions were involved, there were estimated the hydrological impacts of climate changes based on the output of regional climate models for several watersheds/rivers in Romania. REMO5.7-ERA40 (1961–2000) and REMO5.7-A1B (1951–2050), data produced by the Max Planck Institute for Meteorology in Hamburg, were used as climate change scenario for the hydrological models (VITUKI–NHFS and VIDRA). The projections were made for some Romanian rivers draining different geographical regions. Thus, Vișeu (at Bistra g.s.) and Iza (at Vadu Izei g.s.) are tributaries of the Upper Tisa, located in the far north of the country. Mureș River (at Arad and Luduș gauging stations) and its tributaries Arieș (at Turda g.s.) and Târnava (at Mihalț g.s.), drain the central part of Romania. Argeș River (at Malu Spart and Budești gauging stations) and its tributary Dâmbovița (at Lunguletu g.s.) are located in the south of the country (Fig. 10.1).

For rivers in the northern part of the country (Vișeu and Iza), a slight increase of the mean annual discharges (less than 3%) is expected for the period 2021–2050 relative to the reference period (1961–1990). A significant increase (30–40%) is projected in the winter season. For the spring-autumn period, the hydrological simulations indicate a slight variation, more important in spring, when decreases of −4 to −14% are estimated for mean discharges (Fig. 10.7). For the other analyzed rivers, the hydrological simulations indicated a general decrease of the mean annual discharges up to almost −15% by 2050. This decreasing is caused by the pronounced flow reduction during the spring–autumn period, by up to almost −25% in some catchments (a lower decrease of the average flow is expected in summer). In winter, significant increases are projected, which can reach 30% in the south of the country (e.g., Dâmbovița River) (Fig. 10.7) [167]. Concerning floods, the simulations made within CLAVIER project do not indicate obvious changes. However, more frequent floods are expected in winter. Torrential flash floods could occur more frequently, while the floods with long duration and large volume may become rarer. The low flows are expected to last longer [167].

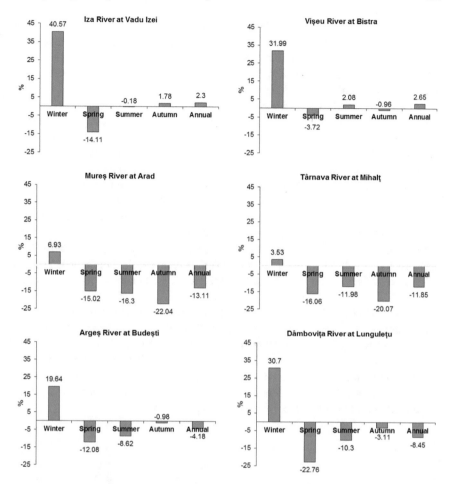

Fig. 10.7 Changes in mean seasonal and annual discharges projected for some rivers in Romania for the period 2021–2050 relative to the period 1961–1990, based on hydrological simulations under the A1B climate change scenario (based on the data from [167])

Within the project CLIMHYDEX, the climate change impacts on extreme flow were estimated in two medium-sized basins in Romania: Crişul Alb (in the western part of Romania) and Bârlad (in the eastern part) (Fig. 10.1). The hydrological simulations were made (using CONSUL and NOAH models) for two future periods 2021–2050 (Scenario S1) and 2071–2100 (Scenario S2). They were compared with the simulations made for the reference period 1976–2005 (Scenario S0), under climate scenarios generated in the CLIMHYDEX Project (8 GCMs have been used, by considering three greenhouse emission scenarios: A1B and E1 for ENSEMBLES GCMs, and RCP4.5 for CMIP5 GCMs) [170].

For the Crişul Alb River basin, the results indicated a general increase of the frequency of flash flood events in the future periods, more pronounced for the second period 2071–2100 (greater than 20%), and especially the increase (by 30–40%) in the frequency of severe flash flood events. In the case of the Bârlad River basin, a general increase of the flash floods frequency was estimated in the upper part of the catchment, and a decrease in the middle and lower basin, for both future periods relative to the reference period [144].

In Bârlad River basin, the scenarios S1 and S2 indicated an increase of the monthly maximum discharges (multiannual averages) during the winter, spring and summer months (with a significant increase in January and February), and a decrease in the autumn months. Regarding the monthly minimum discharges (multiannual averages), an increase was identified in the late winter, early spring and in the summer months, and a decrease in the rest of the year, more pronounced during October–December (Fig. 10.8a). In the future periods (scenarios S1 and S2) compared to the reference period (S0), the simulations indicated for the annual maximum discharges a positive trend (of maximum 15 and 30% respectively) in the upper and lower part of the watershed, and a negative trend in the middle part (of maximum −5 and −15% respectively). Regarding the annual minimum discharges, the simulations showed a decrease of maximum −5 and −15% respectively, for the whole watershed [144, 165, 166].

In Crișul Alb River basin, the analysis of monthly maximum discharges (multiannual averages) showed, for scenarios S1 and S2 relative to S0 scenario, an increase during the winter, spring and summer months, more pronounced from May to August and a decrease in the autumn months. Concerning the multiannual averages of the monthly minimum discharges, an increase was detected in the late winter, early spring as well as in the summer months, and a decrease in the rest of the year, more significant during October–December (Fig. 10.8b).

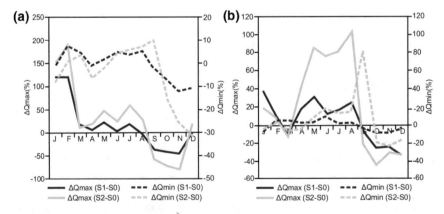

Fig. 10.8 Variation of relative deviation (%) of monthly maximum (Qmax) and minimum (Qmin) discharges (multiannual averages) under S1 and S2 scenarios relative to S0 scenario, at Bârlad gauging station, on Bârlad River (**a**) and at Gurahonț gauging station, on Crișul Alb River (**b**) (adapted from [144])

The simulation of the annual maximum discharges in Crişul Alb River basin, for the scenario S1 compared to S0 scenario, indicated, generally, a slight negative trend in the upper part of the basin (of maximum −3%) and an increasing trend for the lower part (of maximum 7%) while the scenario S2 indicated a significant increase (of maximum 41%) on the entire watershed. Concerning the annual minimum discharges, the simulations indicated for scenario S1 compared to S0 scenario a decrease of up to −8%, and for the scenario S2 an increase of maximum 7% on the entire river basin [144].

A few studies investigated the hydrological alteration induced by climate change in Mureş River basin, the largest catchment in the central part of Romania (Fig. 10.1). Based on scenarios obtained from nine coupled runs of GCM-RCM model driven by A1B socio-economic development scenario and using the SWIM hydrological model, the hydrological response to climate change was simulated in the upper part of the Mureş River basin (upstream Alba Iulia g.s.), for the reference period (1971–2000) and two future periods (2021–2050 and 2071–2100) [173]. The projections indicated a deviation in median monthly discharges, as follows: increase for the winter months for both future periods, and decrease for late spring, summer and early autumn by the year 2100. Over half of the simulations projected a slight increase in median and maximum values for the number of days with high flow, but a more significant increase for the duration of flow events, indicating that it is likely that drought events will become longer by the year [173]. These results are generally consistent with [164], study that investigated the potential hydrological impacts of changes in temperatures and precipitation in the entire Mureş River basin (using the SRES A1B scenario), considering the same future periods (2021–2050 and 2071–2010), compared to the reference period 1961–1990. Based on the results of climate projections (with ALADIN and REMO models), the simulations showed that the annual streamflow will certainly decrease in the long run and early spring snow melt could be more intensive, causing the increase of the river flow. The projections also indicated that the early summer floods might be less significant; summer and autumn low flow extremes may occur more frequent, and severe water scarcity may affect the lower sector of the Mureş River [164].

At national spatial scale, the changes induced by climate change on the mean (monthly and annual) streamflow in 20 large basins (including 275 sections) were investigated for the time horizon 2021–2050, by using the hydrological models WATBAL and CONSUL [174]. The simulations were performed considering two scenarios: Scenario 0, corresponding to the reference period 1971–2000, and Scenario 1, in which the mean monthly discharges for the period 2021–2050 were estimated by using simulations of climate change evolution performed within CLAVIER Project (A1B scenario). The results indicated a general decline of the mean multiannual discharges (of maximum −38% in Vedea River basin), excepting the Someş River basin where an increase (of maximum 23%) was identified. Regarding the monthly mean multiannual discharges, projections showed in most analyzed catchments significant increases in the winter months (December, January, February) and sometimes in March (i.e., in Mureş, Argeş and Siret watersheds) and pronounced decreases in late summer and autumn (August–November) [174].

Researches on future expected impacts of climate changes on water availability and extreme hydrological events in Romania were also carried out within the International Commission for the Protection of the Danube River (ICPDR). The researches are based mainly on IPCC SRES scenarios A1B and A2 and on RCP scenarios, by considering as future periods 2021–2050 and 2071–2100. According to [175], for the entire Danube River basin, seasonal changes in runoff regime are expected in the 21st century, concretized in the increase of the mean discharge in winter and decrease in summer, with local differences. A decline in solid precipitation and consequently in snow cover, together with an earlier snow melt, will lead to the alteration of the runoff regime, by causing in the high parts of the watersheds draining the Carpathians, a shift in peak runoff from early summer to spring.

In the Lower Danube River basin, the studies based on IPCC SRES scenarios show an increase in flood frequency, but the possible changes in flood frequency and magnitude remain uncertain. Flood events are projected to occur more frequently, particularly in winter and spring. For small catchments, an increase in flash floods due to more extreme weather events (torrential rainfall) is expected. Concerning the drought and low flow events, they are likely to become more intense, longer and more frequent. The frequency could increase especially for moderate and severe events. These extreme events will be more severe in summer, due to less precipitation in this season, whereas they will become less pronounced in winter [175].

The impacts of climate change on the flow regime of the Danube River and its tributaries were investigated in [85] by using an ensemble of climate scenarios and the SWIM hydrological model, for two future periods (2031–2060 and 2070–2100) and the reference period 1971–2000. The simulations showed a possible increase in river runoff for the winter and spring months (especially during January–March), while a general decreasing trend is projected in summer runoff for the whole Danube Basin and, additionally, in autumn runoff for the Middle and Lower Danube Basin. The trends are more evident towards the end of the 21st century. At the entrance of the Danube River to its delta (at Ceatal Izmail/Chilia g.s.), the projections based on ENSEMBLES data under A1B SRES show, for the near future period 2031–2060 relative to the reference period, an increase (of 5–40%) in winter and early spring runoff (from January to March). The flow peak specific for May is projected to decrease slightly (up to −5%) and tends to appear earlier (in April). The projections also indicate, in the Lower Danube River, a decrease (up to −20%) of discharges during the warm season (May–October), aggravating the low flow period in late summer and autumn (August–November). For the far future (2071–2100), the simulations indicate increase in runoff for January–April, with rates similar to those projected for the near future period. In the rest of the year, no clear changes were identified, except for a decreasing trend in late summer [85].

10.5   Hydrological Responses to Climate Changes: The Example of the Valea Cerbului River Catchment

This part is focalized on a case study in which the hydrological responses to climate changes were investigated in Valea Cerbului River catchment, based on recorded data and hydrological simulations.

10.5.1   Study Area, Data and Methods

Valea Cerbului River catchment is located in the eastern extremity of the Southern Carpathians, namely in the Bucegi Mountains, a nationally renowned region for tourism related activities (Fig. 10.1). The main stream has almost 10 km length with a mean slope of about 130 m/km. It has its source in the alpine region at 2125 m amsl and the outlet at 870 m amsl. The catchment extends over an area of only 26 km^2 and has a mean altitude of 1500 m amsl (the maximum altitude is 2505 m amsl in Vf. Omu peak) (Fig. 10.9). Its upper part overlaps calcareous conglomerates and has steep slopes (between 45° and 80°) favoring rapid runoff and flash floods occurrence [142, 176]. The hydrographic network is dense, but mainly formed by temporary torrential streams.

The altitude imposes important spatial variations of the climatic parameters. Thereby, the mean multiannual air temperature decreases from 5 to 6 °C at lowest altitudes to −2.5 °C at Vârfu Omu w.s., located at 2505 m amsl, where the frost lasts, in average, 260 days/year and the snow layer has a mean duration of 140–220 days/year, with a monthly average thickness of up to 70–80 cm in March–April [99]. The snow usually lasts until June on the upper part of the northern valleys, two months longer than in the lower basin [177]. The mean annual precipitation in Valea Cerbului catchment varies between 900 and 1030 mm [99]. The wet period is May–August (over 100 mm/month on average), when heavy, convective rains are specific, while in autumn and winter the amounts of precipitation are low.

The mean multiannual discharge of the Valea Cerbului River (calculated for the period 1961–2012, based on data recorded at Bușteni g.s., which controls approx. 96% of the total area of the catchment), is of 0.502 m^3/s, equivalent to a specific discharge of 19.8 l/s/km^2. The maximum flow recorded is of 62.3 m^3/s (on June, 19th 2001) and the minimum discharge is of only 0.030 m^3/s (December, 16th 1963). The flow regime is characterized by high waters in summer (38% from the mean multiannual volume), with the maximum percentages in June and July (13% and respectively 15% of the mean multiannual water volume), due to heavy convective rains specific for this season. An important water volume is flowing in spring (30%) when the runoff is influenced by rainfall cumulated with snowmelt. Low flow is specific for both winter and autumn (13% and respectively 19%) [177].

The Lower Valea Cerbului River basin overlaps the northern part of the Bușteni town, a nationally renowned resort for both, winter and summer related tourism activities. The river is crossed by very important international/national terrestrial

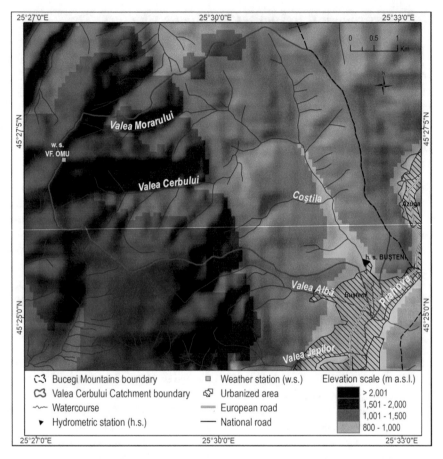

Fig. 10.9 Valea Cerbului River catchment map

routes, namely the European Route E60/National Route 1 (DN1) and the main railway
that links Bucharest with major cities from the central and western parts of the
country.

The analysis is based on three major types of data: hydrological, climatic, and
spatial/cartographic. The hydrological data refers to mean daily discharges measured
at Buşteni g.s., used for both the analysis of the observed changes during 1961–2012
and the calibration of the hydrological model. The source of the hydrological data
series is the "Romanian Waters" National Administration.

The climatic data refers to recorded and emission scenario data on the main parameters, at daily time step: temperature, precipitation, wind speed, relative air humidity
and sunshine duration, respectively solar radiation (substituting the sunshine duration
for the scenario data). The recorded data were obtained from the national weather
stations network (through National Meteorological Administration and European
Climate Assessment and Dataset—ECA&D) and used to simulate the actual hydro-

logical regime of the Valea Cerbului River. The scenario data were obtained within the ESEMBLE project, at Max–Planck Institute (Germany) using the ECHAM5 global climatic model. They were retrieved from the CERA database (Climate and Environmental Retrieval and Archive) and refer to the B1, A2 and A1B scenario families (for the period 2001–2065) and to the reference period C20 (1961–2000) [178]. We considered a medium time horizon which is generally more appropriate to make and implement regional and local policies and adapted management measures. The data on air temperature and precipitation for both the reference (1961–1991) and scenario (2001–2065) periods, were subjected to bias corrections (*delta change method*, as described in [179]).

The spatial data were used as part of the input in the hydrological model and consists of data on altimetry, stream channels, rock and soil types, land cover etc., retrieved from the thematic maps available at different scales (1:25,000, 1:50,000), vectorized, then processed in GIS environment and/or with the specific tools provided by the model's developers.

The changes in the flow data series recorded at Bușteni g.s. were investigated based on frequency and trend analysis for the period 1961–2012. The significance of linear trends was established using the Mann-Kendall test and Sen's slope method, developed on an Excel template [180].

The changes in the hydrological and climatic parameters, as projected by the emissions scenarios, were estimated using WaSiM-ETh (Water Balance Simulation Model), a fully distributed hydrological model that simulates all the components of the water balance. The required input data are climatic data series (e.g., daily data on air temperature, precipitation, sunshine duration, wind velocity, air relative humidity) and gridded spatial data [181]. The model was calibrated and validated using historical data series and with the PEST (Parameter Estimation and Uncertainty Analysis) software [182]. The objective functions used to assess its efficiency were the Nash-Sutcliffe Efficiency Coefficient (NSE = 0.80 in calibration, and 0.60 in validation) and the Pearson Correlation Coefficient ($r = 0.90$ and 0.86 for calibration and validation, respectively).

10.5.2 Observed Changes in Valea Cerbului River Flow

The streamflow changes are mainly caused by climatic and anthropogenic factors. Because the Valea Cerbului Catchment is partially included in the Bucegi Natural Park, the human pressures are quite limited. The intakes for the water supply of Bușteni town are insignificant and do not affect the flow regime [183]. Therefore, the hydrological alteration can mainly be the result of climate changes.

A previous study [177] have shown for the period 1950–2010 a general upward trend in the variability of monthly average discharges. Except for April and May, increasing trends were founded in all the months, statistically significant from August to December (with the level of significance $\alpha = 0.01$ in September and October and $\alpha = 0.05$ in August, November and December). A significant decreasing trend was detected in May (for $\alpha = 0.01$). At the seasonal scale, upward trends were identified in summer, autumn and winter (statistically significant only in autumn, for $\alpha = 0.01$), while in spring the trend is negative but non-significant. Annually, an increasing trend in average discharges was detected, but non-significant statistically.

In this chapter, we focused the investigation on the changes observed during 1961–2012, in high flows (floods) and the low flows of the Valea Cerbului River.

10.5.2.1 Changes in the Magnitude and Frequency of Floods

The annual flood peaks of the Valea Cerbului River recorded at Bușteni g.s. during the period 1961–2012 occurred most frequently during the summer months, each with more than 20% of the cases, while the lowest frequencies were recorded during winter, with no annual floods in January (Table 10.1).

The magnitude of the annual floods (the largest peak flood of each year) has a statistically significant upward trend (with a level of significance $\alpha = 0.1$). The seasonal trends (analyzed based on the maximum monthly discharges averaged for each season) is consistent with the seasonal trends identified using mean daily and monthly discharges: significant positive trends in autumn and winter (for $\alpha = 0.1$) and non-significant in summer; negative trend in spring, non-significant (Table 10.2).

In order to analyze the temporal variability and frequency of floods (during 1961–2012), we established a threshold discharge, namely three times the mean multiannual discharge [142, 184]. Any mean daily discharge exceeding the threshold was considered a flood, except consecutive days with such values, which were counted as single events. This approach allows to include in the analysis of both exceptional and ordinary floods, for more accurate identification of their temporal frequency. The interannual variability of the number of floods shows a slight upward trend, statistically non-significant. The maximum frequency of floods is identified in 1966, 1998, 2005 and 2010, but there are years without floods (defined as mentioned above): 1976, 1986, 1993, 2002 and 2011 (Fig. 10.10a).

At monthly level, the highest frequency of floods in the analyzed period is met in June (over 20% of the total number of floods), followed by July (approx. 16%) and April (almost 12%), when snowmelt is mixed with spring rainfall. The months with the lowest frequency of floods occurrence are January and February, with less than 2% (Fig. 10.10b).

The decadal analysis of the frequency of floods (as defined above) highlights some changes during the period 1961–2010. The total number of floods was smaller in the second and third decades of the analyzed period (1971–1980 and 1981–1990), namely 34 and 31 floods. The maximum number of floods were recorded between 2001 and 2010. The absolute monthly frequency of floods shows a great decadal variability in March, April and August. The maximum frequency of floods is identified in June for all decades (Fig. 10.11a).

Except for the first decade considered in the study, the floods were recorded most often during summer in each decade. In autumn, the frequency of floods for the

Table 10.1 Monthly number of annual flood peaks recorded between 1961–2012 at Bușteni gauging station on Valea Cerbului River (in some cases the maximum annual discharge is recorded twice, in different months)

Month	Jan	Feb	Mar	Apr	May	Jun	Jul	Aug	Sep	Oct	Nov	Dec
No. of annual floods	0	1	2	2	5	11	11	12	4	3	1	1
% of total number of years	0.0	1.9	3.8	3.8	9.4	20.8	20.8	22.6	7.5	5.7	1.9	1.9

Season	Statistical test Z	Level of significance (α)	Sen's slope
Spring (III–V)	−0.49	–	−0.002
Summer (VI–VIII)	0.20	–	0.0011
Autumn (IX–XI)	1.79	0.1	0.0044
Winter (XII–II)	1.81	0.1	0.0027

Table 10.2 Statistical significance of trends in the seasonal variability of the maximum discharge of the Valea Cerbului River at Bușteni gauging station (1961–2010), according to Mann-Kendall test

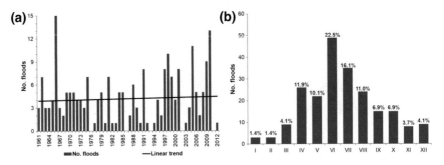

Fig. 10.10 Annual number of floods and the linear trend (**a**) and total monthly number of floods with the equivalent percentages (**b**), recorded at Bușteni gauging station between 1961–2012

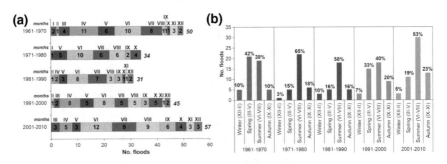

Fig. 10.11 The decadal monthly absolute frequency of floods (**a**) and decadal seasonal absolute and percentage frequency of floods (**b**) at Bușteni gauging station on Valea Cerbului River (1961–2010)

last two decades is increasing and consistent with their magnitude, namely with the
trends identified for the mean and maximum discharges (Fig. 10.11b).

10.5.2.2 Changes in the Magnitude and Frequency of Low Flows

At interannual level, the magnitude of low annual flows (based on mean daily discharges) is decreasing during the analyzed period (1961–2012). Thus, the data show
a positive trend of the annual minimum discharges with a significance level α of 0.05.
The low monthly flows have, in general, the same positive trend, statistically significant for January to March and December (for $\alpha = 0.05$ to $\alpha = 0.01$). The exception
is given by the low flows in April to July, for which a negative trend was identified (statistically non-significant). Seasonally, the minimum discharges have upward
trends in spring (for $\alpha = 0.1$), winter (for $\alpha = 0.01$ in) and autumn (statistically non-
significant), and a downward trend in summer (statistically non-significant). The
trend identified in the low flows, correlated with the seasonal frequency of floods
lead to the conclusion that for the analyzed period, the hydrological regime tends to
become more extreme in the summer for both low and high flows and attenuated in
the winter.

The frequency and the duration of low flows were determined by considering as
a threshold the discharge (Q) with 95% exceedance probability (the 5th percentile),
based on recorded mean daily discharges during 1961–2012 ($Q_{95\%} = 0.092$ m^3/s).
We considered as low flow period any period of at least 10 consecutive days with
discharges below this value. This method helps to identify the driest years for the
study area, regarding hydrological drought duration.

The decadal analysis of low flows shows an increase of their magnitude, especially in the last decade (2001–2010) and an important variation of their frequency
(expressed as a decadal number of days with discharges below $Q_{95\%}$) (Fig. 10.12a).
During the analyzed period, the intervals of dry years are 1963–1968 and 1986–1995
(Fig. 10.12b). The occurrence of low flow periods does not imply that significant

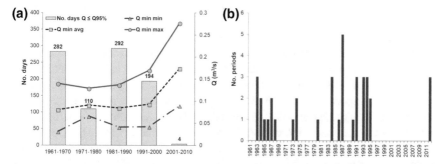

Fig. 10.12 The decadal frequency of discharges below $Q_{95\%}$ and decadal minimum (Q min min),
average (Q min avg) and maximum (Q min max) of mean daily discharges (**a**) and annual number
of low flow periods (**b**) at Bușteni gauging station on Valea Cerbului River (1961–2012)

floods cannot be recorded in the same years, but the interannual distribution of flows can be extremely uneven.

From the total of 38 low flow periods identified during 1961–2012, 63% (24) occurred in winter, 21% (8) in the autumn, 13% (5) in spring and just one case during summer (in July). Monthly, the maximum frequency of the low flow periods is specific in February. On the contrary, in April, May and June, when the snowmelt has an important role in the surface runoff, there were not prolonged intervals with flows below $Q_{95\%}$.

10.5.3 Projected Hydrological Responses to Climate Changes

For Valea Cerbului River catchment the emission scenarios predict a rise in the air temperature of 1.3 °C (the mean value indicated by the emission scenarios used). Compared to the reference period (1961–2000), the scenarios show temperature growths for all seasons, with maximum values in autumn and minimum values in spring. The analysis conducted at the monthly level, reveals upward, significant trends for A2 and A1B scenarios for the future period 2001–2065.

The trends indicated by the simulated precipitation data are not as clear as those of the air temperatures: annually, a slightly descending trend and a higher variability are expected; seasonally, the simulated precipitation shows more important decrease during autumn (from 6.8 to 10.7% depending on the scenario).

The hydrological model used establishes the type of each precipitation event, liquid or solid, by threshold transition temperatures, set in the model optimization stage. Likewise, it can determine the precipitation values and types dissociated for different zones of altitudes, in this case: the zone above 1750 m amsl, the zone between 1250 and 1750 m amsl and the low part, with less than 1250 m amsl. The simulation using the scenarios data as climatic input highlights the diminution of the mean annual snowfall by 5–11% by 2065, relative to the reference period. On the contrary, the projected liquid precipitation slowly increases on the chosen time horizon.

For the winter months, the estimated snowfall is much lower than the simulations for the reference period. Notable differences are estimated for the middle and lower zones of the catchment, namely with approximately 15% less snowfall, but 18–40% more rainfall, as average between the three scenarios (Fig. 10.13a). During spring and summer, projected changes in the precipitation's form have the same tendency: less snow (4% in the spring, 35% in the summer) and more rain (11% in the spring, just 1% in the summer) than during the reference period (as average between the scenarios, for the whole basin). For the autumn, the scenarios show the diminution of precipitation in general, but a strong diminution of snowfall: with 24% in average and over 30% in the middle and lower part of the catchment (Fig. 10.13a). The shift between solid and liquid precipitation during winter and spring entails a significant

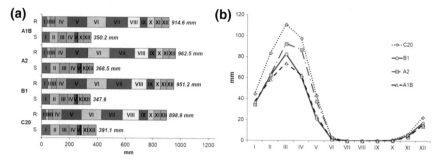

Fig. 10.13 Mean monthly amount of solid (S) and liquid (R) precipitation (**a**) and mean monthly snow depth (**b**) in Valea Cerbului River catchment for the climatic scenarios B1, A2, A1B (2001–2065) and the reference period C20 (1961–2000)

decrease of the snow layer depth and duration, directly affecting the streamflow, because in high mountainous regions, the snowmelt is an important component of river flow.

For the interval November–February, the projections show a uniform decrease of snow depth relative to the reference period. From March to May the simulations are still below the monthly values of the reference period, but for the A2 scenario, the decrease is lower (Fig. 10.13b). The changes in the snow depth are of approximately 27% at annual level (scenarios average) and up to 68% at the monthly level.

The simulations of the hydrological regime under the emission scenarios indicate a lower mean annual discharge of −2% up to −6% (depending on the scenario), but with important monthly and seasonal differences. From January to April, all the three scenarios indicate increased mean monthly discharges, of 4–22%, with the maximum difference in January for the B1 scenario (Fig. 10.14). This is the effect of rising air temperatures, fact that triggers liquid precipitation events, a diminished snow depth and earlier snowmelt. For May–December the average of the scenarios estimates lower mean monthly discharges (between −10 and −29%), excepting several months for the A2 scenario (May–July) and B1 scenario (July), for which the simulated mean monthly discharges exceed those simulated for the reference period (with 2–19%) (Fig. 10.14).

The seasonal projected differences are conclusive for the autumn and the winter, when all three scenarios are consistent with each other and with the changes observed in the climatic parameters (temperature and precipitation): significant decrease during autumn (−13 to −19%) and slight increase during winter (1–7%).

To analyze the impact of climate changes on the magnitude and frequency of extreme flows (high and low), we set two thresholds: the discharge with 95% exceedance probability ($Q_{95\%}$) for low flow and the discharge with 5% exceedance probability ($Q_{5\%}$), for high flow.

The results of the simulations show that the $Q_{5\%}$ threshold does not change significantly: the maximum change is predicted by the A2 scenario (8% higher than

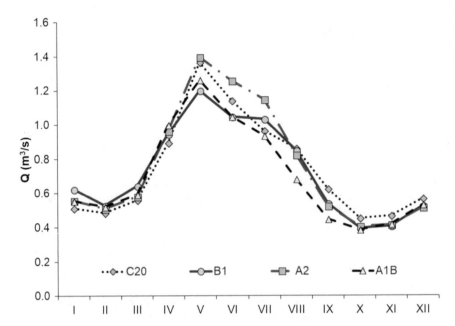

Fig. 10.14 The outcomes of the hydrological model regarding the mean monthly discharge of the Valea Cerbului River at Bușteni gauging station, for the climatic scenarios B1, A2, A1B (2001–2065) and the reference period C20 (1961–2000)

the reference period). The $Q_{95\%}$ decrease up to -12 to -18% relative to the value calculated for the reference period, reflecting the climate's tendency of becoming hotter and drier. Based on the two thresholds, we analyzed the possible change in the mean annual and monthly frequency of days with discharges below the threshold of $Q_{95\%}$ and above the threshold of $Q_{5\%}$. Even though the projected changes in the $Q_{5\%}$ value are not significant, the mean number of days exceeding it is higher: annually with 2–3.7 days and over 3 days for May–July (Fig. 10.15a). A low decrease (by

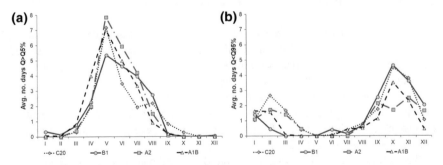

Fig. 10.15 The average monthly number of days with characteristic discharges of the Valea Cerbului River at Bușteni gauging station: greater than $Q_{5\%}$ (**a**) and lower than $Q_{95\%}$ (**b**) for the climatic scenarios B1, A2, A1B (2001–2065) and the reference period C20 (1961–2000)

1–2 days) in the number of days with discharges exceeding $Q_{5\%}$ is expected during August–October.

The simulations show that the annual frequency of discharges below $Q_{95\%}$, projected for the future period will be lower compared to the reference period, up to 18 days. Monthly, the same decreasing trend is expected in February, March, August and November. January is the only month for which all the scenarios indicate a possible low increase, by 1–1.4 days (scenarios average). In May there are not discharges below $Q_{95\%}$ for neither the reference and future scenarios periods (Fig. 10.15b).

The results of our simulations emphasize the fact that the alterations of the climatic parameters, as projected by the IPCC emissions scenarios, would impact on the hydrological regime by shifting water volumes from summer and autumn to winter and early spring, and by reducing the frequency of very low flows. However, the magnitude of both high and low flows could increase, amplifying the torrential flow character of the Valea Cerbului River.

10.6 Conclusions and Recommendations

As the hydrological system is highly sensitive to climate, it is expected that the alteration of climate parameters will impact the water cycle, affecting water availability and human uses, as well as the aquatic ecosystems. This chapter presented a review on the main research findings on hydroclimatic changes (observed and projected) in Romania, in the context of the researches at global and European scale, and provided new results on hydrological responses to climate changes issued from investigations in a small mountainous catchment (Valea Cerbului River basin), located in the Romanian Carpathians. The most important climate changes with hydrological impacts detected in the last half-century (after 1960) are a general warming at annual and seasonal scale (excepting autumn) and increase of the precipitation amount in autumn and of the daily maximum rainfall (mainly in summer and autumn). Important changes also affected the snow-related indices (mean snow depth, number of days with snow cover and with snowfall, continuous snow cover duration), which had significant downward trends. According to the climate simulations under different scenarios, it is likely expected that these trends will continue and intensify in the future.

The changes in the climatic parameters could be held responsible for some alterations and trends in the river flow, detected or estimated in studies conducted at national and regional/local scale. At national level, in Romania, a general decreasing trend in the mean annual streamflow was detected after 1960. Some changes in monthly and seasonal runoff variability were also found: upward trends in winter (related to the increase of the air temperature and the shift of the precipitation from solid to liquid form), downward trends in summer runoff (induced by the general warming and increase of evaporation), upward trends in autumn flow. For the future (by 2050), the simulations under climate scenarios indicate a general decline of the mean multiannual runoff, significant increases of discharges in the winter months and pronounced decreases in late summer and autumn.

In Valea Cerbului River catchment we investigated the streamflow changes based on historical data, as well as the possible hydrological changes, as indicated by simulations using the WaSiM-Eth model, under B1, A2 and A1B climatic scenario, for the time horizon 2001–2065 relative to 1961–2000 period. The simulations indicate a slight decrease in mean annual runoff (of −2% up to −6%) by 2065 and more or less significant changes in the annual flow regime. Thus, from January to April, an increase in mean monthly discharges (of 4–22%) is projected by all considered scenarios, while during May–December the simulations generally show lower mean monthly discharges (between −10 and −29%).

Overall, we highlighted the importance of methodological approaches for estimating climate change and hydrological impacts. Thus, the results of the studies are more or less similar and accurate, depending on the time scale, data used and characteristics of each model. This raise the question of knowing the uncertainties, which is a challenge for interpreting final results. Accordingly, the adaptation policies to climate change and hydrological impacts should take into account this issue of uncertainty, by considering feedback loops, reversible and flexible options, soft strategies for shorter time horizons, or other uncertainty-management methods.

In accordance with the actions taken at international and EU levels, Romania proceeded to the development of strategies and action plans regarding the adaptation to climate changes. The first National Strategy for Climate Change (NSCC) and the related National Action Plan on Climate Change (NAPCC) were designed in 2005 and planned measures and actions for the period 2005–2007. In 2008, the Guide regarding the adaptation to climate change effects was adopted, aiming the identification of necessary measures to limit the negative effects projected by climatic scenarios until 2030 [185]. Later, the NSCC and the related action plan were revised and updated. Currently, in Romania, it is in force the National Strategy on Climate Change and Economic Growth based on Low Carbon Emissions for 2016–2020 and the associated national action plan for its implementation, approved by the Romanian Government in 2016. The strategy addresses two main components: greenhouse gas concentration reduction and adaptation to climate change [186], which has become nowadays a major challenge faced by all, from individuals to local, regional and national authorities. The adaptation is a crucial key to the human existence, enabling the people to survive in changing and critical/extreme conditions, such as those related to climate alteration [187].

According to the NSCC, the process of adaptation to the effects of climate changes must occur at different spatial scales (nationally, regionally, and locally) and in different activity sector, with specific approaches for each location/sector. In the water field, NSCC propose many recommendations and measures to minimize and limit the impacts of the climate changes on water resources and to ensure the adaptation to these changes. A major recommendation concerns the development and updating of scientific knowledge on: water resources availability and requirements for different uses; future impacts of climate changes on rivers flow; water-related risks and their management. The scientific studies must be the basis for the adaptation measures. To this end, we consider that a better collaboration is needed between the research/academic institutions and stakeholders/authorities responsible for designing and implementing adaptation measures to climate changes (at different spatial scales). We also believe that it is necessary a cooperation between the research/academic institution and those holding hydro-climatic database, based on agreements, which facilitate the use of the database for scientific purposes. In this way it will be possible to use more effectively the hydro-climatic database nationwide, for studying climate change and their effects. It is desirable that the research results on this topic be integrated into a national platform accessible to all those interested.

Acknowledgements The authors address special thanks to Prof. Dr. Stuart Lane and Dr. Daniela Balin from University of Lausanne, Faculty of Geosciences and Environment, for their scientific and technical support regarding the application of the hydrological model WaSiM, during a research project funded through the Sciex-NMSch Programme.

Author Contributions All authors contributed equally to this work.

References

1. IPCC (2013) The physical science basis. In: Stocker TF, Qin D, Plattner GK, Tignor M, Allen SK, Boschung J, Nauels A, Xia Y, Bex V, Midgley PM (eds) Climate change 2013. Contribution of working group I to the fifth assessment report of the intergovernmental panel on climate change. Cambridge University Press, Cambridge, pp 1–29
2. IPCC (2014) Summary for policymakers. In: Field CB, Barros VR, Dokken DJ, Mach KJ, Mastrandrea MD, Bilir TE, Chatterjee M, Ebi KL, Estrad YO, Genova RC, Girma B, Kissel ES, Levy AN, MacCracken S, Mastrandrea PR, White LL (eds) Climate change 2014: impacts, adaptation, and vulnerability. Part A: global and sectoral aspects. Contribution of working group II to the fifth assessment report of the intergovernmental panel on climate change. Cambridge University Press, Cambridge, pp 1–32
3. Bates BC, Kundzewicz ZW, Wu S, Palutikof JP (2008) Climate change and water. Technical paper of the intergovernmental panel on climate change. https://www.ipcc.ch/pdf/technical-papers/climate-change-water-en.pdf. Accessed 26 May 2018
4. Kundzewicz ZW, Graczyk D, Maurer T, Pińskwar I, Radziejewski M, Svensson C, Szwed M (2005) Trend detection in river flow series: 1. Annual maximum flow. Hydrol Sci J 50(5):797–810
5. Zhang X, Harvey KD, Hogg WD, Yuzyk TR (2001) Trends in Canadian streamflow. Water Resour Res 37(4):987–998
6. Burn DH, Hag Elnur MA (2002) Detection of hydrologic trends and variability. J Hydrol 255:107–122
7. Burn DH, Sharif M, Zhang K (2010) Detection of trends in hydrological extremes for Canadian watersheds. Hydrol Process 24:1781–1790. https://doi.org/10.1002/hyp.7625
8. Fu G, Chen S, Liu C, Shepard D (2004) Hydro-climatic trends of the Yellow River basin, for the last 50 years. Clim Change 65:149–178
9. Li ZL, Xu ZX, Li JY, Li ZJ (2008) Shift trend and step changes for runoff time series in the Shiyang River basin, northwest China. Hydrol Process 22(23):4639–4646
10. Ma H, Yang D, Tan SK, Gao B, Hu Q (2010) Impact of climate variability and human activity on streamflow decrease in the Miyun reservoir catchment. J Hydrol 389:317–324

11. Piao S, Ciais P, Huang Y, Shen Z, Peng S, Li J, Zhou L, Liu H, Ma Y, Ding Y, Friedlingstein P, Liu C, Tan K, Yu Y, Zhang T, Fang J (2010) The impacts of climate change on water resources and agriculture in China. Nature 467:43–51. https://doi.org/10.1038/nature09364

12. Xu Z, Liu Z, Fu G, Chen Y (2010) Trends of major hydroclimatic variables in the Tarim River basin during the past 50 years. J Arid Environ 74(2):256–267

13. Zhang T, Fang J (2010) The impacts of climate change on water resources and agriculture in China. Nature 467:43–51. https://doi.org/10.1038/nature09364

14. Hu Y, Maskey S, Uhlenbrook S, Zhao H (2011) Streamflow trends and climate linkages in the source region of the Yellow River, China. Hydrol Process 25:3399–3411. https://doi.org/10.1002/hyp.8069

15. Liu Q, McVicar TR (2012) Assessing climate change induced modification of Penman potential evaporation and runoff sensitivity in a large water-limited basin. J Hydrol 464–465:352–362. https://doi.org/10.1016/j.jhydrol.2012.07.032

16. He B, Miao C, Shi W (2013) Trend, abrupt change, and periodicity of streamflow in the mainstream of Yellow River. Environ Monit Assess 185:6187–6199

17. Shen Q, Cong Z, Lei H (2017) Evaluating the impact of climate and underlying surface change on runoff within the Budyko framework: a study across 224 catchments in China. J Hydrol 554:251–262

18. Shen YJ, Shen Y, Fink M, Kralisch S, Chen Y, Brenning A (2018) Trends and variability in streamflow and snowmelt runoff timing in the southern Tianshan Mountains. J Hydrol 557:173–181

19. Rao AR, Azli M, Pae LJ (2011) Identification of trends in Malaysian monthly runoff under the scaling hypothesis. Hydrol Sci J 56(6):917–929

20. Smith LC (2000) Trends in Russian Arctic river-ice formation and breakup: 1917 to 1994. Phys Geogr 21:46–56

21. Peterson BJ, Holmes RM, McClelland JW, Vorosmarty CJ, Lammers RB, Shiklomanov AI, Shiklomanov IA, Rahmstorf S (2002) Increasing river discharge to the Arctic Ocean. Science 298:2171–2173

22. Lins HF, Slack JR (1999) Streamflow trends in the United States. Geophys Res Lett 26:227–230

23. Douglas EB, Vogel RM, Knoll CN (2000) Trends in floods and low flows in the United States: impact of serial correlation. J Hydrol 240:90–105

24. Kahya E, Kalayci S (2004) Trend analysis of streamflow in Turkey. J Hydrol 289(1–4):128–144

25. Cigizoglu HK, Bayazit M, Önöz B (2005) Trends in the maximum, mean, and low flows of Turkish rivers. J Hydrometeorol 6(3):280–290. https://doi.org/10.1175/JHM412.1

26. Yenigün K, Gümüş V, Bulut H (2008) Trends in streamflow of the Euphrates basin, Turkey. Proc Inst Civ Eng Water Manage 161(4):189–198. https://doi.org/10.1680/wama.2008.161.4.189

27. Genta JL, Perez-Iribarren G, Mechoso CR (1998) A recent increasing trend in the streamflow of rivers in southeastern South America. J Climate 11:2858–2862

28. Pasquini AI, Depetris PJ (2007) Discharge trends and flow dynamics of South American rivers draining the southern Atlantic seaboard: an overview. J Hydrol 333:385–399

29. Castino F, Bookhagen B, Strecker MR (2018) Oscillations and trends of river discharge in the southern Central Andes and linkages with climate variability. J Hydrol 555:108–124

30. Sidibe M, Dieppois B, Mahé G, Paturel JE, Amoussou E, Anifowose B, Lawler D (2018) Trend and variability in a new, reconstructed streamflow dataset for West and Central Africa, and climatic interactions, 1950–2005. J Hydrol 561:478–493

31. Svensson C, Kundzewicz WZ, Maurer T (2005) Trend detection in river flow series: 2. Flood and low-flow index series. Hydrol Sci J 50(5):824. https://doi.org/10.1623/hysj.2005.50.5.811

32. Dai A, Qian T, Trenberth KE, Milliman JD (2009) Changes in continental freshwater discharge from 1948 to 2004. J Clim 22(10):2773–2792

33. Blöschl G, Merz R, Parajka J, Salinas J, Viglione A (2012) Floods in Austria. In: Kundzewicz ZW (ed) Changes in flood risk in Europe. IAHS Special Publication 10, pp 169–177
34. Fiala T (2008) Statistical characteristics and trends of mean annual and monthly discharges of Czech rivers in the period 1961–2005. J Hydrol Hydromech 56:133–140
35. Renard B, Lang M, Bois P, Dupeyrat A, Mestre O, Niel H, Sauquet E, Prudhomme C, Parey S, Paquet E, Neppel L, Gailhard J (2008) Regional methods for trend detection: assessing field significance and regional consistency. Water Resour Res 44:W08419
36. Giuntoli I, Renard B, Lang M (2012) Floods in France. In: Kundzewicz ZW (ed) Changes in flood risk in Europe. IAHS Special Publication 10, pp 199–211
37. Petrow T, Merz B (2009) Trends in flood magnitude, frequency and seasonality in Germany in the period 1951–2002. J Hydrol 371:129–141
38. Petrow T, Zimmer J, Merz B (2009) Changes in the flood hazard in Germany through changing frequency and persistence of circulation patterns. Nat Hazards Earth Syst Sci 9:1409–1423
39. Bormann H, Pinter N, Elfert S (2011) Hydrological signatures of flood trends on German rivers: flood frequencies, flood heights and specific stages. J Hydrol 404:50–66
40. Hattermann FF, Kundzewicz ZW, Huang S, Vetter T, Kron W, Burghoff O, Merz B, Bronstert A, Krysanova V, Gerstengarbe FW, Werner P, Hauf Y (2012). Flood risk from a holistic perspective—observed changes in Germany. In: Kundzewicz ZW (ed) Changes in flood risk in Europe. IAHS Special Publication 10, pp 212–237
41. Jónsdóttir JF, Jónsson P, Uvo CB (2006) Trend analysis of Icelandic discharge, precipitation and temperature series. Nord Hydrol 37(4–5):365–376
42. Strupczewski WG, Kochanek K, Feluch W, Bogdanowicz E, Singh VP (2009) On seasonal approach to nonstationary flood frequency analysis. Phys Chem Earth, Parts A/B/C 34(10–12):612–618
43. Demeterova B, Skoda P (2009) Low flows in selected streams of Slovakia. J Hydrol Hydromech 57:55–69
44. Zeleňáková M, Purcz P, Soľáková T, Demeterová B (2012) Analysis of trends of low flow in river stations in eastern Slovakia. Acta Univ Agric et Silvic Mendel Brun 60(5):265–274
45. López-Moreno JI, Vicente-Serrano SM, Moran-Tejeda E, Zabalza J, Lorenzo-Lacruz J, García-Ruiz JM (2011) Impact of climate evolution and land use changes on water yield in the Ebro basin. Hydrol Earth Syst Sci 15(1):311–322
46. Morán-Tejeda E, López-Moreno JI, Ceballos-Barbancho A, Vicente-Serrano SM (2011) River regimes and recent hydrological changes in the Duero basin (Spain). J Hydrol 404(3–4):241–258
47. Lorenzo-Lacruz J, Vicente-Serrano SM, Lopez-Moreno JI, Moran-Tejeda E, Zabalza J (2012) Recent trends in Iberian streamflows (1945–2005). J Hydrol 414–415:463–475
48. Bîrsan MV, Molnar P, Burlando P, Pfaundler M (2005) Streamflow trends in Switzerland. J Hydrol 314:312–329. https://doi.org/10.1016/j.jhydrol.2005.06.008
49. Pellicciotti F, Bauder A, Parola M (2010) Effect of glaciers on streamflow trends in the Swiss Alps. Water Resour Res 46:W10522. https://doi.org/10.1029/2009WR009039
50. Schmocker-Fackel P, Naef F (2010) More frequent flooding? Changes in flood frequency in Switzerland since 1850. J Hydrol 381:1–8
51. Castellarin A, Pistocchi A (2011) An analysis of change in alpine annual maximum discharges: implications for the selection of design discharges. Hydrol Process 26(10):1517–1526
52. Arnell NW, Reynard NS (1996) The effects of climate change due to global warming on river flows in Great Britain. J Hydrol 183:397–424
53. Robson AJ, Jones TK, Reed DW, Bayliss AC (1998) A study of national trends and variation in UK floods. Int J Climatol 18:165–182
54. Dixon H, Lawler DM, Shamseldin AY (2006) Streamflow trends in western Britain. Geophys Res Lett 33:L19406. https://doi.org/10.1029/2006GL027325
55. Hannaford J, Marsh TJ (2006) An assessment of trends in UK runoff and low flows using a network of undisturbed catchments. Int J Climatol 26:1237–1253
56. Hannaford J, Marsh TJ (2008) High flow and flood trends in a network of undisturbed catchments in the UK. Int J Climatol 28:1325–1338

57. Hannaford J, Buys G (2012) Trends in seasonal river flow regimes in the UK. J Hydrol 475:158–174. https://doi.org/10.1016/j.jhydrol.2012.09.044
58. Lindström G, Bergström S (2004) Runoff trends in Sweden 1807–2002. Hydrol Sci J 49(1):69–83. https://doi.org/10.1623/hysj.49.1.69.54000
59. Reihan A, Koltsova T, Kriauciuniene J, Lizuma L, Meilutyte-Barauskiene D (2007) Changes in water discharges of the Baltic States rivers in the 20th century and its relation to climate change. Nord Hydrol 38(4–5):401–412
60. Reihan A, Kriauciuniene J, Meilutyte-Barauskiene D, Kolcova T (2012) Temporal variation of spring flood in rivers of the Baltic States. Hydrol Res 43(4):301–314
61. Korhonen J, Kuusisto E (2010) Long term changes in the discharge regime in Finland. Hydrol Res 41(3–4):253–268
62. Wilson D, Hisdal H, Lawrence D (2010) Has streamflow changed in the Nordic countries? Recent trends and comparisons to hydrological projections. J Hydrol 394(3–4):334–346
63. Stahl K, Hisdal H, Hannaford J, Tallaksen LM, van Lanen HAJ, Sauquet E, Demuth S, Fendekova M, Jódar J (2010) Streamflow trends in Europe: evidence from a dataset of near-natural catchments. Hydrol Earth Syst Sci 14:2367–2382. https://doi.org/10.5194/hess-14-2367-2010
64. Madsen H, Lawrence D, Lang M, Martinkova M, Kjeldsen TR (2014) Review of trend analysis and climate change projections of extreme precipitation and floods in Europe. J Hydrol 519:3634–3650
65. Kundzewicz ZW, Kanae S, Seneviratne SI, Handmer J, Nicholls N, Peduzzi P, Mechler R, Bouwer LM, Arnell N, Mach K, Muir-Wood R, Brakenridge GR, Kron W, Benito G, Honda Y, Takahashi K, Sherstyukov B (2014) Flood risk and climate change: global and regional perspectives. Hydrol Sci J 59(1):1–28
66. Bard A, Renard B, Lang M (2012) Floods in the Alpine areas of Europe. In: Kundzewicz ZW (ed) Changes in flood risk in Europe. IAHS Special Publication 10, pp 362–371
67. Sperna Weiland FC, van Beek LPH, Kwadijk JCJ, Bierkens MFP (2012) Global patterns of change in discharge regimes for 2100. Hydrol Earth Syst Sci 16:1047–1062
68. van Vliet MTH, Franssen WHP, Yearsley JR, Ludwig F, Haddeland I, Lettenmaier DP, Kabat P (2013) Global river discharge and water temperature under climate change. Global Environ Change 23:450–464
69. Arnell NW, Gosling SN (2013) The impacts of climate change on river flow regimes at the global scale. J Hydrol 486:351–364
70. Prudhomme C, Giuntoli I, Robinson EL, Clark DB, Arnell NW, Dankers R, Fekete BM, Franssen W, Gerten D, Gosling SN, Hagemann S, Hannah DM, Kim H, Masaki Y, Satoh Y, Stacke T, Wada Y, Wisser D (2014) Hydrological droughts in the 21st century, hotspots and uncertainties from a global multimodel ensemble experiment. Proc Natl Acad Sci 111(9):3262–3267
71. van Huijgevoort MHJ, van Lanen HAJ, Teuling AJ, Uijlenhoet R (2014) Identification of changes in hydrological drought characteristics from a multi-GCM driven ensemble constrained by observed discharge. J Hydrol 512:421–434
72. Santini M, di Paola A (2015) Changes in the world rivers' discharge projected from an updated high resolution dataset of current and future climate zones. J Hydrol 531:768–780
73. Touma D, Ashfaq M, Nayak MA, Kao SC, Diffenbaugh NS (2015) A multi-model and multi-index evaluation of drought characteristics in the 21st century. J Hydrol 526:196–207
74. Wanders N, Wada Y (2015) Human and climate impacts on the 21st century hydrological drought. J Hydrol 526:208–220
75. Gosling SN, Zaherpour J, Mount NJ, Hattermann FF, Dankers R, Arheimer B, Breuer L, Ding J, Haddeland I, Kumar R, Kundu D, Liu J, van Griensven A, Veldkamp TIE, Vetter T, Wang X, Zhang X (2016) A comparison of changes in river runoff from multiple global and catchment-scale hydrological models under global warming scenarios of 1 °C, 2 °C and 3 °C. Clim Change 141(3):577–595. https://doi.org/10.1007/s10584-016-1773-3
76. Arnell NW, Gosling SN (2016) The impacts of climate change on river flood risk at the global scale. Clim Change 134:387–401. https://doi.org/10.1007/s10584-014-1084-5

77. Alfieri L, Bisselink B, Dottori F, Naumann G, de Roo A, Salamon P, Wyser K, Feyen L (2017) Global projections of river flood risk in a warmer world. Earth's Future 5:171–182. https:// doi.org/10.1002/2016EF000485
78. Strzepek KM, Yates DN (1997) Climate change impacts on the hydrologic resources of Europe: a simplified continental scale analysis. Clim Change 36(1–2):79–92
79. Arnell NW (1999) The effect of climate change on hydrological regimes in Europe: a continental perspective. Global Environ Change 9:5–23
80. Döll P, Müller Schmied H (2012) How is the impact of climate change on river flow regimes related to the impact on mean annual runoff? A global-scale analysis. Environ Res Lett 7(1):1–11. https://doi.org/10.1088/1748-9326/7/1/014037
81. Rojas R, Feyen L, Bianchi A, Dosio A (2012) Assessment of future flood hazard in Europe using a large ensemble of bias-corrected regional climate simulations. J Geophys Res 117:D17109. https://doi.org/10.1029/2012JD017461
82. Alfieri L, Pappenberger F, Wetterhall F, Haiden T, Richardson D, Salamon P (2014) Evaluation of ensemble streamflow predictions in Europe. J Hydrol 517:913–922
83. Forzieri G, Feyen L, Rojas R, Flörke M, Wimmer F, Bianchi A (2014) Ensemble projections of future streamflow droughts in Europe. Hydrol Earth Syst Sci 18:85–108. https://doi.org/ 10.5194/hess-18-85-2014
84. Roudier P, Andersson JCM, Donnelly C, Feyen L, Greuell W, Ludwig F (2015) Projections of future floods and hydrological droughts in Europe under a +2 °C global warming. Clim Change 135(2):341–355. https://doi.org/10.1007/s10584-015-1570-4
85. Stagl JC, Hattermann FF (2015) Impacts of climate change on the hydrological regime of the Danube River and its tributaries using an ensemble of climate scenarios. Water 7:6139–6172. https://doi.org/10.3390/w7116139
86. Donnelly C, Greuell W, Andersson J, Gerten D, Pisacane G, Roudier P, Ludwig F (2017) Impacts of climate change on European hydrology at 1.5, 2 and 3 degrees mean global warming above preindustrial level. Clim Change 143:13–26
87. Marx A, Kumar R, Thober S, Rakovec O, Wanders N, Zink M, Wood EF, Pan M, Sheffield J, Samaniego L (2018) Climate change alters low flows in Europe under global warming of 1.5, 2, and 3 _C. Hydrol Earth Syst Sci 22:1017–1032. https://doi.org/10.5194/hess-22-1017-2018
88. Thober S, Kumar R, Wanders N, Marx A, Pan M, Rakovec O, Samaniego L, Sheffield J, Wood EF, Zink M (2018) Multi-model ensemble projections of European river floods and high flows at 1.5, 2, and 3 degrees global warming. Environ Res Lett 13(1):1–11. https://doi. org/10.1088/1748-9326/aa9e35
89. Naz BS, Kao SC, Ashfaq M, Gao H, Rastogi D, Gangrade S (2018) Effects of climate change on streamflow extremes and implications for reservoir inflow in the United States. J Hydrol 556:359–370
90. Jiang T, Chen YD, Xu C, Chen X, Chen X, Singh VP (2007) Comparison of hydrological impacts of climate change simulated by six hydrological models in the Dongjiang Basin, South China. J Hydrol 336(3–4):316–333. https://doi.org/10.1016/j.jhydrol.2007.01.010
91. Mikhailov VN, Mikhailova MV (2017) Natural and anthropogenic long-term variations of water runoff and suspended sediment load in the Huanghe river. Water Resour 44(6):793–807
92. Reshmidevi TV, Nagesh Kumar D, Mehrotra R, Sharma A (2018) Estimation of the climate change impact on a catchment water balance using an ensemble of GCMs. J Hydrol 556:1192–1204
93. Krysanova V, Vetter T, Eisner S, Huang S, Pechlivanidis I, Strauch M, Gelfan A, Kumar R, Aich V, Arheimer B, Alejandro Chamorro A, van Griensven A, Kundu D, Lobanova A, Mishra V, Plötner S, Reinhardt J, Seidou O, Wang X, Wortmann M, Zeng X, Hattermann FF (2017) Intercomparison of regional-scale hydrological models and climate change impacts projected for 12 large river basins worldwide—a synthesis. Environ Res Lett 12(10):1–12. https://doi.org/10.1088/1748-9326/aa8359
94. Pechlivanidis IG, Arheimer B, Donnelly C, Hundecha Y, Huang S, Aich V, Samaniego L, Eisner S, Shi P (2017) Analysis of hydrological extremes at different hydro-climatic regimes under present and future conditions. Clim Change 141:467–481

95. Vetter T, Huang SH, Aich V, Yang T, Wang X, Krysanova V, Hattermann F (2015) Multi-model climate impact assessment and intercomparison for three large-scale river basins on three continents. Earth Syst Dyn 6:17–43

96. Vetter T, Reinhardt J, Flörke M, van Griensven A, Hattermann F, Huang S, Koch H, Pechlivani-dis IG, Plötner S, Seidou O, Su B, Vervoort RW, Krysanova V (2017) Evaluation of sources of uncertainty in projected hydrological changes under climate change in 12 large-scale river basins. Clim Change 141:419–433

97. Alfieri L, Burek P, Feyen L, Forzieri G (2015) Global warming increases the frequency of river floods in Europe. Hydrol Earth Syst Sci 19:2247–2260. https://doi.org/10.5194/hess-19-2247-2015

98. ICPDR (International Commission for the Protection of the Danube River) (2009) Danube basin: facts and figures. https://www.icpdr.org/flowpaper/viewer/default/files/nodes/documents/icpdr_facts_figures.pdf. Accessed 26 May 2018

99. NMA (National Meteorological Administration) (2008) Clima României. Editura Academiei Române, București

100. Tomozeiu R, Busuioc A, Ștefan S (2002) Changes in seasonal mean of maximum air temperature in Romania and their connection with large-scale circulation. Int J Climatol 22(10):1181–1196. https://doi.org/10.1002/joc.785

101. Croitoru AE, Dragotă CS, Moldovan F, Holobâcă I, Toma FM (2011a) Considérations sur l'évolution des températures de l'air dans les Carpates roumaines. Actes du XXIVème Colloque International de l'Association Internationale de Climatologie, pp 147–152

102. Ioniță-Scholz M, Rimbu N, Chelcea S, Pătruț S (2013) Multidecadal variability of summer temperature over Romania and its relation with Atlantic Multidecadal Oscillation. Theor Appl Climatol 113(1–2):305–315. https://doi.org/10.1007/s00704-012-0786-8

103. Croitoru AE, Piticar A (2013) Changes in daily extreme temperatures in the extra-Carpathians regions of Romania. Int J Climatol 33(8):1987–2001. https://doi.org/10.1002/joc.3567

104. Bîrsan MV, Dumitrescu A, Micu DM, Cheval S (2014) Changes in annual temperature extremes in the Carpathians since AD 1961. Nat Hazards 74(3):1899–1910. https://doi.org/10.1007/s11069-014-1290-5

105. Rîmbu N, Ștefan S, Necula C (2014) The variability of winter high temperature extremes in Romania and its relationship with largescale atmospheric circulation. Theor Appl Climatol 121(1–2):121–130. https://doi.org/10.1007/s00704-014-1219-7

106. Croitoru AE, Piticar A, Ciupertea AF, Roșca CF (2016) Changes in heat waves indices in Romania over the period 1961–2015. Global Planet Change 146:109–121. https://doi.org/10.1016/j.gloplacha.2016.08.016

107. Busuioc A, von Storch H (1996) Changes in the winter precipitation in Romania and its relation to the large-scale circulation. Tellus 48A(4):538–552. https://doi.org/10.1034/j.1600-0870.1996.t01-3-00004.x

108. Tomozeiu R, Ștefan S, Busuioc A (2005) Winter precipitation variability and large-scale circulation patterns in Romania. Theor Appl Climatol 81(3–4):193–201. https://doi.org/10.1007/s00704-004-0082-3

109. Croitoru AE, Toma FM (2010) Trends in precipitation and snow cover in central part of Romanian Plain. Geographia Technica 1:460–469

110. Croitoru AE, Chiotoroiu B, Iancu I (2011) Precipitation analysis using Mann-Kendal test and WASP cumulated curve in Southeastern Romania. Stud Univ Babeş Bolyai Geographia 1:49–58

111. Croitoru AE, Piticar A, Burada DC (2016) Changes in precipitation extremes in Romania. Quant Int 415:325–335. https://doi.org/10.1016/j.quaint.2015.07.028

112. Bojariu R, Dinu M (2007) Snow variability and change in Romania. In: Strasser U, Vogel M (eds) Proceedings of the Alpine Snow workshop. Berchtesgaden National Park Report 52, Munich, pp 64–68

113. Micu D (2009) Snow pack in the Romanian Carpathians under changing climatic conditions. Meteorog Atmos Phys 105(1–2):1–16. https://doi.org/10.1007/s00703-009-0035-6

114. Bîrsan MV, Dumitrescu A (2014) Snow variability in Romania in connection to large-scale atmospheric circulation. Int J Climatol 34(1):134–144. https://doi.org/10.1002/joc.3671
115. Croitoru AE, Piticar A, Dragotă CS, Burada DC (2013) Recent changes in reference evapotranspiration in Romania. Global Planet Change 111:127–136. https://doi.org/10.1016/j.gloplacha.2013.09.004
116. Croitoru AE, Piticar A, Imbroane AM, Burada DC (2013) Spatiotemporal distribution of aridity indices based on temperature and precipitation in the extra-Carpathian regions of Romania. Theor Appl Climatol 112(3–4):597–607
117. Prăvălie R, Bandoc G (2015) Aridity variability in the last five decades in the Dobrogea region, Romania. Arid Land Res Manage 29(3):265–287. https://doi.org/10.1080/15324982.2014.977459
118. Ioniţă M, Scholz P, Chelcea S (2015) Spatio-temporal variability of dryness/wetness in the Danube River basin. Hydrol Process 29(20):4483–4497. https://doi.org/10.1002/hyp.10514
119. Busuioc A, Caian M, Cheval S, Bojariu R, Boroneanţ C, Baciu M, Dumitrescu A (2010) Variability and climate change in Romania. Pro Universitaria, Bucureşti
120. Dumitrescu A, Bojariu R, Bîrsan MV, Marin L, Manea A (2014) Recent climatic changes in Romania from observational data (1961–2013). Theor Appl Climatol 122(1–2):111–119. https://doi.org/10.1007/s00704-014-1290-0
121. Marin L, Bîrsan MV, Bojariu R, Dumitrescu A, Micu DM, Manea A (2014) An overview of annual climatic changes in Romania: trends in air temperature, precipitation, sunshine hours, cloud cover, relative humidity and wind speed during the 1961–2013 period. Carpath J Earth Environ Sci 9(4):253–258
122. Bojariu R, Bîrsan MV, Cică R, Velea L, Burcea S, Dumitrescu A, Dascălu SI, Gothard M, Dobrinescu A, Cărbunaru F, Marin L (2015) Schimbările climatice – de la bazele fizice la riscuri şi adaptare. Editura Printech, Bucureşti
123. Busuioc A, Dobrinescu A, Bîrsan MV, Dumitrescu A, Orzan A (2015) Spatial and temporal variability of climate extremes in Romania and associated large-scale mechanisms. Int J Climatol 35:1278–1300. https://doi.org/10.1002/joc.4054
124. Cheval S, Birsan MV, Dumitrescu A (2014) Climate variability in the Carpathian Mountains region over 1961–2010. Global Planet Change 118:85–96. https://doi.org/10.1016/j.gloplacha.2014.04.005
125. Micu DM, Dumitrescu A, Cheval S, Bîrsan MV (2015) Observed variability and trends from instrumental records. In: Micu DM, Dumitrescu A, Cheval S, Bîrsan MV (eds) Climate of the Romanian Carpathians. Springer, Cham, pp 149–185
126. Micu DM, Dumitrescu A, Cheval S, Bîrsan MV (2015) Changing climate extremes in the last five decades (1961–2010). In: Micu DM, Dumitrescu A, Cheval S, Bîrsan MV (eds) Climate of the Romanian Carpathians. Springer, Cham, pp 187–198
127. Mateescu E (2014) ADER 1.1.1. Sistem de indicatori geo-referenţiali la diferite scări spaţiale şi temporale pentru evaluarea vulnerabilităţii şi măsurile de adaptare ale agroecosistemelor faţă de schimbările globale. Seminar privind diseminarea rezultatelor cercetarilor din domeniul mecanizarii, economiei agrare, pedologiei, agrochimiei si combaterii eroziunii solului, imbunatatirilor funciare, meteorologiei si hidrologiei. http://www.madr.ro/attachments/article/139/ANM-ADER-111.pdf. Accessed 27 May 2018
128. MESD (Ministry of Environment and Sustainable Development) (2008) Ghid privind adaptarea la schimbările climatice. Ministry of Environment and Sustainable Development, Bucureşti
129. NMA (National Meteorological Administration) (2014) Adaptation measures in Romanian agriculture, SEE Project-OrientGate: a structured network for integration of climate knowledge into policy and territorial planning. National Administration of Meteorology, Bucureşti
130. Zaharia L, Perju R, Ioana-Toroimac G (2018) Climate changes and effects on river flow in the Romanian Carpathians. In: Air and water components of the environment, pp 211–218

131. Jacob D, Petersen J, Eggert B, Alias A, Christensen OB, Bouwer LM, Braun A, Colette A, Déqué M, Georgievski G, Georgopoulou E, Gobiet A, Menut L, Nikulin G, Haensler A, Hempelmann N, Jones C, Keuler K, Kovats S, Kröner N, Kotlarski S, Kriegsmann A, Martin E, van Meijgaard E, Moseley C, Pfeifer S, Preuschmann S, Radermacher C, Radtke K, Rechid D, Rounsevell M, Samuelsson P, Somot S, Soussana JF, Teichmann C, Valentini R, Vautard R, Webe B, Yiou P (2014) EURO-CORDEX: new high-resolution climate change projections for European impact research. Reg Environ Change 14(2):563–578

132. Cheval S, Dumitrescu A, Bîrsan MV (2017) Variability of the aridity in the South-Eastern Europe over 1961–2050. CATENA 151:74–86

133. Nistor MM, Ronchetti F, Corsini A, Cheval S, Dumitrescu A, Kumar Rai P, Petrea D, Dezsi Ş (2017) Crop evapotranspiration variation under climate change in South East Europe during 1991–2050. Carpath J Earth Environ 12(2):571–582

134. Bondar C, Buţă C (1995) Trends of water discharges, sediment discharges and salinity of Danube in the Romanian sector. Rom J Hydrol Water Resour 2:61–65

135. Ştefan S, Ghioca M, Rîmbu N, Boroneanţ C (2004) Study of meteorological and hydrological drought in Southern Romania from observational data. Int J Climatol 24:871–881. https://doi.org/10.1002/joc.1039

136. Neculau G, Zaharia L (2009) Tendinţe în variabilitatea precipitaţiilor şi a scurgerii medii în bazinul hidrografic al râului Trotuş. Comunicări de Geografie XIII:249–254

137. Zaharia L, Beltrando G (2009) Variabilité et tendances de la pluviométrie et des débits de crue dans la région de la Courbure de l'Arc carpatique (Roumanie). Geographia Technica, numéro spécial, pp 471–476

138. Neculau G, Zaharia L (2010) Maximum flow variability and flood potential in Trotuş catchment area. Stud Univ Babeş Bolyai Geographia LV(1):87–98

139. Jipa N, Mehedinţeanu L (2012) Trends in variability of water flow of Teleajen river. In: Air and water—components of the environment, pp 535–542

140. Bîrsan MV, Zaharia L, Chendeş V, Brănescu E (2012) Recent trends in streamflow in Romania (1976–2005). Rom Rep Phys 64(1):275–280

141. Bîrsan MV, Zaharia L, Chendeş V, Brănescu E (2014) Seasonal trends in Romanian streamflow. Hydrol Process 28:4496–4505. https://doi.org/10.1002/hyp.9961

142. Perju R, Zaharia L (2014) Changes in the frequency and magnitude of floods in the Bucegi Mountains (Romanian Carpathians). In: Gâştescu P, Włodzimierz M, Breţcan P (eds) 2nd international conference—water resources and wetlands. 11–13 Sept, 2014 Tulcea (Romania). Editura Transversal, Târgovişte, pp 321–328

143. Mitof I, Prăvălie R (2014) Temporal trends of hydroclimatic variability in the lower Buzău catchment. Geographia Technica 1:87–100

144. Mic RP, Mareş C, Corbuş C, Mătreaţă M, Chendeş V, Radu E, Stănescu G, Chelcea S, Teodor S, Mătreaţă S, Adler MJ, Mareş I, Achim D, Preda A, Borcan M, Retegan M, Apostu AD, Brănescu E (2016) Climate change impact on hydrology, CLIMHYDEX—changes in climate extremes and associated impact in hydrological events in Romania. Final report, Bucureşti

145. Mikhailova MV, Morozov VN, Levashova EA, Mikhailov VN (2002) Natural and anthropogenic changes in water and sediment runoff of the Danube at the delta head (1840–2000). In: XXI conference of the Danube countries on hydrological forecasting and hydrological bases of water management, Bucharest, pp 1–7

146. Rîmbu N, Boroneanţ C, Buţă C, Dima M (2002) Decadal variability of the Danube river flow in the lower basin and its relation with the North Atlantic oscillation. Int J Climatol 22:1169–1179. https://doi.org/10.1002/joc.788

147. Gâştescu P, Ţuchiu E (2012) The Danube River in the pontic sector—hydrological regime. In: Gâştescu P, Lewis W Jr, Breţcan P (eds) Water resources and wetlands. Conference proceedings, 14–16 Sept 2012, Tulcea, Romania. Editura Transversal, Târgovişte, pp 13–26

148. Zaharia L, Ioana-Toroimac G (2013) Romanian Danube River management: impacts and perspectives. In: Arnaud-Fassetta G, Masson E, Reynard E (eds) European continental hydrosystems under changing water policy. Verlag Friedrich Pfeil, München, pp 159–170

149. Croitoru AE, Minea I (2015) The impact of climate changes on rivers discharge in Eastern Romania. Theor Appl Climatol 120:563–573. https://doi.org/10.1007/s00704-014-1194-z
150. Prăvălie R, Zaharia L, Bandoc G, Petrișor A, Ionuș A, Mitof I (2016) Hydroclimatic dynamics in southwestern Romania drylands over the past 50 years. J Earth Syst Sci 125(6):1255–1271
151. Șerban P, Corbuș C (1994) Impactul modificărilor climatice asupra bilanțului apei pe un bazin hidrografic. Hidrotehnica 39:3
152. Stănescu VA, Corbus C, Simota M (1999) Modelarea impactului schimbărilor climatice asupra resurselor de apă. Editura HGA, București
153. Șerban AC (2006) Impactul schimbărilor climatice asupra resurselor și sistemelor de gospodărire a apelor. Editura Tipored, București
154. Chirila G, Corbuș C, Mic R, Busuioc A (2008) Assessment of the potential impact of climate change upon surface water resources in the Buzău and Ialomița watersheds from Romania in the frame of Cecilia project. BALWOIS, Ohrid, pp 1–8
155. Corbuș C, Mic R, Mătreață M (2011) Assessment of climate change impact on peak flow regime in the Mureș river basin. In: XXVth conference of Danubian countries, 16–17 June, Budapest, Hungary
156. Corbuș C, Mic RP, Mătreață M, Chendeș V (2012) Climate change impact upon maximum flow in Siret river basin. In: 12th international multidisciplinary scientific geoconference SGEM 2012, conference proceedings, vol III, Albena, Bulgaria, pp 587–594
157. Corbuș C, Mic RP, Mătreață M (2013) Potential climate change impact upon maximum flow in Ialomita river basin. In: National Institute of Hydrology and Water Management—scientific conference, "Water resources management under climate and anthropogenic changes", 23–26 Sept, Bucharest, pp 237–242
158. Adler MJ (2013) Climate change and its impact in water resources. Belgrade AIHS volume. In: Conference proceedings of climate variability and change—hydrological impacts, Belgrade, Oct 2013, pp 117–127
159. Mic RP, Corbuș C, Busuioc A (2013) Climate change impact upon water resources in the Buzău and Ialomița river basins. Rom J Geogr 57(2):93–104
160. Adler MJ, Chelcea S (2014) Climate change and its impact in water resources in Romania. In: Proceedings of XXVI conference of the Danubian countries on hydrological forecasting and hydrological bases of water management, 22–24 Sept 2014, Deggendorf, Germany, pp 83–88
161. Corbuș C, Mic RP, Mătreață M (2014) Estimarea impactului schimbărilor climatice potențiale asupra scurgerii maxime din bazinul hidrografic Olt. Hidrotehnica 59(10–11):28–38
162. Perju ER, Zaharia L, Balin D, Lane S (2014) Changements climatiques dans les Carpates Roumaines et impacts hydrologiques. Étude de cas: les Monts de Bucegi. Actes du XXVIIe Colloque de l'Association Internationale de Climatologie, pp 73–79
163. Retegan M, Borcan M (2014) Assessment of the potential impact of climate change upon surface water resources in the Ialomița River basin from Romania. In: Proceedings of the 14th geoconference on water resources. Forest, marine and ocean ecosystems, 17–26 iunie 2014, Albena, Bulgaria, pp 89–96. https://doi.org/10.5593/SGEM2014/B31/S12.012
164. Sipos G, Blanka V, Mezősi G, Kiss T, van Leeuwen B (2014) Effect of climate change on the hydrological character of river Maros, Hungary-Romania. J Environ Geogr 7(1–2):49–56. https://doi.org/10.2478/jengeo-2014-0006
165. Corbuș C, Mic RP, Busuioc A, Mătreață M (2016) Long-term effects of projected climate change on the extreme flow from Bârlad River basin. Conferința Științifică a INHGA "Apa, resursă vitală și factor de risc – perspective ale unui management integrat", București, pp 53–62
166. Mic RP, Corbuș C, Mătreață M (2016a) Effects of climate change on extreme flow in Romanian River basin Bârlad. In: 16th international multidisciplinary scientific geoconference SGEM 2016, 30 June–6 July 2016, Albena, Bulgaria, conference proceedings, book 3—Water resources forest, marine and ocean ecosystems, vol 1—hydrology and water resources, pp 273–280

167. CLAVIER Project (2007) Climate change and variability: impact on Central and Eastern Europe. Results in work package 3: hydrology. Hamburg, Germany, http://www.clavier-eu. org/?q=node/879. Accessed 25 June 2018

168. CECILIA (2009) Central and Eastern Europe climate change impact and vulnerability assessment. Publishable final activity report. http://www.cecilia-eu.org/restricted/deliverables.php. Accessed 25 June 2018

169. Bojariu R, Papathoma-Köhle M, Wendlová V, Cica RD (2014) Changing risks in changing climate—SEERISK. Report SEE. http://www.meteoromania.ro/anm/images/clima/SEERISKchangingclimate2014.pdf. Accessed 25 June 2018

170. NMA (National Meteorological Administration) (2016) CLIMHYDEX—changes in climate extremes and associated impact in hydrological events in Romania. Final report, București, 87 p

171. Werners SE, Bos E, Civic K, Hlásny T, Hulea O, Jones-Walters L, Kőpataki E, Kovbasko A, Moors E, Nieuwenhuis D, van de Velde I, Zingstra H, Zsuffa I (2014) Climate change vulnerability and ecosystem-based adaptation measures in the Carpathian region. Final report—integrated assessment of vulnerability of environmental resources and ecosystem-based adaptation measures. Alterra Wageningen UR (University & Research centre), Alterra report 2572, Wageningen

172. Stagl JC, Hattermann FF (2016) Impacts of climate change on riverine ecosystems: alterations of ecologically relevant flow dynamics in the Danube River and its major tributaries. Water 8(566):2–55. https://doi.org/10.3390/w8120566

173. Lobanova A, Stagl J, Vetter T, Hattermann F (2015) Discharge alterations of the Mures River, Romania under ensembles of future climate projections and sequential threats to aquatic ecosystem by the end of the century. Water 7:2753–2770. https://doi.org/10.3390/w7062753

174. Corbuș C, Mic RP, Mătreață M, Chendeș V, Preda A (2017) Potential climate change impact on mean flow in Romania. Electronic book with full papers from XXVII conference on the Danubian countries on hydrological forecasting and hydrological bases of water management, 26–28 Sept 2017, Golden Sands, Bulgaria, pp 548–557

175. ICPDR (International Commission for the Protection of the Danube River) (2013) Strategy on adaptation to climate change. http://www.icpdr.org/main/sites/default/files/nodes/documents/icpdr_climate-adaptation-strategy.pdf. Accessed 2 July

176. Perju R (2012) Characteristics of floods in Valea Cerbului catchment. In: Gâștescu P, Lewis W Jr, Brețcan P (eds) Water resources and wetlands. Conference proceedings, 14–16 Sept 2012, Tulcea, Romania. Editura Transversal, Târgoviște, pp 248–253

177. Perju R (2012) Flow controls factors and runoff characteristics in the Valea Cerbului River basin. In: Air and water—components of the environment, pp 503–510

178. IPCC (Intergovernmental Panel on Climate Change) (2000) Special report on emissions scenarios. Cambridge University Press, Cambridge

179. Hawkins E, Osborne TM, Ho CK, Challinor AJ (2013) Calibration and bias correction of climate projections for crop modelling: an idealised case study over Europe. Agr Forest Meteor 170:19–31

180. Salmi T, Määttä A, Anttila P, Ruoho-Airola T, Amnell T (2002) Detecting trends of annual values of atmospheric pollutants by the Mann-Kendall test and Sen's slope estimates—the excel template application MAKESENS. Publications on Air Quality 31, Helsinki

181. Schulla J (2012) Model description WaSiM. Zürich. http://www.wasim.ch/downloads/doku/wasim/wasim_2012_en.pdf Accessed 7 Feb 2017

182. Doherty J (2018) PEST—model-independent parameter estimation user manual part I: PEST, SENSAN and global optimisers, 7th edn. Watermark Numerical Computing. http://www.pesthomepage.org/Downloads.php. Accessed 6 June 2018

183. Chivoiu AD (2010) Valorificarea resurselor de apă din zona orașului Bușteni. In: Gâștescu P, Lewis W Jr, Brețcan P (eds) Water resources and wetlands. Conference proceedings, 14–16 Sept 2012, Tulcea, Romania. Editura Transversal, Târgoviște, pp 250–256

184. Réméniéras G (1999) L'hydrologie de l'ingénieur, 2nd edn. Eyrolles, Paris

185. MMAP (Ministry of Environment, Waters and Forests) (2008) Ghidul privind adaptarea la efectele schimbarilor climatice. http://www.meteoromania.ro/anm/images/clima/SSCGhidASC.pdf; http://www.meteoromania.ro/anm2/clima/adaptarea-la-schimbarile-climatice/. Accessed 10 July 2018
186. MM (Ministry of Environment) (2018) Strategia Națională privind Schimbările Climatice. http://mmediu.ro/categorie/strategia-nationala-privind-schimbarile-climatice-rezumat/171. Accessed 10 July 2018
187. Zaharia L, Ioana-Toroimac G (2016) Developing soft measures for flood risk mitigation and adaptation in Romania: public informing and awareness. In: Sorocovschi V (ed) Riscuri şi catastrofe, an XV, vol 18, nr 1, Ed. Casa Cărții de Știință, Cluj Napoca, pp 7–22

Chapter 11
Monitoring and Management of Water in the Siret River Basin (Romania)

Larisa Elena Paveluc, Gianina Maria Cojoc and Alina Tirnovan

Abstract The Siret hydrographical basin is the largest in the area and holds the highest multiannual average flow rate in Romania (220 m^3/s). The upper basin covers the territory of Ukraine, and it represents 17% of the total volume of water resources. It is also the most inhabited river basin where density is the highest in eastern Romania. The most important hydrographic arteries spring from the Oriental Carpathians. The Moldavian Plateau receives only one important river: Barlad, which has relatively low flow. It has high hydropower potential and important hydro-technical facilities on the Bistrita River. The study approaches the quantitative monitoring of water bodies within the Siret Water Basin Administration and the qualitative monitoring of 353 water bodies (343 bodies of water—natural rivers and 10 water bodies—lakes). There are analyzed all categories of resources: surface (rivers, lakes); groundwater (underground water, deep water). The study approaches the methodology of water resources management, and also the purpose of performing different types of measurements (both quantitative and qualitative) and measures to be taken in order to avoid the negative effects that can result from the exploitation of these resources are presented in detail.

Keywords Hydrological stations · Groundwater · Lakes · Romania · Water quality

11.1 Introduction

The community's attempt to rationally exploit the water resources and reduce or combat pollution has led to the local and international development of networks for tracking qualitative and quantitative water parameters. This information represents the onset of the modeling, and the numerical simulation process of the surface and groundwater flow physical and chemical parameters and their pollution. The river

L. E. Paveluc (✉)
Faculty of Geography and Geology, Department of Geography, Alexandru Ioan Cuza University of Iasi, 20A Carol I 20A Blvd., 700505 Iasi, Romania
e-mail: cristina.despina@ddni.ro

L. E. Paveluc · G. M. Cojoc · A. Tirnovan
Siret Water Basin Administration, Bacau, 1 Cuza Voda Str., 600274 Bacau, Romania

© Springer Nature Switzerland AG 2020
A. M. Negm et al. (eds.), *Water Resources Management in Romania*, Springer Water,
https://doi.org/10.1007/978-3-030-22320-5_11

basin of the Siret River is the most inhabited in Romania, and for this reason, the problem of water resources and its quality is stringent.

Fresh-water, both quantitatively and qualitatively, internationally and implicitly nationally, is extremely low. This is why the food crisis is acute. Most optimistic opinions lead to the idea that it is necessary to obtain an additional quantity of water from desalination. For the mainland regions of the continents, intensive new sources of water are sought, especially by extracting a surplus of groundwater (underground and deep water). Unfortunately, the underground waters in Romania are used in extremely low proportion (up to 6%) and therefore the water resources for the populated regions are reduced. The cause is the high costs of deep-water exploitation.

Global climate change inevitably leads to imbalances. Although the amount of water generated by precipitation has increased, the periods of drought are longer than those with precipitations [39, 41, 58–60]. The cause is the torrential rains when impressive amounts of water accumulate in a few hours [42, 45, 46, 50–53, 57]. These inconveniences have led to accelerating studies on the evolution of the surface and groundwater resources, and at the same time creating an avoidance or loss mitigation model [54–56]. The literature is extremely rich in analysis of water resources for rivers [1, 13–15, 18, 23, 24, 37, 38, 47, 48, 50–53, 65, 69–71], lakes [2, 16, 20, 26–28, 34, 43–56, 64, 66, 68] and groundwater [13, 14, 31–33, 35, 40, 50–53, 61–63], but especially for the problems caused by the pollution and quality mitigation [3–10, 12, 17, 19, 21, 22, 25, 30].

11.2   Study Area

Among the inner water courses, the Siret River is the most important tributary of the Danube River on the territory of Romania. The Siret springs from the Padurosi Carpathians (from the current territory of Ukraine), specifically from Mount Lungul (1382 m), entering Romania in Varcauti, about 5 km NE from the town of Siret, and after a total of 726 km (559 km in Romania), flows into the Danube, near the city of Galati (at Sendreni) [41] (Fig. 11.1).

The Siret River is the most important hydrographic artery in Romania because it has the highest multiannual average flow: 220 m³/s. The hydrographic basin, which covers the territory of Ukraine and Romania, has a total surface area of 42,890 km², from which 44,871 km² in Romania. It springs from the Ukrainian Carpathians, where its upper basin can be identified, and then flows across Romania between the Eastern Carpathians to the west and the Moldavian Plateau to the east [39, 41, 42].

It has the most complex hydrographic network: 1013 rivers, which measure 15,157 km, which means 19.2% of Romania's total. The forest stock includes 15.882 km², which represents 37% of the hydrographic area and 25% of the entire forest stock in Romania [39]. The Siret River Basin has the following mathematical

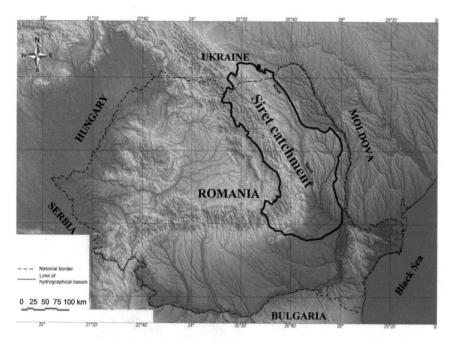

Fig. 11.1 Geographical location of the Siret River Basin on the territory of Romania

coordinates: 47°58′ latitude N; 45°28′ latitude N; 24°49′ longitude E; 28°2′ longitude (Fig. 11.2). The level difference between the spring and the spout is 1236 m.

The hydrographical basin is typically asymmetric, with the right-hand side, in the rainy mountain range. The hydrographic network includes the Siret River and its main tributaries: Siret Mic (on the Ukrainian territory), Suceava (173 km and 2298 km^2), Moldova (213 km and 4299 km^2), Bistrita (283 km and 4490 km^2) and Trotus (162 km and 4456 km^2), Putna (153 km and 2480 km^2), Ramnicu Sarat (137 km and 1063 km^2) and Buzău (302 km and 5264 km^2). The only important left tributary is the Barlad River (207 km and 7220 km^2) [47–56].

11.3   Methodology

The hydrological data was provided by the National Hydrological Institute and by the Siret Basin Water Administration. For the paper, the database was edited using licensed software, both graphically and statistically. The software used for graphic processing of information was Microsoft Excel, and the maps were drawn using ArcGIS 10.3.

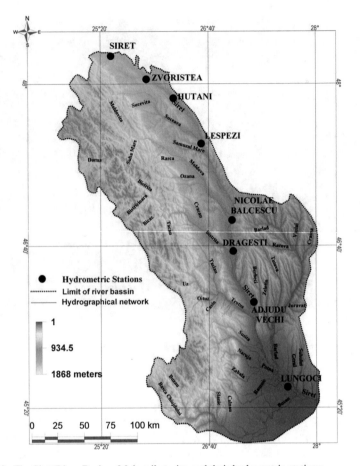

Fig. 11.2 The Siret River Basin—Main tributaries and their hydrometric stations

The hydrological activity of the Siret Water Basin Administration is carried out on an area of 28,878 km^2, managed by four subunits called Water Management Systems (WMS), and each WMS has one or several hydrological stations subordinated to it (Fig. 11.3). In their turn, the hydrological stations manage several hydrometric stations.

The observations and measurements, as well as the hydrometeorological vigilance, which are undergoing within these hydrological stations, are complex and numerous.

The data processing is finalized into hydrological yearbooks for:

- 131 hydrometric stations on rivers and small basins;
- 30 hydrometric stations for the water use;
- 15 hydrometric stations on reservoir lakes;
- 6 evaporimetric stations;
- 88 "villageellite" sections;

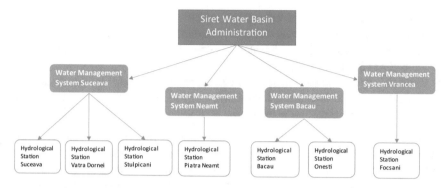

Fig. 11.3 Siret Water Basin Administration (Bacau) structure

- 60 springs;
- 473 of monitored phreatic wells.

The hydrology–hydrogeology activity implies collecting information, validating data and transmitting it from the hydrometric stations in the territory to the main center (Siret Water Basin Administration), representing a number of:

- 106 hydrometric stations in a daily program;
- 133 pluviometric stations with daily transmissions;
- 39 drilling stations for the Hydrogeological Bulletin Transmission.

The purpose of analyzing the Siret river basin evolution and collecting the hydrological data is to manage the water resources, exploit them through various uses, combat the damaging effects of water and maintain the quality of water under conditions appropriate to its use. The water management is the study of how water resources are being exploited in order to achieve a qualitative and quantitative regime of these resources in order to cover the need for water and to eliminate the damages caused by negative actions. The measures and actions that apply to the flood protection are established by the Water Law no. 107/1996 and the instructions arising from it [11, 67].

The main activities ongoing at the hydrometric stations are: level recordings, water and air temperature recordings, precipitation recordings (from specific rain gauge stations indicated by the hydrological stations), meteorological elements recordings (direction and intensity of wind), determination of flows, alluvial sampling, water quality sampling, etc.

In normal situations, the level recordings are registered twice a day (in the morning at 6 o'clock and the evening at 17 o'clock). During high water levels and/or floods, the frequency increases, such as:

– when the attention quota registered (AQ), the observations and measurements are intensified (from 3 to 3 h), and the stakeholders are alerted;

- when the flood quota registered (FQ), which represents the water level at the overflow point on the lowest slopes of the minor bed (there are no floods, only portions of land are filled with water), and the observation and measurements are intensified (from 2 to 2 h);
- when the danger quota registered (DQ), is when the goods and population are evacuated, and decisions are made based on predefined plans (the frequency of observations and measurements is from hour to hour) [11].

Recordings from the staff gauge are done by both the hydrometer and the automatic stations (in order to double the recording and have better accuracy). All information is recorded in the level books and flow measurement flyers. After the most important floods, there is requested an alluvial sampling in order to determine the granulometry [41, 42].

11.4   Results and Discussions

In areas where the rain gauge stations are located, the hydrometer has an obligation to transmit the amount of precipitation to the hydrological stations. Under normal circumstances, the information is transmitted twice a day: at 7 o'clock and 19 o'clock. In exceptional cases, when measured precipitation exceeds 15 mm in 3 h or 25 mm in a 6-h time, the rainfalls are considered dangerous and need to be transmitted to the superior units immediately after registration as a "pluvio warning".

During winter, the snow cover (thickness, density) is measured, and it is determined the water reserve. To achieve these measurements, a series of platforms of different shapes and sizes (which are located in areas that are most suited for the sampling and which do not suffer any influence that could cause erroneous values of the water reserve corresponding to the snow cover) are established. These platforms can be: large, small, and triangular or as a nivometric profile [11].

Within the Siret River Basin, special importance is given to current hydrometric activities. In order to know the spatio-temporal variability of the water resources on a small scale, there were identified four representative basins (monitored by the hydrological stations Vatra Dornei, Suha, Onesti and Bacau).

The evaporimetric network in Romania comprises a total of 55 evaporimetric stations: soil (the evaporation is determined in two types of surfaces free of vegetation soil and covered soil), lake (evaporation at the surface of the water) or experimental. In the Siret River Basin, there are 6 evaporimetric stations. The northernmost Water Management System of the Siret Water Basin Administration is represented by WMS Suceava. It has a number of three hydrological stations (HS): Suceava Hydrological Station, Vatra Dornei Hydrological Station and Stulpicani Hydrological Station.

Each hydrological station has a number of hydrometric stations:

- HS Suceava: 25 hydrometric stations, of which only one with level recordings, 12 with level and flow determinations, 2 with level determinations, adjusted flow rates and reconstituted flow, 6 with level, water flow and alluvial flow and 4 with level determinations, adjusted flow rates, reconstituted flow rate and suspension flow rate;

- HS Vatra Dornei: 17 hydrometric stations, of which the only one with level recordings, 5 with level and flow determinations, 11 with level, water flow and suspension flow;
- HS Stulpicani: 2 hydrometric stations with level and flow determinations.

The hydrological stations within WMS Suceava are among the few in the Siret hydrographic basin that analyze granulometric sampling (8 at HS Suceava and 2 at HS Vatra Dornei).

Within the WMS Suceava were selected two representative river basins: Fundoaia-Cosna Representative Basin at HS Vatra Dornei and Suha Representative Basin at HS Stulpicani. Both representative river basins monitor a number of 6 sections of observations and measurements each.

The location of the evaporimetric stations depends on the physical-geographic factors that influence the evaporation process. Within the Suceava Water Management System there are two evaporimetric soil stations: at HS Suceava, the evaporimetric station with the same name, located in the Suceava-Somuzu Mare River Basin; at HS Vatra Dornei, the evaporimetric station Dorna Giumalau, located in the Bistrita River Basin (Fig. 11.4).

The WMS Neamt has a single hydrological station: the Piatra Neamt Hydrological Station. It has a considerable number of hydrometric stations, namely 29. Most of

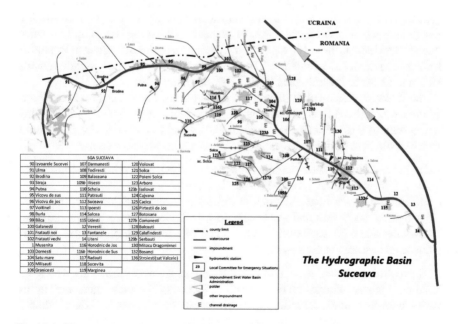

Fig. 11.4 The synoptic scheme of the Hydrographic Basin Suceava

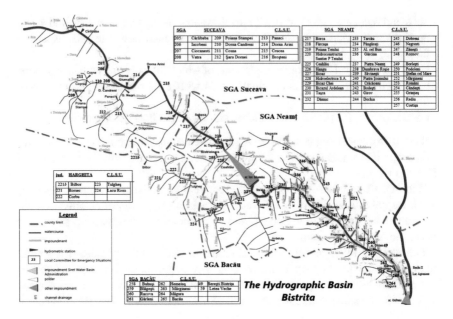

Fig. 11.5 The synoptic scheme of the Hydrographic Basin Bistrita

them are located on the biggest tributary of Siret: Bistrita River. From the total of hydrometric stations, 18 register observations and measurements of level and flow; 9 of the level, water flow and alluvial flow and 2 level, adjusted flows, reconstituted flow (one of which also registers flow of suspensions). At the Hydrological Station Piatra Neamt there is the only evaporimetric station on the lake (Ruginesti) located on the Izvoru Muntelui accumulation. There is also a soil evaporimetric station (Izvorul Alb) in the Bistrita Hydrological Basin (Figs. 11.5 and 11.6).

The Water Management System, located in the same city as the Siret Basin Water Administration (Bacău), is represented by WMS Bacau. It has a number of 2 hydrological stations (HS): Bacau Hydrological Station and Onesti Hydrological Station. Each hydrological station has a certain number of hydrometric stations:

- HS Bacau: 17 hydrometric stations, of which 10 with level and flow determinations and 7 with level, water flow and suspension flow;
- HS Onesti: 17 hydrometric stations, of which 5 with level and flow determinations, 6 with level, water flow and suspension flow, 4 with level determinations, adjusted flow rates and reconstituted flow rates, and 2 with level determinations, adjusted flows, reconstituted flow and suspension flow.

The hydrological stations within WMS Bacau are the only ones in the Siret River Basin that register information from 30 utility waters. In HS Bacau there are 2 utility waters under systematic surveillance. At SH Onesti there is a total of 28, of which 7 are systematically under surveillance, 15 are monitored exceptionally, and the data

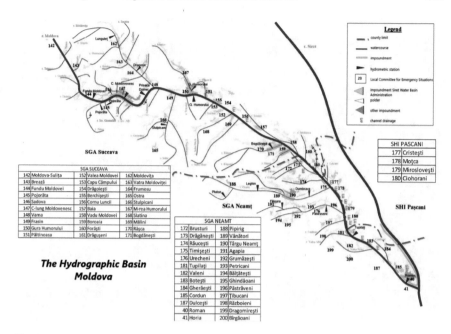

Fig. 11.6   The synoptic scheme of the Hydrographic Basin Moldova

from 6 sections are given by beneficiaries and it is necessary for the hydrological balance calculations.

Each of the two hydrological stations monitors a representative hydrographic basin: The Trebes-Negel Representative Basin—HS Bacau and The Superior Trotus Basin—HS Onesti. The basin with the most sections of observations and hydrometric measurements is the Trebes-Negel Representative Basin that has 11 sections. The Superior Trotus Representative Basin has 7 monitoring sections. At the Bacau Hydrological Station, there are two soil evaporimetric stations: Buhusi, located in the Bistrita River Basin; Racatau in the Siret River Basin (Figs. 11.7 and 11.8).

The last Water Management System that is part of the Siret Water Basin Administration is WMS Vrancea. It has only one hydrological station subordinated, Focsani Hydrological Station. The station has a considerable number of hydrometric stations, respectively 24. From the total of hydrometric stations, only 2 are with level recordings, 9 determine observations and measurements of the level and flow, 11 with level, water flow and alluvial flow determinations and 2 with level measurements, adjusted flow, and reconstituted flow (one also registers suspension flow). From a total of 60 springs monitored, the Focsani Hydrological Station measures only 22 of these (the most of the Siret River Basin) (Fig. 11.9).

The Siret Water Basin Administration (Bacau) has undergone modernization for the hydrometric network by automatic stations. Automated stations are called AHSS (Automated Hydrological Sensor Station) and can transmit hydrological data to the

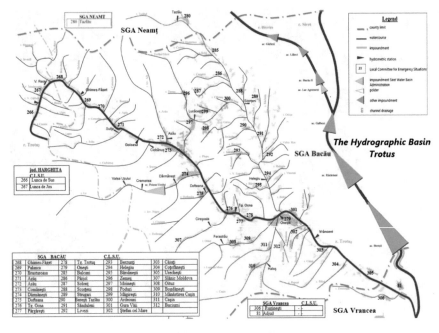

Fig. 11.7 The synoptic scheme of the Hydrographic Basin Trotus

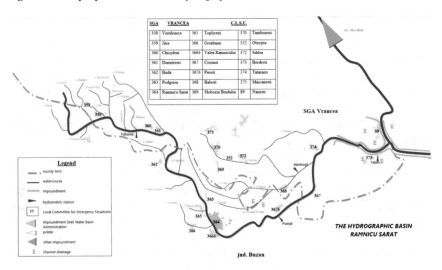

Fig. 11.8 The synoptic scheme of the Hydrographic Basin Ramnicu Sarat

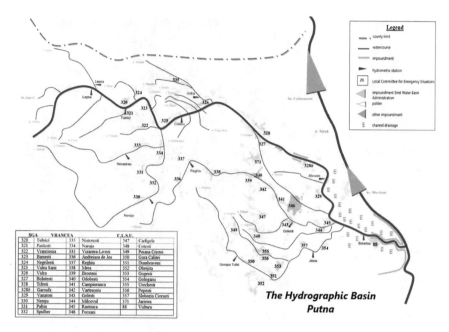

Fig. 11.9 The synoptic scheme of the Hydrographic Basin Putna

hydrological stations within each Basin Administration. The automatic stations are the following:

- Hydrometric stations: transmit daily information on water level, water and air temperature, precipitation and water quality. They are decommissioned during winter because they can not operate at negative temperatures;
- Rain gauges: register only the amount of rainfall throughout the year, even in winter, because they have a heater (which melts the snow);
- Quality stations: these only have sensors that transmit the water quality;
- Accumulations: they are installed in accumulations or polders;
- Snow-gauges: for the daily transmission of the nivometric parameters (layer, density, water equivalent) during the winter period (Fig. 11.10).

In the Siret River Basin, there are 119 automatic stations. Within each hydrological station, there is at least one automatic station (Fig. 11.11). Most stations are located within Suceava Water Management System: 29 at Suceava hydrological station (19 hydrometric stations and 10 rain gauges stations) and at Vatra Dornei hydrological station there are 16 stations (5 hydrometric stations, 10 rain gauges stations and a quality station). At Stulpicani hydrological station there is only one rain gauge station.

The Neamt Water Management System (the Hydrological Station of Piatra Neamt specifically) operates 25 automatic stations, out of which 21 automatic hydrometric stations and 4 automatic rain gauge stations.

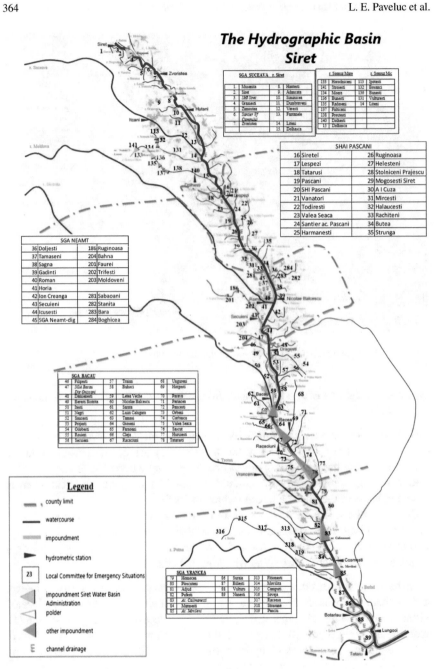

Fig. 11.10 The synoptic scheme of the Hydrographic Basin Siret

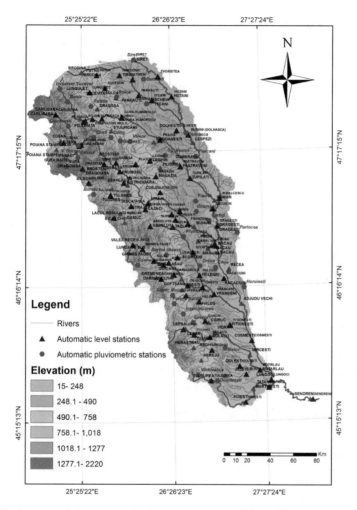

Fig. 11.11 Automatic hydrometric and rain gauges stations within the Siret Water Basin Administration (Bacau)

The Bacau Water Management System has 14 automatic stations at Bacau Hydrological Station (8 hydrometric stations and 6 rain gauges stations) and 15 automatic stations at Onesti Hydrological Station (11 hydrometric stations and 4 rain gauges stations).

The Vrancea Water Management System has under administration 19 automatic stations: 15 hydrometric stations and 4 pluviometric stations (Fig. 11.11).

In addition to the quantitative monitoring of water bodies in the Siret Water Basin Administration, there is also qualitative monitoring for the 353 water bodies: 343 water bodies—natural rivers; 10 water bodies—lakes. Out of the total number of monitored water bodies (55), 50 bodies of water are natural, and 5 bodies of water are heavily modified. The natural and heavily modified water bodies in the Siret River Basin are monitored qualitatively through 87 sections (88 with the Sendreni section belonging to ABA Dobrogea-Litoral). Out of these, there are 50 natural water bodies (rivers) with 76 sections (77 with Sendreni); 6 sections not included in the water body; 5 sections for heavily modified water bodies.

Rivers

1. The Siret water body (accumulation Bucecea—confluence Moldova), code RW12.1.4

It is monitored through 3 control sections:

	Monitoring program
• Siret—Hutani	Operational program, program for ichthyofauna (nutrients)
• Siret—Lespezi	Operational program, program for vulnerable areas, program for ichthyofauna (nutrients)
• Siret—downstream Pascani	Operational program, the best section available, program for ichthyofauna (nutrients)

The water body has the RO05 typology. The average altitude is 511 m, the substrate consists of sand and gravel, the water-course is sinuous, the average width of the bed is 20 m, the slope has an average value of 2.6‰.

2. The Suceava water body (springs—confluence The Ascuns River) + tributaries, code RW12.1.17.1

It is evaluated through two control sections:

	Monitoring program
• Suceava— Brodina	Reference program, program for ichthyofauna, timetable for intercalibration
• Sadau—Sadau	Surveillance program, program for ichthyofauna

The water body has the RO01 typology, the average altitude is 980 m, the substrate consists of stones and gravel, the water-course is tilting, the average width of the bed is 8 m, the slope has a mean value of 23‰.

3. The Putnisoara water body (upstream Strujinoasa confluence), code RW12.1.17.12.1

It is evaluated through a single control section, Putnisoara (Upstream Strujinoasa confluence), with monitoring programs: reference program and program for ichthyofauna. The water body has the RO01 typology. The average altitude is 866 m, the substrate consists of stones and gravel, the water flow is tilting, the average width of the bed is 3 m, the slope has a mean value of 65‰.

4. The Suceava water body (Mihoveni), code RW12.1.17.2

It is assessed through a single control section, Suceava (Mihoveni), with the monitoring programs: the program, the program for vulnerable areas and the program for ichthyofauna. The water body has the RO05 typology. The average altitude is 520 m, the substrate consists of stones and gravel, the water flow is tilting, the average width of the bed is 19 m, the slope has a mean value of 9‰

5. The Suceava water body (Tisauti), code RW12.1.17.3

It is evaluated through a single control section, Suceava (Tisauti), with monitoring programs: operational program and ichthyofauna (nutrients) program. This water body has the RO05 typology. The average altitude is 398 m, the substrate consists of stones and gravel, the course of the water is sinuous, the average width of the bed is 8 m, the slope has an average value of 7‰.

6. The Somuzu Mare water body (Vorniceni), code RW12.1.21.1

It is evaluated through a single control section, Somuzu Mare—Vorniceni, with the monitoring programs: operational program and program for ichthyofauna (nutrients). The water body has the RO04 typology. The average altitude is 398 m, the substrate consists of stones and gravel, the water course is tilting, the average width of the bed is 2 m, the slope has an average value of 12‰.

7. The Moldova water body (springs—confluence Sadova), code RW12.1.40.1

It is assessed through a single control section, Moldova—Fundu Moldova, with monitoring programs: surveillance program, potability program and program for ichthyofauna. This body has the RO01 typology. The average altitude is 1083 m, the substrate consists of stones and gravel, the water course is sinuous, the average width of the bed is 5 m, the slope has an average value of 10‰.

8. The Moldova water body (confluence Sadova—Suha confluence), code RW12.1.40.2

It is evaluated through a single control section, Moldova—upstream Campulung Moldovenesc, with the monitoring programs: surveillance program and ichthyofauna (nutrients) program. This body has the RO01 typology. The average altitude is 1061 m, the substrate consists of stones and gravel, the water course is tilting, the average width of the bed is 13 m, the slope has an average value of 7‰.

9. The Suha water body (Stulpicani) + Brateasa (Ostra) + Botus, code RW12.1.40.25.1

It is evaluated through two control sections:

	Monitoring program
• Suha—Stulpicani	Operational program, program for ichthyofauna
• Baisescu—Ostra	Potability program, program for ichthyofauna

The water body has the RO01 typology. The average altitude is 966 m, the substrate consists of stones and gravel, the water course is sinuous, the average width of the bed is 3 m, the slope has an average value of 30‰.

10. The Moldova water body (Suha confluence—Vier confluence), code RW12.1.40.3

It is rated through 3 control sections:

	Monitoring program
• Moldova—downstream Gura Humorului	Surveillance program, program for vulnerable areas, program for ichthyofauna (nutrients)
• Moldova—Baia	Surveillance program, potability program, program for ichthyofauna
• Moldova—Timisesti	Surveillance program, potability program, program for ichthyofauna

The water body has the RO05 typology. The average altitude is 850 m, the substrate consists of stones and gravel, the water course is sinuous, the average width of the bed is 20 m, the slope has an average value of 8‰.

11. The Ciumarni water body (Gainesti), code RW12.1.40.32.1

It is evaluated through a single control section, Ciumarni—Gainesti with monitoring programs: surveillance program, reference program and program for ichthyofauna. This water body has the RO01 typology. The average altitude is 795 m, the substrate consists of stones and gravel, the water-course is tilting, the average width of the bed is 3 m, the slope has a mean value of 50‰.

12. The Ozana water body (Boboiesti), code RW12.1.40.41.1

It is evaluated through a single control section, Ozana—Boboiesti, with monitoring programs: surveillance program, program for their protection, habitats and species. It has the typology RO01. The average altitude is 999 m, the substrate is formed by stones and gravel, the water flow is tilting, the average width of the bed is 2 m, the slope has an average value of 74‰.

13. The Moldova water body (confluence Vier—Siret confluence), code RW12.1.40.4

It is evaluated through a single control section, Moldova—Roman, with the types of program: operational program, a program for vulnerable areas, a program for ichthyofauna (nutrients). The water body has the RO05 typology. The average altitude is 678 m, the substrate is sandy, the water course is sinuous, the average width of the bed is 18 m, and the slope has an average value of 4‰.

14. The Siret water body (confluence Moldova—accumulation Galbeni), code RW12.1.5

It is evaluated through a single control section, Siret—Dragesti, with the types of programs: the operational program, the best available section, the program for ichthy-ofauna, the program for intercalibration (nutrients). This body has the RO10 typology. The average altitude is 525 m, the substrate consists of sand and gravel, the water-course is sinuous, the average width of the bed is 20 m, the slope has an average value of 1‰.

15. The Bistrita water body (springs—Neagra confluence), code RW12.1.53.1

It is rated through 3 control sections

	Monitoring program
• Bistrita—Carlibaba	Surveillance program, program for ichthyofauna
• Bistrita—Argestru	Surveillance program, potability program, program for ichthyofauna, protection of habitats and species
• Dorna—Dorna Candreni	Surveillance program, potability program, program for ichthyofauna, protection of habitats and species

It has the typology RO01, the average altitude is 1230 m, the substrate is rocky, made of stones and stones, the water-course is tortuous, the average width of the bed is 11 m, and the slope has an average value of 20‰.

16. The Neagra water body (Gura Negrii), code RW12.1.53.17.1

It is evaluated through a single control section, Neagra—Gura Negrii, with the types of programs: the operational program and program for ichthyofauna (nutrients). It has the RO01 typology. The average altitude is 1255 m, the substrate is formed of stones and gravel, the water course is tilting, the average width of the bed is 9 m, the slope has a mean value of 28‰.

17. The Bistrita water body (Neagra confluence—Izvoru Muntelui accumulation), code RW12.1.53.2

It is evaluated through two control sections:

	Monitoring program
• Bistrita—Barnat	Operational program, program for ichthyofauna (nutrients)
• Bistrita—Frumosu	Surveillance program, program for ichthyofauna

It has the typology RO02. The average altitude is 1181 m, the substrate is formed by stones and gravel, the water course is tilting, the average width of the bed is 15 m, the slope has an average value of 10‰.

18.   The Barnarel water body (Crucea), code RW12.1.53.25.1

It is evaluated through a single control section, Barnarel at Crucea, with the types of programs: potability program, program for ichthyofauna. It has the RO01 topology. The average altitude is 850 m, the substrate consists of stones and gravel, the water course is sinuous, the average width of the bed is 20 m, the slope has an average value of 8‰.

19.   The   Bistricioara   water   body   (Bistricioara,   Capu   Corbului),   code RW12.1.53.40.2

It is evaluated through two control sections:

	Monitoring program
• Bistricioara—upstream Capu Corpului	Surveillance program, the best available section, program for ichthyofauna
• Bistricioara—Bistricioara	Surveillance program, program for ichthyofauna

It has the typology RO03. The average altitude is 1066 m, the substrate is formed by stones and gravel, the water course is sinuous, the average width of the bed is 6 m, the slope has an average value of 14‰.

20.   The Putna (Tulghes) water body, codul RW12.1.53.40.11.1

It is evaluated through a single control section, Putna at Tulghes, with the types of programs: potability program, program for ichthyofauna. It belongs to the RO01 typology. The average altitude is 1067 m, the substrate consists of stones and gravel, the water course is sinuous, the average width of the bed is 3 m, the slope has an average value of 33‰.

21.   The Bicaz (Bicaz Chei) water body, codul RW12.1.53.48.1

It is evaluated through a single control section, Bicaz—Bicaz Chei, with the types of programs: surveillance program, program for ichthyofauna, protection of habitats and species. It belongs to the RO01 typology. The average altitude is 1167 m, the substrate is rocky, consists of rocks and gravel, the water course is sinuous, the average width of the bed is 3 m, the slope has an average value of 41‰.

22. The Doamna (Doamna) water body, codul RW12.1.53.56.1

It is evaluated through a single control section, Doamna—upstream village Doamna, with the types of programs: reference program, a program for vulnerable areas, a program for ichthyofauna, the timetable for intercalibration. It belongs to the RO01 typology. The average altitude is 583 m, the substrate consists of stones and gravel, the water course is sinuous, the average width of the bed is 2 m, the slope has an average value of 40‰.

23. The Bistrita (Izvoru Muntelui lake—Pangarati lake) water bodies, codul RW12.1.53.4

It is evaluated through a single control section, Bistrita—Straja, with the types of programs: surveillance program, program for ichthyofauna, knowing the impact of hydromorphological alterations on water. It belongs to the RO05 typology. The average altitude is 1089 m, the substrate consists of stones and gravel, the water course is sinuous, the average width of the bed is 15 m, the slope has an average value of 7‰.

24. The Bistrita (downstream Batca Doamnei lake—Racova lake), codul RW12.1.53.6

It is evaluated through 4 control sections:

	Monitoring program
• Bistrita—Piatra Neamt	Surveillance program, program for vulnerable areas, program for ichthyofauna, knowing the impact of hydromorphological alterations on water
• Bistrita—Roznov	Operational program, program for vulnerable areas, program for ichthyofauna (nutrients)
• Bistrita—Zanesti	Operational program, program for vulnerable areas, program for ichthyofauna, knowing the impact of hydromorphological alterations on water
• Bistrita—Frunzeni	Operational program, program for vulnerable areas, program for ichthyofauna, knowing the impact of hydromorphological alterations on water

It has the typology RO05. The average altitude is 1034 m, the substrate is made of stones and gravel, the water course is sinuous, the average width of the bed is 18 m, the slope has an average value of 6‰.

25. The Cuejdiu water body (Piatra Neamt), code RW12.1.53.57.1

It is evaluated through a single control section, Cuejdiu—Piatra Neamt, with the types of programs: operational program, program for vulnerable areas, a program for ichthyofauna, knowledge of the impact of hydromorphological alterations on water and nutrients. This body has the RO01 typology. The average altitude is 680 m, the substrate consists of stones and gravel, the water course is tilting, the average width of the bed is 4 m, the slope has an average value of 27‰.

26. The Siret water body (Beresti dam—Calimanesti accumulation), code RW12.1.7

It is evaluated through a single control section, Siret—Adjudu Vechi, with the types of programs: surveillance program, program for their protection, protection of habitats and species, and the impact of hydromorphological alterations on water. It has the typology RO10, the average altitude is 647 m, the substrate is sandy, the water course is sinuous, the average width of the bed is 15 m, and the slope has an average value of 1‰.

27. The Ciobanus water body (Ciobanus), code RW12.1.69.17.1

It is evaluated through a single control section, Ciobanus—Ciobanus, with the types of programs: surveillance program, reference program, potability program and program for ichthyofauna. It has the typology RO01. The average altitude is 1052 m, the substrate consists of stones and gravel, the water course is sinuous, the average width of the bed is 2 m, the slope has an average value of 24‰.

28. The Slanic water body (Slanic), code RW12.1.69.27.1.1

It is evaluated through two control sections:

	Monitoring program
• Slanic—upstream Slanic	Potability program
• Slanic—downstream Slanic	Surveillance program, program for ichthyofauna, knowing the impact of hydromorphological alterations on water

It has the typology RO01, the average altitude is 973 m, the substrate is made of stones and gravel, the water course is sinuous, the average width of the bed is 3 m, the slope has an average value of 44‰.

29. The Boulet water body (Mitocu Balan) + Cracau, code RW12.1.53.60.1

It is evaluated through two control sections:

	Monitoring program
• Boulet—upstream village Mitocu Balan	Reference program, program for vulnerable areas, program for ichthyofauna

(continued)

(continued)

	Monitoring program
• Cracau—Slobozia	Surveillance program, program for vulnerable areas, program for ichthyofauna, timetable for intercalibration

This water body has the RO01 typology. The average altitude is 700 m, the substrate consists of stones and gravel, the water course is sinuous, the average width of the bed is 2 m, the slope has a mean value of 27‰.

30. The Trotus water body (confluence Valea Rece—confluence Urmenis), code RW12.1.69.2

It is evaluated through a single control section, Trotus—Ghimes Faget, with the types of programs: surveillance program, the best available section and the program for ichthyofauna. It has the typology RO02, the average altitude is 1116 m, the substrate is rocky, made of rocks and stones, the water course is sinuous, the average width of the bed is 10 m, and the slope has an average value of 22‰.

31. The Asau (Asau) water body, code RW12.1.69.18.1

It is evaluated through a single control section, Asau—Asau, with the types of programs: surveillance program and program for ichthyofauna. It has the typology RO01. The average altitude is 951 m, the substrate is formed of stones and gravel, the water course is tilting, the average width of the bed is 10 m, the slope has an average value of 25‰.

32. The Urmenis water body, code RW12.1.69.20.1

It is evaluated through a single control section, Urmenis—Comanesti, with the types of programs: operational program, a program for vulnerable areas, a program for ichthyofauna and nutrients. It has the RO01 typology. The average altitude is 560 m, the substrate consists of gravel and organic sediments, the water course is sinuous, the average width of the bed is 4 m, the slope has an average value of 40‰.

33. The Uz water body (springs—Poiana Uzului), code RW12.1.69.22.1

It is evaluated through a single control section, Uz-upstream Poiana Uzului Lake, with the types of programs: surveillance program, potability program, program for ichthyofauna and timetable for intercalibration. It has the typology RO01, the average altitude is 1070 m, the substrate is rocky, formed by stones and rocks, the water flow is tilting, the average width of the bed is 5 m, and the slope has an average value of 18‰.

34. The Uz-Poiana Uzului-confluence Trotus water body, code RW12.1.69.22

It is evaluated through a single downstream control section Poiana Uzului, with the types of programs: surveillance program, potability program, program for ichthyofauna and timetable for intercalibration. This water body has the RO01 typology. The average altitude is 1070 m, the substrate is silicate formed by stones and rocks, the water course is sinuous with 1.26 coefficient of sinuosity, the average width of the bed is 2 m, the slope has an average value of 18‰.

35. The Izvorul Alb water body, code RW12.1.69.22.6.1

It is evaluated through a single control section, Izvoru Alb—upstream Poiana Uzului Lake, with the types of programs: surveillance program, program for ichthyofauna and drinking program. This body has the RO01 typology, the average altitude is 910 m, the substrate is formed by stones and rocks, the water course is sinuous, the average width of the bed is 2 m, and the slope has an average value of 120‰.

36. The Plopul water body, code RW12.1.69.21.1

It is evaluated through a single control section, Plopul—upstream Poiana Uzului Lake, with the types of programs: surveillance program, program for ichthyofauna and potability program. This water body has the RO01 typology. The mean altitude is 820 m, the substrate consists of sand and gravel, the water course is sinuous, the average width of the bed is 3–5 m, the slope has an average value of 101‰.

37. The Groza water body, code RW12.1.69.22.7.1

It is evaluated through a single control section, Groza—upstream Poiana Uzului Lake, surveillance program, program for ichthyofauna and drinking program. This water body has the RO01 typology. The average altitude is 791 m, the substrate consists of stones and gravel, the water course is sinuous, the average width of the bed is 4 m, the slope has a mean value of 154‰.

38. The Trotus water body (confluence of Urmenis—Tazlau confluence), code RW12.1.69.3

It is evaluated through two control sections:

	Monitoring program
• Trotus—downstream Darmanesti	Operational program, program for ichthyofauna (nutrients)
• Trotus—upstream Targu Ocna	Operational program, program for vulnerable areas, program for ichthyofauna (nutrients)

It has the typology RO02. The average altitude is 980 m, the substrate is made of stones and gravel, the water course is sinuous, the average width of the bed is 20 m, the slope has an average value of 12‰.

39. The Oituz water body, code RW12.1.69.31.1

It is evaluated through a single control section, Oituz—upstream Onesti, with the types of programs: surveillance program, a program for vulnerable areas and program for ichthyofauna. It has the RO01 typology. The average altitude is 708 m, the substrate is formed by stones and gravel, the water course is tilting, the average width of the bed is 8 m, the slope has an average value of 20‰.

40. The Casin water body (Casin), code RW12.1.69.32.1

It is evaluated through a single control section, Casin—upstream Onesti, with the types of programs: surveillance program and program for ichthyofauna. It has the typology RO01. The average altitude is 620 m, the substrate is made of stones and gravel, the water course is sinuous, the average width of the bed is 10 m, the slope has an average value of 19‰.

41. The Tazlaul Sarat water body (Bolatau, Tescani), code RW12.1.69.33.10.1

It is evaluated through two control sections:

	Monitoring program
• Tazlaul Sarat—upstream Bolatau	Reference program, a program for ichthyofauna, timetable for intercalibration
• Tazlaul Sarat—Tescani	Operational program, program for ichthyofauna (nutrients)

It has the typology RO16, the average altitude is 775 m, the substrate is formed by boulders and gravel, the water course is sinuous, the average width of the bed is 7.5 m, the slope has an average value of 34‰.

42. The Tazlau water body, code RW12.1.69.33.1

It is evaluated through a single control section, Tazlau—Helegiu, with the types of program: surveillance program and program for ichthyofauna. It has the typology RO01. The average altitude is 520 m, the substrate is formed of stones and gravel, the water course is tilting, the average width of the bed is 12 m, the slope has an average value of 13‰.

43. The Trotus water body (Tazlau confluence—Siret confluence), code RW12.1.69.4

It is evaluated through two control sections:

	Monitoring program
• Trotus—Vranceni	Operational program, program for vulnerable areas, program for ichthyofauna (nutrients)
• Trotus—Adjud	Program operaţional, program for ichthyofauna, timetable for intercalibration (nutrients)

It has the typology RO10. The average altitude is 720 m, the substrate is made of stones and gravel, the water flow is sinuous, the average width of the bed is 16 m, the slope has an average value of 8‰.

44. The Putna water body (Tulnici), code RW12.1.79.1

It is evaluated through a single control section, Putna—Tulnici, with the types of program: surveillance program, a program for ichthyofauna, protection of habitats and species, nutrients. It has the typology RO01, the average altitude is 990 m, the substrate is rocky, formed by stones and rocks, the water course is tilting, the average width of the bed is 10 m, and the slope has an average value of 35‰.

45. The Putna water body (Colacu, Botarlau, Podu Zamfirei, Golesti), code RW12.1.79.3

It is evaluated through 5 control sections:

	Monitoring program
• Putna—Colacu	Surveillance program, the best available section, program for ichthyofauna
• Putna—Podu Zamfirei	Surveillance program, program for ichthyofauna (nutrients)
• Putna—Botarlau	Surveillance program, program for ichthyofauna (nutrients)
• Milcov—Golesti	Surveillance program, program for ichthyofauna
• Milcov—Rastoaca	Surveillance program, program for ichthyofauna, knowing the impact of hydromorphological alterations on water, timetable for intercalibration

It has the typology RO08, the average altitude is 635 m, the substrate is made of stones, gravel and sand, the water course is tortuous, the average width of the bed is 10 m, and the slope has an average value of 18‰.

46. The Milcov water body (Reghiu), code RW12.1.79.18.1

It is evaluated through a single control section, Milcov—Reghiu, with the types of programs: surveillance program and program for ichthyofauna. It has the RO01 typology. The average altitude is 595 m, the substrate consists of stones and gravel, the water course is tilting, the average width of the bed is 10 m, the slope has an average value of 30‰.

47. The Ramna water body (Rascuta, Jiliste), code RW12.1.79.19.1

It is evaluated through two control sections:

	Monitoring program
• Ramna—Jiliste	Surveillance program, program for ichthyofauna, timetable for intercalibration
• Ramna—Rascuta confluence	Reference program, a program for ichthyofauna

It has the typology RO01. The average altitude is 383 m, the substrate is formed sand and sludge, the water course is sinuous, the average width of the bed is 2 m, the slope has an average value of 15‰.

48. The Ramnicu Sarat water body (springs—confluence Tulburea), code RW12.1.80.1

It is evaluated through a single control section, Ramnicu Sarat—Tulburea, with the types of program: reference program, program for ichthyofauna. It has the typology RO16. The average altitude is 820 m, the substrate is formed of stones and gravel, the water course is sinuous, the average width of the bed is 4 m, the slope has an average value of 32‰.

49. The Ramnicu Sarat water body (Tulburea, Nicolesti, Maicanesti), code RW12.1.80.2

It is evaluated through two control sections:

	Monitoring program
• Ramnicu Sarat—Nicolesti	Program operaţional, program for ichthyofauna (nutrients)
• Ramnicu Sarat—Maicanesti	Program operaţional, program for ichthyofauna, timetable for intercalibration (nutrients)

It has the RO16 typology. The average altitude is 345 m, the substrate consists of sand and sludge, the water course is sinuous, the average width of the bed is 5 m, the slope has an average value of 10‰.

50. The Siret water body (Calimanesti dam—Danube confluence), code RW12.1.9

It is rated through 4 control sections:

	Monitoring program
• Siret—Cosmesti	Operational program, program for vulnerable areas, program for ichthyofauna (nutrients)
• Siret—Biliesti	Surveillance program, the best available section, program for ichthyofauna, timetable for intercalibration
• Siret—Lungoci	Operational program, program for ichthyofauna (nutrients)
• Siret—Sendreni	Operational program, program for ichthyofauna (nutrients)

It has the typology RO11, the average altitude is 554 m, the substrate consists of sand, the water course is sinuous, the average width of the bed is 34 m, and the slope has an average value of 1‰.

Lakes

In addition to river hydrometry, the Siret Basin Water Administration also monitors the lakes. Each hydrological station has under administration a number of lakes. The Hydrological Station Bacau monitors most of the lakes:

- On Bistrita river

  – Garleni Lake
  – Lilieci Lake
  – Bacau Lake

- On the river Siret

  – Galbeni Lake
  – Racaciuni Lake
  – Beresti Lake
  – Horgesti Lake

Hydrological Station Suceava, which has a total of 4 lakes to be metric:

- Siret:

  – Rogojesti Lake
  – Bucecea Lake

- Dragomirna—Dragomirna Lake
- Somuzul Mare—Somuz II Moara Lake

The Hydrological Station Piatra Neamt monitors three lakes, on the Bistrita—The Izvoru Muntelui Lake, respectively on the Bicaz—The Rosu Lake.

The Poiana Uzului Lake, at the Onesti Hydrological Station, on the River Uz and the Calimanesti Lake at the Focsani Hydrological Station, on the Siret River, complete the list of 15 lakes which are monitored in the Siret River Basin (Fig. 11.12).

The total number of water bodies/lakes within the Siret Water Basin Administration, where is done the qualitative monitoring, is:

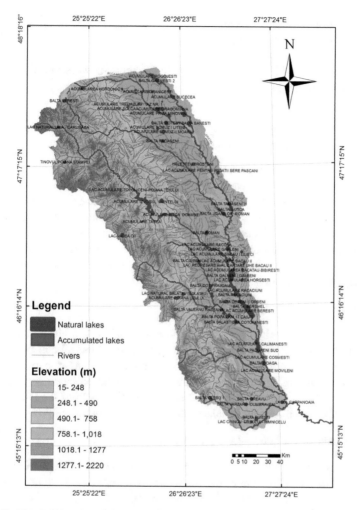

Fig. 11.12 The distribution of the natural and artificial lakes in the Siret River Basin

– 2 water bodies: natural lakes, both of which have a surface area of less than 0.5 km² :

1. The Rosu Lake located in the central part of the Eastern Carpathians;
2. The Lala Lake of glacial origin.

– 10 water bodies (heavily modified lakes):

1. The Rogojesti Lake;
2. The Bucecea Lake;
3. The Dragomirna Lake;
4. The Izvoru Muntelui Lake;

5. The Batca Doamnei Lake;
6. The Agrement Bacau Lake;
7. The Poiana Uzului Lake;
8. The Racaciuni Lake—Siret (upstream Galbeni—downstream Beresti);
9. The Calimanesti Lake;
10. The Solca Lake.

The water bodies represented by the natural lakes of the Siret River Basin are monitored from a qualitative point of view as it follows:

1. The Lala Lake is monitored using an integrated sample consisting of four sub-samples taken from the cardinal points and from the middle of the lake; one sample only is taken for the chemical analysis, from the middle lake;
2. The Rosu Lake is biologically evaluated by means of an integrated sample consisting of sub-samples taken from three monitoring sections: lake entrance, middle of the lake and the lake exit, and the chemical analysis only from the middle of the lake; There are 8 sections where biological samples are taken and 2 sections where chemical samples are taken.

The Lala Lake water body—LW12.1.53.5.1

The Lala water body belongs to the ROLN18 typology. It is a glacial type of lake, located in the ecoregion 10, at an altitude of 1800 m. It has silicon substrate; the average depth is 1.6 m, the area of 0.01 km^2. It is not used in anthropogenic activities. The lake is biologically evaluated through an integrated probe consisting of sub-samples from the 4 cardinal points and one from the middle of the lake. A single sample is taken from the middle lake section for the chemical analysis.

The Rosu Lake water body—LW12.1.53.48.1

The Rosu Lake water body, located in the central part of the Oriental Carpathians, was formed as a result of the landslides that took place at the base of the Ucigasu Mountain and resulted with the blocking of the Bicaz River in 1837 [45, 46]. The lake is supplied with water from the Bicaz River, and its tributaries the Oii River, the Suhard River and 12 torrents. It belongs to the ROLN17 typology, located in ecoregion 10, at an altitude of 980 m. It has a limestone substrate, the average depth is 5.1 m and the maximum is 10.5 m. It has a surface of 0.0126 km^2. The retention time is 3 days. It is used for recreational purposes. The lake is biologically rated through three monitoring sections: the lake entrance, the middle of the lake and the lake exit (integrated sample). The chemical analysis sample is taken from the middle of the lake.

The heavily modified and artificial (lakes) water bodies monitored are the following:

a	The Rogojesti Lake	2	Sections
b	The Bucecea Lake	3	Sections
c	The Dragomirna Lake	2	Sections
d	The Izvoru Muntelui Lake	3	Sections
e	The Batca Doamnei Lake	3	Sections
f	The Agrement Bacau Lake	1	Section
g	The Poiana Uzului Lake	3	Sections
h	The Racaciuni—Siret (upstream Galbeni—downstream Beresti)	3	Sections
i	The Calimanesti Lake	2	Sections
J	The Solca Lake	1	Section

a. The Rogojesti Lake water body

The water body code is LW12.1.1. It belongs to the ROLA10 typology, located at an altitude of 298 m, with an area of 8 km^2, in the hilly area, with a siliceous substrate. The average depth in the middle lake area is 10 m, the retention time is 0.08 years. The main uses of this lake are water supply, flow regulation and hydropower. This lake was characterized by the monitoring of 2 sections: middle lake and dam.

b. The Bucecea Lake water body

The water body code is LW12.1.3. It belongs to the ROLA09 typology, at an altitude of 271 m, with an area of 4.75 km^2, in the hilly area, with a silicon substrate. The average depth in the middle lake area is 8 m, the retention time is 0.01 years. The main uses of this lake are drinking water, flood protection and hydropower. This lake has been characterized by monitoring of 3 sections: middle lake, dam and socket.

c. The Dragomirna Lake water body

The water body code is LW12.1.17.30.2. It belongs to the ROLA10 typology, located at an altitude of 315 m, with an area of 1.8 km^2, in the hilly area, with a silicon substrate. The average depth in the middle lake area is 15 m, the retention time is 0.52 years. The main uses of this lake are flood protection, fish farming and hydropower. This lake was characterized by the monitoring of 2 sections: middle lake and dam.

d. The Izvoru Muntelui Lake water body

The water body code is LW12.1.53.3. It belongs to the ROLA08 typology, located at an altitude of 513 m, with an area of 31 km^2, in the hilly area, with a silicon substrate. The average depth in the middle lake area is 39.67 m, the retention time is 0.84 years. The main uses of this lake are the mitigation of floods on the Bistrita River, the production of electricity, fish farming and recreation. This lake was characterized by monitoring of 3 sections: tail—middle lake, middle lake and dam.

e.   The Batca Doamnei Lake water body

The water body code is LW12.1.53.5. It belongs to the ROLA10 typology, located at an altitude of 324.5 m, with an area of 2.35 km^2, in the hilly area, with a siliceous substrate. The average depth in the middle lake area is 8.1 m, the retention time is 0.12 years. The main uses of this lake are the production of electricity, the supply of water for drinking purposes. This lake has been characterized by monitoring of 3 sections: middle lake, dam and socket.

f.   The Agreement Bacau Lake water body

The water body code is LW12.1.53.7. It belongs to the ROLA02 typology, located at an altitude of 155.3 m, with an area of 0.5 km^2, in the plain area, with a silicon substrate. The average depth in the middle lake area is 2 m, the retention time is 0.0005 years. The main uses of this lake are: water supply for industrial purposes, leisure and flood defense. This lake was characterized by the monitoring of a single section: the middle lake.

g.   The Poiana Uzului Lake water body

The water body code is LW12.1.69.22.2. It belongs to the ROLA08 typology, situated at an altitude of 513.5 m, with an area of 3.34 km^2, in the hilly area, with a silicon substrate. The average depth in the middle lake area is 26.2 m; the retention time is 0.7 years. The main uses of this lake are the production of electricity and the supply of drinking water for drinking purposes. This lake has been characterized by the monitoring of 3 sections: middle lake, dam and Caraboaia socket.

h.   The Racaciuni—Siret (upstream Galbeni—downstream Beresti) water body

The water body code is LW12.1.6. It belongs to the ROLA02 typology, located at an altitude of 129 m, with an area of 20.04 km^2, in the plain area, with a silicon substrate. The average depth in the middle lake area is 5.14 m; the retention time is 0.02 years. The main uses of this lake are power generation and flood mitigation. This lake was characterized by monitoring of 3 sections: tail—middle lake, middle lake and dam.

i.   The Calimanesti Lake water body

The water body code is LW12.1.8. It belongs to the ROLA02 typology, located at an altitude of 75 m, with an area of 7.4 km^2, in the plain area with a silicon substrate. The average depth in the middle lake area is 5.27 m; the retention time is 0.006 years. The main uses of this lake are power generation and flood mitigation. This lake was characterized by the monitoring of 2 sections: middle lake and dam.

j.   The Solca Lake water body

The water body code is LW12.1.17.24. It belongs to the ROLA09 typology, located at an altitude of 485 m, with an area of 166 km^2, in the sub-Carpathian area with a silicon substrate. The main uses of this lake are: water supply for drinking purposes and flood mitigation. This lake was characterized by monitoring a single dam/socket.

Groundwater

In the Siret River Basin have been identified eight groundwater bodies:

- phreatic: ROGWSI01, ROGWSI02, ROGWSI03, ROGWSI04, ROGWSI05 şi ROGWSI06;
- deepwater hydrogeological drillings, managed by the Siret Water Basin Administration, were assigned to two groundwater bodies:

  - Those from the northern and central part (Suceava, Neamt and Bacau counties) to GWPR05;
  - Those in the south (Vrancea and Buzau counties) at GWAG12.

The Siret Basin Water Administration is monitoring a number of 473 groundwater drillings. The information from these drillings can be taken manually by the hydrometer and/or automatically by certain sensors that transmit the water level and temperature. The data is automatically taken from the sensors only from 39 drillings (this data is doubled by the data transmitted by the hydrometer) [29, 36].

A hydrogeological bulletin is edited monthly, comprising 39 drilling stations with two level readings per drilling. Only the hydrological stations Piatra Neamt and Vatra Dornei follow 3 hydrogeological springs.

The monitored groundwater drilling stations are the following:

- The Hydrological Station Bacau: 75 manually monitored drilling stations, of which 4 are also tracked by automatic stations;
- The Hydrological Station Onesti: 20 manually controlled drilling stations, of which 4 are also tracked by automatic stations;
- The Hydrological Station Focsani: 129 manually monitored drilling stations, of which 14 are also tracked by automatic stations;
- The Hydrological Station Piatra Neamt: 119 manually monitored drilling stations, of which 10 are also tracked by automatic stations;
- The Hydrological Station Stulpicani monitors only 2 manually groundwater drilling stations;
- The Hydrological Station of Suceava: 118 manually monitored drilling stations, of which 7 are also tracked by automatic stations;
- The Hydrological Station Vatra Dornei manages only 10 manually drilling stations.

The total number of hydrogeological drilling stations monitored from a qualitative point of view on water bodies is:

- ROGWSI01—2 (Botos-Ciocanesti F1 and Carlibaba 2 Spring);
- ROGWSI02—9;
- ROGWSI03—382;
- ROGWSI04—2 springs (Damuc and Trei Fantani Springs);
- ROGWSI05—115;
- ROGWSI06—4.

ROGWSI01—is located in the Suhard Mountains and Obcina Mestecanisului, with an area of 90.0 km². It is of the fissural type and spring, accumulated in limestone and crystalline dolomites and the crystalline shale of the Tulghes Series. All these types of metamorphic rocks are of pre-Cambrian age and are generally phalanxed in the SV-NE direction. The boundary between Suhard Mountains and Obcina Mestecanisului is marked by the Bistrita River, with the NV-SE oriented flow direction.

ROGWSI03—is a permeable porous type, develops in the deposits of the meadows and terraces of the Siret River and its tributaries and is quaternary of age, and the area is 4256 km². The phreatic aquifer is cantonated in sands and pebbles with stones, covered by deposits made of clays, silt or sandy clays. The permeable layers have average thicknesses of approximatively 5 m. Larger thicknesses of the aquifer formations are recorded in the area of Harlesti and Gheraiesti, Bacau and Sascut hydroelectric stations, where they reach approximately 10 m thick, as well as the Adjud and Ciorani stations where the gravel thicknesses are 20 m.

ROGWSI04—located in the Haghimas Mountains, of the fisural-karst type, being accumulated in Triasic, and Jurassic and Cretaceous deposits with an area of 141 km². The deposits were affected by the alpine orogenesis which led to the formation of a large N-S-oriented sinclinal in which axial and transverse curvatures occur. The Triasic-Cretaceous deposits are generally crossed by many crevice and cracks which, along with the layering and carstice voids, constitute access routes and the flow of meteoric waters into the mass of rocks.

ROGWSI05—a permeable porous body with an area of 2145 km², accumulated in quaternary age deposits growing in the ramble plain. The underground aquifer in the sand and gravel of these deposits is generally located at low depths (1–5 m), except for the areas covered by deluvial-prolual deposits in the Siret Plain, with a piezometric level of 8–10 m deep.

ROGWSI06—is a permeable porous body. It develops in Sarmatian age formations and has a cross-border carater on an area of 3857 km².

ROGWPR05—a permeable porous water body, accumulated in the Sarmatian age deposits that develop on the territory of the Neamt, Bacau and Vaslui counties, belonging to the Moldavian Platform—the Moldavian Plateau, the Barlad Plateau and the Moldavian Subcarpathians, on an area of 12,531 km² the Siret and Prut hydrographic basins.

ROGWAG12—underground deepwater body cantonated in Fratesti and Candesti formations, of the Middle Pleistocene age. The area is of 42,768 km², belonging to the hydroelectric basins Siret, Prut, Argea, Buzau and Ialomita (Fig. 11.13).

The water resources in the management plan of the Siret, surface or underground water basin management are monitored qualitatively and quantitatively conservation and protection. This protects against pollution and floods. Following the collection of information, a database is being gathered, and it is analyzed then after the results are obtained, a series of measures are taken to improve or eliminate the negative effects.

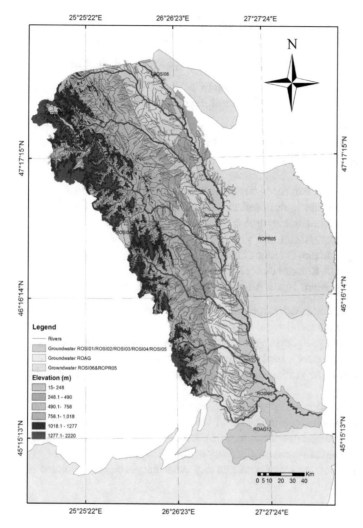

Fig. 11.13 The Siret River Basin—Groundwater bodies

11.5 Conclusions

The surface water resources in the Siret Hydrographic Area account for about 17% of the country's total water resources and are mainly formed by the Siret River and its tributaries (to a very limited extent from lakes and natural ponds). The multi-annual average stock of the Siret basin (activity area of the Siret Basin Water Administration) is approx. 6800 mln. m^3 (the section Lungoci: Q = 215 m^3/s—without the Barlad River, the Buzau River and smaller tributaries from the Lower Siret River).

The total natural resources of the Siret basin are 6868 million m^3 from which:

- Surface resources—6800 million m^3;
- Underground resources—1068 million m^3;

These resources are usable, on average per year—2655 million m^3;

- Surface resources—1955 million m^3;
- Underground resources—700 million m^3.

The groundwater and the deepwater resources are located in the meadows of the Siret, Suceava, Moldavia and Bistrita rivers. The groundwater is estimated at approximately 28 m^3/s from which 16.7 m^3/s are balance sheet resources. In the Siret River Basin there are 30 accumulations with a complex use, the volume of 1847.63 million m^3. The surface water resources of the Siret River Basin include 2 natural lakes. Their water is not used to meet water requirements. From an administrative point of view, the Siret Hydrological Area occupies the entire Suceava County, almost entirely Neamt, Bacau and Vrancea counties, and in part the counties of Botosani, Iasi, Galati, Buzau, Covasna, Harghita, Bistrita Nasaud and Maramures representing 42,890 km^2 of Romania's surface.

The Siret Water Basin Administration considers the following aspects:

(a) Conservation, development and protection of water resources, as well as ensuring the normal flow of water;
(b) Provides protection against pollution and changes in the characteristics of the water resources, banks, beds and cuvettes;
(c) Follows and realizes the restoration of the quality of surface and underground waters;
(d) Conservation and protection of aquatic ecosystems;
(e) The complex capitalization of water as an economic resource and the rational and balanced distribution of this resource, while preserving and improving the natural quality and productivity of the waters;
(f) Mitigation against floods and any other dangerous hydrometeorological phenomena.

11.6 Recommendations

In order to obtain as accurate data as possible it is necessary to multiply the number of the observation points, such as:

- The hydrometric stations on the rivers, in order to observe much more in detail the fluctuations of the level on the rivers, the air and water temperature, and the alluvial deposits.
- The evaporimetric stations both on the ground and on the lake, but also the use of automated equipment series (automatic stations with temperature, humidity, evaporation and radiation sensors).

- The automated rain gauges should also be multiplied as it necessary especially when there are registered local rainfalls.

In the case of representative basins, it is imperative to use modern hydrometrical equipment (level sensors, water and air temperature sensors, automatic rain gauges). Some other recommendations are:

- Preventions, reduction and control of pollution on surface and groundwater and conservation of its living resources in accordance with generally accepted international rules and standards.
- Sustainable and equitable management of the Siret River Waters.
- Conservation, improvement and rational use of surface water and groundwater in the Siret River Waters.
- Control of the hazards caused by accidents with pollutants for surface and groundwater.
- Control of hazards caused by floods and frost on Siret River Waters.

Acknowledgements The authors would like to express their gratitude to the employees of the Romanian Waters Agency Bucharest, Siret Water Direction Bacau, particularly to Dr. Petre Olariu, hydrologist at this research and administration agency, who kindly provided a significant part of the data used in the present study.

References

1. Albu M, Enea A, Romanescu G, Iosub M, Stoleriu CC (2015) Polarization areas of lakes, as quantitative water resources. In: International Multidisciplinary Scientific geoconference SGEM 2015, Water resources. Forest, marine and ocean ecosystem. Conference proceedings, Hydrology & water resources, vol 1, pp 509–516. https://doi.org/10.5593/sgem2015/b31/s12.065
2. Albu M, Stoleriu CC, Enea A, Iosub M, Hapciuc OE, Romanescu G (2016) Geomorphologic risk assessment in Tecucel drainage basin, using GIS techniques. In: Proceedings, 2nd International Scientific Conference GEOBALCANICA 2016, 10–12 June 2016, Skopje, Republic of Macedonia, pp 95–102. https://doi.org/10.18509/gbp.2016.13
3. Andronache I, Fensholt R, Ahammer H, Ciobotaru A-M, Pintilii R-D, Peptenatu D, Drăghici CC, Diaconu DC, Radulović M, Pulighe G, Azihou AF, Toyi MS, Sinsin B (2017) Assessment of textural differentiations in forest resources in Romania using fractal analysis. Forests 8(3):54. https://doi.org/10.3390/f8030054
4. Anton MC, Baltazar-Rojas MM, Aluculesei A, Marguta R, Dorohoi D (2008) Study regarding the water Pollution in Romanian and Spain. Rom J Phys 53(1–2):157–163
5. Bairros da Silva PR, Makara CN, Munaro AP, Schnitzler DC, Diaconu DC, Sandu I, Poleto C (2017) Risks associated of the waters from hydric systems urban's. The case of the Rio Barigui, South of Brazil. Rev Chim (Bucharest) 68(8):1834–1842
6. Briciu AE, Toader E, Romanescu G, Sandu I (2016) Urban streamwater contamination and self-purification in a Central-Eastern European City. Part I. Rev Chim (Bucharest) 67(7):1294–1300
7. Briciu AE, Toader E, Romanescu G, Sandu I (2016) Urban streamwater contamination and self-purification in a Central-Eastern European City—Part B. Rev Chim (Bucharest) 67(8):1583–1586

8. Chirica S, Luca A-L, Lates I (2018) Considerations on drinking water management in the Moldavian plateau and plain region. PESD 12(1):139–147
9. Cical E, Mihali C, Mecea M, Dumuta A, Dippong T (2016) Considerations on the relative efficacy of aluminium sulphates versus polyaluminium chloride for improving drinking water quality. Stud Univ Babes-Bolyai, Chem 61(2):225–238
10. Cirtina D, Capatina C, Simionescu CM (2015) Assessment of Motru and Motru Sec Rivers quality by monitoring of physico-chemical parameters and water quality index. Rev Chim (Bucharest) 66(8):1184–1189
11. Cojoc GM (2016) Analiza regimului hidrologic al râului Bistrita in contextul amenajarilor hidrotehnice. Editura Terra Nostra, Iasi
12. Defo C, Yerima BPK, Noumsi IMK, Bemmo N (2015) Assessment of heavy metals in soils and groundwater in an urban watershed of Yaoundé (Cameroon-West Africa). Environ Monit Assess 187:77–93
13. Diaconu DC, Andronache I, Ahammer H, Ciobotaru AM, Zelenakova M, Dinescu R, Pozdnyakov AV, Chupikova SA (2017) Fractal drainage model—a new approach to determinate the complexity of watershed. Acta Montanistica Slovaca 22(1):12–21
14. Diaconu DC, Peptenatu D, Simion AG, Pintilii RD, Draghici CC, Teodorescu C, Grecu A, Gruia AK, Ilie AM (2017) The restrictions imposed upon the urban development by the piezometric level. Case study: Otopeni-Tunari-Corbeanca. Urbanism Archit Constr 8(1):27–36
15. Enea A, Neamțu D, Stoleriu CC, Romanescu G (2016) Sustainability analysis for building dam lakes in the Oriental Carpathian Mountain, Romania. Case study: Trotus river basin. In: Proceedings, 2nd international scientific conference GEOBALCANICA 2016, 10–12 June 2016, Skopje, Republic of Macedonia, pp 103–110. https://doi.org/10.18509/gbp.2016.14
16. Enea A, Hapciuc OE, Iosub M, Minea I, Romanescu G (2017) Water quality assessment in three mountainous watersheds from eastern Romania (Suceava, Ozana and Tazlău rivers). Environ Eng Manag J 16(3):605–614
17. Iordache M, Popescu LR, Pascu LF, Lehr C, Ungureanu EM, Iordache I (2015) Evaluation of the quality of environmental factors, soil and Water in the Parang Mountains, Romania. Rev Chim (Bucharest) 66(7):1009–1014
18. Iosub M, Iordache I, Enea A, Romanescu G, Minea I (2015) Spatial and temporal analysis of dry/wet conditions in Ozana drainage basin, Romania using the standardized precipitation index. International multidisciplinary Scientific geoconference SGEM 2015, water resources. Forest, marine and ocean ecosystem. Conference Proceedings, hydrology &water resources, vol 1, pp 585–592. https://doi.org/10.5593/sgem2015/b31/s12.075
19. Kowalska JB, Mazurek R, Gąsiorek M, Zaleski T (2018) Pollution indices as useful tools for the comprehensive evaluation of the degree of soil contamination—a review. Environ Geochem Health, 1–26. https://doi.org/10.1007/s10653-018-0106-z
20. Kubiak J, Machula S, Choinski A (2017) Particular example of meromixis in the anthropogenic reservoir. Carpathian J Earth Environ Sci 13(1):5–13
21. Mebirouk H, Boubendir-Mebirouk F, Hamma W (2017) Main sources of pollution and its effects on health and the environment in Annaba. Urbanism Archit Constr 9(2):167–182
22. Merecki N, Agič R, Šunić L, Milenković L, Ilić ZS (2015) Transfer factor as indicator of heavy metals content in plants. Fresenius Environ Bull 24(11c):4212–4219
23. Mic RP, Corbus C, Matreata M (2015) Long-term flow simulation in Bârlad river basin using romanian hydrological model CONSUL. Carpathian J Earth Environ Sci 10(4):147–158
24. Miftode ID, Romanescu G (2017) Land use and superficial runoff in the lower catchment basin of the Uz river (period 1990–2012). Riscuri si catastrofe VI 20(1):101–112
25. Mihai FC (2018) Rural plastic emissions into the largest mountain lake of the Eastern Carpathians. R Soc Open Sci 5:172396. https://doi.org/10.1098/rsos.172396
26. Mihaiescu R, Pop AI, Mihaiescu T, Muntean E, Beldean S, Munteanu N, Alhafez L, Ozun A (2012) Physico-chemical characteristics of the karst Lake Ighiu (Romania). Environ Eng Manag J 11(3):623–626
27. Mihu-Pintilie A, Romanescu G, Stoleriu C (2014) The seasonal changes of the temperature, pH and dissolved oxygen in the Cuejdel Lake, Romania. Carpathian J Earth Environ Sci 9(2):113–123

28. Mihu-Pintilie A, Asandulesei A, Nicu IC, Stoleriu CC, Romanescu G (2016) Using GPR for assessing the volume of sediments from the largest natural dam lake of the Eastern Carpathians: Cuejdel Lake, Romania. Environ Earth Sci 75:710

29. Panaitescu E (2007) Acviferul freatic si de adancime din bazinul hidrografic Barlad. Casa Editoriala Demiurg, Iasi

30. Pantea I, Ferechide D, Barbilian A, Lupusoru M, Lupusoru GE, Moga M, Vilcu ME, Ionescu T, Brezean I (2017) Drinking water quality assessment among rural areas supplied by a centralized water system in Brasov county. Univ Politehnica Bucharest Sci Bull, Ser C—Electr Eng Comput Sci, Seria B 79(1):61–70

31. Papadatu CP, Bordei M, Romanescu G, Sandu I (2016) Researches on heavy metals determination from water and soil in Galati County, Romania. Rev Chim (Bucharest) 67(9):1728–1733

32. Patroescu V, Jinescu C, Cosma C, Cristea I, Badescu V, Stefan CS (2015) Influence of ammonium ions on the treatment process selection of groundwater supplies intended to human consumption. Rev Chim (Bucharest) 66(4):537–541

33. Patroescu IV, Dinu LR, Constantin LA, Alexie M, Jinescu G (2016) Impact of temperature on groundwater nitrification in an up-flow biological aerated filter Using expanded clay as filter media. Rev Chim (Bucharest) 67(8):1433–1435

34. Pop AI, Mihăiescu R, Mihăiescu T, Oprea MG, Tănăselia C, Ozunu A (2013) Physico-chemical properties of some glacial lakes in the Romanian Carpathians. Carpathian J Earth Environ Sci 8(4):5–11

35. Popescu LR, Iordache M, Buica GO, Ungureanu EM, Pascu LF, Lehr C (2015) Evolution of groundwater quality in the area of chemical platform. Rev Chim (Bucharest) 66(12):2060–2064

36. Radevski I, Gorin S (2017) Floodplain analysis for different return periods of river Vardar in Tikvesh valley (Republic of Macedonia). Carpathian J Earth Environ Sci 12(1):179–187

37. Reti KO, Malos CV, Manciula ID (2014) Hydrological risk study in the Damuc village, the Neamt county. J Environ Prot Ecol 15(1):142–148

38. Revuelto J, Lopez-Moreno JI, Azorin-Molina C, Zabalza J, Arguedas G, Vicente-Serrano SM (2014) Mapping the annual evolution of snow depth in a small catchment in the Pyrennes using the long-range terrestrial laser scanning. J Maps 10(3):379–393

39. Romanescu G (2009) Siret river basin planning (Romania) and the role of wetlands in diminishing the floods. WIT Trans Ecol Environ 125:439–453

40. Romanescu G, Cojocaru I (2010) Hydrogeological considerations on the western sector of the Danube Delta—a case study for the Caraorman and Saraturile fluvial-marine levees (with similarities for the Letea levee). Environ Eng Manag J 9(6):795–806

41. Romanescu G, Nistor I (2011) The effect of the July 2005 catastrophic inundations in the Siret River's Lower Watershed, Romania. Nat Hazards 57(2):345–368

42. Romanescu G, Stoleriu C (2013) Causes and effects of the catastrophic flooding on the Siret River (Romania) in July–August 2008. Nat Hazards 69:1351–1367

43. Romanescu G, Stoleriu C (2014) Seasonal variation of temperature, pH and dissolved oxygen concentration in Lake Rosu, Romania. CLEAN—Soil, Air, Water 42(3):236–242

44. Romanescu G, Stoleriu C (2015) Morpho-bathymetry and GIS-processed mapping in delimiting lacustrine wetlands: the Red Lake (Romania). In: Proceedings GEOBALCANICA. International scientific conference, 5–7 June 2015, Skopje, Republic of Macedonia, pp 99–110. http://dx.doi.org/10.18509/GBP.2015.12

45. Romanescu G, Cretu MA, Sandu IG, Paun E, Sandu I (2013) Chemism of streams within the Siret and Prut drainage basins: water resources and management. Rev Chim (Bucharest) 64(12):1416–1421

46. Romanescu G, Stoleriu CC, Enea A (2013) Limnology of the Red Lake, Romania. An interdisciplinary study. Springer-Verlag, Dordrecht, New York LLC

47. Romanescu G, Cojoc GM, Tirnovan A, Dascalita D, Paun E (2014) Surface water quality in Bistrita river basin (Eastern Carpathians). 14th SGEM Geoconference on water resources. Forest, marine and ocean ecosystems, SGEM2014 conference proceedings, June 19–25 2014, vol 1, pp 679–690. https://doi.org/10.5593/sgem2014/b31/s12.088

48. Romanescu G, Sandu I, Stoleriu C, Sandu IG (2014) Water resources in Romania and their quality in the Main Lacustrine Basins. Rev Chim (Bucharest) 65(3):344–349
49. Romanescu G, Tarnovan A, Sandu IG, Cojoc GM, Dascalita D, Sandu I (2014) The quality of surface waters in the Suha hydrographic basin (Oriental Carpathian Mountains). Rev Chim (Bucharest) 65(10):1168–1171
50. Romanescu G, Cojoc GM, Sandu IG, Tirnovan A, Dascalita D, Sandu I (2015) Pollution sources and water quality in the Bistrita catchment (Eastern Carpathians). Rev Chim (Bucharest) 66(6):855–863
51. Romanescu G, Curca RG, Sandu IG (2015) Salt deposits in the Romanian Subcarpathians—genesis, repartition and ethnomanagement. Int J Conserv Sci 6(3):261–269
52. Romanescu G, Jora I, Panaitescu E, Alexianu M (2015) Calcium and magnesium in the groundwaters of the Moldavian Plateau (Romania)—distribution and managerial and medical significance. In: International multidisciplinary scientific geoconference SGEM 2015, water resources. Forest, marine and ocean ecosystem. Conference proceedings, hydrology &water resources, vol 1, pp 103–112. https://doi.org/10.5593/sgem2015/b31/s12.014
53. Romanescu G, Tirnovan A, Sandu I, Cojoc GM, Breaban IG, Mihu-Pintilie A (2015) Water chemism within the settling pond of Valea Straja and the quality of the Suha water body (Eastern Carpathians). Rev Chim (Bucharest) 66(10):1700–1706
54. Romanescu G, Hapciuc OE, Sandu I, Minea I, Dascalita D, Iosub M (2016) Quality indicators for Suceava river. Rev Chim (Bucharest) 67(2):245–249
55. Romanescu G, Iosub M, Sandu I, Minea I, Enea A, Dascalita D, Hapciuc OE (2016) Spatio-temporal analysis of the water quality of the Ozana River. Rev Chim (Bucharest) 67(1):42–47
56. Romanescu G, Miftode D, Mihu-Pintilie A, Stoleriu CC, Sandu I (2016) Water quality analysis in mountain freshwater: Poiana Uzului Reservoir in the Eastern Carpathians. Rev Chim (Bucharest) 67(11):2318–2326
57. Romanescu G, Cimpianu CI, Mihu-Pintilie A, Stoleriu CC (2017) Historic flood events in NE Romania (post-1990). J Maps 13(2):787–798
58. Romanescu G, Hapciuc OE, Minea I, Iosub M (2018a) Flood vulnerability assessment in the mountain-plateau transition zone. Case study for Marginea village (Romania). J Flood Risk Manag 11(S1):S502–S513
59. Romanescu G, Mihu-Pintilie A, Carboni D, Stoleriu CC, Cimpianu CI, Trifanov C, Pascal ME, Ghindaoanu BV, Ciurte DL, Moisii M (2018b) The tendencies of hydraulic energy during XXI century between preservation and economic development. case study: Fagaras Mountains, Romania. Carpathian J Earth Environ Sci 13(2):489–504
60. Romanescu G, Mihu-Pintilie A, Stoleriu CC, Carboni D, Paveluc LE, Cimpianu CI (2018c) A comparative analysis of exceptional flood events in the context of heavy rains in the Summer of 2010: Siret Basin (NE Romania) Case Study. Water 10(2):216
61. Salihou Djari MM, Stoleriu CC, Saley MB, Mihu-Pintilie A, Romanescu G (2018) Ground-water quality analysis in warm semi-arid climate of Sahel countries: Tillabéri region, Niger. Carpathian J Earth Environ Sci 13(1):277–290
62. Sedrati A, Houha B, Romanescu G, Sandu IG, Sandu I, Diaconu DC (2017) Impact of agri-culture upon the chemical quality of groundwaters within the Saharian Atlas steppe. El-Meita (Khenchela-Algeria). Rev Chim (Bucharest) 68(2):420–423
63. Sedrati A, Houha B, Romanescu G, Stoleriu CC (2018) Hydro-geochemical and statistical characterization of groundwater in the south of Khenchela, el Meita area (northeastern Algeria). Carpathian J Earth Environ Sci 13(2):333–342
64. Sevianu E, Stermin AN, Malos C, Reti K, Munteanu D, David A (2015) GIS modeling for the ecological restoration of a nature reserve: Legii lake and valley (NW Romania)—a case study. Carpathian J Earth Environ Sci 10(4):173–180
65. Simić S, Milovanović B, Jojić Glavonjić T (2014) Theoretical model for the identification of hydrological heritage sites. Carpathian J Earth Environ Sci 9(4):19–30
66. Stefan DS, Neacsu N, Pincovschi I, Stefan M (2017) Water quality and self-purification capacity assessment of Snagov Lake. Rev Chim (Bucharest) 68(1):60–64

67. Tîrnovan A (2016) Caracteristicile scurgerii lichide si solide in bazinul reprezentativ Suha (Bucovineana). Editura Terra Nostra, Iasi
68. Tokar A, Negoitescu A, Hamat C, Rosu S (2016) The Chemical and Ecological State Evaluation of a Storage Lake. Rev Chim (Bucharest) 67(9):1860–1863
69. Yang HC, Wang CY, Yang JX (2014) Applying image recording and identification for measuring water stages to present flood hazards. Nat Hazards 74(2):737–754
70. Zelenáková M, Dobos E, Kováčová L, Vágo J, Abu-Hashim M, Fijko R, Purcz P (2018) Flood vulnerability assessment of Bodva cross-border river basin. Acta Montanistica Slovaca 23(1):53–61
71. Zelenáková M, Fijko R, Diaconu DC, Remenáková I (2018) Environmental impact of small hydro power plant—a case study. Environments 5(12):1–10

Chapter 12
Water Resources from Romanian Upper Tisa Basin

Gheorghe Şerban, D. Sabău, R. Bătinaş, P. Breţcan, E. Ignat and S. Nacu

Abstract The Romanian Upper-Tisa Basin identifies itself, from a quantitative point of view, as a space with water resources in excess, due to its place one of the wettest areas of Romania. The aims of the present chapter are to analyse the hydrological regime, which allows the assessment of the water resources of the analysed area. In the first part of the study are presented some introductory elements related to: literature about the study of water resources, general information concerning the analysed territory, information on the monitoring hydrologic activity in the area of study, methods and sources of information that were used for the elaboration of the study. In the second part is presented a brief analysis of the natural conditions (geological, morphological, climatic, hydrographic and soil cover), which conditioning and influence the river runoff. In the third part, it is analysed the actual water resource, being mainly pursued the differences of river runoff, between the two slopes of the basin, one exposed to the east, and the other to the west. Among the various parameters, used to emphasize the resource of water in a territory have been analysed: the

G. Şerban (✉) · R. Bătinaş
Faculty of Geography, Babeş-Bolyai University, 5-7 Clinicilor, 400006 Cluj-Napoca, Romania
e-mail: gheorghe.serban@ubbcluj.ro

R. Bătinaş
e-mail: razvan.batinas@ubbcluj.ro

D. Sabău
"Romanian Waters" National Administration—"Someş-Tisa" Regional Water Branch, 17 Vânătorului, 400213 Cluj-Napoca, Romania
e-mail: danielsabau075@gmail.com

P. Breţcan
Faculty of Humanities, Department of Geography, Valahia University, 35 Lt. Stancu Ion, 130105 Târgovişte, Romania
e-mail: petrebretcan@yahoo.com

E. Ignat
Coţofăneşti Secondary School, DN 11A nr 408, 607130 Coţofăneşti, Bacău County, Romania
e-mail: sisterela@yahoo.com

S. Nacu
"Romanian Waters" National Administration, 11 Ion Câmpineanu, Sector 1, Bucharest, Romania
e-mail: simion.nacu@rowater.ro

© Springer Nature Switzerland AG 2020
A. M. Negm et al. (eds.), *Water Resources Management in Romania*, Springer Water,
https://doi.org/10.1007/978-3-030-22320-5_12

volume of average flow, the average drained layer, the average specific runoff, the coefficient of average river-flow. The fourth part is focused on the analysis of the way in which the population benefits of water resources in terms of the quantity and of the failures existing in the water supply systems. A flow layer composite raster structure was built, for comparing the surface natural resource with used by population water resource, and showing the areas where the amount of water available for the population is in deficit. The study ends with some conclusions related to the analysed subject and with the list of references.

Keywords Romanian Upper Tisa Basin · GIS · Flow indicators · Water resource · Water supply

12.1 Introduction

The water resource is a universal, global, and at the same time delicate, problem, since it proves to be increasingly harder to manage, both quantitatively and especially qualitatively. The numerous complex studies and projects made by specialists find it hard to keep the pace with the fast transformations occurring in the water resources evolution, and hardly contribute to solving the problems related to it. Moreover, the authorities' contribution to the water resource management is sometimes difficult to provide, depending on the level of development of the respective state and on the desire to get involved in the solving of the problems, on the level of endowment of the institutions managing the situation, on the automation implemented for a better decision-making efficiency.

   During the last decennia on the first position have moved the problems related to extreme manifestations of the hydrological phenomena, which come to complicate even more the management of this resource. Whether we talk about massive excess and floods, or about scarcity, the solving of the problems becomes increasingly difficult, especially as these phenomena become more severe and chronic, on the background of the climate changes. Section 12.1.1 will present the related state-of-the-aret review of the water resources.

12.1.1 Literature About Water Resources

Some authors are concerned by issues related to the efficient use and economic component of the water resource [2, 17, 18, 25, 27, 30, 51], or refer to the supplying of water to the population or to other functionalities [58].

   Other authors have turned their attention to the study of the hazards affecting the water resource or the ecosystems [3, 11, 31, 47, 49, 54], or to the study of the physico-chemical components of this resource in protected areas [4, 45, 48].

The study of the water bodies from the framework of different natural units or complex areas is another preoccupation of the researchers [6, 15, 19, 26, 28, 42, 52, 56, 58], since the formation of water drainage and implicitly of water resource depends on natural factors that condition it. In exchange, some researchers correlate, in their study, the water resources with the global or local water circuits from nature, or find alternatives for profitably using the water resources copying processing models or methods based on those from nature [32, 63].

Important are also the feeding sources of some water reservoirs or of the reservoirs themselves, since their monitoring can assure a balance in the profitable resource use [8, 13, 14, 16, 33, 39, 57], and often if these water sources or bodies are affected. This compromises the entire water volume, leading it to a categorization in the low-quality categories.

There are also researchers who try different scenarios, projections for the future, regarding the water resources evolution and management in the short, medium and long run, since for a positive dynamics of the population and of the activities associated to it there may appear even great dysfunctionalities, both quantitatively, and qualitatively [12, 20, 25, 59].

Modelling in the study of the global or local water resources, or the application of different study methodologies, are more recent and foremost preoccupations of the researchers, since it can forecast their dynamics and can allow a real-time monitoring of their evolution [24, 37, 38, 43, 50, 55].

The ground or underground water resource is frequently correlated to the precipitations input, since there are important similitudes and interdependences in the quantities' fluctuation [44, 61, 62]. Moreover, pluvial water can constitute a not at all negligible resource, especially in the areas with water deficit.

A more recent water management branch is dedicated to the study of their drainage inside the human habitats, especially the urban ones, where the impermeable surfaces trigger very significant pluvial water drainage on the ground level [60, 62].

The Romanian Upper Tisza basin is quite well studied, regarding the water resource or the hydric phenomena which occur in this area. The works go from doctoral theses organized on hydrographic basins, to dedicated thematic papers or works derived from large-scale projects [4, 10, 29, 34, 47, 52, 40, 41, 43–50].

The area under analysis has no problems regarding the water resource, being part, according to many studies, of the areas with abundant water balance [10, 34, 40, 52, etc.] yet numerous hazard elements can be identified, either quantitative or qualitative [4, 10, 34, 41, 45, 47, 49].

The fast component of the maximal drainage stage is very frequently present, with its pertaining hazards and risks, as well as winter phenomena along the water courses, and the structural management and protection measures are almost totally missing [10, 41, 47, 49]. Also enough local or areal pollution events damage the water resource of this zone, considering the household residues or those coming from mining exploitation inadequately closed [45].

12.1.2 The Upper Tisa Basin and the Romanian Upper Tisa Basin

The upper basin of Tisa River is mostly overlapping the Basin of historic Maramureş—from which only a third is located on the Romanian territory and on the south-western part of Ukraine—in Transcarpatic Region—respectively the counties of Rahiv, Tiaciv, Hust and Mijgiria (the largest basin in the Carpathians Mountains). This is a well individualized territory, from the geographic point of view, thanks to the outer mountain formations and to the petrographic mosaic, which led to a different and complex evolution of the region [7, 40, 41] (Fig. 12.1).

In the Ukrainian side of the basin, it is better developed. Thanks to many important right affluents (Kisva, Shopurka, Apshytsia, Teresva, Tereblia and Rika), which have created a "gathering water place" at the mountain' base [41] (Fig. 12.1). On the Romanian side of the Maramureş Basin (the left slope of Tisa basin) there are only two important affluents (Vişeu and Iza), while the "Săpânţa" and "Şugătagul Mare"

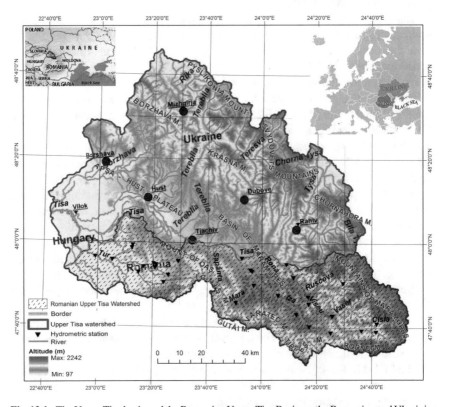

Fig. 12.1 The Upper Tisa basin and the Romanian Upper Tisa Basin on the Romanian and Ukrainian territories (adapted and completed after [29, 41] and from Photograph source 1 & 2—[21, 22])

rivers contribute in a less measure to the development of the basin, because of the existence of the andesitic massif of Igniş Mountains in the south (Fig. 12.2).

Besides, from the two watercourses, only Vişeu river displays similar mountain features like the Ukrainian rivers because Iza collects a good percent of its discharge from the median hilly zone of Maramureş' Basin. Only his left affluent, Mara river, brings a share from the mountain area, though influenced by the hydrotechnical facilities [41] (Fig. 12.2).

In contrast with the Ukrainian side of the basin, which has a larger opening to the Pannonic Basin, the Romanian part is opened only to the north-west. Towards west and south this side is closed by the Neogene volcanic chain of the Oaş-Gutâi-Ţibleş mountains (Fig. 12.1).

The western part of the studied area is defined by different patterns, due to the large opening of Tur catchment towards Pannonic Plain, and by this to the more humid air masses of oceanic origin form Western Europe. The eastern part of Tur catchment (Oaş Depression) is a genuine "gathering water place", with a torrential pattern, morphologically organized under the influence of nearby volcanic mountains, which has closed the area, on a large length of its circumference, towards north, east, south and partially on north-west (Fig. 12.2).

Otherwise, the water resource records here a different spatial distribution, compared to the Maramureş Depression, but also some important differences from the point of view of quantity and quality.

12.1.3   The Hydrometric Monitoring Network

As a basin space with a strong asymmetric character (the Vişeu river on the right and Iza on the left), which drains the slopes with high humidity and with an almost unitary value for circularity degree, the area of the study has been monitored from a hydrometric and pluviometric point of view since the middle of the 20th century, with the purpose of water management and control of the hazardous hydric phenomena. The Tisa River, the main collector, fell under the incidence of complex monitoring only after 2000 (before this year only the water level was tracked at the hydrometric station of Sighetu Marmatiei), when the bases of the cross-border cooperation between Romania and Ukraine were laid in this field [41].

Currently, for the Tisa collector (at the entrance on the Romanian territory and in the urban zone of Sighetu Marmatiei), and as well as for the other two important rivers of Maramureş—Vişeu and Iza, stations for hydric and weather monitoring were organised in key areas. They were placed either in flow formation areas (upper basins) or in water gathering areas (after important confluences with different large tributaries). The quality of the river monitoring, the efficiency of the forecasts and of the population mobilization in emergency situations has increased proportionally with general implementation of flood warning levels on rivers and their afferent alert codes (Fig. 12.3).[1]

---

[1]Records of S.T.W.B.A. ("Someş-Tisa" Water Basin Administration—in Romanian).

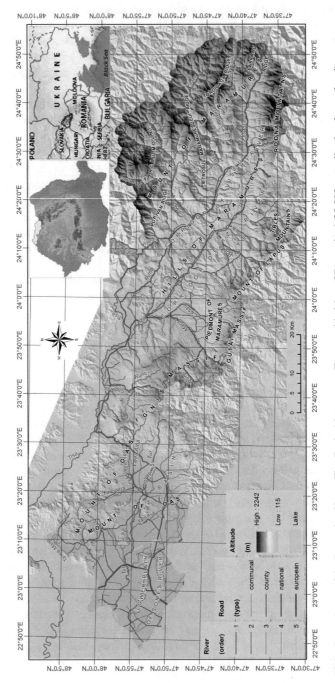

Fig. 12.2 General map of the Romanian Upper Tisa Basin (*data source* Topographic Map of Romania, 1:25,000, actualized and completed)

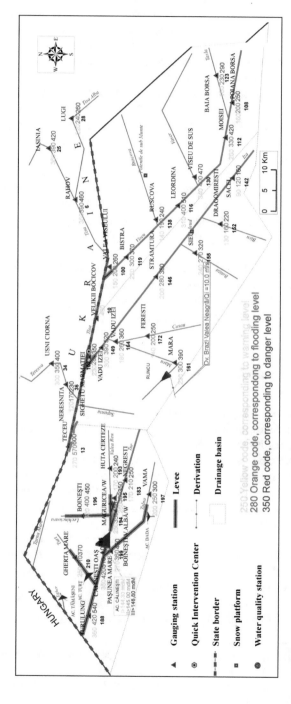

Fig. 12.3 Maximum flow management plan in the Romanian Upper Tisa Basin (after "Someş-Tisa" S.T.W.B.A. 2016)

The observation and gauging program from the river hydrometric stations targets parameters of high importance for the monitoring and management of the water resources and of dangerous hydrological and meteorological phenomena like: water level, air and water temperature, the rainfall amount, the winter phenomena, the depth of river-bed, the speed of water flow, the turbidity, the water pollution, the dynamics of the river-bed in transversal plane etc. [41].

The readings are done frequently (6.00 AM and 6.00 PM, summer time and 7.00 AM and 5.00 PM, winter time), or even more often (hourly) during the development of extreme hydric events (when the water level is surpassing the Warning Level— first defence level used for flood mitigation) in the area for which they are generated. (N.I.H.W.M. 2013).[2]

To ensure a more efficient gauging program, automated hydrometric stations for the basin water balance were introduced, placed on important watercourses or on the ones with a high torrential degree (with the help of DESWAT programs—Destructive Waters and WATMAN—Informational System for the Integrated Water Management). These grant the possibility of live monitoring of some parameters measured by sensors.

The measurements of flow rate or of river alluvial suspended deposits etc., are done less frequently (once every five days), being intensified at hydric events, at even hourly frequency on the smaller rivers. The measurements of winter phenomena are done every five days, and those of submerged vegetation, every ten days [41].

12.1.4   Sources and Technics Used in the Study Development

The text was organised using information gathered through personal investigations, and on the basis of articles, studies and scientific reports produced by other established researchers, or by institutions administrating the hydric environment and natural resources from Romania ("Someş-Tisa" Water Basin Administration—S.T.W.B.A.).

The technical data regarding the basins and watercourses, on which analysis was based, were taken from the Atlas of Water Cadastre of Romania, (1992 edition),[3] partially from the statistics of colleagues from S.T.W.B.A., or the data was generated using ArcGIS 10.x software.

The features regarding the water resources came from the hydrometric activity of S.T.W.B.A and from different reference' sources. They were, either calculated in statistic programs (Microsoft Office, SPSS) or gathered according to syntheses and studies made at the above mentioned regional institution.

The support for GIS modelling was insured from cartographic plans 1:5000, topographic maps 1:25,000, ortophoto plans and other satellite imagery, hazard and risk

---

[2](2013), Guide for the activity of hydrometric stations on rivers, N.I.H.W.M.—The National Institute of Hydrology and Water Management, Bucharest (in Romanian).

[3](1992) The Atlas of Water Cadastre of Romania, Ministry of Environment and Aquaproject S.A., Bucureşti, 683 p (in Romanian).

maps for floods (R.W.N.A.—"Romanian Waters" National Administration), geographic coordinates points and GPS files from different sources (terrain research and bibliographical references). The digital mapping, the files' conversion and the modelling were made using specific software like GPS Utility, Global Mapper, ArcGIS 10.x, AutoCAD.

12.2 Several Natural Components Which Condition the Water Resource

The analysis of natural components, which condition the water resource, has been made strictly for the area of Romanian Upper-Tisza Basin, eventually with some examples from the entire basin, for some aspects' support.

12.2.1 Geological Component

The magma-origin rocks and crystalline schists deposits from Maramureş Basin print the relief massiveness and a peripheral spread, aspect which is also passed in the exaltation of the morphological structures, with altitudes exceeding 2000 m in the south-eastern part of the region (Fig. 12.4).[4]

The sedimentary component characterizes the lower areas inside the basin, and together with the other petrographic formations influence the distinct, quantitative and qualitative features of water bodies in the area (Geological Map of Romania, 1:200,000 scale, Baia Mare and Vişeu sheets).

The river Tur basin has a different configuration, particularly interesting and spectacular, favourable to the accumulation of significant reserves of water. The upper compartment is organized in the form of an amphitheatre, with the opening towards the west, is composed of a consistent peripheral volcanic frame and a central market of gathering waters. The lower compartment is totally open and organized on consistent deposits specific to the Western Plain and allows the accumulation of significant water reserves.

A focused analysis of the petrographic structure of the relief units related strictly to Romanian Upper-Tisza basin, it can be observed important differences, with good conditions for rainwater drainage, but with poor conditions for underground water resources accumulation (Fig. 12.4).

The Pop Ivan Ridge, situated in the northern part of **Maramureş Mountains** (mountain group structured on a veritable mosaic petrographic, comparable maybe with the with very diverse petrographic structure of the Apuseni Mountains Range), is preserved very well in relief thanks to the presence of sericito-chloritic schists and

---

[4](1968), Geologic Map of Romania, 1:200000. Geologic Institute of Romania, Bucharest (in Romanian).

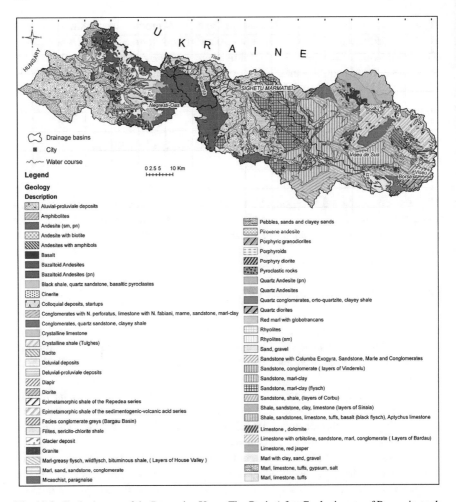

Fig. 12.4 Geologic map of the Romanian Upper Tisa Basin (after Geologic map of Romania, scale 1:200,000—Geologic Map of Romania—Baia Mare & Vişeu sheets 1968)

mica schists with paragneiss in the northern half. While the southern half is organised on a marly-sandstones flysch structure with sandstones, marly clays, menilites and bituminous schists.

On *the Farcău Massif*, it is observed in large areas, the presence of schists, sandstones, limestone, tuffs, basalt rocks (black flysch). While on the main ridge, it rises up to 1954 m, belongs to the basalts and lamprophyres.

The Pietrosul Maramureşului Ridge represents, in terms of petrography, a mosaic of epimetamorphic schists in the east, and sandstones, marly-clays, marly-sandstone flysch in the west, separated by mica schists and paragneiss.

The Toroioaga Massif, the highest in the Maramureş Mountains group (1963 m), is also a petrographic mosaic, where the existence of important metalogenetic deposits, were mined until recently. It identifies two bodies of diorite and quartz diorite in the central axis, which have penetrated a huge plateau of andesites and andesites with amphiboles. Towards the Vaser valley, it can be notice a significant enlargement of epimetamorphic schists with different insertions.

The *south-western part of Maramureş Mountains* belongs almost exclusively to sandstones, marly-clays, menilites and bituminous schists.

The Rodnei Mountains horst is maintained at altitudes above 2300 m due to sericito-chloritic schists, phyllites and meso-metamorphic schists, which compose the central part of the massif, while the eastern half of the massif belongs, primarily, to epimetamorphic schists, sericito-graphitic schists and green tuffogenic rocks. At the north and west extremities marly-sandstone flyschs, wild-flyschs and bituminous schysts, and sporadic colluvial deposits appear.

The corridors of the two main valleys in the Maramureş Basin, **Vişeu** and **Iza** river, but also of **Tisa**, their collector, are characterized by the presence of the sands and fluvial gravel, restricted very much in the gorge sectors by the occurrence of sandstones, marly-clays, menilites and bituminous schysts.

The Maramureş Hills, bounded by the two valleys mentioned above, is one of the narrowest and lowest interfluves (relative altitude) found in the Carpathian Mountains area. Thus, the northern part is defined by sandstones and marly-clays formations, while the southern half is formed on the sandstones and marly-clays deposits with intercalations of marly-sandstones flysch.

An area, partially of accumulative origin, is the left slopes of the Iza River Basin, highly developed, compared with the right side and somewhat in spatial opposition with the Vişeu Basin. The structure is particularly complex, sometimes with piedmont aspect, due to the consistent material coming from the mountain area and he bowed with a moderate slope to the Tisa River corridor, in the north, according to the fluvial drainage.

The sand and gravel deposits occupy the major river beds, both in the case of Iza collector and of its main affluent, Mara River, just as in the case of Vişeu River, as previously presented.

The **Ţibleşului Mountains** are a fairly well preserved volcanic apparatus, in which are found: andesites with pyroxene and amphibole in the main ridge, followed to the base (north-west) by andesites with pyroxene, quartz andesites and massive marly-sandstones flysch, wildflisch, bituminous schysts; secondary follow marls, sandstones and marly-clays and sandstones, marly-clays, menilites and bituminous schysts.

The **Lăpuşului Mountains** (developed on a north-west to southeast direction) are characterized by the presence of andesites, bazaltoide andesites, sandstones, marly-clays (flysch), sandstones, marly-clays, menilites and bituminous schysts.

On the northern and north-eastern limits of the Ţibleşului and Lăpuşului Mountains can be found the same sedimentary components of category marls, clays and sandstones, referred already above, interleaved or in alternation, from one basin to another (between the valleys of Ieud, Botiza and Slătioara rivers).

A very special petrographic configuration is specific to the **Gutâi Massif**, a real edifice, well preserved in relief and a pole of rainfall for the territory of Romania.

From west to east, at the contact with the Igniş Massif, appear in a reduced form clays, marls and sands, followed by andesites with biotite in the Rooster Crest and basaltoide andesites and andesites in the Gutâi Peak. On the northern outskirts starting from the Mara River valley, the petrography is defined by a series of micaceous sandstones and marly-clays (flysch) completed with consistent colluvial and deluvial-coluviale deposits. Same distribution is found on the **Maramureşului Piedmont**, where the mentioned deposits are pierced by Tortonian age formations like marl, limestone, tuffs, gypsum and salt, completed with sands, sandstones and conglomerates.

The huge and massive volcanic plateau of **Igniş Massif** is composed of basaltoide andesites, which have been exploited in the Limpedea quarry and can also be found also found in Piatra Range. To the southeast, of volcanic plateau, on the left slope of the Runcu Valley, appear sporadically andesites with biotite and andesites with amphiboles.

The whole northern (from Piatra Creek) and eastern (up to the Mara River) periphery of the Igniş Massif is organized on the same colluvial and deluvial-colluvial with blocks deposits, continued with a sedimentary structure which consist of limestone, sands, sandstones and conglomerates and one of Tortonian age deposits of marl, limestone, tuffs, gypsum and salt to the periphery. The deluvial-colluvial deposits are pierced in the central part by a few "witnesses" of the same magmatic origin, like of Igniş Massif.

In the western half, the Romanian Upper Tisa Basin is maintaining its petrographic complexity. Thus, basaltoide andesites of Igniş Massif are continued to the periphery with the quartz andesites and coarse pyroclastic deposits, and at the contact with the **Oaşului Depression** with marly clays and sands.

Beyond this depression, organized by a rich fluvial network into a genuine water gathering market, developed on sands, gravel and red clay deposits in the interfluves and sands and gravel on valleys (with limited quantity reserves of underground water), resurface volcanic structures, with insular aspect.

Thus, basaltoide andesites, frequently associated with coarse pyroclastic rocks, appear on considerable surfaces, even if not in the form of a compact plateau, like Igniş Massif. They are found in the two ramifications of the **Oaş Mountains** to the south of the Talna Valley, to the south and the north-west of Călineşti-Oaş water reservoir, continuing north up to the border with Ukraine, and to the north of Rea Valley, on Piatra Vâscului Ridge.

Secondary, andesites with biotite appear in the southern part of the Talna Valley, the rhyolites to the north of the confluence of the Talna Valley with Tur River and porphyric granodiorites in the lobe of the Lechincioara Valley and Tur River and to the east of the same Lechincioara Valley.

The quartz andesites can be observed in the two peaks located in the north-western extremity of the Tur Basin and very frequently andesites with pyroxene intercalated with basaltoide andesites in the whole sector situated between Lechincioara Valley of the Tur River and northern border.

The periphery of the volcanic apparatus related to Oaş Mountains is characterized by the presence of marly clays and sands or by the presence of sands, gravel and red clays on the western outskirts of them.

The entire **Lower Someş Plain**, to the south and the north of the Tur River, belongs to the sand and gravel deposits, who cantonate important groundwater reserves, including in deep aquifers.

12.2.2 Morphological Component

The relief is diversified, ranging from meadows and valley corridors to high mountain ranges [1, 5, 7, 35, 41] (Fig. 12.1 and 12.2).

The mountain component is closing to the eastern, southern and western periphery the Maramureş Depression (Maramureş, Rodna, Ţibleş, Lăpuş, Gutâi mountains, and Igniş Massif) and on the outskirts of the north-eastern, eastern and south-eastern the river Tur Basin (Igniş mountains and Oaş mountains with the two branches).

The piedmont structure is represented by Borşa piedmonts (the southern part of the Maramureş mountains), Văratec piedmont (the northern part of Lăpuş mountains), Gutâi piedmont (northern part of the namesakes' mountains) and Mara-Săpânţa piedmont (located on the eastern and northern periphery of the Igniş Massif). A similar structure is present at the contact between Oaş Depression and the peripheral mountain frame to the north and the east (Oaş Mountains and Igniş Massif).

The glacises component is represented by the Săcel and Vişeu structures, as well as those located on the eastern outskirts (north-south branch) and to the west of Oaş mountains towards Tur Plain, where in the last case the situation is complicated by the numerous volcanic insertions.

The hilly structure is given by the interfluve between the Iza and Vişeu rivers, one of the lowest of its kind in Romania, which determines also an extreme basin asymmetry for the two mentioned watercourses.

The depression component is developed, frequently, in the form of basins, widened in confluence areas: Borşa, Vişeu, Ruscova, Bârsana, Vadu Izei, Mara, Rona, Sighet on which we can add, the gathering water market represented by the Oaş Depression.

Their value, in terms of the *surface water resource,* is extremely important from a quantitative point of view. Qualitatively, however, the problems appear significant, related, either to natural pollution (drainage of heavy metal ions from natural rich deposits exposed to surface erosion, pollution with salts, due to the salt deposits in the diapiric structures, very common throughout the Maramureş Depression, high content of carbonates, because of the petrographic structure, etc.) or by anthropogenic pollution (pollutants resulting from former mining works, organic and inorganic residues derived from the residential areas, residues from logging activities, etc.).

As a *groundwater resource*, their importance is much mitigated except the Oaş Depression and of major confluences or river corridors (ex. Tisa's corridor, where is the water supply source for the Sighetu Marmaţiei municipality). There is a high potential for their pollution, because of the human settlements pressure and their economic activities.

The valley corridors, represented by Vişeu, Iza, Tisa, Tur, Valea Rea, Lechincioara and Talna, have a significant surface and underground potential, as seen previously.

The gorges of Vişeu, Surduc, Tisa, and on the much shorter sections on Tur, Valea Rea and Talna rivers, have rather more significance to the landscape, than as an important water resource reserve. Their significance increases, however, when it comes to maximum leakage, due to the alternation of the narrowing sectors with the slight widening ones. The height of the water layer is having proportions in the first case, while river-shore flooding cannot be neglected in the second case.

The widened meadows (the floodplain of the Tisa River), have high importance, both as surface water resource (sometimes qualitatively modified), and as an underground resource, as has been mentioned previously.

The plains (the Lower Someş Plain and Lower Tur Plain) have a great water resource potential, especially underground, where the thickly settled deposits of sands and gravels allow the accumulation of successive aquifers (phreatic and deep type). The quality of groundwater resource is influenced by the farms or by the presence of human settlements and economic facilities.

The link between Maramureş Depression with other neighbouring regions, in the Romanian sector, is carried out through some high altitude passes: Huta, 587 m, towards the Oaş Depression, Gutâi, 987 m, towards Baia Mare Depression, Neteda, 1040 m, towards Lăpuş Depression, Şetref, 818 m, to the Someş Plateau and Prislop, 1416 m, to the Bucovina's *Obcine*. The huge opening of the Tur basin towards Pannonian Depression prints some water-specific features [40, 41].

By its traits and defining elements (the character of mountain basin, opening to the huge Pannonian Basin, slope orientation, slope degree, the depth of the relief fragmentation, the drainage density, etc.) the relief prints local specificities, with important effects on the climate and hydrological phenomena and processes [40, 41, 52].

The relief influences directly the rivers' runoff through slopes and fragmentation, features which determine the specific water flow speed, and indirectly through the vertical zoning of the climate [10].

12.2.3 Climatic Components (Precipitations and Evapotranspiration)

The landforms structure causes certain climatic features of the studied area, such as cold scandinavian-baltic air masses invasions, doubled by important thermic inversions in the Maramureş Basin, high humidity in the air and significant rainfall as a result of the western oceanic air advection over both the Tur Basin and the Maramureş Basin.

These peculiarities cause specific features of the liquid runoff also, so we considered the analysis of the rainfall spatial distribution of utmost importance for this subchapter, as the amount of precipitations is an essential element of the water resource (Fig. 12.5a, b).[5]

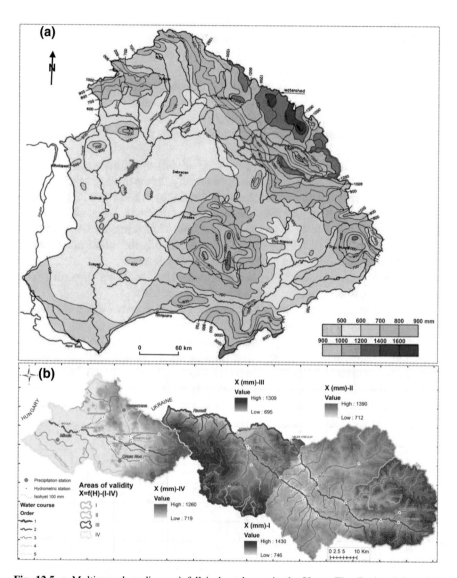

Fig. 12.5 a Multiannual medium rainfall isohyetal map in the Upper Tisa Basin—left and in Romania—right, after Sub-Basin Level Flood Action Plan Tisza River Basin 2009 and Romanian Climate 2008; **b** Areas of rainfall validity in the Upper Tisa Romanian Basin (1965–2015—*data source* "Someș-Tisa" Water Basin Administration—STWBA)

---

[5](2008) Romanian Climate. National Administration of Meteorology, Edit. Academiei Române, București (in Romanian).

According to the maps above, the studied area can be characterized as a high rainfall one, especially in the upper mountain basin that generates the runoff. The rainfall values range from 695 mm in the bottom of the Maramureșului Depression to 1430 mm on the highest peaks of the South and South-East Vișeu River Basin, according to the Sub-Basin Level Flood Action Plan Tisza River Basin (Sub-Basin Level Flood Action Plan in the Tisza River Basin), validated by the National Meteorological Administration (NMA).

For the first area, the rainfall can easily accumulate in the thick deposits of the large Tisa-Iza confluence lane, but for the second, the surface runoff has a significant torrential feature, and drains very rapidly on the steep slopes of Rodnei and Maramureșului Mountains, which is obviously not in favour of the underground water accumulation.

As a matter of fact, the correlations between rainfall and altitude allow the identification of four categories and four areas of validity for the average rainfall, which confirm the gradual reduction of the quantities from East to West. The highest values reduce from about 1430 mm on the northern slopes of the Rodnei Mountains and The Toroiaga Massif Ridge to 1260 mm in the Oaș Low Mountains. As we will further see, these areas of rainfall validity match almost identically the runoff validity areas, which allows us to estimate the balance of the water resource.

The significant rainfall quantities place the studied area in the high humidity zone, or even in the excessive one, in some parts of the river basin [52].

The spatial organization of the relief also leaves its mark on the water lost through evapotranspiration (Fig. 12.6).

The lower areas display high average values of this parameter (Lower Tur Basin, Maramureș Depression), reaching 670 mm at the extreme western limit of the studied area. On the contrary, in the higher mountain areas, the values go below 300 mm (Rodnei Mountains, Toroioaga Massif, Farcău Massif).

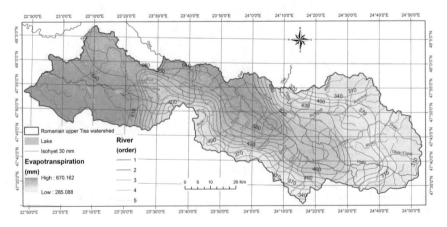

Fig. 12.6 Multiannual medium evapotranspiration isohyetal map in the Upper Tisa Romanian Basin (*source data* Romanian Climate 2008) (See footnote 4)

12.2.4   Hydrographic Component

The hydrographic component is also influenced by the geological substrate, made of hard rocks in the mountain area or clay in the hilly area, and by rainfall, which is of significant quantity.

Three of the Upper Tisa tributaries stand out as far as the development of hydrographic basins and, subsequently, water resource, are concerned: the rivers Vişeu, Iza and Tur (Fig. 12.7). All three of them have more or less pronounced assimetries of the basins, dictated by their petrographic features and the organization of the relief.

Vişeu river has an assimetric basin with an overwhelming development of the right flank, due to some particularly vigorous tributaries (Ruscova, Vaser and Ţâşla), which drain the highly humid and torrential western façade of the Maramureşului Mountains. Its drainage basin has an area of 1581 km² and a total length of the main river of 82 km. The Vişeu drainage basin has an area of 1581 km² and a total length of the main river of 82 km.

The *Iza* river, with an area of 1293 km² and a total length of 80 km, has a left side assimetry of the basin, which develops mostly on the northern slopes of the Rodnei, Ţibleşului, Lăpuşului and Gutâiului Mountains and also on the eastern façade of the Igniş Massif.

These rivers here have an average slope of 0.2–8.9‰ and a sinuosity coefficient of 1.04–2.16 (Atlas of Water Cadastre of Romania 1992) (See footnote 3).

The interfluve between the two collectors of the Maramureş Basin is one of the lowest in the carpathian area and developed through conjugated erosion of the torrential tributaries on the sedimentary clay deposits specific there.

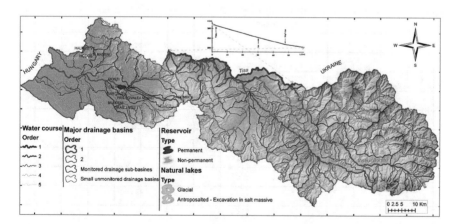

Fig. 12.7 The surface hydrographic network of the Upper Tisa Romanian Basin (*data source* "Someş-Tisa" Water Basin Administration—STWBA and Topographic Map of Romania, 1:25,000)

The Tur river basin has a certain right assimetry, where the vigorous tributaries such as Valea Rea, Lechincioara, Turţ etc., drain the volcanic slopes of Oaş Mountains.

The density of the hydrographic network in the romanian watershed has values that range from 0.3 km/km^2 in the lower parts of the studied area to 0.8 km/km^2 in the mountain area.

According to the GIS calculation algorithm applied in ArcMap 10.x, the drainage density in the studied area reaches up to 6.55 km/km^2 in the mountain area of water courses convergence (Fig. 12.8), which is the case of Upper Ruscova Basin and Median Vaser and Ţâşla Basins. High drainage density values are also specific for the tributaries in the Median Iza Basin and Median and Lower Tur Basin, as well as for the hydrographic crossovers specific for the small depressions (such as the Iza-Mara-Şugău-Rona confluence or the Oaş Depression).

The high drainage density values in the Lower Tur Plain are the result of a dense channel network, that connects the rivers flowing in from the adjacent mountain area.

When studying the shape of the basins and the hydrographic network configuration, we find out that the surface water drainage is rather adapted to the maximum run-off phase than the medium one, which is the basic parameter in estimating the water resource. Some of the basins have an almost circular shape or a more width-developed shape (all of Vişeu Basin, Ruscova Basin, Vaser Basin, Upper Tur Basin etc.), that lead to very powerful flash-floods because of the western display [47, 49].

As previously shown (Fig. 12.3), the monitoring network of the watercourses in the studied area is dense enough, as there are plenty of gauging stations and water quality control sections, as well as hydro-technical structures designed to temper the maximum run-off phase.

Lakes are far more present in the mountain area, although lacustrine basins also appear all along the Maramureşului Depression or in the Tur Basin. Their origins

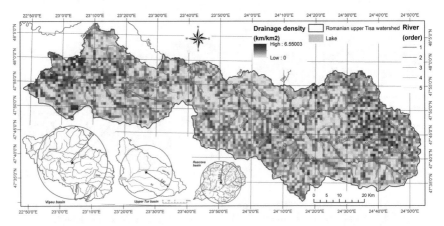

Fig. 12.8 Drainage density in the Upper Tisa Romanian Watershed and the circularity of some basins (*data source* "Someş-Tisa" Water Basin Administration—STWBA and Topographic Map of Romania, 1:25,000)

are complex, depending on the diversity of general and local landform evolution and they affect the studied area not so much in quantity, but more where as quality is concerned, through the mineral substances inflow or by altering the liquid run-off. They often tend to take on a touristic aspect, as they significantly contribute to the annual tourists' number [7], with additions).

Glacial lakes are present in the Rodnei Mountains—lakes Iezer, Buhăiescu I–IV, Repedea and in the Maramureșului Mountains—Tăul Roșu (The Red Tarn).

Periglacial lakes are present in the Maramureșului Mountains—lake Vinderel (0.9 ha) and Gutâi Mountains (Tăul Chendroaiei—Chendroaiei Tarn).

Lakes between landslide waves, linked to deposits structure in the depression area—Tăul Morărenilor (Morărenilor Tarn)—Gutâi Piedmont and lake Hoteni.

Volcanic lakes—Tăul Iezerul Mare (Iezerul Mare Tarn) in Mount Igniș.

Salt lakes (on salt massifs) are prevalent in the Maramureșului Depression, due to its paleogeographic evolution. Nonetheless, their aspect is vestigial, on account of the highly rapid evolution of the diapir structures [48].

There are also *anthropic salt lakes*, formed in collapsed salt mines a few decades ago—at Ocna Șugatag (lakes Gavrilă, Bătrân, Tăul fără fund, Roșu etc.), Coștiui (lake Francisc)—the diapir structure became a ruine itself here, a few years back, when the whole eastern side collapsed over the lake area.

Karstosaline lakes are more of memory themselves because the shallow depths and the rapid evolution of clay ruiniform landforms led to their silting or even their disappearance. Traces of these lakes can be still found at Ocna Șugatag (lakes Vorsing, Pipiriga, Alb—White Lake etc.). This lake category, next to the saline structures that store huge hypersaline water stocks, represent a real danger through their evolution because new subsidences can release from the underground and channel towards water courses thousands of cubic metres of hypersaline water, with a catastrophic effect on the aquatic ecosystems of the collector rivers, felt on tens of kilometres [48].

Anthropic dam lakes in this area are lakes Runcu and Călinești-Oaș. The first one, together with the Brazi-Valea Neagră derivation, which supplies water for the lake Firiza, has a significant ecologic influence over river Mara and its collector, Iza, by reducing the run-off which is very important for the aquatic ecosystem. The second lake, Călinești-Oaș, plays an extremely important role in tempering the flash flood waves formed in the upper Tur basin and concentrated upstream this lake (S.T.W.B.A. 2018).[6]

Lakes formed in former/dry riverbeds (meanders) are Lake Teplița, Paviscu Pond etc. Their ecological impact is quite important.

Underground, the hydrographic component has adapted to the morphological and petrographic conditions, as well as to the influences of rivers major streambeds (Fig. 12.9).

There are four groundwater bodies, two of which phreatic water (ROSO 02 along the rivers Iza and Vișeu, as well as on the interfluve between them, ROSO 17— in the Median and Lower Tur Basin) and two of them deep groundwater, developed in the Median Iza Basin deposits (ROSO 03), but mainly in the thick deposits

---

[6]Records of S.T.W.B.A. ("Someș-Tisa" Water Basin Administration—in Romanian)

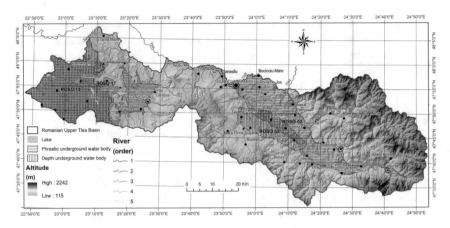

Fig. 12.9 Groundwater bodies in the Upper Tisa Romanian Basin (*data source* "Someş-Tisa" Water Basin Administration—STWBA)

of the northern part of the Western Plain Someş Alluvial Cone–Lower Tur Basin (ROSO 13).

The underground water bodies, along with the significant rainfall in the studied area, contribute to the permanent run-off of the registered water courses.

12.2.5   Pedogeographic and Land Cover Components

The soil represents the surface layer of the Earth's crust where the superficial drainage forms; it influences the run-off through its features: permeability and infiltration capacity [52].

Very often, soils in mountain areas have low permeability because of the high water saturation degree, which ensures a constant supply of underground water for the river.

The soil texture also influences the run-off, as variable or sandy textures allow water to deeply penetrate the soil, thus resulting up to 85% retention of the annual rainfall.

Figure 12.10[7] shows a map of the main classes (categories) and textures of the soils in the Upper Tisa Romanian Watershed.

The map shows the predominance of brown soils, brown luvisols, brown-clay-illuvial soils and brown eumezobazic soils in the Tur Basin and of those brown acid, podzolic and the same last cathegory in the Maramureşului Basin. On the volcanic plateaus we spot andosols and at the base of steep slopes litosols, which have great potential for underground water supplies.

---

[7](1971–1979) Soil Map of S.R. of Romania, 1:200000, ICPA, Bucureşti (in Romanian).

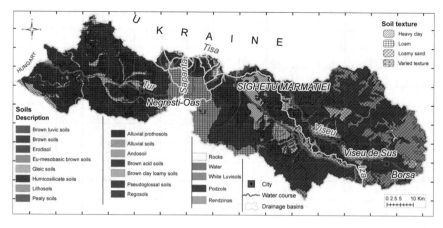

Fig. 12.10 Classes and textures of soils in the Upper Tisa Romanian Watershed (*data source* Soil Map of R.S. Romania, 1971–1979)

Soil texture is mostly loamy (Tur Basin, the volcanic plateau of Mount Igniș, the northern rim of the Gutâi and Țibleș Mountains, northern and eastern Maramureșului Mountains). Then, followed by sand-clay texture (central part of Maramureșului Mountains, the Țibleș Mountains Piedmont and the periferic piedmont-like structure surrounding Mount Igniș) and clay texture (Maramureșului Piedmont, the northern half of the Iza-Vișeu interfluve, the bottom of the Mount Igniș abrupt, the north-eastern and south-western halves of Oaș Mountains and the south-western end of the Lower Tur Plain).

This texture determines the weak to moderate water infiltration in the underground layers and thus the humble underground water stocks.

Other land features, such as land cover altogether and particularly vegetation, play a double role as far as flow forming is concerned (Fig. 12.11).[8]

On the one hand, land cover and vegetation contribute indirectly to the forming of a more loose soil structure (arborescent vegetation) which allows water to infiltrate, and on the other hand, they directly influence the run-off by creating a strong water erosion resistance and maintaining soil moisture.

As far as land use is concerned, we see a real mosaic in the two areas with slightly different morphology and climatic conditions. In the Tur Basin, we see arable lands with complex agricultural use in the gentle slope areas, while in the mountain and pre-mountainous areas the crops are replaced by deciduous forests, pastures and meadows. Human habitats also cover quite a large percentage of the surface, as the zone is highly accessible and favourable to human living.

Land cover in Maramureșului Basin and the surrounding mountain area is, however, far more diversified and tessellated. Most areas are covered by deciduous forests (the bottom of the depression), followed up in the mountains by mixed forests (espe-

---

[8] (2012), Corine Land Cover, https://land.copernicus.eu/pan-european/corine-land-cover/clc-2012

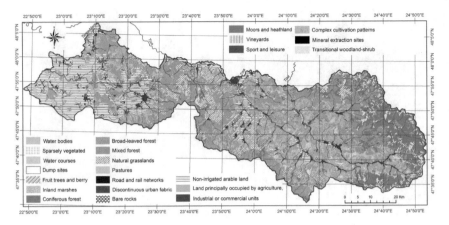

Fig. 12.11 Land cover in the Upper Tisa Romanian Basin (*data source* Corine Land Cover 2012)

cially in the Maramureşului Mountains) and in the upper part by coniferous forests and alpine pastures that is literally basic altitudinal vegetation zoning.

Anthropic areas are located mostly on the three large collector aisles and on the valleys of their tributaries, the latter offering those limited underwater stocks for household use.

Arable lands are quite rare and cover mostly confluence areas, while orchards, especially appletree ones, have a great extension, though scattered through other forms of land use.

12.3 Water Resource

Water resources of technical use in the Upper Tisa basin are important in terms of quantity, but there is the possibility of altering their condition in time, which can render unavailable the stocks of social use [41] (S.T.W.B.A. 2010):

– surface water—250 mil. m^3
– groundwater—50–106 mil. m^3

The most important parameter which characterizes the rivers water resources is the multiannual average stock, expressed as a drained volume or as river flow (Table 12.1),[9] average specific runoff or average flow layer (Fig. 12.12, 12.13 and 12.14).

---

[9]Records of R.W.N.A. ("Romanian Waters" National Administration—in Romanian).

Table 12.1 The distribution of water resources in the Upper Tisa Romanian Basin (multiannual average values, related to a dry year—1990) [41] (R.W.N.A. 2013)

River basin district	Area (km²)	Average conditions		Dry year 1990		
		Annual average flow (m³/s)	Volume (mil. m³)	Average flow (m³/s)	Volume (mil. m³)	Q 1990 (m³/s)
Upper Tisa	4540	66	2082	50	1578	76

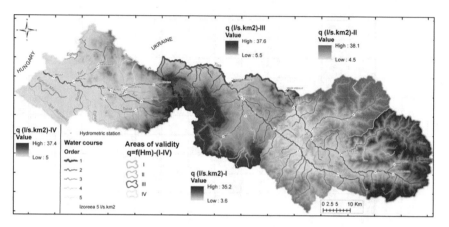

Fig. 12.12 Average specific runoff (q—l/s km²) and the validity areas in the Upper Tisa Romanian Basin (*data source* "Someş-Tisa" Water Basin Administration—STWBA)

12.3.1   Surface Water Resources

When leaving Romania, the Tisza river has a multiannual average flow of 130 m³/s, which includes both the Ukrainian basin (specific flow rate of 20.2 l/s km²) and the contribution of romanian tributaries Vişeu (33.9 m³/s) and Iza (16.6 m³/s). Tisa river has a *specific flow rate* three times higher than the Someş river, although its basin surface is just half the size of the latter's, as a result of the heavy rainfall recorded in the water catchment area [41] (S.T.W.B.A. 2015).

The average specific runoff (*q*) represents the amount of water that drains in a time unit off an area unit. We used the following formula to calculate it:

$$q = Q \times 10^3 / F \left(l/s \ km^2 \right)$$

where

q   specific flow rate (l/s km²)
Q   liquid flow rate/runoff (m³/s)
F   basin surface (km²).

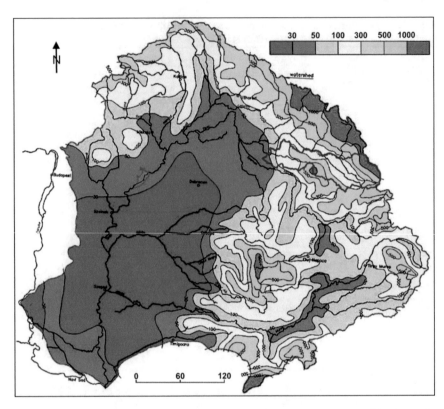

Fig. 12.13 Multiannual values of the average flow layer (Y—mm) in the Upper Tisa Romanian Basin (according to the Sub-Basin Level Flood Action Plan in the Tisza River Basin 2009)

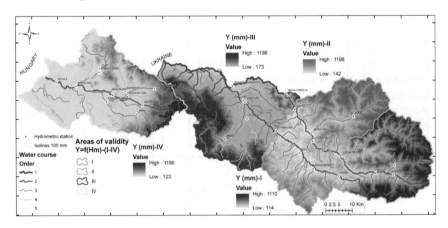

Fig. 12.14 The height of the average flow layer (Y—mm) and the areas of validity in the Upper Tisa Romanian Basin (*data source* "Someş-Tisa" Water Basin Administration—STWBA)

The studied area has a high specific flow rate due to the rich supplies from the mountain area, as a result of heavy rainfall, as previously shown. We identified four validity areas of the relation between the average specific flow rate and the medium altitude of the basins (Fig. 12.12).

In all the four areas, the higher parts of the basins (the catchment areas) show values of over 35 l/s km^2, while the bottom of depressions, the lane areas or the plain ones display much smaller values, up to 3.5 l/s km^2. The highest values that were GIS modeled were also confirmed by the Basin Authority "Someş-Tisa" (S.T.W.B.A.) in 2015 (Fig. 12.12).

The maximum value (38.1 l/s km^2) is recorded in the upper Ruscova basin, which has the smallest circularity coefficient of all the drainage areas in the studied region (1.33) and is one of the most humid ones in the Carpathian Mountains. For that matter, the torrentiality of the mountain Maramureşean area is well-known, a fact also suggested by the high values of the specific medium flow rate.

The height of the flow layer is the thickness of a water layer evenly spread on the whole basin. It is calculated by dividing the volume of water drained in a given period to the surface of the basin that generated it. The drained layer height is obtained in mm using the following formula:

$$Y = W/F \times 10^3 (mm)$$

where

Y the drained layer (mm)
W the drained volume of water (m^3)
F the surface of the basin (km^2).

The height of the flow layer also indicates the abundance of the water resources in the studied area, with values exceeding 1000 mm (Flood Protection Expert Group, Hungary, Romania, Slovakia, Serbia, Ukraine, 2009—Fig. 12.13).

The GIS modeling performed by the authors of this study confirm the issues shown above, based on the survey carried out by the previously mentioned international group (Fig. 12.14).

Just the same as the average specific flow rate, for the flow layer we also found four areas of validity of the relation between the flow layer and the medium altitude of the basins.

Three of the areas of validity show maximum flow layer heights that approach 1200 mm, while the first eastern area of validity barely exceeds this threshold. The minimum values of this parameter do not exceed 200 mm in any of the areas of validity.

12.3.2   Types f Hydrological Regime of the Rivers in the Upper Tisa Romanian Basin

Studying the hydrological regime of a river, aims to discover the laws of water resources time and space variation in a certain area. Knowing this regime has an important practical side, as the local or regional socio-economic development depends both on the quantities and the time-space variation of the water supplies [46].

The research on the runoff regime in the studied area (the spatial component linked to the Tur Basin) confirms it belongs to the *Carpathian type* (C) in the depression and mountain areas and the *Carpathian-periphery type* (Pc) in the plain area. Tisa river and its tributaries from the studied area have a *western Carpathian hydrological regime type*, as their maximum runoff occurs in April and their minimum during winter [50, 52].

12.3.3   Groundwater Resources

Groundwater is an important resource, especially due to its physico-chemical and biological quality. The storage and the movement of groundwater are closely related to the lithology, the spatial arrangement and the supplying conditions, elements that establish the hydraulic conditions of the aquifers, upon which the groundwater layers are refered to as *phreatic* and *deep groundwaters* [41].

Phreatic groundwaters are found in Romania in three macro-regions: the Carpathian orogen, the intra-carpathian depressions and basins and the extra-carpathian ones. The mountain and submountain regions of the Carpathian orogen show little or no interest as far as the borehole water supplies are concerned, therefore are not included in the estimation of groundwater resources (R.W.N.A.) (See footnote 9).

The hydrogeological macro-region of the intra-carpathian depressions and basins includes the Western Plain, the Transylvanian Basin and the Maramureşului Depression, all of them large sedimentary units with precise tectonic, genetic and morphologic borders. Here the average phreatic water flows vary from 0.1 to 1.5 l/s km^2 in the Western Plain and from 0.1 to 1.0 l/s km^2 in the Transylvanian Basin and the Maramureşului Depression [41] (R.W.N.A. See footnote 9).

Deep groundwaters with ascending or artesian features are widespread throughout the sedimentary Carpathian-peripheric regions, reaching depths of several thousand metres, but almost lack in compact rocks areas. The Carpathian regions play a watershed role for the lower lands surrounding them. Inside them, there are, however, tectonic depressions or structures that favor the accumulation of deep groundwaters, with ascending features [41].

The identification and the delimitation of groundwater bodies were carried out only for the areas with important aquifers as water supplies, that is with flows of over

10 m³ per day. The rest of the area, even if it has local underground accumulation conditions, doesn't form a "groundwater body", as stipulated by the Framework Directive 60/2000/EC (S.T.W.B.A. 2010).

The geological criterion influences not only through the age of water bearing deposits but also through the petrographic and structural features or through their capacity of storing the water. Thus, the following groundwater body types were delimitated and characterized: porous, fissural and karstic.

The hydrodynamic criterion is especially linked to the extension of water bodies. Thus, the phreatic water bodies expand only to the limit of the hydrographic basin, which corresponds to their watershed, while the deep-water bodies can extend far outside its boundaries.

The water bodies code (e.g. ROSO02) has the following significance: RO = country code, SO = Water Basin Administration "Someș-Tisa", 02 = number of the water body in the "Someș-Tisa" Water Basin Administration list (S.T.W.B.A. 2010) (Fig. 12.9).

The identified water bodies are of *porous* type, accumulated in Quaternary and Pannonian age deposits and of *fissural* or *mixed—karstic-fissural* or *fissural-porous* type, in Triassic-Cretaceous, Palegene-Middle Miocene and Paleogene-Quaternary deposits (Fig. 12.15).

Three groundwater bodies (ROSO02, ROSO13—pressurized and ROSO17) were delimited in the plain areas, meadows and terraces of several Tisa and Tur tributaries, developed in the alluvial-proluvial permeable porous deposits of recent age, especially Quaternary. As they are close to the surface, they have a free level. The ROSO03

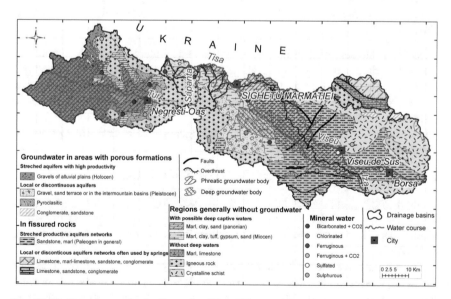

Fig. 12.15 Groundwater in the Romanian Upper Tisa Basin (composite after The Romanian Groundwater Map, 1975 edition and other sources)

body (Maramureşului Depression), although pressurized, is trapped in pannonian or older deposits and has lesser economic importance (S.T.W.B.A. 2010).

The ROSO02 groundwater body—rivers Iza and Vişeu

This groundwater body develops in the Maramureşului Depression, overlays most of the hydrographic basin of river Vişeu and partially the Upper Iza Basin (Fig. 12.9).

The petrography of the area is dominated by sandstones, conglomerates and partially Paleogene sands, with relatively high permeability, which form the base of an extended fissural network [41] (Fig. 12.15).

The average underground flow of the fissural aquifer system is 7–10 l/s km², which means a supply rate of about 250 mm per year (S.T.W.B.A. 2010).

Some springs were identified, with flows varying from 0.2 to 1 l/s in a permanent regime. The phreatic aquifer of the meadows and terraces of rivers Vişeu and Iza is made out of sands and boulders in 4–6 m thick layers, with piezometric levels at 0.1–3.0 m deep and very low flow rates when pumped (under 0.1 l/s/borehole). The only subareas that had more significant flows, from 0.7 to 7 l/s/borehole for level differences of 0.3–1.3 m, are those located at Borşa and Vişeul de Sus (S.T.W.B.A. 2010).

The ROSO03 groundwater body—Maramureşului Depression

This deep groundwater body is of *fissural* type and stationed in Paleogene-Middle Miocene deposits (Fig. 12.15).

In order to evaluate the possibilities of providing drinking water for the settlements throughout the depression, a few boreholes were drilled as follows (S.T.W.B.A. 2010):

– the Călineşti borehole, drilled up to 250 m, encountered a sandstone complex at 135–160 m and 191–216 m depth, has a maximum flow of 0.3 l/s for a level difference of 34 m (the piezometric level is of artesian type, at +0.23 m). The water is potable, of bicarbonate-calcium-magnesium-sodium type, with nitrates in small concentrations, 2 mg/l;
– the drills at Sighetul Marmaţiei, Deseşti and Bogdan-Vodă, at depths of 250–300 m, had no hydrogeological significant results.

The ROSO13 groundwater body—the Someş Cone, Lower Pleistocene

Medium depth groundwater layers of the Someş river alluvial cone and also of the northern part of the Tur river, are stationed in porous-permeable proluvial-alluvial deposits (psefito-psamitic with pellitic intercalations), of Pleistocene age (Fig. 12.15).

The groundwater body is located at depths that vary from 30 m (the lower limit of the dividing layer of clay situated between the phreatic water body and the medium depth water body correspondent to the alluvial cone of river Someş) to 50 m in the eastern side and from 30 to 120–130 m in its western end, towards the border (S.T.W.B.A. 2010).

The groundwater flows from east to west with gradients of about 0002–0.0003, descending in value from east to west.

Hydrochemically, the waters are bicarbonated-calcic type and have a total mineral content of 200–500 mg/l. Some spots have relatively high values of Fe and Mn (iron and manganese) (S.T.W.B.A. 2010).

The deposits that cover it, containing the phreatic groundwater body developed at the top of the alluvial cone of river Someş and especially the dividing clay layer, 3–5 m thick, between the two water bodies, offer this groundwater body some good protection from the surface pollution.

The ROSO13 water body is of transboundary type.

The ROSO17 groundwater body—Upper Tur Plain

The phreatic groundwater body, of porous-permeable type, is stationed in the alluvial, meadow and terrace Quaternary deposits of the Upper Tur and its tributaries (Negreşti-Oaş Depression). Downstream Călineşti, this groundwater body gets in direct contact with ROSO01—The Cone of Someş (S.T.W.B.A. 2010) (Figs. 12.9 and 12.15).

In the meadow and terrace areas, the deposits are made up of sands, siltic sands, sands with gravel, sands with gravel and boulders and lens-shaped clay layers. Some spots have loamy sands with gravel and boulders. The margins of the depression have nested alluvial cones (Fig. 12.15).

In the meadows and terraces the thickness of the alluvial deposits is, generally, of 3–10 m, but can also reach 28 m, in the area of Coca. The bed of the aquifer is made of marl and pannonian clay, and at its top clay, sandy clay and siltic clays develop under the soil, without a continuous extension in the surface. The thickness of these pelitic deposits varies from 0.5 to 4 m (S.T.W.B.A. 2010).

The hydrostatic level is mostly free or can be slightly ascending when loamy formations develop on top of the aquifer. The depth of the hydrostatic level varies largely, from 0.07 to 2.15 m, but its average value is 0.2–1.25 m. The specific flow rate has extremely low values, of 0025–0135 l/s/m (S.T.W.B.A. 2010).

The aquifer gets its water mainly form rainfall, the effective infiltration reaching up to 31.5–63 mm/year. The direction of the groundwater flow is almost always from the aquifer towards the river, but when high flows or floods occur, it can be reversed [41].

The groundwater bodies in the geological layers up to 300 m deep offer extremely reduced water supplies for settlements in the Maramureşului Depression, therefore we can reconfirm the conclusion that this morphohydrographic unit is relatively lacking in groundwater supplies.

Mineral water resources in the Upper Tisa Romanian Basin

The studied area is located under the influence of the volcanic structures of the Oaş-Gutâi-Ţibleş volcanic Mountains and the similar structures from the southern half of the Maramureşului Mountains (Figs. 12.1, 12.2 and 12.4). This situation reflects in the quality of the water and a large number of mineral springs.

Also, the intermontane basin feature of the Maramureşului Depression, which has been separated from the Transylvanian Basin as a result of the Badenian volcanic activity, offers specific properties to the waters that drain it, due to the naturally uncovered salt that appears in certain places. Its effect on the groundwater supplies is the appearance of several chlorosodic sources, which slightly alter the mineral content of the direct collectors.

This basin represents an old Paleogene bay, surrounded by the Maramureşului Mountains and the crystalline ridges of Rodna, Preluca, Dealu Mare, Dealu Codrului. It was, in fact, a continuation of the Transylvanian intra-carpathian basin, where thick layers of Eocene and Oligocene sandstones formed [9].

The forming of the volcanic mountains of Oaş-Gutâi-Ţibleş started in Badenian as a result of the tectonic fragmentation. The initial bay was first divided into two longitudinal compartments, an eastern one, the current Maramureşului Basin, and a western one including Baia Mare, Copalnic and Lăpuş. In the Maramureş Basin, the Mio-Pliocene bay persisted only in the median and lower part of the Iza Basin, where large quantities of sediments laid down, such as marl, clays, sands and also salt, later uncovered by erosion (Coştiui, Ocna Şugatag) [9].

Upper Pliocene marks the third evolutionary stage of the area, when the aerial erosion of the northern Gutâi Mountains glacis began, due to the withdrawal of the marine waters.

As previously shown, the evolution of this zone was complicated by volcanic eruptions and tectonic fragmentation, which resulted in various changes in the quality of surface waters and also in the appearance and the persistence of mineral and even thermal spring water resources, at the western end of the studied area.

Two *areas with carbonated waters in the mofettic aureole* are found here: one in the Oaş-Gutâi-Ţibleş Mountains and one in the central and southern part of Maramureşului Mountains. In these areas and also at their border towards the Maramureşului Depression, most of the springs are *bicarbonated, with a high content of carbon dioxide, bicarbonated, chlorinated and iodinated* [18, 36].

In fact, Artemie Pricăjan confirms the prior mentioned facts in his paper *Mineral and thermal spring waters in Romania*. The complex composition of the mineral waters in the studied area offers large opportunities to use this resource (Fig. 12.16).[10]

The bicarbonated waters (Bixad—carbonated, Certeza, Săpânţa, Crăciuneşti) are exploited by their bottling and offering to human consumption, and the chlorinated and iodinated waters by spa and treatment facilities, some resorts or local centers already functioning nearby (Ocna Şugatag, Coştiui etc.).

12.3.4 Water Balance in the Upper Tisa Romanian Basin

The *water balance* represents the quantitative relationship between the incomes and outcomes, in terms of volumes of water, of a surface, in a given period. The difference

---

[10](1982) Geographic Atlas of Romania. Ed. Didactică şi Pedagogică, Bucureşti (in Romanian).

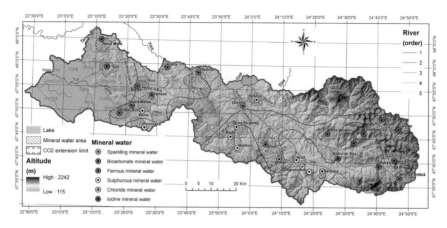

Fig. 12.16 Mineral waters from the Romanian Upper Tisa Basin (after [36]; Geographic Atlas of Romania 1982; [18], with additions)

between the volume of water flowed in and the one flowed out of the studied area is the *water resource*. When calculating the water balance of this area, we used the average values of the equation for the given period, in mm [41].

The first aridity index map for our country was made by Ioan [23]. Later on, Ujvari and Gâștescu [53], quoted by Sorocovschi and Șerban [46], have drawn up the isolines map of the aridity index, where three humidity zones shape up: high humidity ($K_a < 0.8$), variable humidity ($0.8 < K_a < 1.2$) and low humidity ($K_a > 1.2$).

For expressing the elements of the water balance in the PHCTS area we used its simplified equation, formulated by A. Penck: $X = Y + Z$, where X is rainfall, Y the total runoff and Z the evaporation [46].

Besides that, we also calculated the *Aridity Index* (K_a), as to point out the excess of the water balance, the result of the high humidity recorded. This index represents the ratio of the annual evaporated water volume (Z) and the average annual rainfall (X): $K_a = Z/X$ (Table 12.2).

The high values of rainfall and runoff, combined with the low evapotranspiration, place the studied area in the high, excessive humidity zone [10, 52].

12.4 Water Resource Available for the Population

Although the quantity of water from rivers and lakes available for the population is more than enough, its quality threatens sometimes its proper use for a number of reasons. These reasons include dangerous chemicals pollution from former mining areas (river Vișeu and its collector Tisa—[45]), waste substances pollution from residential areas, of high density in the valleys, high levels of turbidity caused by frequent and heavy rainfall, important flows being derived towards other basins—such as river Mara etc. [41].

Table 12.2 The structure of the water balance in the Upper Tisa Romanian Basin (after A.B.A.S.T. and other sources)

Element/area of validity	F (km^2)	Precipitation (X)			Global flow (Y)			Evapotranspiration (Z)			Aridity index
		mm	mil. m^3	%	mm	mil. m^3	%	mm	mil. m^3	%	
I	858.02	1060.92	910.29	21.70	644.86	553.31	26.66	416.06	356.99	16.83	0.39
II	1242.77	975.41	1212.21	28.89	556.35	691.42	33.32	419.06	520.79	24.56	0.43
III	1123.24	939.77	1055.59	25.16	495.62	556.70	26.83	444.15	498.89	23.52	0.47
IV	1324.68	768.28	1017.73	24.26	206.62	273.71	13.19	561.66	744.02	35.08	0.73
Total TISS	4548.72	936.10	4195.83	100	475.86	2075.13	100	460.23	2120.70	100	0.51

The Tisa corridor from the Vișeului Valley and Piatra and the Lower Tur Plain have additional quantities of water due to the underground waterbodies, stored in the fluviatile deposits and those of the Pannonian Basin.

12.4.1 Public Access to Networks or Systems of Water Supply

Public access to local or centralised systems of water supply depends on various factors, such as the position of the locality related to constant abstracted sources, its layout related to the already existing systems, residential neighbourhoods or streets in the peripheral areas, unserved by current systems, isolated households with difficult water network connection etc.

For the studied area, most localities and residential areas have less than **20% of their inhabitants unconnected** to a water supply network of any kind (Fig. 12.17)— from west to east:

- the median Tur basin; Turțului and Tarna Mare basins; Valea Rea upper basin;
- Săpânța river basin;
- the Tisa corridor from Sarasău to Lunca la Tisa;
- Mara and Cosău valleys, except the village of Sârbi;
- the upper Iza basin, except the villages of Șieu, Bogdan Vodă and Săcel;
- the lower Vișeu basin, except Valea Vișeului;
- the upper Vișeu basin, except Moisei.

On the other hand, in the studied area there are localities with a **high percentage of inhabitants unconnected** to water supply networks, neither local nor systems. We identified the following situations, from west to east:

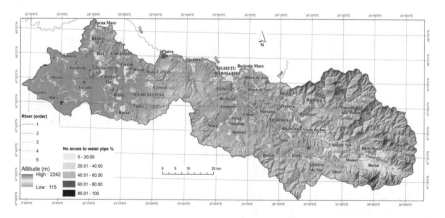

Fig. 12.17 Population with no access to drinking water pipes at locality level (number and percentage of inhabitants)

- the Lower Tur Plain—from multiple causes, one of which being the impossibility to gravitationally organise the systems;
- the Tisa corridor from Piatra to Câmpulung la Tisa, as it is a more remote area towards the main courses of Iza, Mara, Tisa and Lower Rona;
- the median Iza basin and upper Rona basin, due to a more rugged terrain;
- the upper Ruscova basin, due to a very high relief energy and a significant runoff torrentiality (Fig. 12.7 and 12.8);
- the median Vişeu basin and the village of Moisei, due to economic failure after the closure of non-ferrous exploitations in the Southern Maramureşului Mountains.

On counties, localities in Satu Mare have somewhat better percentages than those in Maramureş County, and the situation on river basins has been previously shown.

12.4.2 Required Water for the Population

The required water supplies for the population has been calculated using the specific state standards issued in the years 1989, 1995 and 2006, that is STAS 1343/0-89, SR 1343-1:1995 and SR 1343-1:2006, and the analysis of the two residential areas, urban and rural.

Six different classes have been set accordingly for the specifically required water supply in the studied area (calculation options from minimal to maximal), three each for both rural and urban areas, as follows:

Urban areas

(I)   indoor water installations and sewer—hot water preparation with solid fuel (235 l/person/day);

(II0 indoor water installations and sewer—hot water preparation with gas (280 l/person/day);

(III) indoor water installations and sewer—central heating systems (415 l/person/day).

Rural areas

(I)   areas with water distribution through pumps on streets (100 l/person/day);

(II) indoor water installations and sewer—hot water preparation with solid fuel (235 l/person/day);

(III) indoor water installations and sewer—hot water preparation with gas and electricity (280 l/person/day).

The GIS modelling performed reveals interesting situations, customized for each area of the studied zone (Fig. 12.18).

The highest values of the specifically required water ratio have been determined for the **third calculus option**:

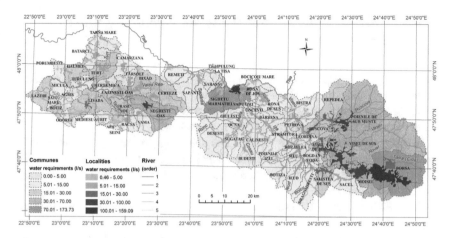

Fig. 12.18 Water requirements for the population in the PHCTS area, determined for the third calculus option

- **on localities**, they range from 0 to 23.17 l/person/day in the rural areas of Satu Mare County and from 0.46 to 32.64 l/person/day in the rural areas of Maramureșului County;
- in the urban areas, values are much higher: 57.00 l/person/day in Satu Mare County—Negrești Oaș and from 60.94 to 159.09 l/person/day in Maramureșului County, the maximum value in Sighetu Marmației;
- **on communes**, values range from 3.09 to 29.49 l/person/day in the rural areas of Satu Mare County and from 3.05 to 47.80 l/person/day in the rural areas of Maramureșului County;
- the urban areas have also much higher values: 61.96 l/person/day in Satu Mare County—Negrești Oaș, and from 68.56 to 173.73 l/person/day in Maramureșului County, with a maximum value in Sighetu Marmației.

As far as the spatial distribution of the specifically required water ratio is concerned, for all three calculus options, some localities—Sighetul Marmației, Borșa, Vișeul de Sus and Negrești-Oaș (urban areas) and Bixad, Trip, Boinești, Certeze, Remeți, Săpânța, Câmpulung la Tisa, Sarasău, Lazu Baciului, Rona de Sus, Bârsana and Petrova (rural areas) stand out with values close to 160 l/person/day first category, respectively pass 5 l/person/day the latter.

The global **water requirement** for the studied area could non be calculated due to the lack of several data categories (industry water requirement, agricultural needs etc.) and their lack of homogeneity.

Altogether, higher water requirements are obvious in the areas with economic activities, either industrial and industry-related (Sighetu Marmației area), tourism-related activities (highly represented by traditional components or local specificity) or due to the significant development of the accommodation base for tourists (in Bixad, Săpânța-Peri, Sighetu Marmației, Rona de Sus, Rona de Jos, Coștiui, Bârsana, Petrova).

12.4.3 Surface Water Resource and the Population Water Requirements

By using different extensions and GIS tools and also the statistical data related to surface runoff and population water requirements, at the end of the study we came up with a composite structured map, which compares both "nature's offer", the surface water resource, and the population water requirements.

In order to properly respect the calculus algorithm, we used the most reduced hydrographic subbasins surfaces (up to 7 km^2) and also didn't overlook the larger basins of the collectors, which crowd together many human habitats and a large number of inhabitants.

The result is very interesting and offers a quite enlightening spatial image (Fig. 12.19).

The features of the drained layer Y (mm)—**component no. 1** of the figure, are already known from the previous chapters. Therefore we will proceed to the explanation of the spatial distribution of the water requirement values.

The **second component** of Fig. 12.19 confirms a diametrically opposed situation to the first component: the areas with high values of the water requirements for population are the corridors, the main collectors valleys and the plains from the western part of the studied area, which are those most favorable to human habitats and their economic and social activities.

The highest values go up to 43.6 mm in the basin areas related to urban agglomerations (Sighetu Marmației) and 0 mm in the runoff formation zone—the mountain area, uninhabited. All along the main courses of Vișeu, Iza, Tur and some of their vigorous tributaries, the values range from 10 to 30 mm.

By using the *Raster Calculator* tool, the difference between the two grids was performed, which enhanced exactly those areas of *deficit*, regarding the water resource versus the water requirements for population (**component 3** of Fig. 12.19). These differences reach up to 17.1 mm near the state frontier with Hungary, where river Tur leaves our country. We are certain that, if we had been calculated and mapped the global *water requirement* of the studied area related to the groundwater supply, the water deficit would have been further accentuated.

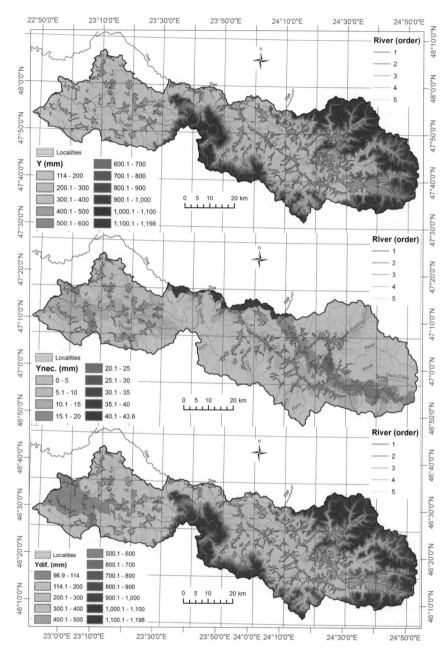

Fig. 12.19 Surface water resource Y (mm), population water requirements Ynec. (mm) and the areas with both excess and deficit of surface water resource Y dif. (mm)

12.5 Conclusions

The Romanian catchment of the Upper Tisa River is far from being considered a poor water resource area, on the contrary, all the analyzed elements, including the balance sheet, indicate a surplus of water. Both average and maximum discharge flow highlight significant volumes of transit water along the basin, while the minimum discharge flow is almost non-existent. The spatial development of the basin within the Carpathian arch, of which only a third is on the Romanian territory, favors the discharge of significant amounts of precipitation and ensures the collection of important water quantities in real "water gathering markets".

The quantitative and qualitative monitoring network is very well developed in the basin, but not the same can be said about the hydrotechnical facilities, able to intervene in the maximum discharge phase and in the regularization of the waterflow, which are completely missing in the Vișeu basin [47].

The natural factors (substratum, relief, precipitation and evaporation processes, pedo-geographic and biogeographical components) are favorable to the development of a significant water resource in terms of quantity, even if there are some qualitatively disturbing elements (the existence of clayey and marlstone substrates, the presence of salt core pits, the existence of rich mineral resources, etc.).

The surface water resource is very consistent, the technical parameters like the average specific flow or leaked layer proving this, a conclusion validated by both national and international studies, as well as by the water management institutions in this area. The underground water resource is slightly less represented on the Maramureș Basin, and better emphasized in the Tisa Corridor or in the lower plain of the West, where the underground deposits allow the cantoning of both groundwater bodies and deep-water reserves. The water balance in the study basin is in excess, is also highlighted by the calculated aridity index.

Due to the presence of the volcanic area in the central part of the studied basin and the salt core pits, there are also important quantities of mineral waters with great potential for consumption and therapeutic use, which are just partially used. Major bottling and spa centers could be arranged, based on these reserves.

Paradoxically, the water resource used by the population has the slightest representation in the mountain range and in the plains, due to various technical or financial considerations. Besides, the lower west area is also slightly deficient regarding the water layer at the disposal of the population, fact validated by the raster composite built at the end of chapter four. However, with the entry of Romania to the European Community, many water-utilities developments projects have appeared and continue to appear, increasingly, elements that increase the capitalization of the water resource and amplify the economic development of the area.

12.6 Recommendations

A minimal and limited anthropogenic intervention in the mountainous (where the surface flow forms) are recommended, in order to maintain the high quality of water resource for the natural and anthropogenic environments.

The future studies may be directed on the environmental features (study of diffuze and areal pollution sources, protection of the natural habitats etc.), but, also, on the hydrological hazards management, because of the very high rate of torrentiality, especially on the western slopes of the basin. Several important surfaces was declared as protected natural areas, without integrated studies related the anthropogenic habitats affected by the flash-floods. For example, along the Vișeu Corridor almost every year destructive floods are present, with a very serious costs for households, infrastructure network, riverbed rehabilitation.

Concerning the water supply, regional and zonal studies may be realized, according the human habitats extension or economic development of the localities. The water sources and traitment plant must be constituted in the mountainous area and connected with the localities from the low altitude spaces through pipelines of large capacity. By this the clean water is assured, and the sustainable development can be made according modern and european reglementations. Regional or zonal water companies can be organized.

Acknowledgements We want to thank to the "Someș-Tisa" Water Basin Administration, Cluj (S.T.W.B.A.), to the Babeș-Bolyai University—Faculty of Geography—Physical and Technical Department for the logistic support, to the "Romanian Waters" National Administration, Bucharest (R.W.N.A.) (See footnote 9) and National Institute of Hydrology and Water Management, Bucharest (N.I.H.W.M.), for the provided data and for all informations that made possible the achievement of this study.

We also thank to all local institutions and to all of those who have offered or will offer suggestions for the improvement of the present paper.

This study has been, also, realized thanks to the "Integrated study on the contribution of the ecosystems from the protected areas Natura 2000: Pricop-Huta Certeze and Upper Tisa to the local communities' sustainable development", international project, Systemic Ecology And Natural Capital Conservation, 2015, Norwegian Institute For Nature Research, Norway.

References

1. Ardelean G, Béreș I (2000) The vertebrate fauna of Maramureș. The Universitaria Collection, Edit. Dacia, Cluj-Napoca, 378 p (in Romanian)
2. Aznar-Sánchez JA, Belmonte-Ureña LJ, Velasco-Muñoz JF, Manzano-Agugliaro F (2018) Economic analysis of sustainable water use: a review of worldwide research. J Clean Prod 198:1120–1132
3. Blaney D (2014) Water resource management in a vulnerable world: the hydro-harzardscapes of climate change by Daanish Mustafa. Glob Environ Polit 14(1):138–139

4. Bătinaş R, Şerban G, Sabău D, Rafan S (2016) Preliminary analysis on some physico-chemical rivers water features in Pricop-Huta-Certeze and Upper Tisa Natura 2000 Protected Areas. In: Şerban G, Bătinaş R, Croitoru A, Holobâcă I, Horvath C, Tudose T (eds) Proceedings of conference air and water—components of the environment. Babeş-Bolyai University, Faculty of Geography, Cluj-Napoca, România, Edit. Casa Cărţii de Ştiinţă 22–23 March, pp 314–319

5. Boar N (2005) The Romanian-Ukrainian cross-border region of Maramureş, 294 p. Edit. Presa Universitară Clujeană (in Romanian)

6. Borsos B, Sendzimir J (2018) The Tisza River: Managing a Lowland River in the Carpathian Basin. https://doi.org/10.1007/978-3-319-73250-3_28. In book: Riverine Ecosystem Management

7. Chiş VT, Kosinszki S (2011) Geographical Introductary Characterization of the Upper Tisa River Basin (Romania-Ukraine), "The Upper Tisa River Basin". Transylvanian Rev Systematical Ecol Res 11:1–14

8. Choiński A, Ilyin L, Marszelewski W, Ptak M (2008) Lakes supplied by springs: selected examples. Limnological Rev 8(4):145–150

9. Ciupagea D, Paucă M, Ichim T (1970) Geology of Transylvanian Basin (in Romanian). Editura Academiei R. S. România, Bucureşti

10. Cocuţ M (2008) The characteristics of water flow in the Basin of Maramureş and in the limitrophe mountain zone. Doctoral thesis—manuscript, Babeş-BolyaiUniversity, College of Geography, Cluj-Napoca, 115 p (in Romanian)

11. Corduneanu F, Vintu V, Balan I, Crenganis L, Bucur D (2016) Impact of drought on water resources in North-Eastern Romania. Case study-the Prut River. Environ Eng Manag J (EEMJ) 15(6)

12. De Marsily G (2007) An overview of the world's water resources problems in 2050. Ecohydrology Hydrobiol 7(2):147–155

13. Diaconu D (2008) The Siriu reservoir, Buzău river (România), Lakes, reservoirs and ponds. Rom J Limnol (1–2):141–149

14. Diaconu DC (2010) Management of storage lakes in Romania, Volumul Conferinţei Aerul şi apa-Componente ale mediului. Presa Universitara Clujeană, pp 149–155

15. Diaconu DC (2013) Water resources from Buzău river watershed. Editura Universitară, Bucureşti, 238 p (in Romanian)

16. Diaconu DC, Mailat E (2010) The management of Reservoir Silting in Romania, vol II. In: 10th International Multidisciplinary Scientific Geoconference SGEM, pp 105–112

17. Gâştescu P (2003) Territorial distribution of water resources in Romania in terms of social-economic demand. Revue roum géogr, tomes 47:48

18. Gâştescu P (2010) Water resources from Romania. Potential, quality, spatial distribution, management. În: Gâştescu P, Breţcan P (eds) volumul "Resursele de apă din România–vulnerabilitate la presiunile antropice", Lucrările primului simpozion naţional de Limnogeografie, 11–13 iunie, Universitatea Valahia, Târgovişte, Edit. Transversal, pp 10–30 (in Romanian)

19. Gâştescu P (2012) Water resources in the Romanian Carpathians and their management. GEO-REVIEW: Scientific Annals of Stefan cel Mare University of Suceava. Geography Series 21(2):9–10

20. Gleick PH (2003) Global freshwater resources: soft-path solutions for the 21st century. Science 302(5650):1524–1528

21. https://urldefense.proofpoint.com/v2/url?u=https-3A__map.viamichelin. com_map_carte-3Fmap-3Dviamichelin-26z-3D4-26lat-3D50.45043- 26lon-3D30.52449-26width-3D550-26height-3D382-26format-3Dpng- 26version-3Dlatest-26layer-3Dbackground-26debug-5Fpattern-3D.-2A& d=DwIDaQ&c=vh6FgFnduejNhPPD0fl_yRaSfZy8CWbWnIf4XJhSqx8& r=4b24WL73wmOUYPR6w18ZP4xywRdjTym_4IWJ945019DTcNUIEC_ nbiAmoMN383wA&m=_NwlsGOQ25siXq99TTNu6I1hFEvSmI8jphH2N8R0G8U&s= cuvEFM21QTbISiFO16gqGOigsmT8K40Cj0u2JP_iNcY&e=.

22. https://urldefense.proofpoint.com/v2/url?u=https-3A__upload.wikimedia.
    org_wikipedia_commons_9_95_Romania-5FUkraine-5FLocator.png&d=
    DwIDaQ&c=vh6FgFnduejNhPPD0fl_yRaSfZy8CWbWnIf4XJhSqx8&r=
    4b24WL73wmOUYPR6w18ZP4xywRdjTym_4IWJ945019DTcNUIEC_
    nbiAmoMN383wA&m=_NwlsGOQ25siXq99TTNu6I1hFEvSmI8jphH2N8R0G8U&s=
    vp_RJzz9Aj_fXJardd9VNf8TL8ZoLr4IwAgfBcmJ79A&e=.
23. Ioan C (1929) Aridity index in Romania. Buletinul meteorologic lunar, Seria II, vol.IX, Aprilie,
    nr. 4 (in Romanian).
24. Kojiri T, Hori T, Nakatsuka J, Chong TS (2008) World continental modeling for water resources
    using system dynamics. Phys Chem Earth, Parts A/B/C 33(5):304–311
25. Konar M, Evans TP, Levy M, Scott CA, Troy TJ, Vörösmarty CJ, Sivapalan M (2016)
    Water resources sustainability in a globalizing world: who uses the water? Hydrol Process
    30(18):3330–3336
26. Lakshmi V, Fayne J, Bolten J (2018) A comparative study of available water in the major river
    basins of the world. J Hydrol 567:510–532
27. Lal R (2015) World water resources and achieving water security. Agron J 107(4):1526–1532
28. Li P, Qian H (2018) Water resource development and protection in loess areas of the world: a
    summary to the thematic issue of water in loess. Environ Earth Sci 77(24):796
29. Lukianets Olga, Obodovskyi I (2015) Spatial, temporal and forecast evaluation of rivers'
    streamflow of the Drainage Basin of the Upper Tisa under the conditions of climate change.
    Environ Res Eng Manag 71(1):36–46
30. Mirchi A, Watkins DW, Huckins CJ, Madani K, Hjorth P (2014) Water resources management
    in a homogenizing world: averting the growth and underinvestment trajectory. Water Resour
    Res 50(9):7515–7526
31. Mustafa D (2013) Water resource management in a vulnerable world: the hydro-hazardscapes
    of climate change. Philip Wilson Publishers
32. Oki T, Kanae S (2006) Global hydrological cycles and world water resources. Science
    313(5790):1068–1072
33. Piasecki A, Marszelewski W (2014) Dynamics and consequences of water level fluctua-
    tions of selected lakes in the catchment of the Ostrowo-Gopło Channel. Limnological Rev
    14(4):187–194. https://doi.org/10.1515/limre-2015-0009
34. Pop OA (2010) Tur Liquid flow study from Tur river watershed. Teza de doctorat (Ph.D. thesis),
    Facultatea de Geografie, Universitatea Babeş-Bolyai, Cluj-Napoca (in Romanian)
35. Posea G, Moldovan C, Posea A (1980) County of Maramureş. Romanian Academy's Publish-
    ing, Bucureşti, p 179 (in Romanian)
36. Pricăjan A (1972) Mineral and thermal waters from Romania. Edit. Tehnică, Bucureşti, p 296
    (in Romanian)
37. Ren C, Guo P, Li M, Li R (2016) An innovative method for water resources carrying capacity
    research–metabolic theory of regional water resources. J Environ Manage 167:139–146
38. Roach T, Kapelan Z, Ledbetter R (2018) A resilience-based methodology for improved water
    resources adaptation planning under deep uncertainty with real world application. Water Resour
    Manage 32(6):2013–2031
39. Romanescu G, Sandu I, Stoleriu C, Sandu IG (2014) Water resources in Romania and their
    quality in the main lacustrine basins. Rev Chim (Bucharest) 63(3):344–349
40. Sabău D, Bătinaş R, Roşu I, Şerban G (2017) Fresh Water Resources in the Natura 2000 Pricop-
    Huta Certeze and Tisa Superioară Protected Areas. In: Şerban G, Croitoru A, Tudose T, Bătinaş
    R, Horvath C, Holobâcă I (eds) Proceedings of conference "air and water—components of the
    environment", 17–19 March, Babeş-Bolyai University, Faculty of Geography, Cluj-Napoca,
    România, Edit. Casa Cărţii de Ştiinţă, pp 166–175
41. Sabău D, Şerban G, Kocsis I, Stroi P, Stroi R (2018) Winter Phenomena (Ice Jam) on Rivers from
    the Romanian Upper Tisa Basin in 2006–2017 Winter Season. In: Zelenakova M (eds) Water
    management and the environment: case studies. WINEC 2017. Water Science and Technology
    Library, vol 86. Springer, Cham, pp 125–174. https://doi.org/10.1007/978-3-319-79014-5_7

42. Savin C (1990) Jiului Water resources from Jiu river major riverbed. Editura Scrisul Românesc, Craiova (in Romanian)
43. Simonovic SP (2002) World water dynamics: global modeling of water resources. J Environ Manage 66(3):249–267
44. Słyś D, Stec A, Zeleňáková M (2012) A LCC analysis of rainwater management variants. Ecol Chem Eng S 19(3):359–372
45. Smical I, Muntean A, Nour E (2015) Research on the surface water quality in mining influenced area in north-western part of Romania. Geogr Pannonica 19(1):20–30
46. Sorocovschi V, Şerban G (2012) Elements of climatology and hydrology. Part II—hydrology. ID Education form. Edit. Casa Cărţii de Ştiinţă, Cluj-Napoca, 242 p (in Romanian)
47. Şerban G, Pandi G, Sima A (2012) The need for reservoir improvement in Vişeu river basin, with minimal impact on protected areas, in order to prevent flooding. Studia Univ. "Babeş-Bolyai", Geographia, LVII, nr.1, Cluj-Napoca, pp 71–80
48. Şerban G, Alexe M, Rusu R, Vele D (2015) The evolution of salt lakes from Ocna Sugatag between risk and capitalisation. În: Sorocovschi V (ed) Volumul "Riscuri şi catastrofe", an XIV, vol. 17, nr. 2. Edit. Casa Cărţii de Ştiinţă, Cluj-Napoca, pp 69–82
49. Şerban G, Sabău A, Rafan S, Corpade C, Niţoaia A, Ponciş R (2016) Risks induced by maximum flow with 1% probability and their effect on several species and habitats in Pricop-Huta-Certeze and Upper Tisa Natura 2000 protected areas. In: Şerban G, Bătinaş R, Croitoru A, Holobâcă I, Horvath C, Tudose T (eds) Proceedings of conference air and water—components of the environment. 22–23 March, Babeş-Bolyai University, Faculty of Geography, Cluj-Napoca, România, Edit. Casa Cărţii de Ştiinţă, pp 58–69
50. Telcean I, Cupşa D (eds) (2015) Methodologic guide for evaluation of rivers from Tisa river watershed. ediţie trilingvă, Ed. Universităţii din Oradea, 624 p (in Romanian)
51. Teodosiu C, Barjoveanu G, Vinke-de Kruijf J (2013) Public participation in water resources management in Romania: issues, expectations and actual involvement. Environ Eng Manag J (EEMJ) 12(5)
52. Ujvári I (1972) The geography of Romanian waters. Edit. Ştiinţifică, Bucureşti, 578 p (in Romanian)
53. Ujvari I, Gâştescu P (1958) Water evaporation to the lakes surface from R.P.R. MHGA 1:49–55 (in Romanian)
54. Woodward RT, Shaw WD (2008) Allocating resources in an uncertain world: water management and endangered species. Am J Agr Econ 90(3):593–605
55. Yan Z, Zhou Z, Sang X, Wang H (2018) Water replenishment for ecological flow with an improved water resources allocation model. Sci Total Environ 643:1152–1165
56. Zaharia L (1999) Water resources in Putna catchment. A hydrological study. Editura Universităţii din Bucureşti, 305p (in Romanian)
57. Zaharia L (2010) The Iron Gates reservoir—aspects concerning hydrological characteristics and water quality. Lakes, reservoirs and ponds. Roman J Limnol 4(1–2):52–69
58. Zaharia L (2005) Study on water resources in Curvature Carpathians and Subcarpathians area to optimize their use for the population supply. In the volume Lucrări şi rapoarte de cercetare. Centrul de cercetare "Degradarea terenurilor si Dinamica geomorfologica", vol I, Ed. Universităţii Bucureşti, pp 137–171 (in Romanian)
59. Zaharia L, Toroimac G-I (2013) Romanian Danube River management: Impacts and perspectives. In: Arnaud-Fassetta G, Masson E, Reynard E (eds) European Continental Hydrosystems under Changing Water Policy. Verlag Friedrich Pfeil, München, pp 159–170
60. Zeleňáková M, Diaconu DC, Haarstad K (2017) Urban water retention measures. Procedia Eng 190:419–426
61. Zeleňáková M, Hudáková G (2014) The concept of rainwater management in area of Košice region. Procedia Eng 89:1529–1536
62. Zeleňáková M, Markovič G, Kaposztásová D, Vranayová Z (2014) Rainwater management in compliance with sustainable design of buildings. Procedia Eng 89:1515–1521
63. Zhang J, Zhang C, Shi W, Fu Y (2019) Quantitative evaluation and optimized utilization of water resources-water environment carrying capacity based on nature-based solutions. J Hydrol 568:96–107

Part VI
Case Studies

Chapter 13
Particularities of Drain Liquid in the Small Wetland of Braila Natural Park, Romania

Daniel Constantin Diaconu

Abstract The aim is to highlight the particularities of the hydrological regime and, especially that of the liquid leakage on the lower course of Danube, within one of the many wetlands along the river to correctly quantify the water intake critical periods, for various uses. The current climatic context leads to changes in both liquid flow and land use, vegetation development as well as the risks associated with these changes. Thus, it was desired to quantify and present the way of using the water resources within these wetlands to ensure water volumes during the hot seasons of the year to ensure the volumes of water required for extinguishing the forest fires that can occur in these areas, many of them being natural reserves. It is also the case of the Small Wetland of Braila, a wetland of international interest, being one of the few natural floodplain areas on the lower course of the Danube after the draining of the former inner Delta (Balta Brailei and Balta Ialomitei), which currently preserves complex aquatic and terrestrial ecosystems in a form close to the original one. The significance of this area results from the recognition as a protected area at a national level, European (SCI and SPA) and international (Ramsar site) but it is also noted by the difficulty to establish the water flow, due to the extremely low slopes and the multitudes of channels on the stream.

Keywords Hydrologic resources · Multi-criteria decision analysis · GIS · Danube river · Wildfire risk

13.1 Introduction

There is more and more clear evidence that the effects of global changes are manifested by a faster and increasingly destructive frequency, the vegetation fires being one of these consequences [1, 2]. Economic losses due to forest fires at European level are estimated at around 13 billion euros [3, 4].

D. C. Diaconu (✉)
Department of Meteorology and Hydrology, Research Center for Integrated Analysis and Territorial Management, Faculty of Geography, University of Bucharest, Nicolae Balcescu Blv. 1, District 1, Bucharest 010041, Romania
e-mail: ddcwater@yahoo.com

© Springer Nature Switzerland AG 2020
A. M. Negm et al. (eds.), *Water Resources Management in Romania*, Springer Water,
https://doi.org/10.1007/978-3-030-22320-5_13

Climatic variability at a large space-time scale, faces vegetation to be in some regions subject to a strong stress that leads to increased flammability [5, 6]. The humidity of vegetal fuel in dry time intervals is closely related to water flows [7, 8].

There is a growing need to apply a centralized/ decentralized solution to disaster management as it develops a scientific approach to help make decisions before, during and after a disaster [9].

In this context, the assessment of the impact of fires and their methods of prevention is required in the activity of managing the fire fires [10].

An important issue of fire management is to decide when and how to intervene so that the loss is minimal. This problem has been studied by several researchers from a centralized perspective, taking into account various assumptions in its formulation and different methods of solution, e.g. simulation and stochastic approach to programming a whole firefighting program [11].

Our objectives are to determine how to intervene on fire and to provide the necessary equipment and water supplies in a scenario with a limited number of specialized personnel and equipment.

The modelling of hydrological processes has the role of predicting their response to the changes induced by human society as well as due to the global changes in the environment. The quantity and dynamics of liquid and sludge leakage, as a result of hydrological processes occurring within a river basin, influence water management activities that need to identify measures of dealing with multiple hydrological risks (floods, droughts). It is, therefore, necessary to achieve a sustainable management of water resources in a changing environment, based on the forecast of water resource dynamics and the associated risk assessment as well as the evolution of habitats [12–15].

The Small Braila Wetlands have a free flooding regime appearing as a largely aquatic area, where there is no strict delimitation between land and water, apart from the highest diking representing positive landforms [16–21]. The entire Small Wetlands is crossed over by streams and natural levees, that connect the Danube channels, but also lakes, moors, puddles. Due to the hydrographical nature of the variability of the water levels, the area of negative and positive landforms also varies, and the delimitation of hydrographic sub-basins becomes more challenging to achieve. In spring, after freshets, when the water levels are high, the low embankments are covered by water, and the small lakes connect. After the seasons with little rainfall, when the water level falls, the morpho-hydrographical configuration is different: the embankments rise, the small backwaters, creeks, small lakes dry up, and the area becomes marshy [22–27].

13.2 Study Area

The Small Island of Braila is located in the Calarasi-Patlageanca sector of the Lower Danube, 365 km long, and more precisely in the Calarasi-Braila sub-sector (195 km long). In this sub-sector, due to the low slope of only 0.02–0.04‰, the Danube makes several meanderings and dislocations (Borcea, Caragheorghe, Cremenea, Macin), thus delimiting two major islands, Balta Ialomitei (Borcei) and Balta Brailei (Balta— stagnant water, not very deep, with specific vegetation and fauna, especially encountered in the floodplain area) [22, 25, 28, 29].

The researched area is located on the western side of the former Balti of Brailei, which in the past had an area of approximately 96,000 ha, being bounded to the west by the Danube and to the east by the Macin channel (Dunarea Veche) (Fig. 13.1). The size of this area is approximately 60 km long and has a maximum width of 23 km. This wet area, due to its extensive area, was able to pick up large volumes of water when the high floods of spring floods reached a high level and retained some of the water even during periods of drought in numerous lakes (ponds) [30].

The water scarcity of these lakes was estimated at over 16,000 ha. The embankment works, made to restrict this wetland to obtain farmland between 1964 and 1970, removed the natural flow of the river and natural regulation of the maximum flow and volumes of water transited on the lower course of the Danube surface over 72,000 ha. Thus, the current enclosure of the Big Island of Brailei, bounded to the east by the Macin channel and west of the Valciu channel (km 236–196). The Danube (km 196–185 and km 175–169) and the Cravia channel (km 185 at 175), as well as the dirty agricultural areas on the left bank of the river, along the line of Gura Calmatui—Stanca—Stancuta—Gropeni—Tichilesti—Chiscani localities.

In the area of the Small Island of Braila, there is a low plain, formed during Holocene, which has altitudes between 8 and 12 m. Exceptions to this rule are the elevations from Blasova (45.2 m) and Piatra Fetei, which are in fact high local blocks (island remnants) of North Dobrogea, remain as erosion witnesses over the holocene alluvium of the plain [24].

The geological analysis shows a succession of alluvial deposits, placed in the semipermeable complex from the surface, up to 10–15 m depth (often with clay), where the groundwater is cantonal, and under it there are depth aquifers, consisting of fine sands up to 25–30 m and then coarse sands and gravels up to 90–100 m. On highly permeable alluvial deposits, with thicknesses of 60–100 m, due to the processes of pedogenesis, young soils, alluvial type, affected by gleizing processes, with particularly high fertility.

Along with this flat and smooth alluvial plain, the river's rich network produces numerous bifurcations of channels and confluences, which close the islands with variable dimensions. Outside the active channels currently, on the surface of the alluvial plains, there are the traces of many abandoned channels of the river, also known as the privy. There are also many canals, lakes, ponds, and marshes along with erosion beams and witnesses such as river island or hillocks.

The lakes in the Braila small wetlands that were left behind after large land improvements appear mostly in the southern part, on a surface of 3500 ha. The largest lakes are Festile, Vulpasu, Sbenghiosu, Beghiu, Sinetele, Balaia, Gasca, Jigna, Rotundu, Manole, Cojoacele Mari, Catinul lui Mos Ion, Catinul cu ventrice, Rastina, Cortinele. All over the world, flow regulation of large rivers has been carried out by the construction of dams, dykes, channelization or floodplain reclamation [12, 31–33].

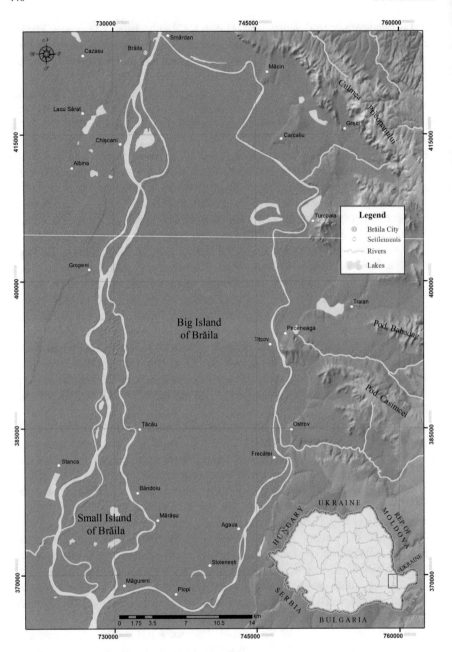

Fig. 13.1 General framing plan for the Braila Small Island

Based on input and as a result of defining the characteristic hydrological units run in the ArcSwat application, all the characteristics of each hydrographical basin have been considered, resulting in an integrated database to which we can generate reports of land-use distribution, the distribution of soils and slopes, as well as the distribution of hydrological units defined at last on each hydrographical sub-basin.

During the next stage of the ArcSwat application, a series of meteorological parameters have been measured on a representative period at the Braila meteorological station for 19 years [243–36]. From the climatic point of view, the analyses natural unit is located in a dry area, with a low precipitation intake, on a multiannual average of 447 mm (Braila Meteorological Station), the average annual temperature of +10.9 °C, the potential evapotranspiration of 650–750 mm recording an annual air humidity deficiency of 250–350 mm [34].

The used meteorological data refer to the following. The temperature data includes: the multiyear maximum, average and minimum temperature of the air (°C), the standard deviation of the maximum and minimum temperature of the air. The rainfall data includes: the average monthly multiyear rainfall and the standard deviation from the average, the asymmetry degree of the rainfall compared to the average, the chances of a rainy day after one with no rainfall or of a rainy day after another one, the average number of rainfall days, the intensity of maximum rainfall in half an hour (mm/hour). The next data was: average multiyear solar radiation, the average monthly multiyear dew point, the average wind speed (m/s), etc.

After identifying the location data and the information about using the meteorological station with specific meteorological parameters in the Small Island of Braila, the tables with entry data were written with the ArcSwat application for the geographical database based on the parameters of the 15 hydrographical basins catalogued under the aspect of integration in the database and referring to the land-use, soil characteristics and the slope side, as well as the meteorological parameters.

Having thus organized a GIS database characteristic for the studied area, The Small Island of Braila, using the ArcSwat application, there have been created layers representing: the basins and the hydrographical network of this area, soil features, land-use characteristics, as well as the land slopes [35, 36].

For this purpose, a database has been developed that includes the necessary information on hydrographic basins, liquid leakage, soils, vegetation, etc. The data model used is a vector type that allows extracting information from scanned and georeferenced maps, or from other raster sources (satellite imagery, orthophotomaps, etc.).

By running the ArcSwat application during the reference period of the meteorological data, we can achieve a variety of data, through which we can undergo statistical and mathematical analysis. The analysis are useful to emphasize the physiogeographical point of view of the studied area according to the necessities of the users, as well as the errors that can render the resulted data inconclusive for the desired study.

13.2.1 Modifications of the Water Course

The Danube channel, or the navigable Danube, is today the entire river sector ranging from km 237 to km 175. During the last century, this main channel undoubtedly suffered the greatest changes in both morphology and the toponymy used for the different sections which make it up. The name under which this channel was known in the past, or at least its southern half, was the Cremenea channel. In 1910, the name of the Cremenea channel referred to the entire section between the bifurcation of the downstream river Giurgeni (km 237) and the confluence in the Gropeni area with the Valciu channel (km 196). But map 28 made by the Romanian Hydrographic Service in 1934 limits the extension to downstream of this channel to its confluence with the secondary channel Pasca (km 210), now known as the Orbu channel. Today this toponym refers exclusively to the secondary channel bounded between km 226 and km 216.

Another particularly important hydromorphological process took place between 1934 and 1963, and that is the increase in the liquid and obvious flow rate of the former Pasca channel which, on the section between km 226 and km 216, became navigable and is currently associated with the hydronim Danube waterway. The geomorphological phenomena are also of great importance, taking place along this channel and having important consequences on river navigation. From those, special attention should be given to the accumulation of sediment produced in the Danube bed in the critical area of Gropeni (Km 196), at the confluence of the Danube and Valciu branches and further downstream in the Tichilesti area (km 190–191). Due to these phenomena, in the last decades, at least during periods of low water levels of the river, it was necessary to direct the river navigation on the secondary channel Caleia (Calia). The predominantly cumulative sedimentary regime on the Cravia branch led to the appearance of the small island, Chiciul Cucului, alongside the channel, before 1963 (and especially after 1934). Sequentially it led to the disappearance of the little Brancusi's secondary channel after 1963, which led to the affixation of the island to Mount Fundu Mare, as well as to the small island of Popa Vasile on the right bank of the river.

The Valciu channel represents the second main branch of the studied hydrographic network. Removed from the Macin channel just 1.5–2 km downstream of the Vadu Oii bifurcation, this channel is transported over a length of approximately 40 km to its confluence with the channel called the navigable Danube located in the area of Gropeni (km 196) approx. 21% of the total flow of the river. The width of the Válciu river bed varies between 160 and 440 m, and the bank shares have values between 5 and 6 m determined in the Black Sea—Sulina (MN-S) altimetric system. An exception is the upstream section of its right bank, where the intense alluvial activity of the channel has led to the formation of several tall beams which have allowed the settlement of the population and the development of several localities: Marasu, Bandoiu, and Tacau, whose existence is still registered from the beginning of the last century. Also on the left bank of the Valciu branch are the tributaries Plange Bani and Noua Nebuni, which represent the natural channels of communication and water supply of the lakes Plange Bani and Jepşile de Jos, located in the middle section of the wetland of Braila (Fig. 13.2).

The embankment of the Great Island of Braila has determined that the natural reservoir, the Small Island of Braila, to present a new distribution of the stream flow in the downstream sector, where the stream flows and the maximum levels have higher values. The embankment of the Braila Wetlands has eliminated the

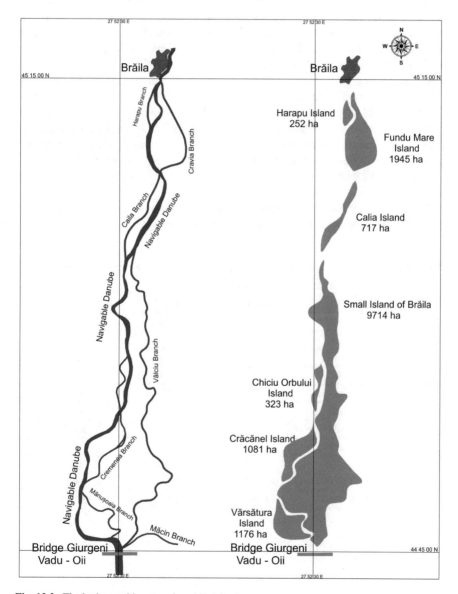

Fig. 13.2 The hydrographic network and its islands

mitigation of the stream flows. It is determining at the downstream point of the embanked sector (Braila) an increase of the stream flows in comparison to the period prior to the embankment. As a multiyear mean, the increase was one of 290 m³/s (the multiyear average stream flow before the embankment was 5980 m³/s, and after the embankment it rose to 6280 m³/s). The maximum average stream flow prior to the embankment, 10,170 m³/s, has risen as a result of the embankment to 11,030 m³/s. The maximum average stream flow, 15,800 m³/s, was registered on the 28th of May 1970 and on the 26th April 2006 (Table 13.1, Fig. 13.3) [28, 29]. At the other measuring sections the maximum stream flows were registered at Vadu Oii 14,960 m³/s (22nd April 1970), at Gropeni 9330 m³/s (31st May 1980), at Balaia 2910 m³/s (4th March 1980) and at Smardan 2080 m³/s (30th May 1980). The average annual flows in the sections of quantitative monitoring within the Braila Wetlands for the period of time 1965–1999 are 6270 m³/s—Vadu Oii, 4180 m³/s—Gropeni (The Navigable Danube Way), 1320 m³/s—Balaia (The Valciu Waterway), 729 m³/s—Smardan (The Macin Waterway) and 6280 m³/s. Starting with October (when we register minimum stream flows–4400 m³/s la Braila) the liquid drains start to grow gradually until they reach a maximum peak in April–May (in Braila, there is an average of 8670 m³/s in April and of 8690 m³/s in May) and they start to fall until October.

The flow variability, on the lower course of the Danube, explains periods with excess or deficiency of humidity at the level of the meadow.

Table 13.1 The maximum and minimum discharges, for the Danube river in the period 1931–2010

Year	Discharges in the control section (m³/s)			
	Bazias	Giurgiu	Vadu Oii	Braila
1940	13,520	14,970	14,950	15,020
1942	14,020	15,370	14,750	14,820
1947	1280	1560	1540	1550
1949	1040	1820	1910	1920
1953	1360	1650	1620	1630
1954	1200	1485	1720	1460
1970	13,040	14,930	14,790	15,000
1985	1400	1800	1880	2030
2003	1470	1690	2080	2100
2006	15,800	16,300	16,200	15,800
2010	13,200	14,340	15,410	15,150

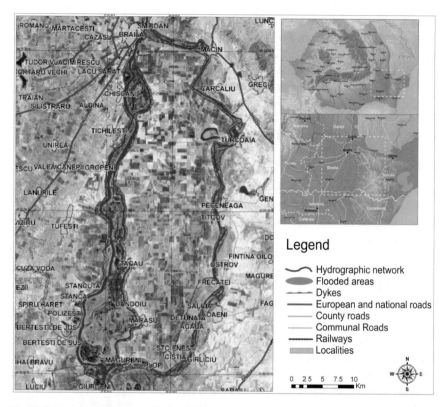

Fig. 13.3 The flooded areas in the Small Wetlands of Braila, during the biggest floods 04 April 2006

13.2.2   The Distribution of Stream Flows on the Channels

In the Natural Park of the Braila Small Wetland, upstream, we meet a number of islands such as the Varsatura Island, with a length of 5.5 km and a maximum width of 4 km, has a surface of approx. 1200 ha, being delimited by the Danube branches (km 236 to 228) and Manusoaia; The small island of Braila, formed by the merging of the former islands of Constantin (Iapa) and Popa, has a length of 33 km, a maximum width of 8.5 km and the surface of approx. 9800 ha, bounded to the east by the Valciu branch and to the west by the Manusoaia, Cremenea and Danube branches. The Cracanel Island (Caras-Chiciu) has a length of 8 km, a maximum width of 2 km and an area of approximately 1130 ha, being bordered by the Danube branches (km 225 + 500 to 216 + 500) and Cremenea. The (Chiciul) Orbului Island (5.5 km long), a maximum width of 1 km and an area of 360 ha, is located between the Danube branches (km 215 + 500 to 210) and Orbu. The Island of Lupul (Calia or Caleia) is about 9.5 km long, with a maximum width of 1 km and an area of approximately 755 ha, being delimited by the Danube branches (km 196 to 186) and Caleia; Fundu

Mare Island with a length of 9.5 km, a maximum width of 4 km and an area of approx. 1950 ha is delimited by the Danube branches (185 to 175 km) and Cravia; The Arapu Island has a length of 3.5 km, a maximum width of 1.5 km and an area of approx. 250 ha, being between the Danube branches (km 180 to 176) and Arapu.

The embankment has maintained broadly the same situation of stream flow on the channels, but it has modified the actual value of the stream flows so that the percentages of the drains are almost the same as the ones before the embankment. On average, the Macin Channel transports 11.70% of the stream flow from Vadu Oii, the Navigable Danube 67.11% and the Valciu Waterway 21.19% of the stream flow at Vadu Oii:

- The Macin Channel—transports 9.63% at the shallow waters from Vadu Oii (in 1990–8.47%) and 13.18% at the deep waters (in 1970–13.17%);
- The Navigable Danube Channel—transports at the deep waters 65.41% (May) the stream flow at Vadu Oii (in 1970–64.81%) and 69.54% (in 1990–70.43%);
- The Valciu Channel—transports 20.79% at the shallow waters (in 1990–21.10%) and 21.43% at the deep waters (in 1970–21.52%). In 1970 the stream flow at Vadu Oii was 13,590 m^3/s, and in 1990–2530 m^3/s.

In order to achieve the objective undergone in the Small Island of Braila to identify and manage the water resources in order to develop some bodies of water as reservoirs, there have been two stages as follows. A terrain one during which there have been identified the current geographical conditions of stream flow and of terrain usage and an official one that had to do with the building up of a database and the integration of all the hydro-meteorological information by using the ArcSwat application.

Hydrological parameters, presented as multiannual environments after 1994, according to data taken from the Braila Hydrological Station, synthetically describe the present state of the Danube waters in this sector. The turbidity of 0.051 g/l, the average solid flow of 10.4 million tons, the variance amplitude the 4.5 m level and the water drainage rate that records annually the following average water velocity values: 0.6, 0.8 m^3/s during average water and 1.1 m^3/s during high water.

It is worth mentioning that the hydrological regime of the lower course of the Danube is marked by the moment when the indigestion of some precincts was made in order to obtain agricultural land within the Danube meadow, starting with 1904, but we can significantly mention the year 1962 when the afforested area 106,233 ha to reach the maximum of 430,000 ha in 1987. As a result of this, on many river sections, the active section of the river was reduced, thus increasing the leakage velocity at high levels, the entrainment of sedimentary alluviums and the recording of higher maximum downstream flows than naturally up to now. But the second important moment is the construction of the energy and navigation system from the Iron Gates I (Portile de Fier I) (1964–1972) and the Iron Gates II (Portile de Fier II) (1985–1986), which formed two huge reservoirs with a huge volume, in which more than 80% of the transported river is deposited. If, for the period 1931–1970, the average annual flow rate of alluvia transported on the Vadu Oii—Braila sector, had values ranging from 1640 to 1680 kg/s, between 1971 and 1992 it decreases to average values between 710 and 820 kg/s. in the period 1992–2003, it would be only

320–340 kg/s, i.e. one-third of the flow of alluviums that flowed on the Danube in natural conditions [37, 38].

A key element for successful management of the computing program is the availability of the required data in the study area, especially in electronic format. For this purpose, a spatial database is created. The spatial data collected were: topography, liquid flows, river basin boundaries, land uses/cover and soil types. For the intended area, there have been determined three stages of slope delimitation, the distance error being from 0 to 2%; 2 to 4%; > 4% (Fig. 13.4). The low slope of the terrain leads, especially when the water level is low, to a slow flow of water, which on some channels has a direction opposite to normal upstream flow.

There have been outlined 15 hydrographical reception sub-basins feeding the Danube and a series of channels of the Danube, along with channels belonging to each reception sub-basin (Fig. 13.5). It was done based on the digital model of the terrain, with a resolution 1 at 25 m, and on the predefined data regarding the hydro geographical basins and the hydrographical network in the studied area, using the ArcSwat application. To each defined hydrographical sub-basin, there has been attributed a closing section resulting, thus, in 15 control sections of the main characteristic elements (Fig. 13.6). The problem with this area is the delimitation of the hydrographical basins on land with small slopes, entirely different than the hydrographical basins from the hills or the mountains.

A feature of the lower course of the Danube is the fact that the amounts of sludge alluviums have small values, which vary between 0 and 40 kg/s and represent about 5% of the total alluvial transport. In the sections Vadu Oii and Braila, which limit downstream Balta Ialomitei and Balta Braila respectively, the flows of dragged alluviums have values of 1–3% of the total quantity of alluviums. In the sections on the main branches of the Balta Brailei hydrographic system, there were no flows of dragged alluviums.

In order to define the hydrological units, there have been introduced data about the usage of the terrain, soils, and slopes. The hydrographical basins thus determined have been classified into 5 subunits of agricultural use, considering the usage of the land: agricultural, forestry of the wetlands and the surfaces exclusively covered by water or buildings (Fig. 13.7). Of the ecosystems identified here, 50% are natural, 30% are semi-natural and 20% are man-made [39]. Because the territory is subjected to flood periods and water withdrawal periods each year, the two types of ecosystems, terrestrial and aquatic, are interdependent, creating a specific Danube biomass. There is no strict territorial and temporal delimitation between these types of ecosystems, with a periodic succession and replacement.

A number of about 150 plant species have been identified in the wetland system in the Small Wetland of Braila. Among the most common wood species are: the willow (*Salix alba, Salix cinerea, Salix fragilis*), the poplar (*Populus alba, Populus nigra*), the ulmus (*Ulmus foliacea*), the myricaria (*Myricaria germanica*), the blackberry (*Rubus caesius*). The most common wet plant species are: reed (*Phragmites australis*), cattail (*Typha latifolia, Typha angustifolia*), *Scirpus lacustris, Lythrum salicaria, Galium palustre, Euphorbia palustris, Solanum dulcamara, Sium latifolium, Glyceria maxima, Stachys palustris, Butomus umbellatus, Iris pseudacorus.*

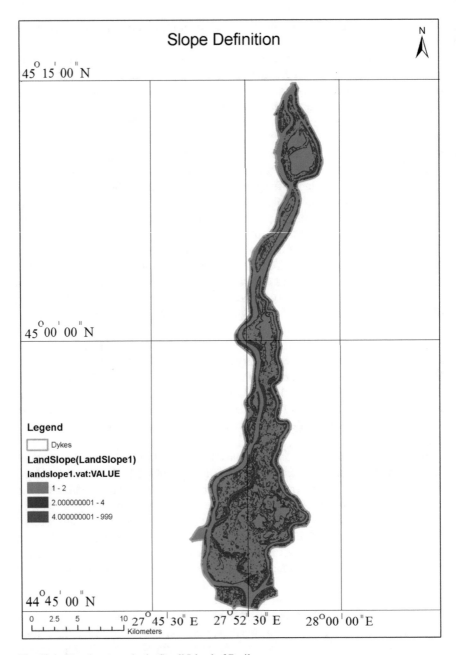

Fig. 13.4 The slope map in the Small Island of Braila

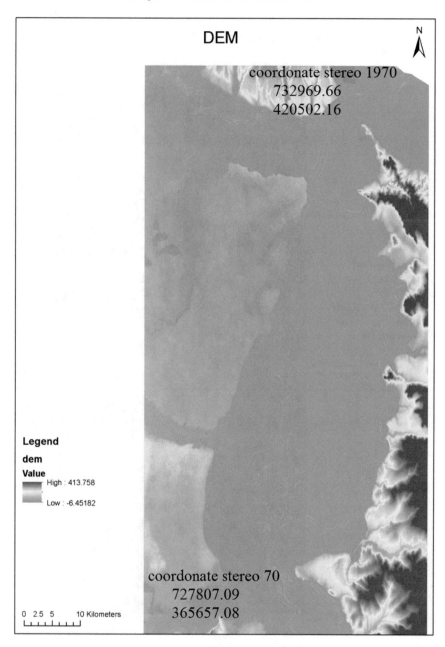

Fig. 13.5 The digital model of the terrain Small Island of Braila

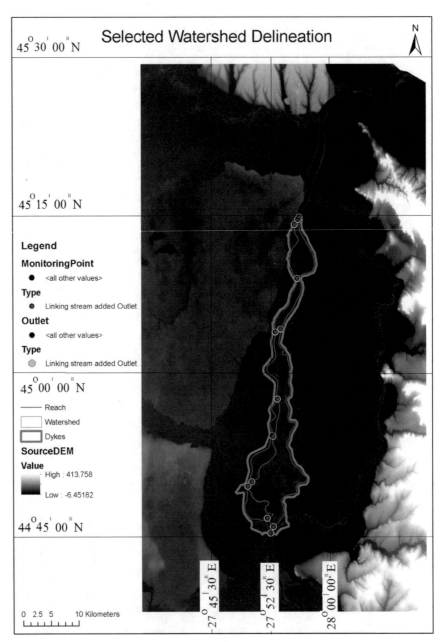

Fig. 13.6 Closing section of the hydrographical basins

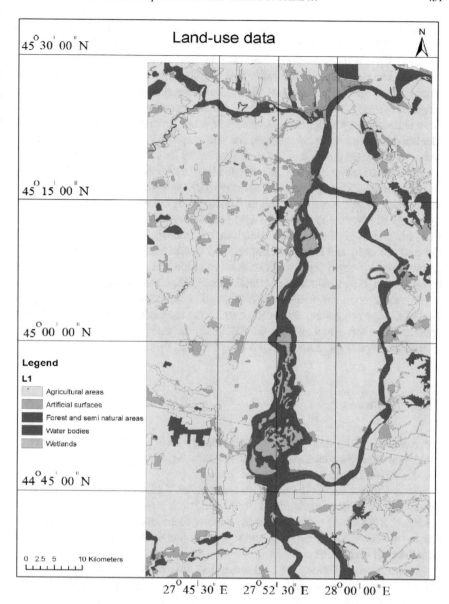

Fig. 13.7 Land-use map Small Island of Braila

Special vegetation is fixed in the sandy soils (*Tragus racemosus, Cynodon dacty-lon,* etc.). The most important aquatic associations are *Myriophyllo-nupharetum, Hydrocharitetum morsus-ranae,* especially located on canals and lakes; *Salvinio-Spirodeletum polyrrizae,* in the water from the marshes with a reed, and Trapetum natansis, in deep waters, together with *Lemna minor, Potamogeton perfoliatus, Pota-mogeton crispus, Potamogeton luncens, Potamogeton pectinatus.* There have also been introduced hydrodynamic features of the soil, characteristic for the 5 classes identified as share and presented on the soil map: chernozems, luvisols, gleysol, wetlands and areas covered in water all year round (Fig. 13.8) [40].

The Danube sector between Harsova and Braila, a section with a length of 83 km, shows an average slope of the water surface of 0.02‰, which represents a decrease of the Danube water's share by 2 cm/km. The hydrological model used to estimate the hydrological potential is a physical model, which calculates the discharge from the available data (e.g. climate, soil, vegetation, land management practices) and allows the study of short and long term impact, and data processing continuously. SWAT simulates hydrology using a water balance equation that allows estimation of water budget and hydrographic flow potential. ArcSWAT, which is the GIS interface for SWAT, was used in this study. ArcSWAT requires information on topography, soils, land use, slope categories, and weather data from the study region. Actual data was gathered to calibrate and validate the SWAT for the study area, resulting in a series of specific maps (Figs. 13.5, 13.6, 13.7, 13.8 and 13.9).

13.3 Methods

Hydrological modeling is a method that employs an array of interconnected environmental variables with associated mathematical equations to model a hydrological process within an area over a period of time.

The SWAT model (Soil and Water Assessment Tool) is a public model jointly developed by USDA Agricultural Research Service and Texas A&M AgriLife Research. SWAT is a physically-based, spatially-distributed model which can simulate the effects of soil, vegetation, and topography on the movement of water at and near the land surface [41].

Four required variables were used to run the SWAT model: digital elevation model (DEM), land cover raster, soil raster and meteorological data. These variables were entered into ArcSWAT, which is an ArcGIS interface for the SWAT model.

13.4 Applications of the Liquid Leak Study

Mediterranean Europe benefits from extensive experience and the ability to fight forest fires, but the same can't be said for the northeastern part of Europe. This region is not properly prepared for such large fires, the intervention in such situations having

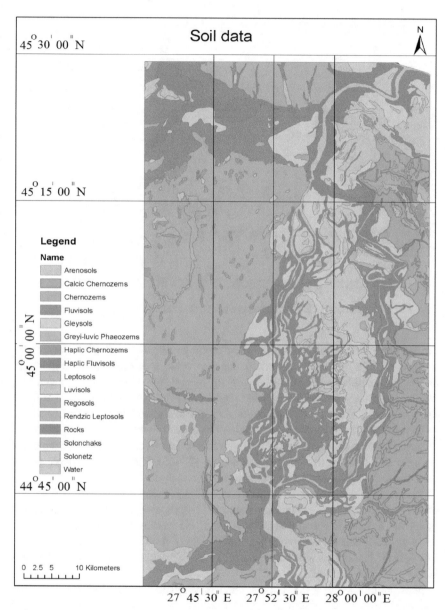

Fig. 13.8 Soil type map

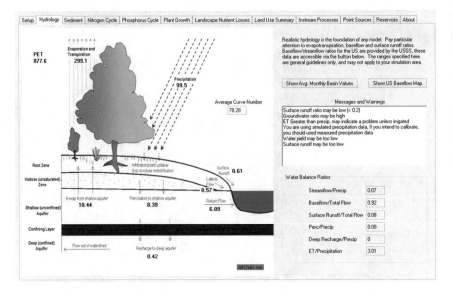

Fig. 13.9 Representation of the hydric balance within the studied area

limited capacity. Determination of water drainage directions, clotting rate/erosion rate provides the possibility of anticipating the evolution of the hydromorphological area of the Small Island of Braila, to optimally allocate the water resources needed for different purposes. One of these goals is the scenario of intervention in case of forest fires within this natural park in the context of global climate change [42]. Leading to an increase in extreme temperatures and a drop in rainfall during the warmer times of the year. Climate change is one of the main factors influencing global environmental changes and leading to major consequences [43–45]. The water resource of the Danube is closely related to the climatic variations in the basin area.

The average surface temperature of the planet has increased at the end of the 19th century by about 1.1 °C, a change determined largely by the increase in the amount of carbon dioxide in the atmosphere. Most of the heating took place over the last 35 years, 2016 was the warmest year, where eight of the 12 months that make up the year—from January to September, except June—were classified as warmer months [46].

Most studies suggest that climate change, whether direct or indirect, will pose challenges for the European society, namely economic, environmental, social, geopolitical and technological risks. Changing the land cover or its use by deforestation or extension of agricultural crops can aggravate the effects of floods or heat "islands" which in turn lead to the emergence of another wave of extreme phenomena with negative consequences for the environment [47–49].

However, natural systems provide vital ecosystem goods and services in many human activities, including agriculture, forestry, fishing, tourism and the provision of clean water and air [50]. The impacts of climate change on natural systems are expected to be far-reaching—for example, biodiversity loss for species, habitats, and functions of ecosystems and services.

In this context, given the hydrological regime and the physical-geographic characteristics of the studied area, it was desired to identify water reservoirs capable of storing a significant volume of water that could be used in extinguishing vegetation fires. Choosing the locations of the reservoirs necessary for the extinction of the vegetation fire has taken into account the physical-geographical features of the terrain (the type of combustible forest material, the type of soil, the position of the underground waters, the direction of their flow as well as the position of the surface water, the slope, the access paths and the restrictions given by the fact that the studied area has been declared a Natural Park) [51, 52].

13.4.1 Distance to Roads and Accessibility

Small wetland of Braila is isolated from the road transport, due to its location between the branches of Danube, and the access is only possible by ferry or by other small ships. The road access is possible all the way to the vicinity of the Stancuta town DJ 212, for a distance of four kilometers, Gropeni, Chiscani DJ 212, Giurgeni or Harsova E60. Inside the natural park, the routes are difficult because there are no functional roads, but paths or, at the most, earth roads (Figs. 13.10, 13.11, 13.12 and 13.13). The type of terrain, as well as the development of the vegetation, makes the

Fig. 13.10 The isolated area and the lack of road access

Fig. 13.11 The isolated area and the lack of road access

Fig. 13.12 The terrain and vegetation on the islands

Fig. 13.13 The terrain and vegetation on the islands

researched area inaccessible to fire trucks. Under these circumstances, we must find a solution based on faster and smaller vehicles capable of heaving on rough ground. At the same time, the building restrictions generated by the rules of a natural park make it almost impossible to build roads or a permanent infrastructure for putting out fires.

13.4.2 The Building Restrictions

The building restrictions are generated by the features of the foundation ground, first of all of some classical areal reservoirs used for depositing water to put out vegetation fires, due to the elevated position of the underground water and because the terrain is made up of alleviations mixed with organic materials. Using materials such as reinforced concrete, iron, and other building materials may have an impact on the environment and are restricted in this area. Also, the access to building machines is forbidden and inaccessible (excavators, cranes, concrete pumps, concrete mixers etc.).

13.4.3 Optimal Solution

The optimal solution in achieving the objective is to build water sources in the areas with a major risk of fires, deriving from the existing lakes. To this purpose, wooden roads will be built, as well as the protection of the banks and that of the lake basin, by putting up flexible geogrids to protect the banks from the variations of the water levels but also by fixating the alluvial material and impeding it from being sucked in by the pumps that extract the water.

The lakes defined as water sources intended for the fire extinction will be analyzed from the point of view of their depth and of the level variations, but also from the point of view of accessibility by road, as well as their position in regard to the areas with greater fire risk. The corresponding pumps and power generators will be mobile, and thus they will be towed by ATVs to the areas affected by the fire. Thus, the necessary logistic support will be ensured with a minimum effect on the environment. A minimal arrangement of natural materials (wood, stone) that are not polluting and which visually fall into the landscape, without altering the landscapes or the environment, can be achieved. The volume curves of the water accumulations (basins, enclosures) will be made in advance and the minimum levels of water exploitation will be indicated (Fig. 13.14).

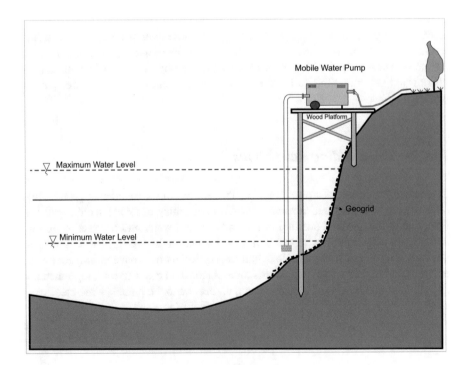

Fig. 13.14 Construction model of a water source for fire risks

Fig. 13.15 Anti-erosion network of synthetic material

Banks and channel drains will need to be protected against erosion due to the level changes occurring during pumping as well as to prevent clogging of the water pump sorb with alluvial or vegetal materials. Anti-erosion networks are provided as a system for preventing and controlling inclined surfaces, made up of tear-resistant, low-elasticity polyester stacked meshes, arranged between them to obtain a three-dimensional volumetric configuration.

These networks have the role of fixing the soil and fostering the development of a continuous vegetal carpet. They can be made of high-density polyethylene (PEID), polypropylene or polyester, 2% carbon black for ultraviolet protection, antioxidant additives, and 0.25–0.75% lubricants. They generally have large chemical inertia, which is practically inert and does not affect the hydric or terrestrial environment in any form. The open and volumetric structure allows fixing of the slopes, ensuring an increase in the friction angle. Also, high tensile strength and low creepage provide long-term reinforcement of the slope (Fig. 13.15).

13.4.3.1 Reservoirs Locations

Taking into account the potential fire risk in the Small Wetland of Braila, seven locations for some lakes have been identified, which can be used as reservoirs for depositing and delivering water for putting out the vegetation fires. These locations will be named *reservoirs* and they are numbered from 1 to 7 from upstream to downstream. They can also be turned into fixed sources of collecting and pumping the water with minimum building works (deepening the lake basins if such is the case, putting up a motor-pump, building access roads used for extinguishing fires). At the same time, in choosing the location, we took into account the strict protection of the two areas from the Small Wetlands of the Braila reservation, and they have not been equipped with reservoirs (Fig. 13.16).

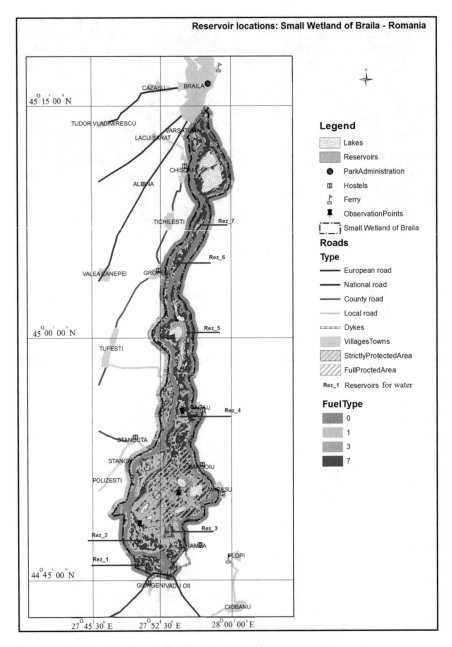

Fig. 13.16 Reservoir locations in the Small Wetland of Braila, Romania

Table 13.2 The features of the reservoirs suggested for use in case of a vegetation fire

No.	The name of the reservoirs	Potential water volume (mil. mc.)	Area (ha)	Perimeter (m)
1.	Rez_1	0.070	4.66	1141
2.	Rez_2	0.266	17.70	2502
3.	Rez_3	0.226	15.06	1817
4.	Rez_4	0.358	23.84	2850
5.	Rez_5	5.78	385.2	14,102
6.	Rez_6	0.079	5.29	1850
7.	Rez_7	0.227	15.16	1744

Considering the great resource of water in the studied area, the number of water sources used for extinguishing the possible fires can be supplemented with mobile motor-pumps. The motor-pumps can collect the water straight form the Danube branches or from the numerous lakes or ponds, which have been identified as being 20, the most important of them being: the Lupascu, the Begu and the Toplosca Ponds, the Melinte, Sbenghiozul, Lupoiu and Curcubeu Lakes, The Zaina, Cucova de Jos Mountain Lakes, as well as the Iezerul Popii, Misailla, and Fundul Mare Lakes. The water volume of the seven permanent reservoirs has been estimated to a total volume of 7 million m^3 (Table 13.2).

13.5 Conclusions

Running the ArcSwat application emphasizes the local characteristics of wetlands with a negative balance of the water flow taking into consideration important surfaces occupied by water, which determines important evaporation between the 1st of March and the 30th of October. Difficulties in applying this method include the difficulty of accurately determining hydrographic basins in the context of an extremely low slope of the terrain and a high density of the hydrographic network (branches, canals, lakes) [53].

In a thematic context, the hydrological information determined for this territory must be aggregated with data on forest vegetation, i.e. the development of fuel map and fire risk mapping [54]. Considering the area specific, we believe that, from the point of view of the water resource, necessary for fire extinguishing there should not be imposed special storage conditions, and we suggest the building of reservoirs (basins) of the earth with the depth of approx. 4 m, supplied with the water from the phreatic aquifer in the areas with greater fire risk, from where there can be supplied the necessary water for extinguishing vegetation fires in the Small Braila Wetlands, by pumping with individual groups of pumps and portable electrical generators [55].

13.6 Recommendations

Romania, compared to Europe, does not face large-scale vegetation fires. But it is necessary to prepare an adequate infrastructure in order to prevent and mitigate its effects. Global climate change by emphasizing extreme phenomena and protecting natural areas is the premise of these new approaches.

The implementation of these projects can be done gradually, especially within the areas that are difficult to access. Staging development will offer the opportunity to optimize the technical solutions implemented.

Acknowledgements This research is being developed by the project "Utilizing stream waters in the suppression of forest fires with the help of new technologies" acronym *Streams-2-Suppress-Fires"*, *cod 2.2.2.73841.323_MIS*-ETC 2666.

References

1. Trigo RM, Pereira J, Pereira MG, Mota B, Calado TJ, Dacamara CC, Santo FE (2006) Atmospheric conditions associated with the exceptional fire season of 2003 in Portugal. Int J Climatol 26:1741–1757
2. Bowman DMJS, Williamson GJ, Abatzoglou JT, Kolden CA, Cochrane MA, Smith AMS (2017) Human exposure and sensitivity to globally extreme wildfire events. Nat Ecol Evol 1:58. https://doi.org/10.1038/s41559-016-0058
3. Cogeca C (2003) Assessment of the impact of the heat wave and drought of the summer 2003 on agriculture and forestry. Comm Agric Organ. Eur Union Gen Comm Agric Coop Eur, Union Bruss, p 15
4. De Bono A, Peduzzi P, Kluser S, Giuliani G (2004) Impacts of Summer 2003 Heat Wave in Europe
5. Blarquez O, Ali AA, Girardin MP, Grondin P, Fréchette B, Bergeron Y, Hély C (2015) Regional paleofire regimes affected by non-uniform climate, vegetation and human drivers. Sci Rep 5:13356
6. Turco M, Bedia J, Di Liberto F, Fiorucci P, von Hardenberg J, Koutsias N, Llasat M-C, Xystrakis F, Provenzale A (2016) Decreasing fires in mediterranean Europe. PLoS ONE 11:e0150663. https://doi.org/10.1371/journal.pone.0150663
7. Ruffault J, Martin-StPaul NK, Rambal S, Mouillot F (2013) Differential regional responses in drought length, intensity and timing to recent climate changes in a Mediterranean forested ecosystem. Clim Change 117:103–117
8. Zvijáková L, Zeleňáková M (2013) The proposal of procedure used in the process of environmental impact assessment for water management. In: Public recreation and landscape protection—with man hand in hand…. Mendel University in Brno, Czech Republic, Brno, pp 205–221
9. Galindo G, Batta R (2013) Review of recent developments in OR/MS research in disaster operations management. Eur J Oper Res 230(2):201–211
10. UNISDR (2015) Sendai framework for disaster risk reduction 2015–2030 (Report)
11. Ntaimo L, Gallego-Arrubla JA, Gan J, Stripling C, Young J, Spencer T (2013) A simulation and stochastic integer programming approach to wildfire initial attack planning. For Sci 59(1):105–117
12. Buijse AD, Coops H, Staras M, Jans L, Van Geest G, Grift R, Ibelings BW, Oosterberg W, Roozen FC (2002) Restoration strategies for river floodplains along large lowland rivers in Europe. Freshw Biol 47(4):889–907

13. Romanescu G, Stoleriu CC, Enea A (2013) Limnology of the Red Lake. An Interdisciplinary Study. Springer-Verlag, Dordrecht, New York,LLC, Romania
14. Romanescu G, Sandu I, Stoleriu C, Sandu IG (2014) Water resources in Romania and their quality in the main lacustrine Basins. Rev Chim (Bucharest) 65(3):344–349
15. Zelenakova M, Zvijakova L, Hlavata H (2017) Risk analysis in environmental impact assessment, Public Recreation and Landscape Protection, pages: 317–322
16. Kahit FZ, Zaoui L, Danu MA, Romanescu G, Benslama M (2017) A new vegetation history documented by pollen analysis and C14 dating in the alder of Ain Khiar—El Kala wet complex, Algeria. Int J Biosci 11(6):192–199. doi:http://dx.doi.org/10.12692/ijb/11.6.192-199
17. Romanescu G, Cojocaru I (2010) Hydrogeological considerations on the western sector of the Danube Delta—a case study for the Caraorman and Saraturile fluvial-marine levees (with similarities for the Letea levee). Environ Eng Manag Journal 9(6):795–806
18. Romanescu G (2013) Geoarchaeology of the ancient and medieval Danube Delta: modeling environmental and historical changes. A review. Quatern Int 293:231–244
19. Romanescu G (2013) Alluvial transport processes and the impact of anthropogenic intervention on the Romanian Littoral of the Danube delta. Ocean & Coastal Manage 73:31–43
20. Romanescu G (2014) The catchment area of the Milesian colony of Histria, within the Razim-Sinoie lagoon complex (Romania): hydro-geomorphologic, economic and geopolitical implications. Area 46(3):320–327
21. Romanescu G, Stoleriu C, Lupascu A (2012) Biochemistry of wetlands in barrage Lacul Rosu catchment (Haghimas - Eastern Carpathian). Environ Eng Manage J 11(9):1627–1637
22. Romana Academia (2006) Geografia Romaniei, volumul V—Campia Romana, Dunarea, Podisul Dobrogei. Litoralul romanesc al Marii Negre si Platforma Continentala, Editura Academiei Romane, Bucuresti
23. Conea I (1957) Din geografia istorica a baltilor Ialomitei si Brailei. Probleme de Geografie 4
24. Dimitriu RG, Melinte-Dobrinescu MC, Pop IC, Varzaru C, Briceag A (2010) Geomorfologia Dunarii in arealul Parcului Natural Insula Mica a Brailei. Editura Eokon, Cluj-Napoca
25. Ielenicz M, Patru I (2005) Geografia fizica a Romaniei. Editura Universitara, Bucuresti
26. Romanescu G (2005) Morpho-hydrographical evolution of the Danube Delta, II, Management of water resources and coastline evolution. Land use and the ecological consequences. Editura Terra Nostra, Iasi
27. Romanescu G (2006) Complexul lagunar Razim-Sinoie. Editura Universitatii Alexandru Ioan Cuza, Iasi, Studiu morfohidrografic
28. Burtea MC, Sandu IG, Cioromele GA, Bordei M, Ciurea A, Romanescu G (2015) Sustainable Exploitation of Ecosystems on the Big Island of Braila. Rev Chim (Bucharest) 66(5):621–627
29. Burtea MC, Ciurea A, Bordei M, Romanescu G, Sandu AV (2015) Development of the potential of ecological agriculture in the Village Ciresu. County of Braila Rev Chim (Bucharest) 66(8):1222–1226
30. Romana Academia (1969) Geografia Vaii Dunarii Romanesti. Editura Academiei Romane, Bucuresti
31. Gore JA, Shields FD (1995) Can large rivers be restored? BioScience 142–152
32. Romanescu G (1996) Delta Dunarii. Studiu morfohidrografic, Editura Corson, Iasi
33. Romanescu G, Pascal M, Pintilie-Mihu A, Stoleriu CC, Sandu I, Moisii M (2017) Water quality analysis in wetlands freshwater: common floodplain of Jijia-Prut Rivers. Rev Chim (Bucharest) 68(3):553–561
34. Stagl JC, Hattermann FF (2016) Impacts of climate change on riverine ecosystems: alterations of ecologically relevant flow dynamics in the Danube River and its major Tributaries. Water 8(12):566. https://doi.org/10.3390/w8120566
35. Lévesque É, Anctil F, Griensven AV, Beauchamp N (2008) Evaluation of streamflow simulation by SWAT model for two small watersheds under snowmelt and rainfall. Hydrol Sci J 53(5):961–976
36. Diaconu DC, Tiscovschi A, Mailat E (2013) GIS application in the management of reservoir lakes case study: the series of lakes Valcele, Budeasa, Bascov, Arges river. Geoconference on Water Resources, Forest, Marine and Ocean Ecosystems Book Series: International Multidisciplinary Scientific GeoConference-SGEM, pp 165–171

37. Mierla M, Romanescu G, Nichersu I, Grigoras I (2015) Hydrological risk map for the danube Delta-a case study of floods within the fluvial Delta. IEEE J Selected Topics Appl Earth Obs Remote Sens 8(1):98–104. https://doi.org/10.1109/JSTARS.2014.2347352

38. Romanescu G (2009) Siret river basin planning (Romania) and the role of wetlands in diminishing the floods. WIT Trans Ecol Environ 125:439–453. https://doi.org/10.2495/WRM090391

39. Vadineanu A, Adamescu MC, Vadineanu RS, Cristofor S, Negrei C (2003) Past and future management of Lower Danube Wetlands System: a bioeconomic appraisal. J Interdisciplinary Econ 14:415–447

40. WWW Danube-Carpathian Programme (1999) Evalution of wetlands and foodplain areas in the D.R.B. (final report 27 May 1999); Danube Polution Reductioon Programme. WWW Danube-Carpathian Programme

41. Arnold JG (2010) SWAT: model use, calibration, and validation. Trans ASABE 55(4):1491–1508

42. Chilikova-Lubomirova, M (2016) River systems and water related extremes with respect to drought. SGEM: water, resources, forest, marine and ocean ecosystems conference proceedings, vol I, International multidisciplinary scientific GeoConference-SGEM, pp 621–628

43. Houghton JT, Ding Y, Griggs DJ, Noguer M, van der Linden PJ, Dai X, Maskell K, Johnson CA (2001) Climate change 2001: the scientific basis. Contribution of working group I to the third assessment report of the intergovernmental panel on climate change. Cambridge University Press, Cambridge

44. Kundzewicz Z (2008) Climate change impacts on the hydrological cycle. Ecohydrol Hydrobiol 8(2–4):195–203. https://doi.org/10.2478/v10104-009-0015-y

45. Nasta P, Palladino M, Ursino N, Saracino A, Sommella A, Romano N (2017) Assessing long-term impact of land-use change on hydrological ecosystem functions in a Mediterranean upland agro-forestry catchment. Sci Total Environ 605:1070–1082. https://doi.org/10.1016/j.scitotenv.2017.06.008

46. https://climate.nasa.gov/evidence/. Accessed 07 Nov 2015

47. Pintilii RD, Andronache I, Diaconu DC, Dobrea RC, Zelenáková M, Fensholt R, Peptenatu D, Draghici CC, Ciobotaru AM (2017) Using fractal analysis in modeling the dynamics of forest areas and economic impact assessment: Maramures, County, Romania, as a Case Study. Forests 8(25). https://doi.org/10.3390/f8010025

48. Rigden AJ, Li D (2017) Attribution of surface temperature anomalies induced by land use and land cover changes. Geophys Res Lett 44(13):6814–6822. https://doi.org/10.1002/2017GL073811

49. https://www.epa.gov/heat-islands/using-trees-and-vegetation-reduce-heat-islands. Accessed 07 Nov 2015

50. Romanescu G (2016) Tourist exploitation of archaeological sites in the Danube Delta Biosphere Reserve area (Romania). Int J Conserv Sci 7(3):683–690

51. International Commission of River Danube Protection (ICPDR) (2009) Danube River Basin District Management Plan

52. International Commission of River Danube Protection (ICPDR) (2013) Assessment report on hydropower generation in the Danube basin. Vienna, 2013. http://www.icpdr.org/main/sites/default/files/nodes/documents/hydropower_assessment_report_danube_basin_-_final.pdf.140pp. Accessed 23rd Dec 2014

53. Mitsopoulos I, Mallinis G, Zibtsev S, Yavuz M, Saglam S, Kucuk O, Bogomolov V, Borsuk A, Zaimes G (2016) An integrated approach for mapping fire suppression difficulty in three different ecosystems of Eastern Europe. J Spatial Sci 1–17:00001. https://doi.org/10.1080/14498596.2016.1169952

54. Zaimes GN, Tufekcioglu M, Tufekcioglu A, Zibtsev S, Corobov R, Emmanouloudis D, Uratu R, Ghulijanyan A, Borsuk A, Trombitsky I (2016) Transboundary collaborations to enhance wildfire suppression in protected areas of the Black Sea region. J Eng Sci Technol Rev 9(1):108–114

55. Water Framework Directive (2000) Water Framework Directive 2000/60/EC of European Parliament and European Commission. European Community Official Journal

Chapter 14
Assessment of Some Diurnal Streamwater Profiles in Western and Northern Romania in Relation to Meteorological Data

Andrei-Emil Briciu, Dinu Iulian Oprea, Dumitru Mihăilă, Liliana Gina Lazurca (Andrei), Luciana-Alexandra Costan (Briciu) and Petruț-Ionel Bistricean

Abstract Water and air measurements were conducted in river valleys of Romania to detect the shapes of diurnal profiles and their spatial variations. The studied river water parameters are pressure/level, temperature, and electrical conductivity. The air parameters, used for understanding the diurnal water profiles are pressure, temperature and relative humidity. Time intervals used in this study vary from few weeks to few months and sites are grouped depending on common time intervals for comparison purposes. The selected water monitoring sites have similar diurnal shapes of the studied parameters in areas with the natural flow (afternoon maximum water level and temperature and minimum electrical conductivity; the opposite events occur early in the morning) and disturbed evolutions in areas where dams and hydroelectric plants exist. The natural particular monitoring sites characteristics can also significantly impact the results of measurements. The mean diurnal water profiles, obtained from detrended time series, can be used for theoretical models.

Keywords Natural flow · Human impact · Thermal water input · Detrending

14.1 Introduction

Rivers are a very dynamic part of the hydrosphere. They have many oscillations (of their discharge/level and other parameters) caused by the Earth revolution (that generates the annual cycle with its spatio-temporal variations), climate teleconnections and actual climate changes. However, beyond these well-known variations, rivers also show a cyclic diurnal evolution that partly overwrites every major trendline.

A.-E. Briciu (✉) · D. I. Oprea · D. Mihăilă · L. G. Lazurca (Andrei) · L.-A. Costan (Briciu)
Department of Geography, Ștefan cel Mare University of Suceava (USV), Suceava, Romania
e-mail: andreibriciu@atlas.usv.ro

P.-I. Bistricean
Suceava Weather Station, National Meteorological Administration (ANM), Suceava, Romania

© Springer Nature Switzerland AG 2020
A. M. Negm et al. (eds.), *Water Resources Management in Romania*, Springer Water,
https://doi.org/10.1007/978-3-030-22320-5_14

The diurnal cycle in rivers is a well-known feature that is studied for a long time, but it is often revealed only by high temporal resolution measurements done with instruments able to record fine changes in the studied water parameter (e.g. millimetric variation in water level). Measurements of water parameters done multiple times per day (at least hourly) are a must because low temporal resolution measurements (such as those done once or twice per day) do only record moments of the day that represents only a small length on the curve line of the diurnal cycle. Such low-resolution measurements generate daily values on the longer time series that may be far from being representative for the recorded day. Even the average value of the diurnal evolution of a streamwater parameter measured with high temporal resolution does not offer more than a (reasonably) distorted interpretation of the selected time interval.

Because of the permanent mixing of the diurnal cycle with the (periodic or aperiodic) multi-diurnal evolutions, every day has a unique streamwater evolution in the 24 h cycle. Groups of days or separate days may record similar diurnal shapes in the evolution of a studied streamwater parameter, just like in the case of the evolution of an atmospheric parameter. However, unlike the air, the streamwater recorded in a section tells the story of the entire catchment upstream, which is a much stronger control factor for a river than for the air above.

Not only consecutive days on a river have different diurnal shapes, but also different rivers have different diurnal evolutions in the same day, as a mark of their different catchments. As every particular diurnal evolution is unique and heavily influenced by other evolutions, a question arises: which is the real/denoised shape of the diurnal cycle of a river at given point in space? Because of the Heisenberg's uncertainty principle, a pure diel signal may never be guessed (in our case: if all supradaily/macro-oscillations are removed from time series, which mean value remains to be changed by the daily cycle (and will it still exist)?).

The purpose of this chapter is to extract abstract/mean diurnal cycles of various streamwater parameters in different catchments, river sections and time intervals in western and northern Romania. The selected territory does not have studies that try to obtain relevant shapes of water parameters of the selected rivers. Our study is useful for those that undertake case studies about stream waters (by using our methodology) and for researchers interested in Romanian waters (by using our examples and results).

The meteorological data in this study can be used to understand the water behavior. Summary studies such as those of Poole and Berman [23] or Nimick et al. [21] indicated that the diurnal cycle of temperature in stream water is linked to the air temperature and relative humidity which are, amongst others, control factors.

14.2   Review of the Literature Describing Diurnal Streamwater Profiles

There are multiple diurnal cycles in stream water, corresponding to the multiple parameters that may be taken into account, and all cycles are interdependent [21]. For example, the streamwater temperature, which has higher amplitudes during the summer, baseflow and in shallow streams [22, 30], is higher during the daytime and decreases the water viscosity. The decreased viscosity generates improved water infiltration into the ground, leading to decreased streamflow during this time of the day [14]. This effect is similar to that generated by the increased evapotranspiration during the daytime [2].

The most numerous studies about diel cycles in rivers are about river level and/or discharge. This is due, from a historical point of view, to the earlier availability of the instruments needed for measuring these parameters and to the strict necessity to measure them firstly. Secondly, studies about diel thermal cycles follow, mainly due to the ease of technical access to the required instruments (specific thermometers and loggers). Other physical, chemical and even biological properties of river waters were also measured to observe if diurnal periodicities exist (e.g. turbidity, electrical conductivity, pH, oxidation-reduction potential, dissolved oxygen, chlorophyll).

The diurnal cycle of a streamwater parameter is similar to a sinusoid [9] or a smoothed saw-teeth evolution of data values (this is due to the asymmetric behavior in the increase and decrease processes). Troxell [27], Wicht [31], Burger [11], Rycroft [25] and Tschinkel [28] observed the diel variations in river level and linked them to the similar evolution in the groundwater nearby. Callède [13] observed that, in the absence of atmospheric precipitation, the diurnal evolution of river discharges is modelled by the diurnal cycle of the air temperature and sunshine, mainly through evapotranspiration. High waters do often miss diurnal oscillations, which are almost always detectable during base flow (between rain events).

The diurnal evolution of a river parameter can also exhibit natural periodicities due to direct water evaporation, atmospheric precipitation, thawing, freezing, tides [13, 15, 18] and even due to occult precipitation [13]. Oceanic tides are causing oscillations in tidal rivers. Earth and atmospheric tides cause semidiurnal oscillations in rivers by using the groundwater tides as a proxy [3, 5, 8, 16].

Not only the natural factors but also the anthropogenic ones are the cause of diurnal and subdiurnal oscillations in the evolution of a streamwater parameter. Water abstraction from groundwater is one of the most widespread human indirect intervention on river flow characteristics [20, 29], in our case because of the daily cycle of human activities. Dams have the most evident impact on rivers, especially due to regular flow change and hypolimnetic water releases, which affects temperature [24]. Regulated rivers do often have human-induced periodic oscillations that override entirely the natural oscillations. Wastewaters from urban areas influence water flow, temperature and chemistry on a daily cycle [4, 17].

To explain the evolution of a streamwater parameter, meteorological measurements are involved. Benyahya et al. [1] observed that microclimate data collected near the river gauge is more effective in explaining the river behavior than the regional climate data from an official meteorological network.

Lundquist and Cayan [18] showed the wide variety of streamwater flow profiles across USA caused by diverse natural factors. Gribovszki et al. [15] have written a comprehensive review on the literature about diurnal oscillations in streamflow rates. Studies on the river temperature are also reviewed in recent papers [12, 30].

The diurnal evolution of the rivers analyzed in this chapter was previously directly or indirectly discussed in scientific articles of Briciu and Oprea-Gancevici [9] and Briciu et al. [6–8, 10]; rivers fed by thermal groundwaters have a different local thermal regime due to the quasi-constant temperature of such waters.

The good understanding of the diurnal cycle is useful for good prediction of a streamwater parameter, such as temperature [26].

14.3 Geographic Description of the Study Area and Measurement Sites

14.3.1 General Description of the Study Area

The study area is located in the Romanian Carpathians and the Moldavian Plateau (Fig. 14.1). 2/3 of the 21 sites used for measuring rivers are placed in the mountain area. The terrain elevation of the chosen sites ranges from 118 m above sea level (a.s.l.) (minimum, Cerna River downstream of Băile Herculane) to 810 m a.s.l. (maximum, Dorna River upstream of Vatra Dornei). In general, the selected sites can be found in river valleys, at elevations between 300 and 400 m a.s.l.

The geology of the studied sites is dominated by carbonaceous rocks in the studied areas of western Romania and site number 21 (Şugău Gorge). In sites 1–8, volcanic rocks are also present. Sites 9–20 are dominated by sedimentary and metamorphic rocks; friable sedimentary rocks are present in the plateau sites.

The climate of Romania is temperate continental with more precipitation in the sites from Apuseni Mountains (Western Romanian Carpathians, sites 1–5) and Southern Carpathians (sites 6–8) than in Outer Eastern Carpathians (sites 9, 10, 16, 18, 20, 21) and Moldavian Plateau (sites 11–14, 15, 17, 19). The rainfall amount also greatly varies with terrain elevation, slope aspect and the presence or absence of foehn; thus, the annual sum of precipitation ranges from around 550 mm (some plateau areas, mountain valleys with foehn) to approximately 800 mm (higher valleys, on slopes with western aspect).

The air thermal amplitude is generally higher in the areas with fewer atmospheric precipitations. In the area around Geoagiu-Băi settlement (sites 1–3), the mean annual air temperature was 10.6 °C during 1961–2010 [7]. In the area around Moneasa settlement (sites 4 and 5), the mean air temperature was 7.9 °C during 1961–2010 [6], while, in the area around Băile Herculane settlement (sites 6–8), the mean air temperature was 8 °C during the same time interval [10]. The sites in northern

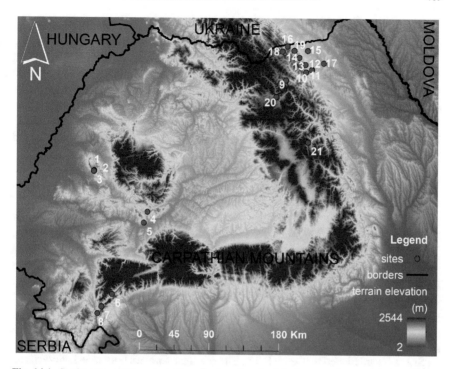

Fig. 14.1 Study area and the selected monitoring sites: 1—Megheş upstream spring, 2—Megheş downstream spring, 3—Băilor Creek, 4—Cib before Băcâia, 5—Bobîlna at Feredee, 6—Cerna below Cerna Sat, 7—Cerna before Seven Hot Springs, 8—Cerna downstream Băile Herculane, 9—Moldova at Pojorâta, 10—Humor near Mănăstirea Humor, 11—Soloneţ at Cacica, 12—Blândeţu at Cacica, 13—Solca at Solca, 14—Sucevița before Bercheza, 15—Pozen at Horodnic, 16—Suceava at Straja, 17—Suceava at Dărmănești, 18—Brodina before Brodinioara, 19—Vițău before Putna, 20—Dorna before Vatra Dornei, 21—Şugau before Bicaz

Romania (9–21) have, in general, mean air temperature of 5–6 °C in the selected mountain areas and of 7–8 °C in the plateau areas [19].

All sites in Romania have multiple consecutive days with negative air temperature during winter, especially those in the northern studied territory, and this has an important impact on diurnal streamwater profiles because of the ice formation. Winter in western Romania is warmer, and rivers in sites of this part of Romania are often receiving thermal waters (as thermal springs or groundwater seepage).

14.3.2   Measurement Sites

A site is composed of the space where water measuring sensor was placed and the space nearby where the air measuring sensor(s) was/were placed too. Usually, the two types of sensors are placed within 20 m of each other, but, in some few cases, an air sensor placed much farther recorded more relevant data and was used instead of the closer air sensor (which may have been missing or malfunctioned).

The most important rivers (based on their discharge in the last 60 years) measured for this study are Suceava (at Dărmănești, ~16.5 m³/s) and Cerna (at Pecinișca (Băile Herculane), ~14 m³/s). Most studied rivers have mean annual discharges under 1 m³/s at the selected measuring sites. On some small rivers (such as Dorna and Bobîlna, in the selected sites), during some winters or summers, the water flow almost ceases because of the water freezing or, respectively, because of the missing rainfalls and the high evapotranspiration.

Sites 1 and 2 were established on Megheș River, a tributary of Moneasa River. The two sites are at only 0.33 km from each other and downstream water sensor was meant to measure the impact of a hypothermal spring found between the two sites.

Site 3 was placed on Băilor Creek, another tributary of Moneasa, as it is a mix of cold waters from Grota Ursului Cave and thermal waters/springs from Băilor Valley.

Site 4 is represented by Cib River upstream of Băcâia settlement and downstream of Cib Gorge. Air pressure was measured few meters away from the water monitoring point, but air temperature and relative humidity were selected from inside the gorge upstream (~1 km away) to be not affected by the nearby settlement.

Site 5 is Bobîlna River at Feredee, where hypothermal springs can be found; air data was completed (when missing) with measurements done at Rapolțel (4.5 km away).

Sites 6–8 are placed on Cerna River. Site 6 is upstream of Herculane Dam/Prisaca Lake and is not influenced by its flow regulation, but has flow regulated by Valea lui Iovan Lake and Dam, found 15 km upstream. Site 7 is placed in Seven Hot Springs area where Cerna River receives water from hyperthermal springs. Site 8 measures the effect of Băile Herculane city (including the numerous thermal springs of the urban area) on the temperature of the homonymous river.

Site 9 measures the thermal regime of Moldova River, being found between run-of-the-river power plants and having heavily modified discharge. Site 10 is placed on Humor, a tributary of Moldova River with a natural flow.

Sites 11–13 represent measurements done on rivers with significant salt input and the surrounding air. 11 is represented by Soloneț River, while 12 is represented by Blîndețu, a tributary of Soloneț. Site 13 is found in Solca River Valley.

Sites 14, 15, and 18 are found on other rivers that, like Soloneț and Solca, are tributary of Suceava River: 14—Sucevița, 15—Pozen, 18—Brodina. Site 19 is placed on Vițău before it flows into Putna, another tributary of Suceava River. Suceava River is measured in sites 16 and 17.

Site 20 is placed on Dorna River before Vatra Dornei city and Roșu Dam (also before flowing into Bistrița River). Site 21 was meant to measure Șugău River inside Șugău Gorge; Șugău is a tributary of Bicaz, which is a tributary of Bistrița River.

14.4 Instruments, Data and Methods

14.4.1 Instruments

Water parameters used in this study are temperature, electrical conductivity and pressure/level. Air parameters are temperature, relative humidity, and atmospheric pressure. Air and water temperature measurements exist for all sites, while other parameters were used preferentially (due to the limited number of instruments).

In order to measure water temperature, we used iButton (DS1922L-F5#—$\pm$0.5 °C accuracy, 0.0625 °C resolution) and TruBlue (585 CTD—$\pm$0.2 °C accuracy, 0.01 °C resolution) sensors and loggers. The TruBlue 585 CTD model was also used for measuring the electrical conductivity (1% of reading accuracy, 1 μS/cm resolution) and water pressure ($\pm$0.05% full-scale accuracy, 0.01% full-scale resolution; range: 0–10.8 m H_2O). The electrical conductivity values are automatically corrected for a reference temperature (25 °C). The pressure recorded by TruBlue was total pressure (air + water column) and water pressure created by the water column above the pressure sensor was obtained by using the atmospheric pressure data (difference).

Air temperature was measured with DS1921G-F5# ($\pm$1 °C accuracy, 0.5 °C resolution) and DS1923-F5# iButtons ($\pm$0.5 °C accuracy, 0.0625 °C or 0.5 °C resolution (depending on settings, in order to gain logger memory or measurement resolution) also used for recording the air relative humidity with a resolution of 0.6%) and with TruBlue 275 Baro instruments ($\pm$0.1 °C accuracy, 0.01 °C resolution). TruBlue 275 Baro was used for logging the atmospheric pressure too ($\pm$0.05% full-scale accuracy, 0.01% full-scale resolution; range: 8–16 psia).

14.4.2 Data

Water temperature measurements were done in all 21 sites of this study (17 rivers). Air temperature measurements were also conducted at all sites (Fig. 14.2). Only some selected rivers and sites in western Romania have water electrical conductivity and pressure measurements (1, 4, 6, 7). Air relative humidity measurements were conducted in all sites of the western part of Romania (excepting site 7) and also in site number 9.

Measurements in our sites were done during different time intervals, which sometimes are partly overlapping. In order to create groups of sites with common time intervals (usually sites that are in the same area), we cut the original time series to shorter, but common time series. All time intervals begin at 00:00 and end before midnight. All data values were recorded at regular intervals, that vary from every 10 min to every hour.

The studied time interval of sites 1–5 is October 5th, 2016–July, 17th, 2017 (site no. 4 has two time intervals with missing data: Nov. 21st–Dec. 1st, 2016; Jun.

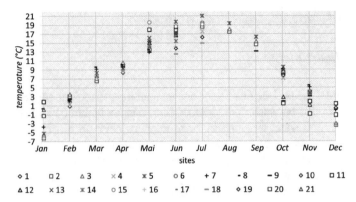

Fig. 14.2 Variability of average monthly air temperatures of the studied sites (obtained from days with measurements): 1—Megheș upstream spring, 2—Megheș downstream spring, 3—Băilor Creek, 4—Cib before Băcâia, 5—Bobîlna at Feredee, 6—Cerna below Cerna Sat, 7—Cerna before Seven Hot Springs, 8—Cerna downstream Băile Herculane, 9—Moldova at Pojorâta, 10—Humor near Mănăstirea Humor, 11—Soloneț at Cacica, 12—Blândețu at Cacica, 13—Solca at Solca, 14—Sucevița before Bercheza, 15—Pozen at Horodnic, 16—Suceava at Straja, 17—Suceava at Dărmănești, 18—Brodina before Brodinioara, 19—Vițău before Putna, 20—Dorna before Vatra Dornei, 21—Șugau before Bicaz temperature (°C)

22nd–26th, 2017). Data in sites 1 and 4 were recorded every 15 min, while data in the other sites were recorded every 30 min.

The time interval of sites 7 and 8 is shorter than that of the previously mentioned sites: October 5th, 2016–May 1st, 2017. Data were recorded every 15 min at site 7 and every 30 min in site 8. Data in site 6 were recorded every 15 min during May 3rd, 2017–October 6th, 2017 time interval.

Data in site 9 were recorded every hour in the following interval: September 3rd, 2017–February 2nd, 2018.

Sites 10–15 have a common time interval (May 18th, 2011–June 13th, 2011) and sampling rate (every 10 min). However, some sites have common time intervals with other sites too. Thus, site 15 has common time interval with sites 16–19 (June 29th, 2011–July 22nd, 2011) and a sampling rate of 1 measurement every 10 min. Site 11 also has common time interval (October 27th, 2011–January 4th, 2012) and sampling rate (every 30 min) with sites 20 and 21, having measurements done during the same measurements campaign.

The 2011–2012 campaign was the only one that used a resolution of 0.0625 °C for the air temperature measurements. The air temperature resolution of 0.01 °C was used in sites that had TruBlue sensors (1, 4, 6 and 7).

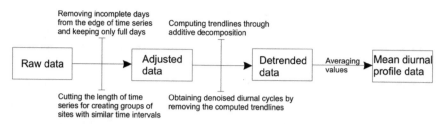

Fig. 14.3 Flowchart depicting the general steps used in obtaining the mean diurnal profiles

14.4.3   Methodology

The water sensors were placed in rivers with good water mixing at distances of up to 1 m from banks and under water columns that usually varied from 0.1 to 1 m thickness. The air sensors were generally placed at 2 m above terrain in shadowed places and recipients made for avoiding raindrops and sunlight.

Raw data consist of consecutive days with measurements made during all 24 h of a day. The general steps used for obtaining mean diurnal profiles are described in Fig. 14.3. All data are set to EET—Eastern European Time (Standard Time); measurements done during EEST—Eastern European Summer Time (Daylight Saving Time) had the time converted to EET. A common and unique time standard across all sites is necessary in order to avoid the mismatch of processes, such as the moment when peaks in the diurnal evolution occur.

In order to observe an abstract diurnal evolution of a parameter, we applied a detrending to the time series by using XLSTAT software. Detrending is done by, firstly, obtaining a relevant trend of the entire time series through an additive seasonal decomposition which uses a period equal to the length of a day and by, secondly, removing this trend from the raw time series. Additive decomposition was used instead of the multiplicative one because the diurnal cycle has not amplitudes that increase over time.

As a technical note, because of the length used as period for the detrending process, the trend and detrended time series are shorter than raw data with the equivalent of 1 day. This happens because the first and last 12 h of the selected time interval are removed during the detrending process. Trend obtained through this method cannot be detected at the ends (12 h) of time series (at least not with high significance).

Figure 14.4a shows the trend data close following of raw data so that every increase of water level/pressure of water column due to rainfall or every decrease due to groundwater depletion is removed for insulating the diurnal signal. As a result, detrended data are a succession of positive and negative values because of the mathematical operation of differencing (Fig. 14.4b). Detrended data lose their absolute values, but retain amplitudes (however, they are neither standardized nor normalized time series). Detrended data are used for obtaining mean diurnal evolution of a parameter. In order to preserve the original shape of data, time series smoothing

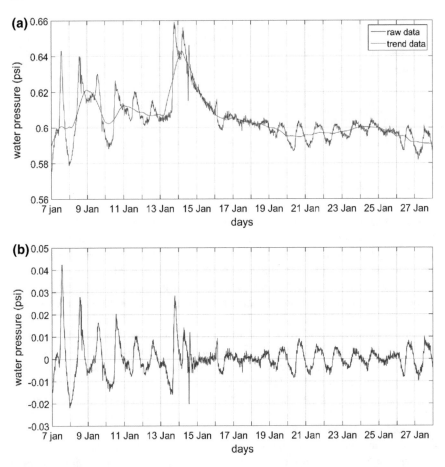

Fig. 14.4 Example of products used in the process of obtaining mean profiles, **a** raw and trend time series of water level in site 1 of Megheş River during 7–27 January 2017; **b** the detrended data resulted as trend is removed from raw data

was applied only indirectly by averaging all days for the final profile. This study is focused on analyzing diurnal profiles obtained from detrended data.

For standardization and comparison purposes in the discussion sections, the moments when maxima and minima occur are rounded, in some cases with higher frequency of measurements, with maximum ±15 min and the chosen time step is the half hour.

The mean profile relative values obtained from a detrended time series can be added to the mean of the selected raw time series (singular value added to a vector of values) to obtain a diurnal profile of absolute values that has a 99% fit to the raw diurnal profile. This operation can be applied to the entire time series that have been detrended or to sections of it (in this case, the mean of the raw time series section will be added to the profile of the same section).

The relative diurnal profiles are useful for comparison of different rivers (because, in raw format, they have different raw and amplitude values) and for using them as a theoretical mean evolution in various simulation models.

14.5   Diurnal Streamwater Profiles

Water temperature of Megheș River increased from 8.23 °C, at site 1, to 10.41 °C, at site 2, a big increase of 2.18 °C despite the short distance between the two sites (0.33 km; increase caused by thermal water input). The water temperature shape is similar in both sites (Figs. 14.5 and 14.6a), with minima at 7:30–8:00 and maxima at 15:00–15:30 (the times of maxima and minima are similar in air; mean air temperature is 6.52 °C in site 1 and 9.69 °C in site 2, closer to Moneasa settlement). The earlier minimum and maximum is recorded upstream (site 1), and there is a higher thermal amplitude in site 2 (in air and water). One supplementary/partial explanation for

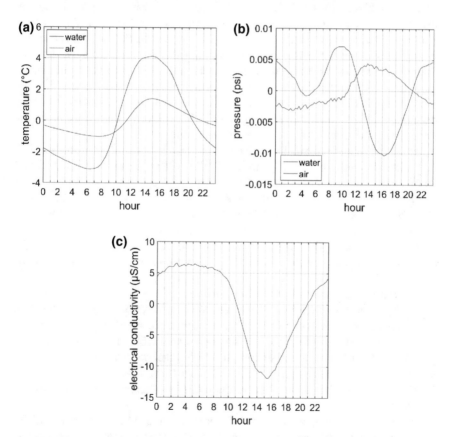

Fig. 14.5 Mean profiles of parameters in site 1: **a** temperature, **b** pressure, **c** electrical conductivity

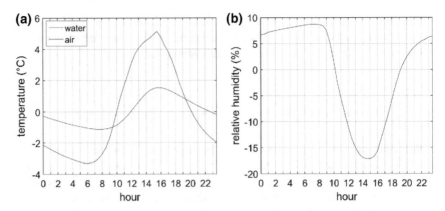

Fig. 14.6 Mean profiles of parameters in site 2: **a** temperature, **b** relative humidity

the higher amplitude and temperature in water is the placement of the water sensor near the right bank (lower depth; site 2). The water sensor in site 1 was placed near the center of stream, on the line with the maximum flow speed, at higher depth. However, streambed has a maximum 2 m width, a maximum 0.75 m depth and water is very well mixed due to the high streambed slope and vertical irregularities. The shape of the diurnal air profile is different in the selected sites because only the air temperature of the left bank (closer to site 2) was used for the downstream site, while an air temperature average from both banks was used in the analysis of site 1 (these banks have different forest coverage and type). The observed differences indicate the importance of choosing the proper measurement site.

As the temperature rise in the afternoon, the air relative humidity (mean: 90.02%) and water electrical conductivity (mean: 489.39 µS/cm) decrease (Figs. 14.5 and 14.6), as there is an inverse relationship between these parameters and the air/water temperature. Because conductivity values are already corrected for temperature changes, the observed diurnal oscillation is caused by variations from groundwater flow, which is influenced by evapotranspiration in the studied catchment. The water level (mean: 0.608 psi) has a maximum after midday, uncorrelated with the atmospheric pressure (mean: 14.245 psi), which shows evident atmospheric tides (Fig. 14.5b).

The diurnal oscillation of water temperature in site 3 is much weaker than in sites 1 and 2 because of the cold waters that spring nearby, from Grota Ursului Cave (karst groundwater with weak diurnal oscillation) and because of the thermal springs that fed the stream water with waters having no detectable diurnal oscillation [6] (Fig. 14.7). The maxima of temperatures in air and water occur at 14:00, and minima at 6:30 (mean water temperature is 15.32 °C, while the equivalent air temperature is 10.21 °C); the air relative humidity (mean: 89.07%) is in almost perfect antiphase with air temperature.

Water minimum temperature of Cib River (site 4, Fig. 14.8) is recorded at 7:30 and the maximum temperature is recorded at 15:30 (mean temperature: 7.61 °C). The

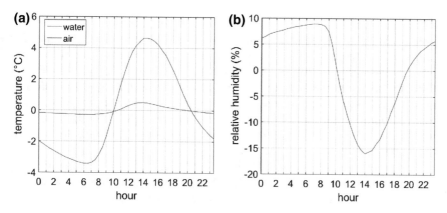

Fig. 14.7 Mean profiles of parameters in site 3: **a** temperature, **b** relative humidity

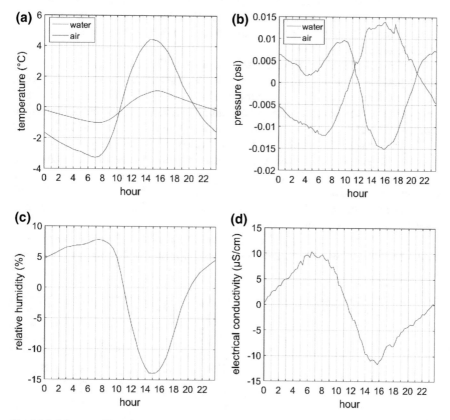

Fig. 14.8 Mean profiles of parameters in site 4: **a** temperature, **b** pressure, **c** relative humidity, **d** electrical conductivity

equivalent air temperatures occur 1 h earlier (mean temperature: 6.03 °C). Water level (mean: 0.517 psi), atmospheric pressure (mean: 14.161 psi), air relative humidity (mean: 89.51%) and water electrical conductivity (mean: 588 μS/cm) have the same behavior previously observed in sites 1 and 2. However, beyond the common behavior shown by the electrical conductivity diurnal profile, Briciu et al. [8] revealed that, on some short intervals of consecutive days, significant semidiurnal oscillations could be found, with a period equal to those of the atmospheric tides. This phenomenon is explained as being caused by the thermal karst waters that Cib River collects inside Cib Gorge.

Water maximum temperature of Bobîlna River (site 5, Fig. 14.9) occurs at 14:30, half an hour later than in the air, while the minimum temperature occurs at 7:30, 1.5 h later than in the air. The mean water temperature was (9.72 °C), while the mean air temperature was almost equal (9.74 °C).

Cerna River in all studied sites (6–8; Figs. 14.10, 14.11 and 14.12) shows different mean profiles from one site to another because of the human impact. Site 6 lies below Valea lui Iovan Dam, while site 7 is below Herculane Dam; flow is strongly regulated for hydroelectrical purposes.

Minimum water temperature in site 6 occurs at 8:30, 2 h later than the minimum air temperature from the sensor located in the wooded bank nearby. In site 7, water and air minimum temperatures occur at the same time, 7:30. The maximum water temperature in site 6 occurs at 16:30 (the same 2 h difference from the equivalent air temperature like in the case of minimum temperatures), but occurs at 14:30 in site 7. This difference is explained by the morphology of Cerna Valley, which is very narrow and deep at site 7. The air temperature maximum at site 7 was recorded at 16:00, but the diurnal peak has an inflexion at 14:30 corresponding to the temperature maximum recorded in site 6. Mean water temperature in site 6 was 14.94 and 6.76 °C in site 7, while the equivalent air temperatures were 16.85 and 4.51 °C respectively (note the different time intervals).

The water level and electrical conductivity in sites 6 and 7 exhibit very altered values (deviation from a normal state). The water level has 2 diurnal peaks instead

Fig. 14.9 Mean profiles of water and air temperature in site 5

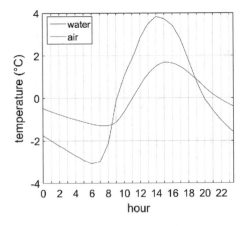

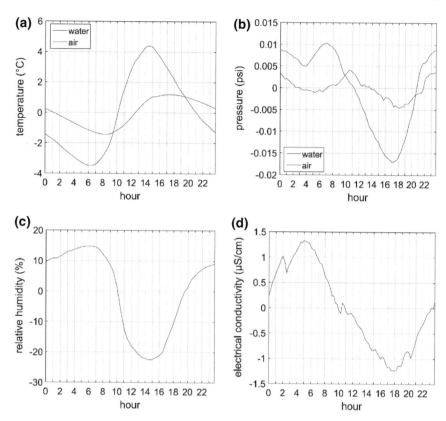

Fig. 14.10 Mean profiles of parameters in site 6: **a** temperature, **b** pressure, **c** relative humidity, **d** electrical conductivity

of one, at 10:30 and 23:30 in site 6 and has a quasi-flat peak during the 9:30–12:15 hourly interval at site 7.

Means of water pressure, air pressure, air relative humidity and water electrical conductivity in site 6 are as follow 0.694 psi, 14.099 psi, 81.77% and 135.57 μS/cm. Means of water pressure, air pressure and water electrical conductivity in site 7 are as follow 0.837 psi, 14.525 psi, and 217.66 μS/cm.

Mean water temperature of Cerna in site 8 (Fig. 14.12) was 7.18 °C, a clear increase from site 7, that recorded the same time interval. This temperature, also greater than that of the mean air temperature (4.89 °C) is obtained through the numerous hyperthermal springs that feed the river. The minimum water temperature is recorded at 8:30, 1 h after the minimum in the air temperature, and the maximum are recorded at 16:00 (the same 1-h delay from the atmospheric maximum).

Site 9 is a case where water regulation for hydroelectric plants is strongly influencing the diurnal thermal water regime (Fig. 14.13). Moldova River exhibits at this site 5 peaks per day in an average day, but there are also days with a singular diurnal peak in temperature. The lowest valley in water temperature is recorded at 8:00,

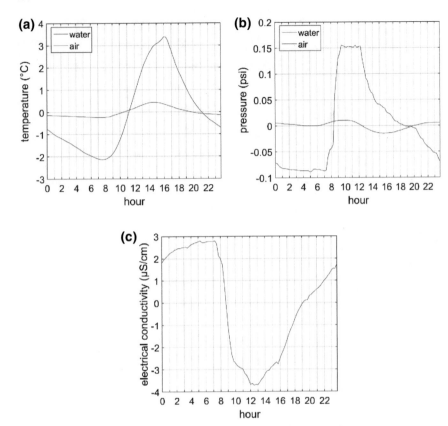

Fig. 14.11 Mean profiles of parameters in site 7: **a** temperature, **b** pressure, **c** electrical conductivity

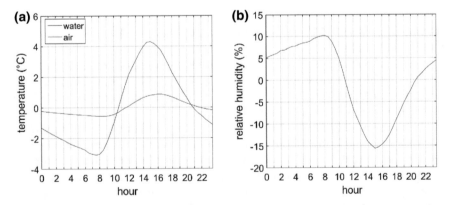

Fig. 14.12 Mean profiles of parameters in site 8: **a** temperature, **b** relative humidity

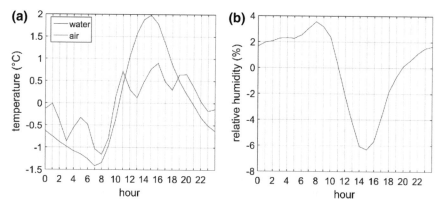

Fig. 14.13 Mean profiles of parameters in site 9: **a** temperature, **b** relative humidity

while the highest peak is recorded at 16:00 (min and max have an exact 1 h delay from the equivalent moments in the air temperature). The mean water temperature in the studied time interval was 6.47 °C and the mean air temperature was 4.57 °C. The recorded mean air relative humidity was 87.7%.

All sites in Fig. 14.14 have air temperature maxima at 14:30, excepting the site on Sucevița River, where the eastward exposition of the sensor moved the maximum towards the morning (10:30) and the site on Soloneț River, where the westward exposition of the river bank moved the maximum at 15:30. The air temperature minima generally occur at 5:00. The water temperature maxima occurred at the same time with the virtual moment when the air temperature maxima occur in this part of Romania (14:30), except Soloneț and Pozen rivers (15:30), Soloneț River (14:00) and Humor River (15:00). Water temperature minima range from 5:00 to 6:00.

Mean water temperature for Humor, Pozen, Solca, Blândețu, Soloneț, and Sucevița were: 16.39, 15.39, 18.53, 17.96, 19.1 and 16.21 °C. The equivalent air temperatures in the same places were: 16.33, 19.51, 16.8, 16, 18.37 and 15.08 °C.

It is to be noted the water thermal amplitude that is greater than the air thermal amplitude in some sites because of the direct solar irradiation of stream water and pebbled streambed during multiple consecutive hours.

The evolutions shown in Fig. 14.15 are similar to those in Fig. 14.14. The water temperature maximum and minimum are more constant across sites than their equivalents in air temperature, indicating that streamwater is less susceptible than surrounding air to local influences. The water temperature minimum occurs at 7:30 in sites no. 15, 18 and 19. At the same sites, the air temperature minima occur at distinct hours: 4:30, 7:00 and 6:30 respectively. The water temperature maximum occurs at 17:00, while the air temperature maxima range from 12:30 to 13:30. The terrain morphology and aspect of these sites influences the air temperature and the diurnal profiles record two peaks: the chronologically first peak is caused by direct solar irradiation of local surfaces, while the second peak is generated by the air temperature of the greater surrounding area.

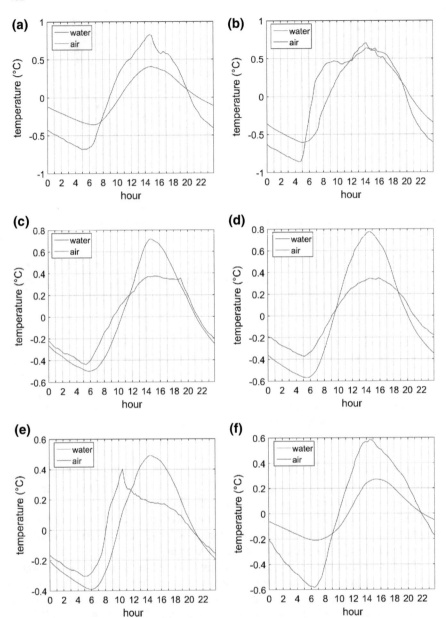

Fig. 14.14 Mean profiles of water and air temperature in the following sites: **a** site 10, **b** site 11, **c** site 12, **d** site 13, **e** site 14, **f** site 15

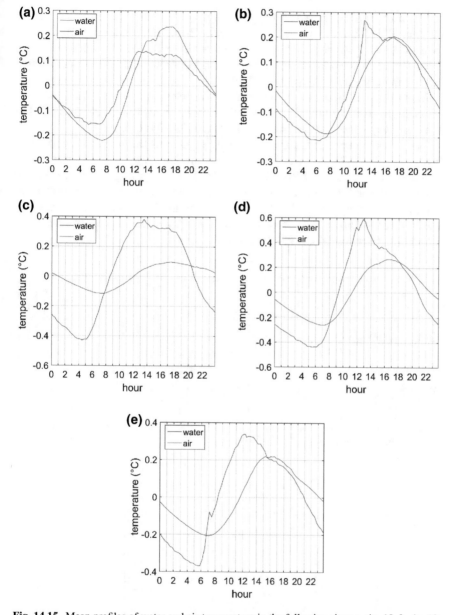

Fig. 14.15 Mean profiles of water and air temperature in the following sites: **a** site 18, **b** site 19, **c** site 15, **d** site 16, **e** site 17

The profiles of water and air temperature at sites 16 and 17 are similar because they are placed in the same valley, of Suceava River. The water temperature minimum occurs at 7:00, preceded by the air minimum, which occurs at approximately 5:30. The water temperature maximum occurs at 16:30, while the air temperature maximum occurs at approximately 12:30.

Mean water temperature of Brodina, Vițău, Pozen and Suceava (at Straja and Dărmănești) rivers in July 2011 were: 15.86, 16.6, 15.92, 17.06 and 19.85 °C. The mean air temperatures were: 14.86, 16.18, 18.98, 17.58 and 20.96 °C.

The mean water temperature of Solonet, Şugău and Dorna rivers described in Fig. 14.16 were: 2.72, 2.31 and 0.23 °C. The equivalent air temperatures were: 1.78, 1.09 and −0.7 °C. Soloneț River (Fig. 14.16a) had a surprisingly perfect match of minima (8:00) and maxima (15:00) times in water and air temperature during October 27th, 2011–January 4th, 2012 time interval. The minimum water temperature in site 20 (Fig. 14.16b) was recorded at 7:30 (7:00 for the air temperature), while the maximum was recorded at 15:30 (air—14:30).

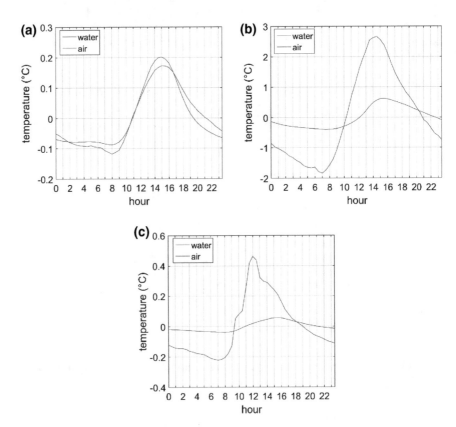

Fig. 14.16 Mean profiles of water and air temperature in the following sites: **a** site 11, **b** site 20, **c** site 21

The air temperature recorded inside Şugău Gorge has a special shape because of the canyon-type morphology of the river valley in the studied area. The air temperature minimum occurs at 7:00, while the maximum is at 12:00 (Fig. 14.16c). Şugău River is fed by karst groundwaters which impose a weak mean diurnal variation of water temperature (diurnal minimum occurs at 8:00; diurnal maximum occurs at 15:30).

Water time series that span over winter often has days with a very flat diurnal peak of temperature or an irregular diurnal increase in temperature because of the reduced flow during winter, correlated with ice formation. Such days are usually those when water temperature falls below 1 °C. In opposition, a very preeminent diurnal peak is often present during summer. Similar behavior applies to the evolution of other water parameters. Diurnal peaks of any parameter are disturbed/erased during and immediately after rainfalls.

14.6   Conclusions

The water monitoring sites have revealed similar diurnal shapes of the temperature, level and electrical conductivity in areas with the natural flow: afternoon maximum water level and temperature and minimum electrical conductivity; the opposite events occur early in the morning. For both water and air parameters, the moments when minima occur are less temporally variable than those of maxima, at all sites. The water parameter which is common to all sites in this study is the temperature. The water temperature recorded diurnal cycles with amplitudes, when using the detrended time series, ranging from 0.095 °C (site 21, Şugău River) to 2.978 °C (site 5, Bobîlna River).

The natural particularities of the monitoring sites can greatly impact the results of measurements. The evolution of air parameters can explain the evolution of the water parameters in natural areas, but it is only partly useful in areas with intense human activity, such as the areas where dams and hydroelectric plants exist. Smoothing techniques may be applied to all mean profiles in order to obtain more abstract representations of diurnal evolution in every site.

14.7   Recommendations

The analysis of a physical, chemical or biological parameter of a river should take into account the diurnal cycle. The cycle may be useful for estimating a representative average of a day, for including a daily variation into analysis or for removing the diurnal oscillation/noise from the long term evolution of a time series. The figures in this study can be used for estimating absolute or percentage changes of a parameter relative to its average value, during a day, for rivers in the study area or nearby.

This also applies to parameters that were not included in this analysis, but whose relationship with one of the studied parameters is well known.

Acknowledgements Some measurements of this study were conducted within the research project entitled "Field studies in orthotidal potamology". This work was supported by a grant of the Romanian National Authority for Scientific Research and Innovation, CNCS – UEFISCDI, project number PN-II-RU-TE2014-4-2900.

References

1. Benyahya L, Caissie D, El-Jabi N, Satish MG (2010) Comparison of microclimate vs. remote meteorological data and results applied to a water temperature model (Miramichi River, Canada. J Hydrol 380:247–259
2. Bond BJ, Jones JA, Moore G, Phillips N, Post D, McDonnell JJ (2002) The zone of vegetation influence on baseflow revealed by diel patterns of streamflow and vegetation water use in a headwater basin. Hydrol Process 16:1671–1677
3. Briciu A-E (2014) Wavelet analysis of lunar semidiurnal tidal influence on selected inland rivers across the globe. Sci Rep 4:4193. https://doi.org/10.1038/srep04193
4. Briciu A-E (2017) Studiu de hidrologie urbană în arealul municipiului Suceava (Urban hydrology study in Suceava municipality area). Ştefan cel Mare University Publishing House, Suceava. ISBN 978-973-666-506-6
5. Briciu A-E (2018) Diurnal, semidiurnal, and fortnightly tidal components in orthotidal proglacial rivers. Environ Monit Assess 190(3):160. https://doi.org/10.1007/s10661-018-6513-x
6. Briciu A-E, Mihăilă D, Oprea-Gancevici DI, Bistricean P-I (2016) Analysis of surface thermal waters in Baile Herculane area. In: SGEM2016 conference proceedings, vol 1, pp 63–70. ISSN 1314-2704
7. Briciu A-E, Mihăilă D, Oprea-Gancevici DI, Bistricean P-I (2017) Some aspects regarding the thermal water temperature of some sites in Băile Felix, Geoagiu-Băi and Hârşova areas, Romania. In: SGEM2017 conference proceedings, vol 17, no 31, pp 601–608. ISBN 978-619-7408-04-1/ISSN 1314-2704
8. Briciu A-E, Mihăilă D, Oprea DI, Bistricean P-I, Lazurca LG (2018) Orthotidal signal in the electrical conductivity of an inland river. Environ Monit Assess 190(5):282. https://doi.org/10.1007/s10661-018-6676-5
9. Briciu A-E, Oprea-Gancevici DI (2015) Diurnal thermal profiles of selected rivers in Romania. In: SGEM2015 conference proceedings, vol 1, pp 221–228. ISSN 1314-2704
10. Briciu A-E, Oprea-Gancevici DI, Mihăilă D, Bistricean, P-I (2016) Analysis of surface thermal waters in Moneasa area. In: SGEM2016 conference proceedings, vol 1, pp 71–78. ISSN 1314-2704
11. Burger H (1945) Einfluss des Waldes auf den Stand der Sewasser. Tech. rep., IV Mittlg. Der Wasserhaushalt im Valle di Melera von 1934/35 bis 1943/44 – Mitt.d. Schweiz. Anstalt f. forstl. Versuchsw., 25 Bd. 1
12. Caissie D (2006) The thermal regime of rivers: a review. Freshw Biol 51:1389–1406
13. Callède J (1977) Oscillations journalières du débit des rivières en l'absence de precipitations. Cahier ORSTOM, série Hydrologie 14:219–283
14. Constantz J, Thomas CL, Zellweger G (1994) Influence of diurnal variations in stream temperature on streamflow loss and groundwater recharge. Water Resour Res 30:3253–3264
15. Gribovszki Z, Szilágyi J, Kalicz P (2010) Diurnal fluctuations in shallow groundwater levels and streamflow rates and their interpretation—a review. J Hydrol 385:371–383

16. Jasonsmith JF, Macdonald BCT, White I (2017) Earth tide-induced fluctuations in the salinity of an inland river, New South Wales, Australia: a short-term study. Environ Monit Assess 189(4):188. https://doi.org/10.1007/s10661-017-5880-z
17. Kinouchi T, Yagi H, Miyamoto M (2007) Increase in stream temperature related to anthropogenic heat input from urban wastewater. J Hydrol 335:78–88
18. Lundquist JD, Cayan DR (2002) Seasonal and spatial patterns in diurnal cycles in streamflow in the western United States. J Hydrometeorol 3:591–1603
19. Mihăilă D, Briciu A-E (2012) Actual climate evolution in the NE Romania. Manifestations and consequences. In: 12th international multidisciplinary scientific geoconference, SGEM2012 conference proceedings, vol 4, pp 241–252. ISSN 1314-2704
20. Morgenschweis G (1995) Kurzzeitige vorhersage der wasserentnahme aus einem flussgebiet. Vortragsmanusskript zur 8. Wiss. Tagung Hydrologie und wasserwirtschaft zum Thema Verfügbarkeit von Wasser vom 22/23. Marz 1995 in Bochum, 16 Seite
21. Nimick DA, Gammons CH, Parker SR (2011) Diel biogeochemical processes and their effect on the aqueous chemistry of streams: a review. Chem Geol 283(1–2):3–17
22. Nimick DA, Cleasby TE, McCleskey RB (2005) Seasonality of diel cycles of dissolved trace-metal concentrations in a Rocky Mountain stream. Environ Geol 47:603–614
23. Poole GC, Berman CH (2001) An ecological perspective on in-stream temperature: natural heat dynamics and mechanisms of human-caused thermal degradation. Environ Manag 27:787–802
24. Prats J, Val R, Armengol J, Dolz J (2010) Temporal variability in the thermal regime of the lower Ebro River (Spain) and alteration due to anthropogenic factors. J Hydrol 387:105–118
25. Rycroft HB (1955) The effect of riparian vegetation on water-loss from an irrigation furrow at Jonkershoek. J South Afr For Assoc 26:2–9
26. Smith K (1981) The prediction of river water temperatures/Prédiction des températures des eaux de rivière. Hydrol Sci J 26(1):19–32. https://doi.org/10.1080/02626668109490859
27. Troxell HC (1936) The diurnal fluctuation in the ground-water and flow of the Santa Anna River and its meaning. Trans Am Geophys Union 17(4):496–504
28. Tschinkel HM (1963) Short-term fluctuation in streamflow as related to evaporation. J Geophys Res 68(24):6459–6469. https://doi.org/10.1029/JZ068i024p06459
29. Verma RD (1986) Environmental impacts of irrigation projects. J Irrig Drain Eng 112:322–330
30. Webb BW, Hannah DM, Moore RD, Brown LE, Nobilis F (2008) Recent advances in stream and river temperature research. Hydrol Process 22:902–918
31. Wicht CL (1941) Diurnal fluctuation in Jonkershoeck streams due to evaporation and transpiration. J South Afr For Assoc 7:34–49

Chapter 15
Drought and Insolvency: Case Study of the Producer-Buyer Conflict (Romania, the Period Between the Years 2011–2012)

Gheorghe Romanescu and Ionuţ Minea

Abstract The important hydroelectric potential of Romania has been tapped on most of the mountainous rivers: Bistrita (along with the Siret and Pruth), Olt, Jiu, Arges, Lotru, Somes, Raul Mare. The Danube, with its potential of 2100 MW, has also been harnessed, with the Iron Gates I as the only hydropower plant in Romania that functions permanently. The hydroelectric power potential of Romania is important and it has been exploited on most mountainous rivers: Bistrita (along with Siret and Pruth), Olt, Jiu, Arges, Lotru, Somes, Raul Mare; in addition to them, we must include the Danube, which holds a potential of 2100 MW (Iron Gates I is the only hydropower plant in Romania that functions permanently). The regional climate is increasingly unpredictable, and the hydrological risk events (droughts and tidal waves) occur more frequently. In this case, infrastructures capable of water intake are required (either to compensate for the droughty periods or to mitigate floods). For the past 25 years, droughty periods have increased in length and severity, though, the mean amount of precipitations has augmented. The drought recorded in the autumn of the year 2011 and the spring of the year 2012 entailed a drastic reduction in power production provided by hydropower plants, reasons for which the company Hidroelectrica S.A. became unable to distribute power to beneficiaries. On an average hydrological year, Romania produces 17.33 TWh, which means 35% of the consumption. In this case, Hidroelectrica S.A. was sued by a series of partners who were no longer satisfied with the quality of the distribution.

Keywords Average hydrological year · Forecasting error · Hydrological forecasting · Hydrological drought · Hydropower

G. Romanescu · I. Minea (✉)
Department of Geography, Faculty of Geography and Geology, "Alexandru Ioan Cuza" University of Iasi, Bd. Carol I 20A, 700505 Iasi, Romania
e-mail: ionutminea1979@yahoo.com

© Springer Nature Switzerland AG 2020
A. M. Negm et al. (eds.), *Water Resources Management in Romania*, Springer Water,
https://doi.org/10.1007/978-3-030-22320-5_15

489

15.1 Introduction

The economic damage caused by a lack of water in some regions is extremely high. For this reason, certain developed States implemented programs for water storage in special tanks, with multiple usages. Water and power demand is acute for most underdeveloped or developing countries. From this perspective, the States in Africa and Asia is worth noting [9]. As for Europe, an acute lack of water and too little storage investments are real issues in the East and in the South. This category also includes Romania, which does not own many storage facilities in areas most affected by droughts, in the autumn and particularly in the winter. Water storage can have both positive and negative effects (lowering the phreatic level downstream, and so on) [32, 33]. The purpose of water storage include water supply for irrigations, for power production, for flood mitigation, and so on. Water is stored in mountainous areas, and it is used for mitigating floods and for producing electric power [2, 12, 16, 17, 34, 36, 42, 55].

Global climatic changes have also affected the territory of Romania [6, 23–26, 43, 48]. The mean precipitations of the last 25 years have increased, and so have the droughts [8]. Unfortunately, the intensity of precipitations has increased at an alarming rate, leading to heavy rains falling within a very short time span: 100–200 mm/24 h [18–20, 40, 44]. On a local level, infrastructures capable of water intake are required, to ensure it meets up to the protection levels necessary for a viable economy. Over 1.6 billion people are affected by a lack of water worldwide.

On a global level, hydroelectric power represents 19% of the entire amount of electric power. The exploitable potential is six times higher. Romania produces, on the average, 17.33 TWh per year, which means 35% of the consumption. However, Romania's hydroelectric power potential is double. Hidroelectrica S.A. is the main holder and operator of the hydroelectric power in Romania (Fig. 15.1). The most important facilities for hydropower production in Romania are as follows: 26 hydro-units for regulating the secondary frequency-power; 95 hydro-units for rapid tertiary reserve; 5 hydro-units; 28 hydro-units for regulating reactive power in the secondary tension-regulating band; and 5 hydro-units for restoring the National Energetic System [47] (see Figs. 15.2, 15.3, 15.4, 15.5, and 15.6).

Over 450,000 MW was generated worldwide by the hydropower facilities in the year 2001; half of them are located in Europe and half in North America. Besides conventional hydropower, more than 80,000 MW of power comes from the pumped-hydro capacity [21]. Currently, 1.6 billion people worldwide have limited access to electricity: most of them live in Africa and Asia.

Hydrocarbon resources of Romania are limited, reason for which, efforts have been made to exploit the hydropower resources as much as possible. The hydro-electric power potential of Romania is estimated at 5900 MW for the rivers, and at 2100 MW for the Danube. Under these circumstances, the total estimated hydro-electric power potential reaches 8000 MW (1134 MW being represented by micro hydropower plants that currently produce 400 MW only) (Table 15.1). Romania has 296 small and large hydropower plants (in function); furthermore, other 411 micro

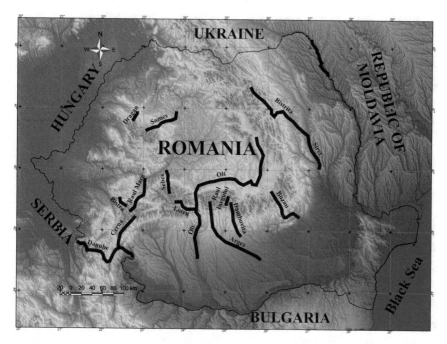

Fig. 15.1   The main rivers developed for producing hydroelectric power in Romania

Fig. 15.2   Siriu Reservoir in the Buzau catchment

Fig. 15.3 The dam at Stanca-Costesti on the Prut river

Fig. 15.4 The Iron Gates I Hydropower Plant, the largest in Romania (2100 MW)

Fig. 15.5 Izvorul Muntelui Reservoir, built in 1961 on the Bistrita River, the third largest in Romania (after the Iron Gates I and Stanca-Costesti)

Fig. 15.6 Vidra Reservoir, which supplies the Ciunget Hydropower Plant on the Lotru River

Table 15.1 The developable and developed technical potential of Romania

Catchment basin district	Developable technical potential (TWh/year)	Developed potential (TWh/year)	Current potential use degree (%)
Somes	2.20	0.63	28.64
Crisuri	0.90	0.37	41.11
Mures	4.30	1.47	34.19
Banat rivers	0.65	0.60	92.31
Jiu	0.90	0.52	57.78
Olt	5.00	4.20	84.00
Arges	1.60	0.94	58.75
Ialomita	0.75	0.30	40.00
Siret	5.50	2.00	36.36
Danube	12.00	6.45	53.75
Total	33.80	17.48	51.72

hydropower plants are currently a project in progress or they are under construction. The plan is to build, 550 micro hydropower plants by the year 2020. Compared to a large number of installations, electric power production is far from lucrative.

The Danube basin comprises of 8647 small and large hydropower plants, and it produces 99,473 GWh/year. In the entire catchment basin of the Danube, 88.4% of the hydropower is produced by only 3.4% of the hydropower plants, mostly by those on the main course of the river. In this case, only a minimal number of large hydropower plants are productive (for example, Iron Gates I built by Romania and Serbia, which function permanently). Scientific studies in Romania [13, 30, 39] and abroad [3–5, 7, 27, 28, 38, 49] have treated in details or sometimes tangentially, some topics related to the dysfunctionality of electric power distribution caused by extreme events, such as droughts and floods [10, 29, 35, 37, 41, 45, 46]. In this sense, it is worth underscoring the hydrological drought that affected Romania in the period between the year 2011 and 2012, and the subsequent insolvency of Hidroelectrica S.A. Therefore, natural hazards have been proven to affect economic activity, mostly in economically struggling countries.

15.2 Materials and Methods

The hydrological data were provided by the National Institute of Hydrology and Water Management in Bucharest for the mean multi-annual flows (for hydrographical arteries producing electric power) and the mean monthly flows within the period between the years 2000 and 2012. The other data were collected from each hydroelectric unit and from the individual watershed management agencies across Romania.

This study has taken into account only the mean flows of tributaries recorded at the hydrometric stations, situated upstream from the most important dams. For electric power production, the data provided by the Hidroelectrica S.A. were used. The conflict between Hidroelectrica S.A. as debtor and other institutions that purchased electric power has been the subject of ongoing court cases within the Bucharest Court (Bucharest Court Civil Section VII).

The data were processed in the Laboratory of Hydrology of the Faculty of Geography and Geology from the "Alexandru Ioan Cuza" University of Iasi. The graphical material was processed in CorelDRAW X3. The field measurements were conducted during and immediately after the described hydrological phenomena occurred.

15.3 Results

The joint stock company that specialized in electric power production within the Hydropower plants named "Hidroelectrica," was established under the Government Decision no. 627/13 July 2000, following the reorganization of the National Electricity Company, and is registered in the Trade Register of the Bucharest, under number J40/7426/2000 Series A no, 733887 issued on 10.08.2000. Hidroelectrica S.A. has a registered office in Bucharest (the address is sector 2, Strada Constantin Nacu, nr. 3), and a fiscal identification number, 13267213. The subscribed and paid in the share capital of the company is 4,475,643,070 RON, owned by the Romanian State represented by the Ministry of Economy, Commerce and Business Environment with a percentage of 80.0561% and 19.9439%, by SC Fondul Proprietatea SA.

Following is the production capacity of the branches. The 140 micro hydropower plants with installation capacities <4 MW comprise a total of 287 hydropower groups with a total installation capacity of 111.86 MW. The 23 hydropower plants with installation capacity ranging between 4 and 10 MW, where there are 46 hydropower groups with a total installed capacity of 165.68 MW (Table 15.2); 106 hydropower plants with installation capacities exceeding 10 MW, comprising a total of 247 hydropower groups with an installation capacity of 6074.27 MW; 5 pumping stations with 91.5 MW installation capacity (Table 15.3). The production capacity of branches is provided by the following: 140 micro hydropower plants with installation capacities <4 MW, comprising a total of 287 hydropower groups with a total installation capacity of 111.86 MW; 23 hydropower plants with installation capacity ranging between 4 and 10 MW, where there are 46 hydropower groups with a total installed capacity of 165.68 MW (Table 15.2); 106 hydropower plants with installation capacities exceeding 10 MW, comprising a total of 247 hydropower groups with an installation capacity of 6074.27 MW; 5 pumping stations with 91.5 MW installation capacity (Table 15.3).

The hydrological context for the period between 30th September 2011 and 30th April 2012, requires collecting data on the mean annual and multi-annual liquid discharge for all hydrographical arteries, comprising the hydropower units. The main point of interest is represented by stations that record the flows of reservoirs and the

Table 15.2 Distribution per branches of micro hydropower plants with installation capacities under 4 MW and of hydropower plants with installation capacity, ranging between 4 MW and 10 MW

Branch	CHEMP (MHC)				CHE			
	Pi ≤ 4 MW				4 MW < Pi ≤ 10 MW			
	Pi (MW)	Ep (GWh/year)	No. of plants	No. of groups	Pi (MW)	Ep (GWh/year)	No. of plants	No. of groups
Bistrita	23.76	75.47	30	56	4.10	14.00	1	2
Buzau	6.55	23.38	9	16	33.15	121.65	4	7
Cluj	11.19	34.61	19	55	20.44	36.00	3	9
Curtea de Arges	9.75	36.34	14	34	66.74	167.60	10	18
Hateg	6.03	18.48	12	23	0.00	0.00	0	0
Oradea	13.16	47.92	10	23	20.00	41.10	2	4
Iron Gates	0.00	0.00	0	0	0.00	0.00	0	0
Ramnicu Valcea	3.83	11.89	4	7	0.00	0.00	0	0
Sebes	0.25	1.90	2	2	4.25	6.00	1	3
Slatina	0.00	0.00	0	0	0.00	0.00	0	0
Targu Jiu	3.37	8.10	3	5	10.00	20.00	1	1
Caransebes	6.94	24.20	11	17	7.00	25.00	1	2
Sibiu	27.04	86.82	26	49	0.00	0.00	0	0
Total	111.86	369.11	140	287	165.68	431.35	23	46

CHEMP Low-capacity hydropower plants; *MHC* Micro hydropower plants; *CHE* hydropower plants; *Pi* installation capacity; *Ep* theoretical precipitation potential

corresponding catchment basins. The National Institute of Hydrology and Water Management in Bucharest provided the entire set of data.

The mean tributary flow for the year 2011 and 2012 was lower than the mean multi-annual tributary flow for all rivers, with plants for electric power production (Table 15.4). The lowest values were recorded on the rivers of Lotru, Sebes, Cerna, Bistra, Targul, and Pruth; their values were about 50% of the mean multi-annual tributary flow. The other rivers recorded values of about 60–70% of the mean multiannual tributary flow. The values for both years are close. The Danube—the most important hydropower provider—reached about 76.38% in the year 2011 and 79.28% in the year 2012 (see Table 15.4).

Drought does not influence directly, the production of electric power per se. However, it may entail repercussions if it spans over a long period. Meteorological drought is represented by the lack of precipitations, while hydrological drought by a reduction of flows. Socioeconomic drought occurs when precipitations drop below the normal values, thereby, affecting the economic activities. The elements of hydrological risk define a state of uncertainty specified in the contract between the Supplier and the Buyer. If the previous years recorded rich flows, it is possible to produce

Table 15.3 Distribution per branches of hydropower plants with installation exceeding 10 MW, and of pumping stations

Branch	CHE				Pumping stations		
	Pi > 10 MW						
	Pi (MW)	Ep (GWh/year)	No. of plants	No. of groups	Pi (MW)	No. of stations	No. of groups
Bistrita	598.00	1568.35	18	43	0.0	0	0
Buzau	177.25	459.40	8	19	0.0	0	0
Cluj	298.50	537.40	4	7	0.0	0	0
Curtea de Arges	525.10	972.15	18	38	0.0	0	0
Hateg	507.52	883.02	13	27	0.0	0	0
Oradea	194.00	390.00	4	8	10.00	1	2
Iron Gates	1462.80	6561.00	3	16	0.0	0	0
Ramnicu Valcea	1109.90	2737.00	14	29	61.5	3	7
Sebes	342.00	600.00	3	6	20.0	1	2
Slatina	379.00	889.00	8	26	0.0	0	0
Targu Jiu	179.60	442.60	4	10	0.0	0	0
Caransebes	181.00	330.10	2	4	0.0	0	0
Sibiu	119.60	294.40	7	14	0.0	0	0
Total	6074.27	16664.42	106	247	91.5	5	11

CHE hydropower plants; *Pi* installation capacity; *Ep* theoretical precipitation potential

electric power in the droughty periods as well, if the water was stored (and only if the demand was low). Not all dams benefit from water intake facilities (on the Danube, the entire amount of water goes through turbines). Such amounts of water were used at HPP Marisel on Somes, at HPP Stejaru on Bistrita, at HPP Corbeni on Arges, and at HPP Calceag on Lotru. Flows higher than the tributary flow went through turbines in the period between October 2011 and March 2012. The normal flows recorded in April 2012 did not lead to the resumption of the normal turbine process for tributary flows, because reservoirs had to be filled. Nonetheless, Hidroelectrica S.A. managed to supply 90% of the electric power stipulated by the contract. The power was tapped from the Iron Gates I and II, the Olt River hydrotechnical works, units downstream from the HPP Stejaru on the Bistrita, the Siret River hydrotechnical works, and others (1070 MWh of the 1500 MWh mentioned in the contract).

Therefore, it is possible to assess the previous hydrological situation (one or two years earlier) to get better insight into the hydrological regime. In this case, the year 2010 was very rich compared to the mean multi-annual tributary flow; substantial excess flows were recorded on all the hydrographic arteries (except for Sebes, but its electric power production potential is low) (Table 15.2). In the year 2010, the Danube recorded a mean tributary flow of 7602.00 m^3/s, compared to a mean multi-annual

Table 15.4 Comparison between the mean tributary flow in the year 2011 and 2012, and the mean multiannual tributary flow for rivers, comprising the hydropower units

River/section	Year						
	2000	2001	2002	2003	2004	2005	2006
Arges/Vidraru	13.52	17.21	14.77	13.90	20.20	25.28	19.56
Bistrita/Izvorul Muntelui	38.87	48.51	61.82	30.48	44.63	59.96	66.23
Lotru/Vidra	10.06	13.40	11.70	12.94	15.27	21.00	18.46
Somes/Fantanele	10.21	15.73	13.64	8.57	14.42	13.96	16.26
Dragan/Dragan	6.13	9.99	8.22	5.29	9.41	9.45	9.70
Sebes/Oasa	3.80	4.51	4.43	4.23	5.61	7.84	6.55
Cerna/Valea lui Iovan	6.08	7.33	6.46	8.30	10.94	11.76	10.59
Raul Mare/Gura Apelor	8.19	9.87	8.31	8.09	12.07	12.68	12.10
Buzau/Siriu	6.74	7.30	8.45	4.76	7.53	19.02	10.15
Bistra/Poiana Marului	5.78	7.56	5.79	4.46	6.56	8.69	8.55
Dambovita/Pecineagu	2.39	2.83	2.28	2.12	2.89	4.65	3.38
Raul Targului/Rausor	2.03	2.90	1.90	2.09	3.00	4.46	4.02
Cerna/Herculane	5.55	4.22	4.92	5.19	7.66	8.91	11.83
Danube/Iron Gates	5448.6	5463.6	5618.2	3930.3	5451.8	6345.4	6437.6
Siret/Dragesti	54.53	57.71	98.21	48.86	56.46	104.18	151.59
Pruth/Stanca Costesti	58.73	84.08	110.27	70.30	74.09	98.90	114.42
Olt/Ramnicu Valcea	101.22	99.55	111.97	84.06	121.99	194.98	148.40

(continued)

Table 15.4 (continued)

| River/section | Year | | | | | | Qmm | % |
	2007	2008	2009	2010	2011	2012		Qmm 2011/2012
Arges/Vidraru	20.18	17.62	18.69	24.74	13.58	11.73	19.76	68.72/59.36
Bistrita/Izvorul Muntelui	49.86	54.88	34.85	74.36	38.58	32.71	48.36	79.77/67.63
Lotru/Vidra	18.50	14.14	17.37	22.01	10.42	10.17	19.59	53.19/51.91
Somes/Fantanele	11.92	11.08	11.98	16.74	7.85	8.69	11.97	65.58/72.59
Dragan/Dragan	7.45	7.62	7.54	10.39	4.68	6.05	8.10	57.77/74.69
Sebes/Oasa	7.87	6.81	6.27	7.87	5.37	4.29	8.28	64.85/51.81
Cerna/Valea lui Iovan	9.07	8.16	10.44	12.64	4.89	6.17	10.31	47.42/59.84
Raul Mare/Gura Apelor	12.55	10.30	12.79	14.65	6.76	7.66	10.37	65.18/73.86
Buzau/Siriu	8.51	6.87	7.89	13.18	7.30	7.42	9.80	74.48/75.71
Bistra/Poiana Marului	6.81	5.89	5.53	7.40	3.64	4.80	7.24	50.27/66.29
Dambovita/Pecineagu	3.17	3.18	3.36	3.76	2.56	2.06	3.39	75.51/60.76
Raul Targului/Rausor	2.50	2.94	3.53	4.26	1.82	2.10	3.64	50.00/57.69
Cerna/Herculane	6.23	5.58	7.34	9.95	2.89	4.18	5.79	49.91/72.19
Danube/Iron Gates	4696.2	4826.4	5421.6	7602.0	4226.3	4387.1	5533.0	76.38/79.28
Siret/Dragesti	80.69	133.91	72.60	167.35	49.36	40.76	75.54	65.34/53.95
Pruth/Stanca Costesti	72.78	142.44	59.44	183.54	53.77	40.01	84.12	63.92/47.56
Olt/Ramnicu Valcea	118.99	121.91	121.99	178.50	119.77	80.60	127.60	93.86/63.16

Qmm mean multiannual tributary flow

tributary flow of 5533.0 m^3/s (137.39%). The highest increases were recorded on the rivers of Siret (221.53%) and Pruth (218.18%) (Table 15.5). The fact that some plants had higher amounts of water passing through the turbines was due to the excess water recorded in the previous year.

For the mean multi-monthly tributary flow, data corresponding to over 13 years was analyzed (2000–2012). It is worth stating that, in ensuring a low degree of error, hydrological data must be analyzed for, at least, 30 years. The mean monthly tributary flow for the period between October 2011 and April 2012 (7 months) was low, compared to the mean multi-monthly tributary flow for rivers comprising hydropower units (Table 15.6). Most values ranged between 60 and 70% of the mean multi-monthly tributary flow. In the period October 2011–April 2012 (7 months), the Danube recorded a mean flow of 3789.3 m^3/s. This is lower compared to the mean multi-monthly flow of 5664.41 m^3/s (66.89%) for the period 2000–2012.

The period between the month of October 2011 and April 2012 (7 months), the Danube recorded a mean flow of 3789.3 m^3/s, compared to the mean multi-monthly flow for the period between the year 2000 and 2012 (for the months of October 2011, November 2011, December 2011, January 2012, February 2012, March 2012, and April 2012)—5664.41 m^3/s (66.89%). The period between the year 2000 and 2010 recorded higher mean values for flows throughout the entire country (in the year 2005, 2006, 2008, and 2010, historic floods occurred on the Romanian territory). Even under these circumstances, the mean monthly tributary flows recorded in October 2011–April 2012, represented only 60–70% (for some arteries, values were even lower). Over a longer period, values can be slightly different, and they feature a descending trend (concerning values of 40–50%).

The interval, 30.09.2011–30.04.2012 recorded low amounts of precipitations for the entire river network in Romania. The lowest values of mean flows were recorded in November 2011 and January–February, 2012, because precipitations were reduced and temperatures were very low (February). Mean flows ranged between 30 and 50% of the multi-annual monthly means. The lowest values (<30% of the normal monthly values) were recorded in November, December, and January, on the rivers situated in the southwest of Romania, and in February, on the rivers situated in the west and in the eastern hill and plain areas. The month of April recorded values close to the normal monthly values in Banat, Transylvania, Crisana, and Maramures.

At the hydrometric station of Bazias, on the Danube (entry into Romania), the mean monthly flows, which is lower than the multiannual monthly means by 53–78% were recorded. The lowest values were recorded in the months of November–December, 2011 and in February 2012 (53–57% of the monthly means). The lowest daily value was 2150 m^3/s, and it was recorded within the period, 22–23.02.2012 [22]. For the entire recording period—between the year 1921 and 2014 (94 years), the interval between September and March (7 months)—at the hydrometric station of Orsova (upstream from HPP Iron Gates), droughts were recorded between the years, 1921–1922, 1946–1947, 1953–1954, 1990–1991, and 2011–2012. The droughtiest periods turned out to be between the years, 1953–1954 and 2011–2012. The mean multi-monthly flow for the period, September–March (7 months) was 2480 m^3/s (1953–1954) and 3340 m^3/s (2011–2012). In this case, the flow of 3340 m^3/was

Table 15.5 Comparison between the mean tributary flow for the year 2010, and the mean multi-annual tributary flow for rivers comprising the hydropower units

River/section	Year						
	2000	2001	2002	2003	2004	2005	2006
Arges/Vidraru	13.52	17.21	14.77	13.90	20.20	25.28	19.56
Bistrita/Izvorul Muntelui	38.87	48.51	61.82	30.48	44.63	59.96	66.23
Lotru/Vidra	10.06	13.40	11.70	12.94	15.27	21.00	18.46
Somes/Fantanele	10.21	15.73	13.64	8.57	14.42	13.96	16.26
Dragan/Dragan	6.13	9.99	8.22	5.29	9.41	9.45	9.70
Sebes/Oasa	3.80	4.51	4.43	4.23	5.61	7.84	6.55
Cerna/Valea lui Iovan	6.08	7.33	6.46	8.30	10.94	11.76	10.59
Raul Mare/Gura Apelor	8.19	9.87	8.31	8.09	12.07	12.68	12.10
Buzau/Siriu	6.74	7.30	8.45	4.76	7.53	19.02	10.15
Bistra/Poiana Marului	5.78	7.56	5.79	4.46	6.56	8.69	8.55
Dambovita/Pecineagu	2.39	2.83	2.28	2.12	2.89	4.65	3.38
Raul Targului/Rausor	2.03	2.90	1.90	2.09	3.00	4.46	4.02
Cerna/Herculane	5.55	4.22	4.92	5.19	7.66	8.91	11.83
Danube/Iron Gates	5448.6	5463.6	5618.2	3930.3	5451.8	6345.4	6437.6
Siret/Dragesti	54.53	57.71	98.21	48.86	56.46	104.18	151.59
Pruth/Stanca Costesti	58.73	84.08	110.27	70.30	74.09	98.90	114.42
Olt/Ramnicu Valcea	101.22	99.55	111.97	84.06	121.99	194.98	148.40

(continued)

Table 15.5 (continued)

River/section	Year						Qmm	% Qmm 2010
	2007	2008	2009	2010	2011	2012		
Arges/Vidraru	20.18	17.62	18.69	*24.74*	13.58	11.73	*19.76*	*125.20*
Bistrita/Izvorul Muntelui	49.86	54.88	34.85	*74.36*	38.58	32.71	*48.36*	*153.76*
Lotru/Vidra	18.50	14.14	17.37	*22.01*	10.42	10.17	*19.59*	*112.35*
Somes/Fantanele	11.92	11.08	11.98	*16.74*	7.85	8.69	*11.97*	*139.84*
Dragan/Dragan	7.45	7.62	7.54	*10.39*	4.68	6.05	*8.10*	*128.27*
Sebes/Oasa	7.87	6.81	6.27	*7.87*	5.37	4.29	*8.28*	*95.04*
Cerna/Valea lui Iovan	9.07	8.16	10.44	*12.64*	4.89	6.17	*10.31*	*122.59*
Raul Mare/Gura Apelor	12.55	10.30	12.79	*14.65*	6.76	7.66	*10.37*	*141.27*
Buzau/Siriu	8.51	6.87	7.89	*13.18*	7.30	7.42	*9.80*	*134.48*
Bistra/Poiana Marului	6.81	5.89	5.53	*7.40*	3.64	4.80	*7.24*	*102.20*
Dambovita/Pecineagu	3.17	3.18	3.36	*3.76*	2.56	2.06	*3.39*	*110.91*
Raul Targului/Rausor	2.50	2.94	3.53	*4.26*	1.82	2.10	*3.64*	*117.03*
Cerna/Herculane	6.23	5.58	7.34	*9.95*	2.89	4.18	*5.79*	*172.74*
Danube/Iron Gates	4696.2	4826.4	5421.6	*7602.0*	4226.3	4387.1	*5533.0*	*137.39*
Siret/Dragesti	80.69	133.91	72.60	*167.35*	49.36	40.76	*75.54*	*221.53*
Pruth/Stanca Costesti	72.78	142.44	59.44	*183.54*	53.77	40.01	*84.12*	*218.18*
Olt/Ramnicu Valcea	118.99	121.91	121.99	*178.50*	119.77	80.60	*127.60*	*139.89*

Qmm mean multi-annual tributary flow

Table 15.6 (Qm) Mean multi-monthly tributary flow for the period between the year 2000 and 2012, compared to the mean multi-monthly tributary flow for the period October, 2011–April, 2012

| River/section | Month | | | | | | | | Multi-monthly Qm 2000–2012 | Multi-monthly Qm Oct. 2011–Apr. 2012 | % Multi-monthly Qm 2000–2012 |
	Oct. 2011	Nov. 2011	Dec. 2011	Jan. 2012	Feb. 2012	Mar. 2012	Apr. 2012				
Arges/Vidraru	5.23	3.71	3.78	3.07	3.30	5.37	25.79		*11.35*	*7.17*	*63.17*
Bistrita/Izvorul Muntelui	15.45	10.83	12.76	9.06	8.64	25.91	86.08		*39.00*	*24.10*	*61.79*
Lotru/Vidra	4.03	3.83	3.05	2.77	2.83	3.64	26.63		*10.92*	*6.68*	*61.17*
Somes/Fantanele	2.96	2.02	7.44	3.15	2.55	7.05	36.51		*11.81*	*8.81*	*74.59*
Dragan/Dragan	1.05	0.78	7.04	2.19	1.26	5.42	26.66		*8.01*	*7.47*	*93.25*
Sebes/Oasa	2.87	2.10	2.07	1.84	1.54	1.89	8.85		*4.19*	*3.02*	*72.07*
Cerna/Valea lui Iovan	2.01	1.41	1.95	1.47	1.71	5.78	19.83		*8.32*	*4.88*	*58.65*
Raul Mare/Gura Apelor	3.06	2.28	2.46	2.02	1.76	2.72	22.23		*7.52*	*5.21*	*69.28*
Buzau/Siriu	2.92	2.25	2.29	2.01	1.90	10.40	25.23		*8.79*	*6.71*	*76.33*
Bistra/Poiana Marului	1.31	0.94	1.63	1.29	1.26	3.83	17.05		*5.76*	*3.90*	*67.70*
Dambovita/Pecineagu	1.19	1.00	1.00	0.85	0.71	1.04	3.51		*1.97*	*1.32*	*67.00*
Raul Targului/Rausor	1.14	0.87	0.95	0.80	0.73	1.01	3.96		*2.26*	*1.34*	*59.29*
Cerna/Herculane	0.93	1.04	1.14	1.03	1.65	4.81	13.12		*7.81*	*3.38*	*43.27*

(continued)

Table 15.6 (continued)

River/section	Month							Multi-monthly Qm 2000–2012	Multi-monthly Qm Oct. 2011–Apr. 2012	% Multi-monthly Qm 2000–2012
	Oct. 2011	Nov. 2011	Dec. 2011	Jan. 2012	Feb. 2012	Mar. 2012	Apr. 2012			
Danube/Iron Gates	2925.1	2358.9	2751.8	3751.6	3171.2	5504.9	6061.6	*5664.41*	*3789.3*	*66.89*
Siret/Dragesti	24.70	22.30	40.00	14.94	9.72	50.86	107.23	*64.25*	*38.53*	*59.96*
Pruth/Stanca Costesti	20.90	16.60	21.71	16.58	19.76	57.87	89.63	*66.67*	*34.72*	*52.07*
Olt/Ramnicu Valcea	43.39	46.40	37.81	34.48	30.97	78.45	208.40	*115.71*	*68.55*	*59.24*

Table 15.7 Percentage of electric power estimated for an average hydrological year and obtained in the year 2011 and 2012

Catchment basin	Percentage of estimated electric power for an average hydrological year (%)	Percentage of electric power produced in 2011 (%)	Percentage of electric power produced in 2012 (%)
Bistrita (including Siret and Pruth)	10.37	11.16	7.91
Buzau	2.35	3.05	2.67
Somes	3.39	3.24	3.19
Arges	6.70	6.62	5.34
Raul Mare	5.15	2.84	2.81
Crisul Repede	2.60	2.26	2.18
Danube	37.66	39.16	51.23
Olt	23.43	25.31	18.98
Sebes	3.49	3.46	2.37
Cerna (including Motru and Jiu)	2.70	1.91	2.02
Bistra Marului	2.16	0.99	1.30

recorded in the probability zone, which is not taken into account when calculating electric power provision (2–3%). Such a period when minimum flows reoccur comprises around 50 years.

In Romania, in the last two years (2011 and 2012), the low amounts of precipitations entailed a decrease in hydropower by 4 and 5 TWh, respectively, compared to an average hydrological year, which led to a loss of 1200 million EUROS in the hydroelectric sector (Table 15.7). The tributary flows on national rivers, which produce electric power, are lower by 40–60%, than the flows estimated for a normal hydrological year.

Globally, 2.1 million GWh are produced annually, which represents between 16 and 18% of the global electricity consumption. The exploitable potential is 6–7% higher. Within an average hydrological year, Romania produces 17.33 TWh, which means 35% of the consumption. The potential in the domain is almost double [47]. In the drought period of the year 2011–2012, Hidroelectrica S.A. produced only 13 TWh.

15.4   Discussions

For a correct understanding of the phrase "normal hydrological year", it is necessary to review several hydrological terms accepted by the international forums and by the specialized institutions in Romania [14, 50–52, 54]. A hydrological year represents a continuous 12-month period, selected in such a way that the overall changes in the storage are minimal so that carryover is reduced to a minimum. The average year (normal year) represents the year for which the observed hydrological or meteoro- logical quantity, approximately equals the long-term average of that quantity. The normal hydrological year is the year when the mean annual flow has a value close to the multiannual mean for this variable ($\pm 10\%$) [53]. For it to be representative, the mean multi-annual value of mean annual flows is calculated for a long period, gen- erally covering, at least, thirty consecutive years of observations. Such forecasting is highly valid if the period analyzed is long. If the string of data is reduced, the degree of error increases. The estimated flow may be higher or lower than 1: supraunitary value if a certain interval is richer (>1); subunitary value if the interval is poorer (<1).

Runoff coefficient (η_o) represents the ratio of runoff depth (Y_0) to a precipitation depth (P_0), both expressed in equivalent heights of water, within the same period [31]:

$$\eta_0 = \frac{Y_0}{P_0} * 100 \, [\%]. \tag{15.1}$$

This parameter measures the hydrological productivity of transforming precipita- tions into the mean total precipitations. The module coefficient of mean total runoff (K_i) represents the ratio of the mean annual runoff to the mean multi-annual runoff:

$$K_i = \frac{Q_{01}}{Q_0} = \frac{q_{0i}}{q_0} = \frac{V_{0i}}{V_0} = \frac{Y_{0i}}{Y_0} [-]. \tag{15.2}$$

The module coefficient of mean total runoff may exceed or may be lower than one. Thus, expressing whether a certain interval is richer (>1) or poorer (<1), compared to the average (normal) hydrological year. Depending on necessities, the relations can be established between the different parameters. The most common of such relation is determined between Y_0 and q_0:

$$Y_0 = \frac{V_0}{10^3 * F} = \frac{31.56 * 10^6 * Q_0}{10^3 * F} = 31.56 * 10^3 * \frac{Q_0}{F} = 31.56 * q_0 \tag{15.3}$$

To increase the estimated degree, one can consider the value of the minimum flow (low-water mark). In this case, errors are reduced to around 10%. Unfortunately, the model only approximates reality, the reason for which, it is not being considered.

Forecasting error represents the difference between a forecast and the observed value. The probability event 1/n corresponds to an average return period of n years. The drought index is the computed value which is related to some of the cumulative effects of a prolonged and abnormal moisture deficiency. An index of hydrological drought corresponding to the levels below the mean, in streams, lakes, reservoirs, and the likes. However, an index of agricultural drought must relate to the cumulative effects of either an absolute or an abnormal transpiration deficit. The probability of forecast is the chance of occurrence of a quantified assessment (in real time) of future events. The long-term hydrological forecast represents the forecast of the future value of an element of the regime, of a water body, for a period extending beyond ten days, from the issue of the forecast. Medium-term (extended) hydrological forecast represents the forecast of the future value of an element of the regime of a water body, for a period ending between two to ten days from the issue of the forecast. The short-term hydrological forecast represents the forecast of the future value of an element of the regime of a water body, for a period ending up to two days from the issue of the forecast. Net volume is the value between the minimum and the maximum exploitation levels.

Most of the time the hydrological forecasts concern rapidly occurring events (tidal waves) or long-term events (seasonal). The very long ones (taking 12 months, for instance) comprise a high degree of error (30–40%). The accuracy and opportunity of the hydrological forecasts depend on several factors. Notably are the viability and on the amount of hydrological and meteorological information, on the response time of the catchment basin, on the rapidity of basin status at a certain point, on the forecasting techniques or on the models used, and on how quickly beneficiaries receive the forecast [11]. Regardless of the forecasted element, the forecasts issued involve a degree of error, expressed in percentage of the real value or absolute values (m^3/s, m^3, days, months, and so on).

The Centre of Hydrological Forecasts of the National Institute of Hydrology and Water Management in Bucharest issues short-term (1–2 days), medium-term (3–7 days) and long-term (monthly, seasonal and annual) hydrological forecasts. This is done for a series of countrywide representative forecast sections, which are also disseminated by Hidroelectrica S.A

The National Institute of Hydrology and Water Management in Bucharest— through the National Centre of Hydrological Forecasts—elaborates a short-term (1–2 days), medium-term (3–7 days), and long-term (monthly, seasonal, and annual) hydrological forecasts for a series of forecast sections representative nationwide, forecasts that were also disseminated by Hidroelectrica S.A. The precision of hydrological forecasts decreases, as the anticipation time increases; for this reason, the long-term forecasts elaborated (mostly the seasonal and annual forecasts), are used for only information and research purposes. They must be updated periodically by medium- and short-term forecasts. Generally, long-term hydrological forecasts refer to the mean monthly or annual flow regime, and they are 60% confirmed (40% error).

The National Institute of Hydrology and Water Management develops research studies and provides services in the fields of hydrology, hydrogeology, and in the management of water resources. They aim to support activities and decisions related to the effective management of water resources, during hydrological risk events (floods, droughts), and under normal situations, by the decision makers in the field: "Romanian Waters" National Administration and the Ministry of Environment and Forests.

The main fields of activity are the following:

– short, medium-, and long-term diagnoses and the hydrological forecasts of a national and trans-border interest;

- mathematical modelling of surface and underground water runoff;
- warnings in case of extreme hydrological phenomena (floods, droughts, frost) for crisis prevention and management.

Forecasts always involve probability. A physical model features the values that appear rarely, and that favours extreme events, such as droughts or tidal waves. The forecasting of hydrological elements is only possible under the conditions of cyclic repeatability, over a long period. Unfortunately, the hydrological information is on a recent date. The future cannot be an replica of the past. From both a theoretical and a practical perspective, it is impossible to regulate completely, the mean multi-annual flow used. The reservoirs capable of such annual or multi-annual regulation are situated on the rivers of Bistrita (HPP Stejaru), Arges (HPP Corbeni), Lotru (HPP Ciunget), Somes (HPP Mariselu), and Sebes (HPP Calceag). On the Danube and on the Olt River, storage is not possible, though, this river does hold 33% of Romania's hydroelectric power potential.

The most common long-term—up to 12 months—forecast method used, is the Extended Streamflow Prediction (ESP). The ESP system is that part of the NWS-RFS (National Weather Service River Forecast System) that provides the possibility of making long-term forecasts for streams. The ESP uses conceptual hydro-logic/hydraulic models to forecast future streamflow, using the current snow, soil moisture, river, and reservoir conditions with historical meteorological data. The ESP procedure assumes that some meteorological events that occurred in the past are representative of events that may occur in the future. Each year of historical meteorological data is assumed to be a possible representation of the future and is used to simulate a streamflow trace. The simulated streamflow traces can be scanned for maximum flow, minimum flow, the volume of flow, reservoir stage, and so on, for any period in the future. ESP produces a probabilistic forecast for each streamflow variable and the period of interest. Historical simulation is generated as part of a 12-month interval. The error depends on the type of flow: for minimum flow, 10–15%; for average flow 50%. For a long period (up to 6 months), another viable method to apply is the Multi-model forecast technique (applied by the Romanian experts Adler Mary-Jeanne and Valentina Ungureanu in 2006) [1].

Four models (Deterministic statistical model; Hydro-meteorological analogues; Stochastic models; Deterministic model) have been used for low flow forecasting in the extreme droughty period between the months of June and November, 1994. The models from which the best data were obtained are the Stochastic model and the Stanford model. The use of the Multi-model improves the forecast; during these months, the forecasting errors represent 10%. Therefore, the multi-model technique is highly effective. The use of the Multi-model technique increases the forecasting performances, and this is why it is recommended for operational activity (see Table 15.8) [15].

Hidroelectrica S.A. declared insolvency and made a change in the Energetic Strategy of Romania for the period between the year 2011 and 2035. In this case, there is the possibility of compromising the objectives assumed by Romania according to Annex I of the Directive, 2009/28/EC, which stipulates that the objective set for

Table 15.8 Measured flow, calculated flow (Q_c), forecasted value (Q_f), and errors (Er), in case of each model and of the Multi-model used for 6 months, June–November, 1994 (Oltet River)

Month	Measured flow (m³ s⁻¹)	Deterministic statistical model (1)		Hydromet. analogues method (2)		Stochastic model ARMA type (3)		Stanford type model (4)		Multi-model procedure (5)	
		Q_c m³ s⁻¹	Er (%)	Q_c m³ s⁻¹	Er (%)	Q_c m³ s⁻¹	Er (%)	Q_c m³ s⁻¹	Er (%)	Q_f m³ s⁻¹	Er (%)
VI	3.61	3.03	−16.07	3.12	−13.57	3.80	5.26	3.50	−3.05	3.36	−6.86
VII	3.18	3.36	5.66	3.8	19.50	3.26	2.52	3.20	0.63	3.41	7.08
VIII	1.19	1.09	−8.40	0.727	−38.91	1.26	5.88	1.50	26.05	1.48	24.42
IX	0.474	0.220	−53.59	0.624	31.65	0.550	16.03	0.900	89.87	0.46	−3.14
X	2.98	1.78	−40.27	2.5	−16.11	3.20	7.38	3.00	0.67	2.79	−6.50
XI	1.71	1.02	−40.35	1.49	−12.87	2.02	18.13	1.59	−7.02	1.59	−6.82

Romania for the year 2020, is to get 24% of the energy from renewable sources [47]. The hydrological deficit that Hidroelectrica S.A. had to face in the year 2011 and 2012 entailed a severe drop in the production capacity: 13 TWh, is comparable to the historic minimum recorded in the year 2003. Prolonged drought inevitably led to the insolvency of Hidroelectrica S.A. [47]. The debtor Hidroelectrica S.A. can be subjected to a reorganization plan, which can be successful if rapid and effective measures are taken, in terms of restructuring the activity; these measures are included in the company's reorganization plan.

For the period 2011–2035, an Strategic Energy Project was elaborated for Romania, which posits that the hydro potential represents a sustainable development alternative for the energy sector. Authorities are allowed to construct hydropower plants with an installed capacity of 1400 MW by the year 2035 when the national hydroelectric power potential will have been used up by 67% (59% by the year 2020) [47]. In such a scenario, the Hidroelectrica S.A. will consider increasing by 2035 the production capacity from 17.33 TWh (in a normal hydrological year) to 32.20 TWh. Several future projects aim to raised the capacity by 7.99 TWh. These include the 2.28 TWh representing ongoing projects that are to be finalized by 2015 (according to the Development Strategy of Hidroelectrica S.A.), a number of units on the Danube (totalling 4.66 TWh), and others with a micro-potential of 2.22 TWh.

"The energy production and resources of a country represent the strategic fields, of which the public has the legitimate right of being transparently and completely informed." Consequently, it is admitted that "Hidroelectrica is a vital company in a strategic sector, with national security implications, on which the evolution of the Romanian economy and the society as a whole depend" (Report of the insolvency administrator cited by [47]). The debtor, Hidroelectrica S.A. must be reorganized using rapid and effective restructuring methods. It is highly necessary to implement a national program for highlighting the developable hydroelectric power potential in the context of the effective management of water resources. This should be reinforced by a specific national legislation stipulating that all the administrative and economic entities that act as beneficiaries of the hydroelectric power potential must contribute to the investments made in this field. Electric power production management is based on maximizing the production obtained, using the available flow. The exploitation plans take into account the average statistical levels, which entail maximum insurance.

Unfortunately, they fail to consider the minimum levels, which are far more, accurate in terms of forecasting. Exceptional occurrences of minimum flows are assumed by the producer (Hidroelectrica S.A.) and by the buyers. It is worth mentioning that the peak electric power demand often coincides with the minimum water resources available. All developed States need electric power resources the most, during summer, when air conditioning devices are used. It is impossible to take into account the exceptional situations when planning the power and energy, to deliver from the hydropower resources nationwide. The streams on which the power plants could be constructed hold low energetic potential, which means that the investments made would not be productive. From this perspective, they are economically ineffective.

In the future, Hidroelectrica (S.A.) must include, in its portfolio, only the 129 power plants with 293 groups (with the installation capacity exceeding 4 MW), and the five pumping stations with an overall installed capacity of 6331 MW (98.1% of total installation capacity). Great-capacity power plants may also generate an optimal cost per MWh, thus, leading to increased profitability for the company. The selling of Micro Hydropower plants (MHC) will bring an infusion of about 120–150 million Euros, and it will entail a decrease in expenses for staff, maintenance, and exploitation (of around 20 million Euros/year). Hidroelectrica S.A. must focus its financial resources on updating the technology of the great-capacity power plants, which have been exploited for over 30 years [55].

15.5   Conclusions and Policy Implications

Romania's mountainous landforms favour a relatively significant hydroelectric power potential (8000 MW). In extra-mountainous areas, there are hydropower plants, situated only on Siret and Pruth, which provide important liquid discharge. Most hydropower plants in Romania are distributed on mountainous rivers: Bistrita (along with Siret and Pruth), Olt, Jiu, Arges, Lotru, Somes, Raul Mare, and so on. The largest hydrotechnical work is the one done on the Danube (Iron Gates I, with a potential of 2100 MW, shared with Serbia). Iron Gates I is the only hydropower plant in Romania, which functions permanently, and that contributes to the largest amount of hydropower.

The global climatic changes also affect Romania. For the past 25 years, drought periods are longer and more severe, though, the mean amount of precipitations has also increased. Unfortunately, heavy rains (100–300 mm/24 h) causing large amounts of water within a short time span have become more frequent and more significant; they lead inevitably, to catastrophic floods. The drought recorded in the autumn of the year 2011, and in the spring of the year 2012, entailed a drastic reduction in power production provided by the hydropower plants, the reason for which the company Hidroelectrica S.A. became unable to distribute power to the beneficiaries. Drought affected the Danube to the same extent; the river is the main hydropower producer. In this case, Hidroelectrica S.A. was sued by a series of partners who were no longer satisfied with the quality of the distribution. The contracts signed between Hidroelectrica S.A., and the partners depend on the forecast of energy production in the context of a normal hydrological year. A forecasting error may represent 10%. The drought comprising the entire Romanian territory, entailed a drastic drop in production: many hydropower plants produced only 60–70% of their capacity. Some hydropower plants were able to produce more energy because of the water storage programs conducted in the year 2010 (a year with rich rains). Unfortunately, on the Danube and on the Olt, it is impossible to store water in the context of significant flow.

The period of drought recorded between the year 2011 and 2012, entailed the insolvency of Hidroelectrica S.A. and the progress of ongoing civil cases within the Bucharest Court. Unfortunately, one must consider reassessing the hydrological forecasts, probably by taking into account, the minimum mean flows or other conditions accepted by all parties: producer, distributor, recipient. It is obvious that it would have been impossible to take into account the extreme hydrological drought recorded within the period, 30th September 2011–30th April 2012, upon signing the contract (for delivering electric power) between the Hidroelectrica S.A. and the buyers.

Acknowledgements The author would like to express his gratitude to the employees of the Romanian Waters Agency, Bucharest and Hidroelectrica S.A. Bucharest. This work was financially supported by the Department of Geography from the "Alexandru Ioan Cuza" University of Iasi, and the infrastructure was provided through the POSCCE-O 2.2.1, SMIS-CSNR 13984-901, No. 257/28.09.2010 Project, CERNESIM.

References

1. Adler MJ, Ungureanu V (2006), Multi-model technique for low flow forecasting. In: Climate variability and change—hydrological impacts (Proceedings of the fifth FRIEND world conference held at Havana, Cuba, November 2006), vol 308, IAHS Publ. pp 151–157
2. Ashaary NA, Wan Ishak WH, Ku-Mahamud KR (2015) Forecasting the change of reservoir water level stage using neural network. In: The 2nd international conference on mathematical sciences and computer engineering (ICMSCE 2015), pp 103–107
3. Bartholmes JC, Thielen J, Ramos MH, Gentilini S (2009) The European flood alert system EFAS-Part. 2: Statistical skill assessment of probabilistic and deterministic aplicational forecasts. Hydrol Earth Syst Sci 13:141–153
4. Bauzha T, Gorbachova L (2017) The Features of the Cyclical Fluctuations, Homogeneity and Stationarity of the Average Annual Flow of the Southern Buh River Basin. Ann Valahia Univ Targoviste, Geogr Ser 17(1):5–17. https://doi.org/10.1515/avutgs-2017-0001
5. Blumer VB, Mühlebach M, Moser C (2014) Why some electricity utilities actively promote energy efficiency while others do not—a Swiss case study. Energ Effi 7(4):697–710
6. Cheval S, Baciu M, Dumitrescu A, Breza T, Legates DR, Chendes V (2011) Climatologic adjustments to monthly precipitation in Romania. Int J Climatol 31(5):704–714
7. Christensen NS, Wood AW, Voisin N, Lettnmaier DP, Palmer RN (2004) The effects of climate change on the hydrology and water resources of the Colorado River basin. Clim Change 62:337–363
8. Croitoru AE, Minea I (2015) The impact of climate changes on rivers discharge in Eastern Romania. Theoret Appl Climatol 20(3–4):563–573. https://doi.org/10.1007/s00704-014-1194-z
9. DFID (2009) Water storage and hydropower: supporting growth, resilience and low carbon development, A. London: DFID evidence-into-action paper, Department for International Development
10. Dascalu SI, Gothard M, Bojariu R, Birsan MV, Cica R, Vintila R, Adler MJ, Chendes V, Mic RP (2016) Drought-related variables over the Barlad basin (Eastern Romania) under climate change scenarios. CATENA 141:92–99
11. Diaconu CD, Jude O (2009) Prognoze hidrologice. Editura Matrix Rom, Bucuresti (in Romanian)
12. Dimitrovska O, Radevski I, Gorin S, Taleska M (2015) Quality of the environment in mountain areas and sustainable use of mountain resources. In: International scientific conference geobalcanica, pp 1193–1198
13. Drobot R, Serban P (1999) Aplicatii de hidrologie si gospodarirea apelor. HGA Press, Bucuresti (in Romanian)

14. EPC, Consultanta de Mediu (2013) Raport de Mediu Strategia Energetica a Romaniei pentru perioada 2007–2020 actualizata pentru perioada 2011–2020, pagina 143–144, iulie 2012. EPC, Consultanta de Mediu. Link: http://www.mmediu.ro/beta/wp-content/uploads/2012/07/2012-07-31_evaluare_impact_planuri_raportmediustrategiaenergeticaromania.pdf, 10 Oct 2015. Accessed 02 Dec 2015 (in Romanian)

15. Euro Insol (2013) Plan de reorganizare a Activitatii Debitoarei Societatea Comerciala de Producere a Energiei Electrice in Hidrocentrale Hicroelectrica S.A. Tribunalul Bucuresti, Sectia a VII-a Civila, Dosar nr. 22456/3/2012 (in Romanian)

16. Gleick PH (2015) Impacts of California's ongoing drought: hydroelectricity generation. Pacific Institute, Oakland

17. Guegan M, Madani K, Uvo CB (2012) Climate change effects on the high-elevation hydropower system with consideration of worming impacts on electricity demand and pricing. University of California, Riverside

18. Hapciuc OE, Minea I, Romanescu G, Tomașciuc AI (2015) Flash flood risk management for small basins in mountain-plateau transition zone. Case study for Suceviţa catchment (Romania), International multidisciplinary scientific geoconference—SGEM din Albena, Bulgaria, Conference proceedings, hydrology and water resources, pp 301–308. https://doi.org/10.5593/sgem2015/b31/s12.039

19. Hapciuc OE, Romanescu G, Minea I, Iosub M, Enea A, Sandu I (2016) Flood susceptibility analysis of the cultural heritage in the Sucevita catchment (Romania). Int J Conserv Sci 7(2):501–510

20. Hapciuc OE, Iosub M, Tomașciuc AI, Minea I, Romanescu G (2016) Identification of the potential risk areas regarding the floods occurrence within small mountain catchments. In: Geobalcanica 2nd international scientific conference, 177–183

21. IEA (2009) Integration of wind and hydropower systems. International Energy Agency (IEA), R&D, Task 24

22. INHGA (2015) Date hidrologice. Institutul National de Hidrologie si Gospodarire a Apelor, Bucuresti (in Romanian)

23. Iordache I, Ursu A, Liviu M, Iosub M, Istrate V (2016) Using MODIS imagery for risk assessment in the cross-border area Romania-Republic of Moldova. In: 6th international multidisciplinary scientific geoconference SGEM 2016, SGEM2016 conference proceedings, vol 2, Photogrammetry and remote sensing, pp 1075–1082. https://doi.org/10.5593/sgem2016/b22/10.137

24. Iosub M, Iordache I, Enea A, Romanescu G, Minea I (2015) Spatial and temporal analysis of dry/wet conditions in Ozana drainage basin, Romania using the standardized precipitation index, international multidisciplinary scientific geoconference—SGEM, Albena, Bulgaria, pp 585–592. https://doi.org/10.5593/sgem2015/b31/s12.075

25. Iosub M, Tomasciuc AI, Hapciuc OE, Enea A (2016) Flood risk analysis in Suceava city, applied for its' main river course. In: Proceedings, 2nd international scientific conference GEOBALCANICA 2016, 10–12 June 2016, Skopje, Republic of Macedonia, pp 111–118. https://doi.org/10.18509/gbp.2016.15

26. Irimia LM, Patriche CV, Quenol H, Sfica L, Foss C (2018) Shifts in climate suitability for wine production as a result of climate change in a temperate climate wine region of Romania. Theoret Appl Climatol 131(3–4):1069–1081. https://doi.org/10.1007/s00704-017-2033-9

27. Liu P, Nguyen TD, Cai X, Jiang X (2012) Finding multiple optimal solutions to optimal load distribution problem in hydropower plant. Energies 5:1413–1432

28. Mantel SK, Hughes DA, Slaughter AS (2015) Water resources management in the context of future climate and development changes: a South African case study. J Water Clim Change 6(4):772–786

29. Marengo JA, Tomasella J, Alves LM, Soares WR, Rodriguez DA (2011) The drought of 2010 in the context of historical droughts in the Amazon region. Geophys Res Lett 38(12):L12703. https://doi.org/10.1029/2011GL047436

30. Matreata M (1997), Dinamico-statistic model for low flows. Studii de hidrologie, INMH, pp 24–37 (in Romanian)

31. Minea I, Romanescu G (2007) Hidrologia mediilor continentale. Aplicaţii practice, Demiurg Press, Iaşi (in Romanian)
32. Minea I, Croitoru AE (2015) Climate changes and their impact on the variation of groundwater level in the Moldavian Plateau (Eastern Romania). In: International multidisciplinary scientific geoconferences, SGEM 2015, 15th geoconference on water resources, forest, marine and ocean ecosystems, conference proceedings, vol I, Hydrology and water resources, pp 137–145
33. Minea I, Croitoru AE (2017) Groundwater response to changes in precipitations in north-eastern Romania. Environ Eng Manag J 16(3):643–651
34. Mukheibir P (2007) Possible climate change impacts on large hydroelectricity schemes in Southern Africa. J Energy South Afr 18(1):4–9
35. Murarescu O, Muratoreanu G, Frinculeasa M (2014) Agrometeorological drought in the Romanian plain within the sector delimited by the valleys of the Olt and Buzau Rivers. J Environ Health Sci Eng 12(1):152
36. Obaid RR (2015), Seasonal-water dams: a great potential for hydropower generation in Saudi Arabia. Int J Sustain Water Environ Syst 7(1):1–7. https://doi.org/10.5383/swes.7.01.001
37. Potopová V, Cazac V, Boincean B (2017) Flood and drought risk management at the catchment level: a case study in the Republic of Moldova management. Rizik sucha a záplav na úrovni povodí: případová studie v moldávii. Vliv abiotických a biotických stresorů na vlastnosti rostlin (Influence of abiotic and biotic stresses on properties of plants 2017), Czech University of Life Science Prague 12–14(9):170–173
38. Radevski I, Gorin S (2017) Floodplain analysis for different return periods of river Vardar in Tikvesh valley (Republic of Macedonia). Carpathian J Earth Environ Sci 12(1):179–187
39. Romanescu G, Stoleriu C, Romanescu AM (2011) Water reservoirs and the risk of accidental flood occurrence. Case study: Stanca–Costesti reservoir and the historical floods of the Prut River in the period July–August 2008, Romania. Hydrol Process 25(13):2056–2070
40. Romanescu G, Zaharia C, Stoleriu C (2012) Long-term changes in average annual liquid flow river Miletin (Moldavian Plain). Carpath J Earth Environ 7(1):161–170
41. Romanescu G, Tirnovan A, Cojoc GM, Sandu IG (2016) Temporal variability of minimum liquid discharge in Suha basin. Secure water resources and preservation possibilities. Int J Conserv Sci 7(4):1135–1144
42. Romanescu G, Cimpianu CI, Mihu-Pintilie A, Stoleriu CC (2017) Historic flood events in NE Romania (post-1990). J Maps 13(2):787–798
43. Romanescu G, Stoleriu C (2017) Exceptional floods in the Prut basin, Romania, in the context of heavy rains in the summer of 2010. Nat Hazards Earth Syst Sci 17:381–396
44. Romanescu G, Hapciuc OE, Minea I, Iosub M (2018) Flood vulnerability assessment in the mountain-plateau transition zone. Case study for Marginea village (Romania). J Flood Risk Manag 5502–5013. https://doi.org/10.1111/jfr3.12249
45. Rosu L, Zagan R (2017) Management of drought and floods in Romania. In: Natural resources management: concepts, methodologies, tools, and applications. Information Resources Management Association, IGI Global, Hershey PA, USA, p 2063. https://doi.org/10.4018/978-1-5225-0803-8.ch002
46. Stagge JH, Kohn I, Tallaksen LM, Stahl K (2015) Modeling drought impact occurrence based on meteorological drought indices in Europe. J Hydrol 530:37–50
47. Stematiu D (2013) Hidroenergia – componenta esentiala in cadrul surselor de energie regenerabila. Lucrarile celei de-a VIII-a editii a Conferintei anuale a ASTR, Sectiunea Energie Durabila, pp 273–280 (in Romanian)
48. Sfica L, Croitoru AE, Iordache I, Ciupertea AF (2017) Synoptic conditions generating heat waves and warm spells in Romania. Athmosfere 8(3):50. https://doi.org/10.3390/atmos8030050
49. Tsakiris G, Vangelis H (2004) Towards a drought watch system based on spatial SPI. Water Resour Manag 18(1):1–12. https://doi.org/10.1023/B:WARM.0000015410.47014.a4
50. UNESCO (2015) http://webworld.unesco.org/water/ihp/db/glossary/glu/RO/GF0195RO.HTM, 1 Sept 2015. Accessed 5 Dec 2015

51. UNESCO (2015) http://webworld.unesco.org/water/ihp/db/glossary/glu/aglo.htm, 1 Sept 2015. Accessed 8 Dec 2015
52. UNESCO (2015) Glosar International de Hidrologie. http://webworld.unesco.org/water/ihp/db/glossary/glu/HINDRO.HTM, 1 Sept 2015. Accessed 9 Dec 2015
53. WMO (2012) International glossary of hydrology. World Meteorological Organization, Geneve, WMO-No. 385
54. WWF (2015) http://d2ouvy59p0dg6k.cloudfront.net/downloads/wwf_7_mituri_despre_ hidroenergie.pdf, 10 Oct 2015. Accessed 9 Dec 2015
55. Zelenáková M, Fijko R, Diaconu DC, Remenáková I (2018) Environmental impact of small hydropower plant—a case study. Environments 5(12):1–10

Chapter 16
Water Resources from Apuseni Mountains—Major Coordinates

Răzvan Bătinaş, Gheorghe Şerban and Daniel Sabău

Abstract The strong precipitation and cool climate render the Apuseni Mountains a real "water castle" with a complex, radial development of the river network. The density of the river network has an average value between 0.6 and 1.0 km/km^2, higher than the values found in the Eastern and Southern Carpathians; a deviation from these values occurs in the karstic areas where the surface drainage is replaced by an underground one: 0.4–0.5 km/km^2 in the Pădurea Craiului Mountains, Vaşcău Plateau and Trascău Mountains (Pascu 1983). The hydrographic network is a result of the *climate*; therefore, the quantitative precipitation differences between the two slopes of the Apuseni Mountains are also manifested in the specific average flow rate: 20–40 l/s km^2 on the west side of the Apuseni Mountains and 10–20 l/s km^2 on the eastern side. According to this parameter, the entire mountain region is among the country's water-surplus areas. The areas with moderate water resources belong to the peripheral and depression areas, with values of 5–10 l/s km^2. The water resources evaluation is done by analyzing the hydrological regime, respectively by establishing the water balance and calculating the aridity index, rated for Romania by I. Ujvari, since 1972. The introductory chapter presents aspects related to the general organization of the available water resources, followed by an assessment of the factors determining the water flow. The observations related to the water flow parameters were analyzed in the third part of the study, including details about the parameters of liquid flow.

Keywords Apuseni mountains · Water resources · GIS · Discharge

R. Bătinaş · G. Şerban (✉)
Faculty of Geography, Babeş-Bolyai University, 5-7 Clinicilor, 400006 Cluj-Napoca, Romania
e-mail: gheorghe.serban@ubbcluj.ro

R. Bătinaş
e-mail: razvan.batinas@ubbcluj.ro

D. Sabău
"Romanian Waters" National Administration—"Someş-Tisa" Regional Water Branch, 17 Vânătorului, 400213 Cluj-Napoca, Romania
e-mail: danielsabau075@gmail.com

© Springer Nature Switzerland AG 2020
A. M. Negm et al. (eds.), *Water Resources Management in Romania*, Springer Water,
https://doi.org/10.1007/978-3-030-22320-5_16

16.1 Introduction

Being a very sophisticated and comprehensive term, the water resource is a particularly important, even decisive, component for the existence of natural and anthropic habitats. Whether we are talking about a surplus, balanced or deficient water resource in a territory, it is obvious the necessity for its management, in the context of increasing vulnerability and the aggravation and chronicisation of dangerous meteorological and hydrological phenomena. Frequently, some seasons require compensation in terms of the distributed water quantities to the two environments (natural and anthropic). Thus, the needs of human population can be achieved through hydrotechnical facilities which can redistribute the water resource in a uniform manner throughout the year. A deficit in the natural environment can have serious repercussions in the life cycle of plants and, indirectly, to the animal organisms.

The mountainous areas in the Romanian Carpathians still maintain a less aggressive hydrosphere environment by the anthropic factor. This is often due to the establishment of protected natural areas, which impose stricter rules on the exploitation of water resources and human intervention in nature.

The high degree of anthropology in the Apuseni Mountains, with the exceptions of the eastern, northeast and northern parts, due to the special fragmentation of this mountain group, has left its mark on the water resource, especially from a qualitative point of view. The effects of mining and residues from households have affected the water resources, making them frequently unusable and unpurified.

Apuseni Mountains are situated on the movement trajectory of the western air masses, which generated a heterogeneous water resources distribution in the mountainous area. The steep western flank (compared with the altitudes of Dealurile de Vest/Western Hills and Câmpia de Ves/Western Plain), is defined by a higher discharge value, resulted from the elevated rainfall amounts as well. Moreover, many scientific meteorological studies grant the Stâna de Vale resort, from the Vlădeasa Massif, the status of "*the pole of precipitation at the national level*", with values exceeding the threshold of 1500 mm per year.

The eastern flank, marked by the development of the Someşul Mic and Arieş basins and the direct tributaries of Mureş River drained from Trascău and Metaliferi Mountains is defined by a lower water budget. This is the consequence of a "*precipitation shadow*" formation due to foehnal wind circulation, which moves to the East towards the Transylvanian Depression.

16.1.1 Literature About Water Resources in Apuseni Mountains

The study of the natural or anthropogenic hazards affecting the water resources or the ecosystems is one of the main purposes for numerous authors [2, 4, 6, 8, 14, 41, 61, 64, 65, 68] etc.

Generally, talking about water resources, many authors refers to the efficient use and economic component of this [3, 25, 26, 32, 34, 39, 66, 71].

The water bodies analysis from different natural units or extended areas is another preoccupation of the researchers [9, 19, 27, 33, 35, 54, 67, 69, 71].

Important are also the feeding sources of some water reservoirs or of the reservoirs themselves, since their monitoring can assure a balance in the profitable resource use [12, 17, 18, 20, 45, 52, 63, 70].

The ground or underground water resource is frequently correlated to the precipitations input since there are important similitudes and interdependences in the quantities' fluctuation [59, 72, 73].

The study of the water resource available within the western mountainous area has started consistently along with the detailed research carried out by various authors for the hydrological analysis of the large basins spread in this region. Complex analyzes of the entire area have been undertaken by several authors and approached the issue of liquid discharge [31], hydrographic regionalization [40] and alluviums transported by watercourses [10, 43]. The maximum discharge was tackled considering its effects on human communities and the environment in general, by analyzing flash floods [2, 4, 61], for defining the estimation and modelling methodologies for flash floods [16, 28, 29].

Thus, the area of Criş rivers has been analyzed from the water budget perspective [11] and through several complex hydrological syntheses [51], or by the influence of reservoirs on the flow transition [23, 30, 44, 47, 48]. Hydrological approaches have also studied extreme phenomena such as river drought [1] or rivers' winter phenomena [50]. Topics related to the existing hydrotechnical facilities were discussed for Someşul Mic basin, with analysis of the liquid flow evolution [60, 63].

The underground water reserves were also a subject that was analyzed in conjunction with the heterogeneous geology of the mountainous area, with a particular emphasis on the karstic sectors [42, 53]. The vastly varied and complex metallogenic potential also led to studies that approached the impact of mining activities on water resources [5, 7, 15, 22, 24].

16.1.2 Apuseni Mountains in Carpathian and National Context

The Carpathians, the immense and complex transnational orogenetic structure (crossing through Czech Republic, Slovakia, Poland, Ukraine, Romania and Serbia—Figs. 16.1 and 16.2) are stretching along 1500 km, between Vienna and the Timoc River. They are exceeding in length other mountain range from Europe like the Alps (1000 km), the Dinaric Alps (800 km), the Stara Planina Mountains (500 km) and the Pyrenees (450 km). The area occupied by the Carpathian Mountains is about 170.000 km^2, higher than that occupied by the Alps (about 140.000 km^2) [80]. The major mountain groups are divided as follow: the northern part is represented by

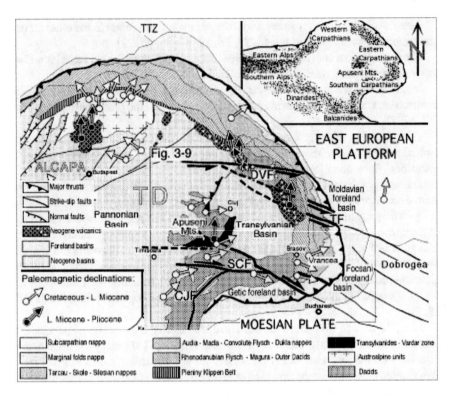

Fig. 16.1 Tectonic map of Carpatho-Pannonian system. Adapted after Linzer et al. [36]

Tatra Massif, followed to the south by Beschids Mountains and then the Romanian Carpathians.

The Tatra Massif presents itself as a block more isolated than the "arch itself" but very well individualized due to the central-southern granite platoon, which gives it massiveness and high altitudes. The edifice complicates and differs greatly with the transition to its southern half due to the multiple contacts between the lithospheric plates on which it has formed and numerous faults and secondary fractures resulted due to the internal pressure and processes produced between these plates (Fig. 16.1).

Of all the Carpathian subunits, the Romanian ones seem to have a very special structure and complexity, determined by the contact between the Eastern European Platform and the Moesian Platform, which induced a spatial deformation of the mountain chain in the form of a vast and sophisticated curve. To the west of this curve, beyond the submerged basin of the Transylvanian Depression, another Carpathian subunit was identified, built on the tectonic connections with the main chain and the fractures produced at the western boundary of the mentioned basin. This unit is known as the Western Carpathians, which from north to south, are divided into Apuseni Mountains, Poiana Ruscă Mountains and Banat Mountains (Fig. 16.2).

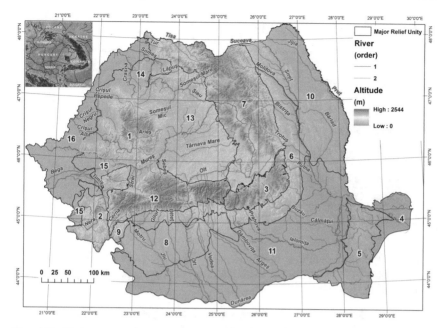

Fig. 16.2 The Apuseni Mountains in Romanian and Carpathian context. In inset all Carpathian Mountain Range (after https://upload.wikimedia.org/wikipedia/commons/thumb/5/5a/Carpathians_dem.jpg/580px-Carpathians_dem.jpg)

The Apuseni Mountains are the largest component of this Western Carpathian unit, covering an area of about 11,420 km^2 and a maximum altitude of 1849 m at the Cucurbata Mare Peak in the Bihor massif. Their spatial delimitation is quite clear, being separated from the neighboring natural units by various distinctive elements present in relief (Fig. 16.2):

- to the South and South-East, the Mureș valley separates this mountainous sector between the towns of Păuliș, in the West and Alba Iulia, in the East, towards Lipovei Hills, Poiana Ruscă Mountains and Orăștie Corridor (Table 16.1);
- to the West, the boundary is again quite clear towards the Western Hills (Dealurile de Vest), as the mountain range penetrates digitally to the West Plain, forming a few elongated depressions (Zarand, Beiuș and Vad-Borod), former bays of the Pannonian Sea. The differences are more than visible on digital elevation model map (DEM) on the alignment between Păuliș in the south and Brusturi & Suplacu de Barcău settlements, in the north.
- to the North, on the line between the localities of Suplacu de Barcău, Cizer, Moigrad-Porolissum, Călățele, Căpușu Mic and Gilău, the limit is sufficiently obvious; the separation is made against the Șimleului Depression, Almaș-Agrij Depression, Huedin Depression and Păniceni Plateau.
- to the East, even if the transition is made at a much higher altitude, it is quite clear by the specificity created by hilly and corridor spaces; between Gilău, Iara,

Table 16.1 Major relief units of Romania

No.	Name	No.	Name
1	Apuseni Mountains	9	Mehedinţi Plateau
2	Banat Mountains and Poiana Ruscă Mountains	10	Moldavian Plateau
3	Carpathians of Curvature	11	Romanian Plain
4	Danube Delta and Razim-Sinoe lagoon system	12	Southern Carpathians and Orăştie Corridor
5	Dobrogea Plateau	13	Transylvanian depression
6	Eastern and Southern Subcarpathians	14	Western Hills (North of Mureş)
7	Eastern Carpathians	15	Western Hills (South of Mureş)
8	Getic Piedmont	16	Western Plain

Copăceni, Poiana Aiudului, Ighiu and Alba Iulia, the separation takes place across Depression of Iara, Dealul Feleacului, Măhăceni Plateau, Aiud Hills and Turda—Alba Iulia Depression.

The altitude, defined by the presence of three higher areas (Bihor, Vlădeasa, Muntele Mare and Gilău massifs) and completed by the peripheral areas of lower altitudes, profoundly marks the distribution of the bio-edaphic components under the decisive influence of the climatic characteristics (Geografia României—Vol. I, 1983).

Besides, the potential of mountain habitat is determined by the morphological conditions mentioned, being influenced, among other things, by the petrographic variety, which at the level of the Apuseni Mountains is an unrepeatable mosaic within the Carpathian arch.

The mountainous area of the Apuseni Mountains is characterized by a hydrographic network with a development dictated by the position of three major hydrographic areas: Someşului Mic basin, Crişuri basin and Mureş basin to which are added some basins of lower hydrological relevance, Upper Barcău and Middle Someş (through the Almaş and Agrij upper basins) (Geografia României—Vol. I, 1983).

16.1.3 Sources and Technics Used in the Study Development

For the hydrological analysis we used information available online on the sites of the Mureş, Someş-Tisa and Crişuri basins Water Administration. An additional source of data was the River basin management plans, updated on the national level. The interpretation of the data regarding the liquid flow took into account the cartographic interpretation of some thematic resources, made available from the competent authority of Hungary (for the Tisa basin, as the rivers in the Apuseni River are tributary to it). The raw data interpolation, the calculation of the hydrological balance items, were performed using ArcGIS 10.4, respectively MS Excel 2016, available on the labs of Faculty of Geography, from Babeș-Bolyai University.

16.2   Natural Components Which Condition the Water Resource

The approach of this chapter will be done in a somewhat atypical sequence, depending on the production of natural phenomena and processes, ranging from the precipitation component to the development of the surface and underground water resource. The analysis of each natural component conditioning the liquid flow will be made on the basis of the major hydrographic basins that collect the affluents related to the mountainous area of the Apuseni Mountains.

16.2.1   Climatic Components (Precipitations and Evapotranspiration)

From the climatic point of view, the Apuseni Mountains area is defined by the positioning on the trajectory of the western air circulation, the relatively modest altitudes and the fragmentation of the relief. The monitoring of climatic parameters is carried out by eight meteorological stations, located both at altimetric superior levels (north-central area) and at the periphery of the mountainous area (Fig. 16.3).

The climate is continental moderate with western influences. The development of the central high-altitude zone on the north-south direction, as a genuine orographic barrier, led to an asymmetric distribution of the rainfall budget in favor of the western slope. Thus, at similar altitudes, precipitation discharges record sensitively different values: at Stâna de Vale rains 1571 mm (at an altitude of 1110 m), whereas on the eastern slope at Băişoara station was recorded only 847 mm (at an altitude of 1384 m).

The comparison can also be applied to low altitude stations located at the periphery of the mountainous area, in the marginal depressions (Table 16.2). At Ştei, on the western flank, precipitation accumulates a layer of 681 mm (278 m altitude), while in Turda, on the eastern slope, rains only 552 mm (at 318 m altitude). The foehn processes are occurring mainly on the eastern slope, leading to an increase in the air temperature, the decrease of air humidity and the nebulosity, and implicitly a diminished contribution to watercourses budget.

In the depressions, there is a sheltered climate with higher frequencies of atmospheric calm, but also with frequent thermal inversions. Developments over the last decades suggest changes in climate peculiarities, as winters become warmer and dry, and snow days are fewer.

Regarding the impact of climate change on liquid flow, a possible extension on a long period of the western hydrological regime, specific to the Crişuri rivers, with

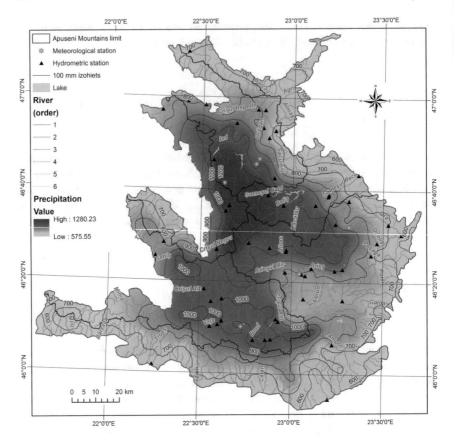

Fig. 16.3 Multiannual average rainfall isohyetal map in the Apuseni Mountains. Adapted after ICPDR (2009), NMA (2008)

high water at the beginning of the spring and frequent rainfalls could spread to other rivers in Apuseni, currently with a stable winter hydrological regime. Another aspect of climate changes with significant effects on the rivers is the increase of the rainy torrentiality, a phenomenon that can have devastating hydrological effects on the small water courses in the mountain area, causing dangerous flash floods [21].

High precipitation values are concentrated on the central axis (Bihor and Vladeasa massifs), over 1000 mm, but also to the southwestern part (ridge of the Codru-Moma massif). At the mountain periphery, the rainfall decreases to just below 600 mm. The precipitation map was generated on the basis of materials developed by the ICPDR— International Commission for Protection of Danube River (Sub-Basin Level Flood Action Plan in the Tisza River Basin, 2009), validated by the National Meteorological Administration [91].

The combined loss of soil water is achieved by direct evaporation, or by plant sweating. Thus, evapotranspiration at the level of the studied area presents values between 669 and 376 mm (Fig. 16.4). The minimum values are associated with the

Table 16.2 Main climatic features of the meteorological stations within Apuseni Mountains

No	Meteorological station	Altitude (m)	Precipitation (mm)	Temperature (°C)			Humidity (%)
				Average	Max.	Min.	
1	Băișoara	1360	847.4	4.8	29.6	−25.3	75
2	Câmpeni	611	738.0	7.3	35.8	−32.7	81
3	Roșia Montană	1198	739.0	5.4	34.7	−29.7	76
4	Stâna de Vale	1108	1570.7	3.9	30.6	−28.9	89
5	Vlădeasa 1400 m	1404	1360.5	4.9	27.6	−23.8	80
6	Vlădeasa 1800 m	1836	1151.3	1.0	25.2	−30.0	85
7	Huedin	560	596.7	7.9	38.4	−26.3	77
8	Ștei	278	681.0	9.8	37.2	−24.1	78

Source Gaceu (2006)

high mountainous areas (Bihor, Vlădeasa, Muntele Mare, Metaliferi), while the maximum values are found on western apophyses of the Apuseni Mountains (Zarandului, Codru-Moma) and in the northern part (Plopiș, Meseș) with low altitudes.

16.2.2   Geological and Morphological Components

The Apuseni Mountains represent a true petrographic mosaic, where the granite platoons alternate with large expansions of crystalline shale—both offering in the relief, large and extensive massiveness, maintained at a high altimetric level, with calcareous plateaus, marble structures, rocky clays peripheral to the mountainous area often broken by fault lines [92] (Fig. 16.5).

Under the impulse of the Meso-Cretaceous orogenesis, through the bending of mountain side, sinking of the Transylvanian Depression and then of the Pannonic Depression, the Apuseni Mountains remained exonerated, being fragmented into a number of compartments defined by the fault lines, which later become supporting axes for marginal and contact depressions, which surrounds the mountain ensemble. Wide depression sectors have occurred, alternating with narrow stretches of the gorge' sectors as they appear along the Arieș River, Ampoi, Geoagiu or Criș rivers. The individual hydrographic basin analysis revealed specific features that highlight the heterogeneity of petrography and morphology of the studied area. Large petrographic diversity, tectonic and morphological fragmentation determined by river valleys has generated an inventory of 75 relief units represented by mountains (ridges and massifs), depressions, hills and corridors (Tables 16.3 and 16.4).

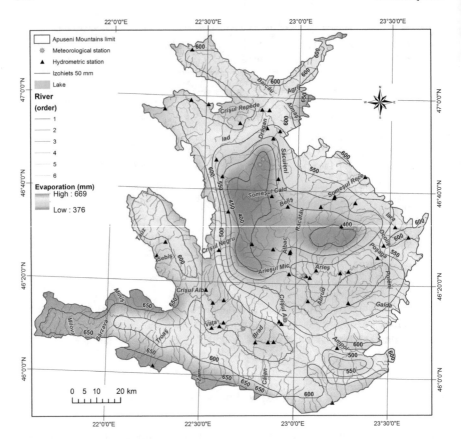

Fig. 16.4 Multiannual average evapotranspiration isohyetal map in the Apuseni Mountains. *Source data* [91]

Mureş (Arieş) basin

In the Arieş basin are included geological formations that are extremely diverse as composition, which is reflected in the physiognomy of the relief and in the chemical characteristics of springs, rivers and lakes. The petrographic mosaic of different geological ages represents the proof of uncontrollable movements of the earth's crust in the region, where almost all the geological events specific to the Carpathian domain can be found [92].

The crystalline foundation is developed between the Arieş valley and the Crişul Repede River, occupying entirely Muntele Mare massif and the southern part of the Bihor Mountains, being the largest crystalline island in the Western Carpathians [92]. The area of the crystalline also extends to the south of Arieş, having an island character in the Metaliferi Mountains and in the Trascău Mountains. Pre-Cambrian and Cambrian crystal formations are penetrated into the Muntele Mare massif by granitic Hercynian age intrusions, to which are added crystalline limestone strips and crystalline dolomites, especially in the Trascău Mountains (Fig. 16.5).

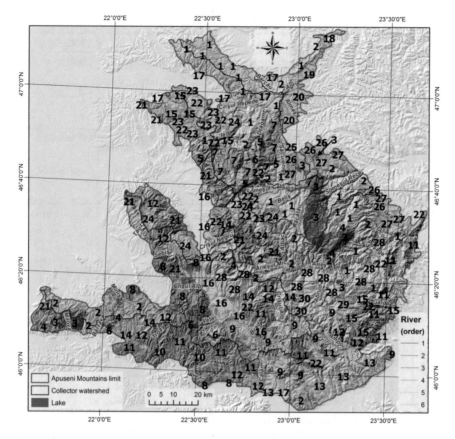

Fig. 16.5 Major categories of rocks in the Apuseni Mountains petrographic mosaic (adapted after Geologic map of Romania, scale 1:200.000—Cluj & Brad sheets, 1968)

Sedimentary formations include petrographic complexes from paleozoic to quaternary, determined by an extremely long period of sea transgressions, which led to the formation of the most diverse rocks, as a result of the varied deposition conditions [92]. Thus, the sedimentary formations of paleozoic age are represented by the sandstone and conglomerates forming the *Arieşeni cloth*. Mesozoic age sediments are found in Bihor, Muntele Mare and Trascău massifs, and are represented by conglomerates, limestones and triassic dolomites.

On the karst plateau of Bihor, but also in the Bedeleu and Petreşti massifs of the Trascăului ridge appear Jurassic formations made of massive limestone, placed transgressively over the crystalline substratum, alternating with ophiolites deposits.

The Cretacean age formations occupy large areas of Găina massif, Metaliferi Mountains and Trascău Mountains, being represented by sandstone, conglomerate and marls (Table 16.3). The tertiary sediments are located at the mountain contact with depression area, respectively in the eastern part of Muntele Mare and in the northeast of the Trăscau Mountains, towards the Plateau of Transylvania. Dominant

Table 16.3 The petrographic map' legend of Apuseni Mountains

No.	Petrography	Geologic age	No.	Petrography	Geologic age
1	Mica-shale, paragnaise	Anteproterozoic	16	Marl-clay, sand, gravel	Pliocene
2	Sericito-chlorite shale	Proterozoic	17	Sandstone, limestone, marl, gravel	Miocene
3	Granits	Paleozoic & Precambrian	18	Sandstone, marl-clay	Inf. Miocene
4	Gnaise	Paleozoic & Precambrian	19	Limestone, marl, gypsum, sandstone, clay	Sup. Eocene
5	Andesite	Paleogen & Cretaceous	20	Clay, sand, sandstone, bituminous marly-limestone	Mid. Oligocene
6	Diorite	Paleogen & Cretaceous	21	Conglomerates, tuff sandstones, tuffs, purplish clayed shales	Permian
7	Dacite	Paleogen & Cretaceous	22	Massive limestone	Sup. Jurassic
8	Andesite	Neogene	23	Micro conglomerate, sandstone, clayed shale, limestone, marl	Inf. Jurassic
9	Quartz andesite	Neogene	24	White and pink massive limestone, dolomite, clayed shale	Sup. Triasic
10	Porphyritic granodiorites	Paleogene & Sup. Cretaceous	25	Limestones, marls, gypsum, sandstone, clay	Sup. Eocene
11	Basalt (melafire and diabase)	Mezozoic & Permian	26	Marl, gypsum, clay	Mid. Eocene
12	Calcarenites, mudstone, marly-limestone, conglomerates, silicolites	Inf. Cretaceous	27	Continental red clay (Inferiour striped clay)	Paleocene
13	Tuff, conglomerates, sandstone, marl	Sup. Cretaceous	28	Conglomerate, sandstone, marl	Sup. Cretaceous

(continued)

Table 16.3 (continued)

No.	Petrography	Geologic age	No.	Petrography	Geologic age
14	Sandstone, clay-shale, conglomerates, limestone	Inf. Cretaceous	29	Aleuritic flysch + micro conglomerate and quartz sandstone	Mid. Cretaceous
15	Sandstone, clay-shale, marl-shale, conglomerates, limestone + conglomerate, calcarenite	Inf. Cretaceous	30	Conglomerates, calcarenites, marl shale	Inf. Cretaceous

formations are those of paleogenic age (located in the Iara Depression), followed by neogene ones (limestone, sands and Pannonian gravels).

The eruptive (magmatic) formations are chronologically found in the Muntele Mare massif, represented by hercynian age granitic intrusions and by eruptive formations in the Arieșeni area [92]. In Trascău Mountains, there are Jurassic age ophiolitic magmatic rocks while in Găina massif and in some areas of Muntele Mare, can be spotted banatitic Cretacic age magmatic rocks (granites, andesites). Also, in Metaliferi Mountains, can be found quaternary volcanic rocks (basalt and andesite), which were the object of the first geological observations, explained by their metalogenetic interest associated to gold-silver deposits (the northern part of the "Golden Quadrillater").

The relief of the basin is described by a general altitude decrease in the West-East direction with local peculiarities imposed by the presence of limestone masses, which appear as a discordant note in the landscape (the abrupt of Bedeleu, Petreștilor ridge, both located in Trascău Mountains Fig. 16.6).

An important role in the development of the hydrographic network and implicitly in the formation of superficial flow has a number of elements specific to the relief: the orientation of the slopes (in correlation with the general circulation of the air masses), the fragmentation density and the depth of fragmentation (expressed by relief energy), slopes' tilting.

Thus, the slope orientation favors the ones exposed to the dominant western direction air masses, and implicitly in providing additional water from precipitation. In the area of tributaries' springs, the relief has high slopes that favor the rapid flow of water and the development of some fluvial modeling processes, ranging from torrents and ravines on the slopes to the strong erosion of the river bed.

Crișul Alb basin

The morphology of the Crișul Alb basin, as an integrant part of the Apuseni Mountains, is defined by the southern flanks of the Bihor Mountains (Găina massif), the western sector of Metaliferi Mountains, Brad and Halmagiu depressions and a part of the southern slopes of the Codru-Moma Massif. The area associated with the Găina

Table 16.4 The occupied surface of relief morphological units of Apuseni Mountains

No.	Relief unit	Area (km²)	No.	Relief unit	Area (km²)	No.	Relief unit	Area (km²)
1	Hills of Hălmaşd	77.34	26	Abrud depression	39.09	51	Zlatna depression	26.62
2	Hills of Meseş	155.84	27	Albac depression	57.01	52	Bihor Massif	277.33
3	Fangs of Trascău	60.21	28	Almaş-Agrij depression	609.86	53	Drocea Massif	287.56
4	Bedeleu Ridge	421.76	29	Almaş-Balşa depression	50.91	54	Gilău Massif	698.96
5	Hăjdate Ridge	57.51	30	Depression of Băiţa	28.56	55	Highiş Massif	453.35
6	Henţ Ridge	159.84	31	Depression of Beiuş	351.29	56	Moma Massif	256.05
7	Corridor of Deva	97.83	32	Borod depression	47.76	57	Vlădeasa Massif	357.44
8	Corridor of Vinţ	242.12	33	Brad depression	268.67	58	Găina Massif	524.17
9	Hill of Agrij	44.23	34	Câmpeni depression	21.62	59	Muntele Mare Massif	752.35
10	Hill of Căpuş	142.20	35	Câmpia Turzii Depresion	175.36	60	Mountains of Ampoi	378.49
11	Hill of Cued	204.25	36	Conop depression	89.73	61	Codru Mountains	352.83
12	Hill of Lugaş	129.23	37	Gurahonţ depression	90.41	62	Mountains of Detunate	590.12
13	Hill of Mădrigeşti	99.62	38	Hălmagiu depression	151.66	63	Hountains of Huş	708.28
14	Hills of Agrij	178.28	39	Huedin depression	178.55	64	Mountains of Iada	396.58

(continued)

Table 16.4 (continued)

No.	Relief unit	Area (km²)	No.	Relief unit	Area (km²)	No.	Relief unit	Area (km²)
15	Hills of Aiud	507.90	40	Depression of Iara	75.35	65	Măgureaua Mountains	585.62
16	Hills of Barcău	296.17	41	Depression of Lupşa	111.49	66	Meseş Mountains	224.03
17	Hills of Budureasa	140.49	42	Meteş depression	20.62	67	Mountains of Meseş	19.74
18	Hills of Dobreşti	161.54	43	Ocoliş-Poşaga depression	38.72	68	Mountains of Miseş	71.71
19	Hills of Mărăuş	179.78	44	Pleşcuţa depression	33.04	69	Pădurea Craiului Mountains	504.35
20	Hills of Moma	126.65	45	Sălciua depression	27.98	70	Plopiş Mountains	383.40
21	Hills of Nucet	71.25	46	Depression of Şimleu	172.49	71	Mountains of Săcărâmb	725.39
22	Hills of Sălaj	310.25	47	Depression of Trascău	30.36	72	Piedmont of Şimleu	171.79
23	Hills of Tărcăiţa	108.06	48	Vad-Borod depression	271.55	73	Plateau of Besna	55.85
24	Hills of Tăşad	352.93	49	Vlaha-Hăjdate depression	188.47	74	Plateau of Padiş	488.30
25	Hills of Teuz	139.67	50	Depression of Zalău	75.49	75	Păniceni Plateau	243.40

massif is dominated by the main ridge, with altitudes that oscillate around 1400 m, which is defined by conglomerates, quartz sandstone and marl with limestone intercalations, above which are placed sandstones miocene in alternation with shale clay. Locally, the flysch deposits are crossed by granodiorithic eruptive bodies, andesite, dacite or riolite strands. Also, under the insular aspect, can be found Jurassic limestone formations (Măgura Vulcan, 1260 m and Piatra Bulzului, 963 m). To the south, monoclinic structures connected with the piedmont hills of the Hălmagiului Depres-

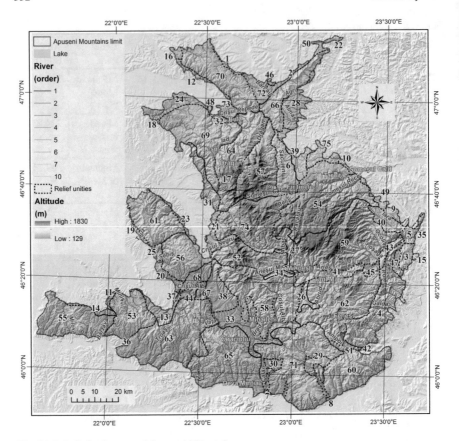

Fig. 16.6 Relief units map of Apuseni Mountains

sion [80], are detached from the main ridge. Smaller altitude values reduce the role of the orographic barrier in the path of air masses arrived from the western, which led to an average rainfall budget of about 1200 mm.

The south-western part of the Găina massif continues with the modest and elongated structures of the piedmountain Hills of Hălmagiu, arranged in the form of ridges separating the right tributaries of the Crişul Alb. Pliocene-age sedimentary deposits are petrographically expressed through muddy clays, sands and gravel. Locally, the volcanic intrusions on the southern side, uncovered by differential erosion, led to the emergence of some hillocks in the landscape (Măgura Ociului, 439 m). The southern area of Bihor is continued by Brad Hills, which compared to the Hills of Halmagiu no longer presents a piedmont structure, being defined by intrusive volcanic structures.

The tilting towards north of volcanic and sedimentary facies impose to the Crişul Alb tributaries, which drains to the south, an obsecvent character, materialized in relief by a symmetrical shape of the valleys and frequently riverbed break-points. The valley of Crişul Alb flows perpendicularly on the inclination direction of strata, favoring the formation of a cuesta with a length of about 25 km [80].

Crișul Alb flow direction in the upper sector is made on a north-south trajectory, while reaching Brad Depression area, changes towards west-northwest. The geology of the depression is marked by Tortonian sedimentary formations, peripherally filled with volcanic or limestone deposits. At the level of the volcanic hills, located in the vicinity of Brad city, mining activities have been carried out over time, especially for gold-silver deposits. Nowadays, the works are much diminished, being rather associated with private initiatives for prospecting purposes.

Crișul Repede basin

The relief of the Crișul Repede river basin, corresponding to the analyzed area, is associated with the morphological units of Vlădeasa and Pădurea Craiului mountains, with small peripheral depression units, to which we can add the spectacular defile sector, created by the main collector. The mountain sector is dominated by the Vlădeasa massif (1836 m), which appears as a series of ridges, which descends northwards to the Crișul Repede River. Besides, the upper elevations in the massive are also associated with the catchment limit lines established between Someșul Mic, Crișul Repede and Crișul Negru basins.

The main ridges developed on a north-south direction became hydrographic boundaries which delimit the Săcuieu, Drăgan and Iada valleys, tributaries which also provide the main hydrological budget of Crișul Repede. To the west, the basin is morphologically associated with the Pădurea Craiului Mountains, which have lower altitudes, predominantly between 400 and 600 m. The highest peaks are, from west to east, the following: Măgura Dadului (945 m), Hodrinașa Hill 1014 m), Dealul Mare (957 m), Măgura Beiușele (1003 m).

The Crișul Repede Defile, a landmark of major importance of climbing tourism, links the Huedin Depression to the Vad-Borod Depression. It consists of an alternation of broader areas with basin-like appearance and narrower areas, with steep walls, often geomorphologically marked, by endo- and exocarst forms. The narrowest part is between Șuncuiuș and Vadu Crișului (about 4 km). The enlargement zones, starting from the east to the west, are the following: Drăganului Valley, Ciucea, Bucea, Bratca and Șuncuiuș (Horvath 2009).

The Vlădeasa Massif consists of a petrographic mixture consisting of eruptive rocks, crystalline shale and mesozoic sedimentary deposits. The eruptive sector has a north-south direction of development, having over 45 km in length and a maximum width of 30 km between the Săcuieu and the Iadei valleys. The petrographic structures have accumulated in different geological periods, the oldest ones being the andesites, which were later penetrated by the riolites (which cover the largest surface today) and the dacite, diorites, etc.

The crystalline schist develops from Drăgan Valley to the valley of Iada (Stâna de Vale), to the east, where they come into contact with the eruptive sector. Sedimentary deposits of permian-mesozoic age are represented by conglomerates, sandstone, limestone. Thus, Permian age deposits occur in two areas: one at Remeți and another at the springs of Iada Valley, formed by conglomerates within which there are fragments of the crystalline foundation. The Triassic age includes a northwestern strip that reaches Iada valley up to Bulz village and the entire area of the Crișul Repede defile. The most frequent rocks are limestones, to which are added conglomerates, schist and sandstone. The Jurassic age, with calcareous formations, extends especially in the eastern area of the Pădurea Craiului Mountains with insular occurrences on Iada valley at Remeţi, having direct contact with the Vlădeasa Mountain' eruptive deposits on Şuncuiuş-Vad defile. The Cretaceous is restricted to only a few spots on both sides of the Iada Valley (Horvath 2009).

Crişul Negru basin
The river basin of Crişul Negru is morphologically associated with the western flanks of the Bihor Mountains, the eastern slopes of the Codru-Moma Massif and the slopes of Pădurea Craiului Mountains (Fig. 16.6). The depression area in the upper basin is defined by the Beiuş Depression, where the main course has formed a dendriform fluvial network with a right asymmetry, towards Bihor mountainous area, where springs the most vigorous tributaries [80]. The highest morphological units drained by the main course and the right tributaries of the Crişul Negru, in the upper section of the hydrographic basin, belong to the Biharia Mountains and to the Bătrâna (Old Lady) Massif. Thus, at the on the first mentioned massif, the altitudes rise to values of more than 1700 m, which allowed the conservation of the most significant traces of the cryonivale modeling processes in the Apuseni Mountains. The highest ridge has a broadly curved appearance, with small heights associated with the Cucurbata Mare (1849 m) and Cucurbata Mica (1770 m) peaks.

Someşul Mic basin
The relief of this heavily hydrotechnically aranged area is formed mainly on metamorphic and magmatic rocks and lesser on sedimentary rocks (mesozoic limestone in the Someşul Cald springs region, Tertiary age limestone deposits at the edge of the mountainous area, lower stripped clays in some sectors—Agârbiciu river basin, etc.).

The natural potential of the area derives from the support provided by the three units: Gilău—Muntele Mare Mountains, Vlădeasa Massif and Bătrâna Mountains as part of the Bihor Mountains. These units come into contact with each other along the Someşul Mic and its tributaries [63].

Between Bătrâna Mountains (1579 m), the Gilău Mountains (1404 m) and Muntele Mare (1825 m), the tectonic processes caused an obvious axial loss. As a result, an "orthographic amphitheater" was formed on a vast morpologically oriented west-east area, consisting of the storied interfluves, all descending to the east (depression corridor Gilău-Căpuş-Huedin). This elongated amphitheater has been occupied by the Someşul Mic hydrographic system, contributing to the diversification of the geographic landscape, but also to the increase of the economic potential of the territory. As geographic phenomena act synergistically and consistently, the morphological and hydrographic components mentioned, by their specificity, closely cooperate with the meteo-climatic component [37].

This explains the fact that along the morphohydrographic corridor of Someşul Mic, there is a high occurrence of the western air masses, with typical adjacent processes, like humidity discharges and weather disturbances, which will generate a genuine valley topoclimate [63].

Bătrâna Mountains, a toponym from the dominant peak (1579 m), represents the orographic point in the central part of the Apuseni Mountains, but also a hydrographical boundary because it separates the basin of Crişul Pietros, Someşul Cald and Arieş river. The fundamental characteristics are imposed by the geological constitution marked by the intercalation of crystalline rocks, with sedimentary rocks, especially limestone and dolomite (Fig. 16.5). The alternation in stripes of non-carstificable rocks (75.3%) with carstifiable rocks led to the emergence of seamless peaks, depressions and relatively low plateaus, the latter dominated by endo and exocartst shapes (caves, avenues, valleys, etc.) [13].

Viewed as a whole, the mountain ridges are arranged on different levels: at 1500–1600 m there are residual peaks from an old peneplena (Măgura Vânătă 1641 m, Bătrâna peak 1579 m), followed by a lower level 1400–1500 m (Piatra Burned 1488 m, Moţului Church 1466 m), but related to the same period of carving, the Măguri-Mărişel polyclinic complex. The carstifiable formations have enriched the old surface by generating depressions and plateaux considered as carstoplen remains (Padiş plateau). At 1000–1100 m there are extensive heights, shaped on crystalline schist and granite. This step represents the mountain that has entered strong under the human impact: settlements, pastures, roads, etc. [46].

Muntele Mare Mountains are based on an old hercynian age chain, cut and then peneplenized, covered with large waters in the paleogene and raised again into neogen. Built like a monolithic giant block through the center of its granite platoon, it features a dominant relief of rounded ridges with smooth or slightly curled surfaces, contrasted with the depths of the valleys and the high slopes of the valleys' sides. It is the area known in the literature as the Mărişel Platform (1200–1600 m), which after deforestations were established crops, the *crâng* households, resulting in an active human life [46].

Gilău Mountains are considered an extension to the north of Muntele Mare massif, at a lower altitude (1600–1200 m). The area of their maximum deployment is represented by the interfluvium separating the water catchment area of Someşul Cald from Crişul Repede one [63].

Also constituted from resistant rocks (crystalline shale), the relief of the Gilău Mountains has the same physiognomy of slightly wavy and smooth ridges. The north-eastern tilting direction of the whole mass is underlined by the decrease of the altitudes from 1692 m (Cârligaţi peak—hydrographic crossing point in the Apuseni Mountains), 1649 m (Cumpănăţelu peak), 1404 m (Călăţele Hill), 1099 m (Dealu Negru), near the village of Beliş and only 900 m on the ridges that surround the town of Râşca.

As a result of external agents action started in Danian age, three levels of relief modeling were finalized. The upper level (called Fărcaş-Cârligaţi) is situated at 1400–1800 m, was modeled in Danian-Oligocen period. The middle level (Maguri-Mărişel), more fragmented than the first one, is situated at 1200–1400 m and was modeled in Sarmatian-Meotian period. The lowest level of Pliocen-Quaternary age (called Feneş-Deva), appears in the form of *shoulders* along the valleys entering the mountainous area. The situation of the leveling platforms should not, however, be absolutized from the point of view of the height limits, because sometimes, due to the increased tectonization, they are much lower (especially in the case of low altitude levels) [46].

In the northern and eastern parts of the Gilău—Muntele Mare area, the foundation consisting of crystalline rocks sinks under the sedimentary (paleogenic) formations of the border area. As a result, the smooth ridge landscape is replaced by scenery of steep slopes that are aligned with the lower slopes towards Transylvanian Depression [37].

Between the mentioned mountain ranges, the valleys were inserted; of these, the hydrographic system of Someşul Mic drains about two thirds of the entire complex area. The valleys in this hydrographic system have high depths (300–400 m) and accentuated longitudinal slope (25–30 m/km) resulting in a great hydropower potential, used by the built facilities from Fântânele, Tarniţa, Someşul Cald and Gilău.

Along these valleys, there are very steep sectors with impressive gorges, such as Someşul Cald (its upper sector and the one at Lăpuşteşti), Someşul Rece, Răcătău and Irişoara, etc.

On the geological-geomorphologic background a diversified landscape was composed made of [63]:

- high karst plateau (1400 m) particularly attractive for speoturism and hiking;
- smooth or slightly curved mountain ridges with scattered settlements among meadows and forest clusters;
- valleys with steep slopes well forrested, with small meadows occupied by sheepfolds or small settlements (Giurcuţa de Sus, Smida, Ciurtuci etc.).

The crossing of the granite intrusions, turn the valleys into narrow corridors which become favorable to the construction of dams, while depression zones formed by erosion, at the confluence of the watercourses, provide the basins necessary for the development of lacustrine cuvettes.

In the case of Fântânele reservoir, it is noted the contact of Muntele Mare granite with the crystalline deposits of Gilău mountains, both extended on large surfaces. They are, in some places, penetrated by andesitic and dacite veins and apophyses [49].

A gravity dam was built at the entrance of Someşul Cald in the defile of Beliş, in the sector with hard rocks represented by granodiorites and granites, rocks that caused the Someşul Cald valley to make a double detour to the southeast and then again to its initial direction [47, 56–58]. Besides, these rocks were also exploited in the quarry for building the edifice on the left bank of the river, a few hundred meters downstream of its location.

The double arch dam of *Tarniţa* reservoir was located into a narrow sector of Someşul Cald Valley. Here the slopes have a thin layer of deluvium and detritus covering the crystalline shale from the foundation (amphibolites, quartz amphibolo-clorite), whose cracks and fractures were naturally filled with quartz and carbonate. The reservoir cuvette is formed on the same crystalline shale, covered by sands and quaternary gravels [57].

For *Someşul Cald* reservoir the dam was built on a lithology given by crystalline shale of Gilău, where the basic rock is represented by cracked amphibolites. There are breached fractures, and at depths of over 15 m, the cracks are closed by diaclase of calcite and quartz [38].

The reservoir cuvette extends on the basin generated by the confluence of the Someşul Cald River with the Agârbiciu Valley. The gorge petrography on the reservoir site is dominated by the sericitoaceous quartz shale and the sericito-chlorite shale. Towards the upstream end of the reservoir, Someşul Cald valley cuts an orthoamfibolite intercalation [63].

In the case of *Gilău* reservoir, the dam was located into a narrowing point of Someşul Mic valley at the exit of the mountainous area, determined by the presence of a 140–200 m thick andesitic body of vein form. This andesite with amphibols and biotite has a grayish- sometimes reddish colour, being heavily cracked and easily disaggregated near the surface. On the altered scattered rock, the rock has the appearance of gravel sand, almost entirely lacking in cohesion, but where hard, consistent, less altered areas appear. At a depth of 3–5 m, the eruptive body consists of compact, unaltered gray-green andesite [38]. The reservoir cuvetteation is situated in an erosion basin contoured by the confluence of Someşul Cald with Someşul Rece creek.

Midlle Someş basin is dominated by the main ridge of the Meseş Mountains, situated in the north-eastern extremity of the Apuseni area. Their southern part draws tributaries of low importance belonging to Crişul Repede (Poicu valley), while their northern part is associated with the Almaş and Agrij rivers, tributaries of the Someş, in its middle sector.

The Meseş Mountains, like the Plopiş Mountains, have a crystalline foundation, covered by neogene sedimentary deposits, with maximum altitudes not exceeding 1000 m.

Barcău basin is morphologically dominated by the Plopiş (Şes) Mountains, which appear as a horst with a smooth appearance, with peaks that have altitudes around 750 m.

*

All these geological and morphological components create very diverse conditions for the existence and circulation of water resource, with somewhat lesser possibilities of underground storage, due to the low permeability (Fig. 16.7) but with very good water drainage conditions at the ground (Fig. 16.8).

16.2.2.1 Rocks' Permeability

As mentioned at the beginning of this study in the Apuseni, although we encounter the most extensive endoreal areas at the national level, surface drainage values are quite important.

Except the predominantly karstic areas, the hydrographic network density values range from 0.8 to 1.2 km/km^2, with local convergence zones, sometimes exceeding 1.8 km/km^2 (Câmpeni and Brad Depression).

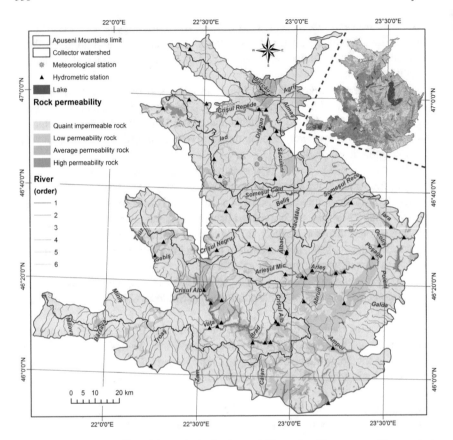

Fig. 16.7 The permeability of the rocks in the Apuseni Mountains. The petrographic mosaic, in the upper right side (adapted after Geologic map of Romania, scale 1:200000—Cluj & Brad sheets, 1968)

Generally, areas with high permeability correspond to the main valley corridors, depression basins or torrential basins in the higher area, characterized by the presence of quaternary pebbles and sands in thick packs, or by the presence of sandy petrographic structures.

On the opposite end, there are granite plateaus, crystalline massifs and clay basin areas, which due to the specificity of the rocks can't accumulate underground water.

In the collector basins (**Mureş**), the relatively high permeability of the rocks is identified *in the middle basin of Arieş river*. Also, similar features can be found in an area characterized by the presence of sedimentary formations mosaic with non-cohesive rocks, *in the lower basin of the Iara river*. Similar behaviour is found also on a diverse sedimentary substrate *in the corridor of the Ampoi River* but also in the valleys of the Mureş right tributaries between the Vinţu de Jos and Mintia villages (Fig. 16.7).

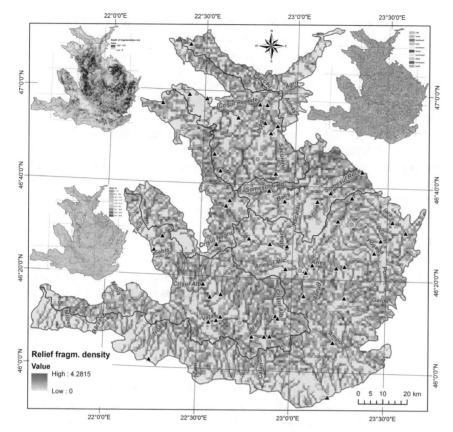

Fig. 16.8 Drainage density—big map, fragmentation depth—up left, aspect—up right and slope—middle left, in the Apuseni Mountains (height source: Topographic Map of Romania, 1:25.000)

The upper basin of **Crişul Alb** River is best represented regarding its rock permeability. Practically, from its source to the exit from the mountain area, both on the main course and on the right tributaries, the constituent rocks have high permeability and favorability for underground drainage. A high rock permeability is also presented in the basins of the right tributaries at the entrance of Gurahonţ Depression.

In the **Crişul Repede basin**, the collector's corridor with the main left-hand confluences of Săcuieu, Drăgan and Iada Valley, plus the southern compartment of Huedin Depression, are characterized by the presence of relatively high permeability rocks.

The **Crişul Negru basin** as far as it belongs to the Apuseni Mountains, overlaps surfaces with lesser rocks permeability. However, there are some areas with a high underground drainage located in *the upper periphery of Beiuş Depression* made of consistent sedimentary deposits. Other areas include *some right tributaries basins on the upper course of the main collector,* and *a lower sedimentary area, situated between the Bihor and Pădurea Craiului massifs.*

The upper basin of **Someşul Mic** is almost devoid of high permeability rocks due to its overlapping over the granite platoon of Muntele Mare and over the crystalline shale in the Gilău Mountains. Only the *sedimentary periphery of the Gilău reservoir area* has a sedimentary substrate favorable to underground water infiltration.

The **Middle Someş basin** is just the rest of drainage area, deployed at a very low altimetric level, an area that drains the north-eastern low apophysis of Apuseni Mountains—the Meseş Mountains.

Its central axis, built of crystalline shale of poor consistency compared to the Gilău, Vlădeasa massifs, etc., is quasi-impermeable. Its adjacent areas, on the western and eastern slopes, are occupied by sedimentary basins (Şimleu, with Zalău micro-depression to the west and Almaş-Agrij to the east). The rocks have a high permeability and allow the creation of moderate underground water deposits—on the hearth of the Zalău basin there is even an artesian aquifer.

The **Barcău basin** is formed on the northern slope of the Plopiş Mountains (Şes), the other low apophysis, situated in the northwest of the Apuseni Mountains.

The only space with some permeability of the component rocks is the rest of the Şimleu Depression, which partially overlaps the Barcău basin. The entire mountain area is quasi-impermeable due to the same crystalline shale which is found underground.

16.2.2.2 Drainage Density (Density of the Relief Fragmentation)

As we mentioned at the beginning of this study, in the Apuseni, although we encounter the most extensive endoreal areas in Romania, the morphometric parameters related to surface flow are significant. Excepting the karst areas, river network density values fall within the gap of 0.8–1.2 km/km^2 with local convergence areas, where they sometimes exceed 1.8 km/km^2 (Câmpeni and Brad depression).

Drainage density, as a result of relief dynamics, records significantly higher values compared to the density of the permanent network and gives a relevant image on the surface liquid flow areas. The statistic analysis, from the information presented, shows that the high values of this parameter are observed within the river basins exposed to the west. In three of the four cases (Barcău, Crişul Alb and Crişul Negru) the *average values of the drainage density* exceed 1 km/km^2, while in the case of Crişul Repede the value is very close to the unit (Table 16.5).

The values recorded on other basins is below 1 km/km^2, in the context of a different orientation, or due to smaller amounts of precipitation within the context of lower altitude (Middle Someş basin).

The maximum drainage density is reached in Someşul Mic basin, almost 4.3 km/km^2, developed on a petrographic structure, particularly resistant, of metamorphic-magmatic type, which prevents ingress of water in underground layers. Sedimentary rocks, tougher or friable, cover areas that are much smaller compared to other basins (Mureş basin). In fact, the maximum values of the parameter in all analyzed collectors reach or exceed 3 km/km^2, which confirms the term established in the literature as "water castle" attributed to the Apuseni Mountains.

Table 16.5 Drainage density values in the Apuseni Mountains collecting basins

No.	Collector basin/element	Surface (km²)	Drainage density (km/km²)	
			Maximum	Average
1	Mureş	5064.06	3.795	0.852
2	Crişul Alb	2247.29	3.861	1.009
3	Crişul Repede	1444.83	3.792	0.969
4	Crişul Negru	1081.16	3.836	1.007
5	Someşul Mic	988.50	4.282	0.973
6	Middle Someş	358.92	2.994	0.682
7	Barcău	235.40	3.435	1.160
8	Total Apuseni Mountains	11,420.16	4.282	0.950

Referring to the **spatial distribution** of the parameter, important concentrations of the primary drainage network are observed in the upper basins of watercourses, where the leakage is formed as well as along the valleys in general. In the vicinity of interfluves density values is exceeding frequently 2.5 km/km² while to the morphological contact with collectors' meadows they are significantly reduced by 1 km/km² or less.

Flanks with western orientation have high values for the drainage network. There is a noticeable difference of 0.5–1 km/km² between basins exposed mainly to the west compared to those exposed mainly to the east (Fig. 16.8).

In other words, there is a **visible cartographic correlation** between the distribution *of high drainage density values* and the similarities of the *depth fragmentation of the relief (dark brown), the northwestern, western and southwest slopes (blue shades), and the high relief slopes (yellow and red)* (Fig. 16.8). These elements can only lead to the conclusion that, from a morphometric and morphohidrographic point of view, the Apuseni Mountains have a remarkable potential for water resources, in general, and for hydropower in particular. Besides, the latter has also been redeemed with very good efficiency in the Someşul Cald basin, with additional water supply from neighboring basins.

16.2.3 Pedogeographic and Land Cover Components

Soils play an important role both in the formation of superficial runoff and in the underground supply of rivers. Due to their friability characteristics, soils have a direct influence on the formation of alluvial spillage.

Stream run-off, as a hydrological process, depends on soil permeability. Thus, *permeable soils* favor the storage of water and its transfer to the underground porous environment, and as a result, they lead to a more uniform feeding of the rivers, making their hydrological regime less dependent on liquid precipitation or melting snow times (Fig. 16.9).

Edaphical blanket of Apuseni Mountains has a smaller variety compared with petrographic diversity. Thus, large areas are occupied by cambisoils (about 70% of the total), followed by clay-alluvial soils (17%), while other soils are represented in lower proportions. Altitude and climatic conditions combined with lithological features, specific relief forms and land cover (Fig. 16.10), lead to a differentiation between the western part and the rest of the whole mountain area.

In the high central part, brown soils of lower alpine meadows are characteristic, along with iron-illuvial podzols. At lower altitudes on the fields covered with juniper, spruce or mixed forests, from the eastern slopes of Bihor and Gilău-Muntele Mare, brown acid soils and brown podzols are the main occurring types. Their thickness is generally less than 1 m, having a texture favorable to the infiltration of water from the surface to the underground.

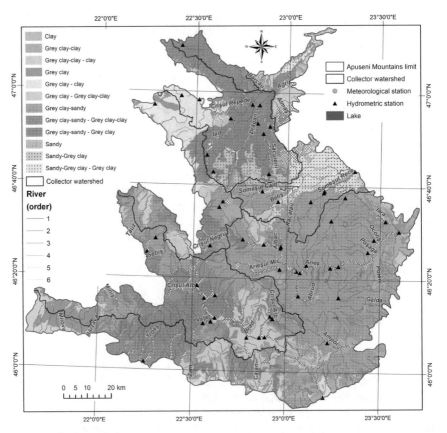

Fig. 16.9 Texture of soils categories in the Apuseni Mountains (source data: Soil Map of R.S. Romania, 1971–1979)

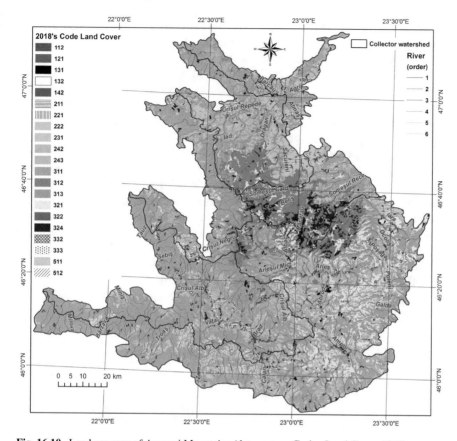

Fig. 16.10   Land use map of Apuseni Mountains (data source: Corine Land Cover, 2018)

On the lower altimetric mountain area, associated with the Zarand, Metaliferi, Trascău, Codru-Moma, Plopiş and Meseş ridges, are occupied by brown soils and eutrophic forest brown soils. Under the influence of parental rocks, the limestone rocks develop rendzinic soils. Insular, also on the limestone rocks or connected to the peripheral deposits of the Transylvan Basin may appear terra rosa soil formations.

Figure 16.9 shows a map of soil texture in the Apuseni Mountains, a parameter which can determine the favorability of the infiltration process in the stream run-off system. From soil texture analysis, the largest weights are associated with sandy loam (36%), followed by the loamy sandy soil (16%). Other types have proportions of less than 10%.

Spatial distribution highlights the fact that soils with sandy loam texture occupy significant areas associated with the Bihor Mountains, the Muntele Mare Mountains, Gilău and Trascău Mountains and partly the Codru-Moma Mountains. Important areas in the Vlădeasa, Meseş and Plopiş Mountains present a loamy sandy clay texture.

Regarding the **land cover**, in relation to water resources, we notice that in Apuseni Mountains a big a share is associated with natural surfaces, especially with deciduous forests, coniferous forests and natural pastures (Table 16.6).

Massive forest cover is favorable to water resources, especially in the case of their capitalization, due to a significant reduction in slurry spillage, but also to the mitigation of torrential run-off. More than half of Apuseni Mountains are covered with broad-leaved forest (52.7%), followed by three different types of surface with similar percentages: natural grasslands (9.2%), coniferous forest (9%) and pastures (8.4%). Significant surfaces are associated with land principally occupied by agriculture, with significant areas of natural vegetation, complex cultivation patterns, transitional woodland-shrub, mixed forest and non-irrigated arable land. Smallest areas belong to human settlements, mineral extraction and dump sites. The last ones provide significants sedimend loads to watercourses and important changes in water quality parameters, especially regarding heavy metals content and pH values.

A detailed analysis of the spatial distribution of the land cover typology using the data provided by the Copernicus Land Monitoring Service, derived from Corine Land Cover 2018, is presented in the map of Fig. 16.10.

The most hazardous threats for water quality are the tailing ponds and dump mining sites, exposed to meteorisation processes.

Table 16.6 Size of different land use types in Apuseni Mountains using Corine Land Cover—2018

Code	Name	Area (Ha)	Code	Name	Area (Ha)
112	Discontinuous urban fabric	17,309.5	243	Land principally occupied by agriculture, with significant areas of natural vegetation	79,215.2
121	Industrial or commercial units	320.4	311	Broad-leaved forest	602,423.1
131	Mineral extraction sites	2227.4	312	Coniferous forest	103,062.6
132	Dump sites	235.7	313	Mixed forest	30,320.1
142	Sport and leisure facilities	54.5	321	Natural grasslands	105,241.1
211	Non-irrigated arable land	24,234.7	322	Moors and heathland	2038.2
221	Vineyards	69.8	324	Transitional woodland-shrub	34,518.8
222	Fruit trees and berry plantations	4565.7	332	Bare rocks	654.0
231	Pastures	96,600.8	333	Sparsely vegetated areas	823.3
242	Complex cultivation patterns	36,856.7	511	Water courses	91.5
			512	Water bodies	1528.1

16.2.4 Hydrographic Component

The hydrographic analysis was done on three types of water structures: the underground environment, rivers and lakes.

16.2.4.1 The Hydrogeological Component

Of the Apuseni Mountains is determined by a varied petrographical context, with contradictory alternations, defined by the impermeability of some mountainous masses (Muntele Mare, Vlădeasa, Gilău) or rapid drainage in the context of extensive karst areas (Padiş Plateau—Bihor Mountains, Trascău Mountains, Pădurea Craiului Mountains).

The presence of quite extensive endoreal areas further complicates water drainage at the local level.

However, *the mineral* and *thermal springs* have a poor distribution, compared with the area of the Eastern Carpathians and are spread mainly around the eruption mofetic region of Săcărâmb—Deva, within a territory with deep volcanic intrusions, or in the western part of the region (mesothermal waters from Vaţa de Jos and Moneasa).

Some are valued for *spa cures* (the baleno-climaterical resort of Geoagiu-Băi, Moneasa, Vaţa de Jos), others by *bottling the water from springs* (Izvorul Minunilor/Wonders Spring—Stâna de Vale, Apa Cezara—Băcâia, Aquavia—Săcuieu), while hot water in Beiuş depression are intercepted by drilling and are used for heatig houses purposes [21].

In the northern part of the mountain group, there are some ferruginous springs located in Pădurea Neagră and Meseşenii de Sus settlements.

Underground water bodies in the Apuseni Mountains were identified by consulting the Management Plans of Major River Basins, taking into account only aquifers with importance for water utility system and exploitable discharge values higher than 10 m³/day (Fig. 16.11).

The studied area includes eight inventoried bodies of underground water, most of them are trans-basinal:

- *one* waterbody on the basins of Someşul Mic, Arieş and Crişul Negru identified in the national inventory with the following code: **ROSO04**—Bihor Mountains—Vlădeasa;
- *one* in the Mureş basin **ROMU06**—Brădeşti—Trascău Mountains and another *two* common with the Crişul Alb Basin **ROMU09**—Poieni—Metaliferi Mountains and **ROMU10**—Abrud—Metaliferi Mountains);
- *one* underground water body belongs to the Crişul Negru basin—**ROCR03** Dumbraviţa de Codru—Moneasa, Codru Moma Mountains, *two* are common to Crişul Negru, and Crişul Alb basin—**ROCR04** Clăptescu, Codru-Moma Mountains, **ROCR05** Vaşcău, Codru-Moma Mountains, and *one* belongs to basins of Crişul Negru and Crişul Repede—**ROCR02** Zece Hotare, Pădurea Craiului Mountains.

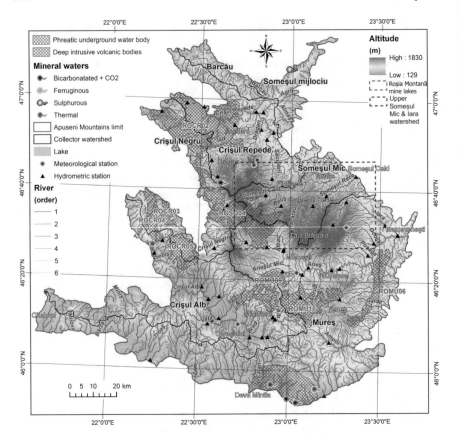

Fig. 16.11 Groundwater bodies, hydrographic, hydrometric and meteorological monitoring networks in the Apuseni Mountains area (data source: Mureş, Crişuri and Someş-Tisa Basin Water Administration)

16.2.4.2 The Watercourses

In the Apuseni Mountains have a radial distribution, imposed by the orographic node of Bihor and Vlădeasa. Altitude analysis has highlighted a number of distinct peculiarities. Most watercourses exhibit, in the upper sectors, tiny slopes with shallow depths and smoothly interfluves. The rich flow of the rivers and the leakage permanence are determined by the high degree of impermeability of the subsoil rocks, the reduced thickness of alteration crust and the intense exchange with the underground groundwater.

Besides, the causes above, together with rich humidity, allow the development on appreciable surfaces for ridge marshes near the summits, called *"tinoave"* (Muntele Mare, Călineasa area, Someşul Cald spring area). To the periphery the valleys deepen and adjust their longitudinal profile through erosion or structural steps, generating rapids and cascades. Along the valleys, many sectors of the gorges and defiles, epigenetic or antecedents (especially on limestone) have evolved. The most representative hydrographic landmarks are defined by the valley of Mureş, the collector of all rivers that drain to the south and east (except Someşul Cald and Someşul Rece creek, collected by Tisa on the territory of Hungary through Someşul Mic, and from Dej of Someş) and Almăjului and Agrijului stream collected by Someş, respectively the three Crişuri drained to the west (towards Tisa), after passing the Hills and the Western Plain.

The case studies presented below will refer to the watercourses located in these hydrographic areas. By the relatively central position of the high-altitude areas, the prerequisites conditions for river network are created, leading into a veritable "water castle edifice" with radial drainage, similar to volcanic areas.

The most important watercourses that cross the Apuseni Mountains are Arieşul, Someşul Mic, Crişul Repede, Crişul Negru and Crişul Alb. The northern, northeastern and southern flanks are defined by lesser tributaries, drained towards Someş or Mureş river (Fig. 16.11).

The river network of the **Mureş** basin, with its quasi-covering tributary of the eastern flank *Arieş*, is characterized by an asymmetry of its catchment associated with the studied area. In Apuseni, Arieş River creates a great asymmetry as it penetrates its springs close to their western edge, and as a result, traverses almost the entire mountain edifice. Having its springs in the Bătrâna massif, at an altitude of 1108 m, Arieş is composed of two hydrographic branches that drain, the eastern flank of the Bihor Mountains—Arieşul Mare and Arieşul Mic. They join together into a confluence upstream of Câmpeni, in Mihoieşti reservoir [5].

Arieşul Mare has an asymmetric basin, with superior development on the left, marked by strong tributaries: Cobleşul, Gârda Seacă with Ordâncuşa, Albacul (Lenght, 19 km, Drainage basin, 95 km^2), while on the right it receives shorter tributaries, like Cepelor Valley, Bucura and Neagra creeks. *Arieşul Mic* (34 km, 160 km^2) has its springs at an altitude of 1450 m, in the southern slope of Cucurbăta Mare peak and drains the area adjacent to the Biharia massif, the northern slopes of Găina Mountain and Arieşului Mic Mountains. Most of its tributaries have a small length (less than 10 km).

After the union of the two main branches, Arieş receives, downstream of Câmpeni, the most important tributary on his right: *Abrud* (24 km, 223 km^2). Developed at the contact between the Bihor massif and the Metalifers, the Abrud is characterized by a strong asymmetric basin, in favor of the right tributaries, with a high coefficient of sinuosity, imposed by the neogene hills [5].

The middle part of the Arieş River is dominated by the important contribution of left tributaries, which drain the southern flank of Muntele Mare Mountain: Bistra (18 km, 43 km^2), Bistrişoara, Lupşa, Valea Caselor, Sălciuţa, Poşaga, Ocolişul and Ocolişelul. To this is added the most important left tributary of Arieş, *Iara* (48 km, 321 km^2). From the right side, Arieş receives in the middle sector the tributaries that drain the northern slope of the Metaliferi Mountains, respectively the north-western flank of Trascăului Mountains: Ştefanca, Valea Muşcanilor, Valea Şesei and Râmetea (17 km, 42 km^2).

The southern region of the Apuseni Mountains, morphologically expressed by the Zarand ridge and Metaliferi Mountains, is characterized by a hydrographical inventory, defined by short rivers, with small flows that are directly tributary to Mureș river.

The eastern and south-eastern flank dominated by the calcareous summit of Trascău, the peaks of the Southern Metalifers, are marked by hydrographic courses with lengths exceeding 20 km, but with moderate or even modest flows: Valea Galda, Ampoiul and Valea Geoagiului.

Someșul Mic basin develops on the north-eastern slope of the Apuseni Mountains, overlapping the morphological units associated with three mountain massifs: Gilău— Muntele Mare Mountains, Vlădeasa Massif and Bătrâna Mountains component of the Bihor Mountains.

As in the Arieș basin, Someșul Mic is defined by two branches, relatively parallel, as a general direction of flow, but with very different peculiarities, including from the point of view also the hydrotechnical arrangements. *Someșul Cald* springs from the Bihor Mountains, from the karst region, situated at the edge of Padiș Plateau, which gives it a drainage alternation either on the surface or underground. The most important tributaries are the Bătrâna Valley, Firii Valley, Belișul, Valea Neagră, Râșca and the Agârbiciu valley. The course of the river is strongly modified by the presence of four storage lakes, equipped with hydropower plants. Moreover, from this point of view, only the northern sector of the Apuseni is well represented hydro-technical. The course of the river closes in Gilău, where it merges with the other hydrographic branch Someșul Rece [63].

Someșul Rece springs from Muntele Mare massif, through the Zboru stream, with its origin under the Balomireasa Peak, at almost 1600 m altitudes. The area of origin is defined by a flat relief, with large marshes, called here *molhașuri*, some of them being preserved as natural reservations (Căpățânii Molhaș). The tributaries are represented by short watercourses with the relatively rich flow: Fieșul, Ursul and Negruța. The natural drainage is influenced by a transverse system of underground water adductions, which complements the flows used by hydropower plants located on the Someșul Cald basin. In the upper sector there is a single reservoir, with a name quasi-identical to the river (Someșul Rece I), which takes over the captured waters of the Iara River tributaries, and then sends them through Răcătău valley towards Beliș-Fantanele reservoir [63]. The lower sector closes by a flanking delta at the upstream end of Lake Gilău.

The hydrographic system of the western flank of the Apuseni Mountains is represented by a triad of Crișuri (although there are five hydronimes with the component "Criș"—respectively Crișul Negru, Crișul Alb, Crișul Repede, Crișul Pietros and Crișul Băița) whose drainage direction is predominantly western. From north to south, the hydrographic network on the western side of the Apuseni is collected by Crișul Repede, Crișul Negru and Crișul Alb.

Crișul Repede drains the northern slopes of the Gilău-Vlădeasa and Pădurea Craiului massifs. Asymmetry of the basin is very strong, being in favor of the left tributaries, which have a generous development, especially in the upper sector. The river springs from the altitude of 670 m near Izvorul Crişului, situated in a hilly area on the northern edge of the Huedin Depression.

The upper sector is marked by the alternation of some narrow sectors, gorge-type with larger tectono-erosive depression arreas, which has generated contradictory slopes along the river profile. Thus, if the river has large slopes near it springs, in the Huedin depression these are considerably reduced. The most important tributaries that come from the mountain range are Călata, Săcuieu, Drăganul and Iada. Moreover, the last two, due to a rich leakage, have been enriched with big reservoirs, supplemented with a system of adductions in order to capitalize the generous hydropower potential [30].

Crişul Negru has a hydrographical basin, located further south, having its origin in the northern slope of Cucurbăta Mare Peak, the highest altitude of the Apuseni Mountains. The most representative tributaries in the upper sector are Crişcior, Tărcaiţa, Finişul and Crişul Băiţa (23 km). Downstream, towards Beiuş city, the river receives other tributaries, which drain the karst areas: Valea Neagră and Valea Chişcăului, after which receives its most important tributary, Crişul Pietros (Atlas of the Water Cadastre of Romania 1992).

Crişul Alb, spring from the western slope of Bihor Mountains from an altitude of 900 m, near Certezul peak (1184 m). In the upper course, the slopes are steep, so that at Crişcior, after 31 km, the river already reaches the altitude of only 292 m. The most important tributaries in the mountain sector are Valea Satului (18 km), Bucuresci, Luncoi, Chişindia, and the largest tributary on the left is Cigherul (56 km) (Atlas of the Water Cadastre of Romania 1992).

The northern apophyses of the Plopiş and Meseş Mountains are dominated by the upper sections of **Barcău**, to which is added a small area belonging to the **Middle Someş** basin, represented by the tributaries Agrij and Almaş.

16.2.4.3   Lakes

Of the Apuseni Mountains are relatively few in number, being dominated by those belonging to the category of accumulation lakes/reservoirs, frequently used for hydropower purpose (Fig. 16.11).

In the **Arieş** basin, the largest lake unit is the *Mihoeşti* reservoir, located at the confluence of Arieş Mare with Arieşul Mic. In the vicinity of Roşia Montană village, remains of a mining past can be identified, expressed by the presence of several lakes of anthropogenic origin, known locally as *tăuri* (Fig. 16.12). Their water was used in the past to wash gold-silver ores (Tăul Mare, Corna, Brazi, Anghel, Cartuş, Găuri, Ţarinii, Ţapului and Muntari).

In the upper section of the Iara River, four small hydro-technical facilities are identified, whose water volumes are diverted through adductions pipes to Someşul Cald basin (Iara, Şoimu, Calul and Lindrul) for hydropower purposes.

Of course, the largest hydropower development in the Apuseni Mountains, but also from Transylvania is that of **Upper Someşul Mic**, which combines about 280 MW of installed power, besides the four major reservoirs: Fântânele Lake (Beliş)—

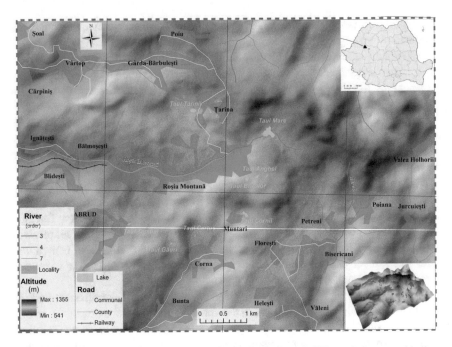

Fig. 16.12 Anthropic lakes in the vicinity of Roşia Montană, one of Europe's largest gold-silver reserves

795.67 ha, 18.95 km long, Tarniţa Lake—206.99 ha, 8.40 km, Someşul Cald Lake—81.27 ha, 4.25 km and Gilău—66.15 ha, 2.34 km (Fig. 16.13).

A special category is the former mining deposits with sterile flotation, developed very recently in the communist period. They belong to the category of the tailings ponds (Gura Roşiei, Seliştei, Ştefăncii, Şesei, Brăzeşti, Cuţii, Sartăş and Băişoara) and are located approximately in the same central area of the Apuseni Mountains, but are found in different hydrographic collectors basins. Some are in conservation and have a high degree of clogging, others are still active and represent a potential source of pollution for the aquatic environment.

In the **Crişul Repede** basin can be identified the two large reservoirs on the northern slope of Vlădeasa (Drăgan and Leşu), complemented by other small anthropic lakes, with the function of hydrotechnical water intakes (Fig. 16.14).

On **Crişul Alb** can be found the non-permanent reservoir of Mihăileni, located in the upper sector, in the vicinity of Brad town.

The great extension of **karst** in the Apuseni Mountains has favored the genesis of **typical lakes** (formed on dissolution rocks) with temporar or permanent character and small surfaces. Thus, we can mention the temporary lake of Ponor Poiana, Vărăşoaia Lake and Ighiu Lake.

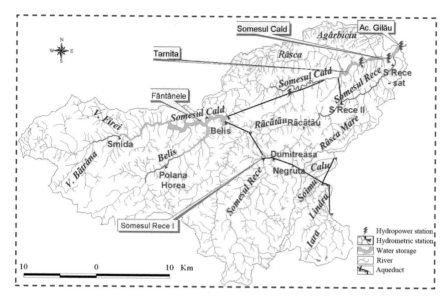

Fig. 16.13 Hydrotechnical works of Upper Someşul Mic—Iara basins

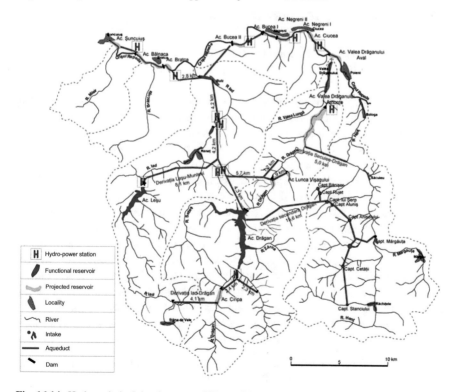

Fig. 16.14 Hydrotechnical development of Upper Crişul Repede basin

16.2.4.4   The Hydrometric Monitoring Network

The hydrological monitoring activity of the river network from Apuseni Mountains is done through a number of 58 hydrometric stations (Fig. 16.11). Observation programs involve day-by-day level observation and indirect flow monitoring operations to ensure run-off forecasts, respectively for rational management of available water resources.

- Most hydrometric stations are associated with the Arieş river basin (18 stations), then Crişul Alb basin (13 stations), Crişul Repede (11), Someşul Mic (6) and Crişul Alb (4). The hydrological oriented activity is based also on nine meteorological stations, which provides additional background data related to precipitation amounts and type, air temperature, air mass movement, wind conditions etc. (NIHWM 2013).

16.3   Water Resource in the Apuseni Mountains

16.3.1   Underground Water Resource

A detailed analysis of the distribution and capacity of aquifers/groundwater bodies has led us to identify significant differences induced by conditional factors previously studied (Fig. 16.15).

Thus, **in areas with porous formations**, *local aquifers* are distinguished *or discontinued* in *pyroclastic rocks*. Those underground deposits can be found in specific areas belonging to the South-East and north-western parts of Metaliferi Mountains, the North-East side of Zarand Mountains, Southern area of Codru-Moma and as an insular appearance in the western side of Trascău.

Local aquifers in **carstiferous rocks** occur mainly in *limestone and dolomite* in the southern and central part of Bihor Mountains and in the South-East of Codru-Moma Mountains. Also, *local aquifers* are found in *the limestone formations of the Transylvanian Basin*, related to the Northern side of Gilău Mountains, the North-East area of Trascău Mountains and the South-East and North-East of Meseş Mountains.

In cracked rocks, there is *an extensive network of productive aquifers* found in *sandstone and marl*, located at the contact between mountain groups Bihor, Muntele Mare and Gilău, in the Southern Codru-Moma Mountains and sporadically in the Southern Plopiş Mountains, as well as in the North side of Meseş Mountains.

Also, *local aquifer networks or discontinued aquifers, often drained by springs* are found in the following formations and locations:

- *limestone and calcareous sandstone* in the Center of Metaliferi Mountains and in the extreme North-East part of Trascau Mountains;

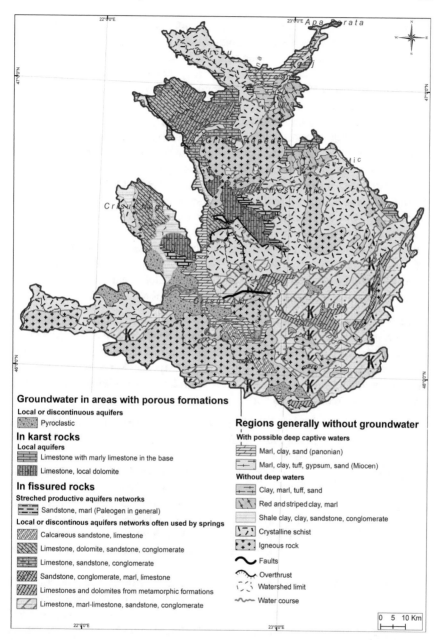

Fig. 16.15 Groundwater in Apuseni Mountains (composite after The Romanian Groundwater Map, 1975 edition and other sources)

- *limestone, dolomite, sandstone and conglomerates* on extended areas from Central and Nordic part of Codru-Moma Mountains, the western half of Pădurea Craiului Mountains and North-Western side of Bihor Mountains;
- *limestones, sandstones and conglomerates* in extended areas, exclusively related to the extreme East of Pădurea Craiului Mountains and in the North and Northwest sides of Vlădeasa Mountains; they appear like islands in the West and extreme South of Pădurea Craiului Mountains;
- *sandstones, conglomerates, marls and limestones* extended in the Central-Eastern part of Pădurea Craiului Mountains and less extensive but significant in the South-Eastern of Vlădeasa Mountains (Someşul Cald springs), the western half of the central ridge of Trascău and sporadically in a central area of Metaliferi Mountains;
- *limestones and dolomites from metamorphic formations* on extended areas in the South and West of Trascau Mountains, as well as on less extensive areas in the northern half of the central ridge of the same mountain group;
- the largest area occupied by ***local aquifers or discontinuous drainage springs*** is found on almost the entire surface of Trascău Mountains, in the eastern half of Metaliferi Mountains, in the eastern half of Zarandului Mountains, in the southern half of Bihor and on smaller surfaces in the South of Metaliferi Mountains.

Underground captive aquifers *developed in marl, panonian clays and sands* can be found in the South-West of Meseş Mountains, South-East of Plopiş Mountains and in the western extremities and especially on South-West part of Bihor Mountains.

However, appreciable landscape surfaces in the Apuseni Mountains are less represented by groundwater (***practically without underground aquifers***), due to their quasi-waterproof rocks' composition:

- *clays, marl, tuffs and derived sands* are present in the Southern Meseş Mountains and the Southeast side of Plopiş Mountains, as well as isolated spots in the South-East of Metaliferi Mountains;
- the *red and striped clays*, occur especially at the periphery of Transylvanian Basin, and can be found in the Southwest of Almaş-Agrij Depression, South part of Huedin Depression, North (the Râşca Basin) and North-East sides of Gilău Mountains, and on some small areas in Northwest part of Trascău Mountains;
- *clayey shale, clays, sandstones,* abundant *conglomerates* in the East and Northwest sides of Codru-Moma Mountains, as well as at the eastern periphery of Beiuş Depression;
- the *crystalline shale* present on particularly extensive surfaces in Muntele Mare, as well as in Gilău Mountains, the central and southern parts of Bihor Mountains, the northern half of Vlădeasa Massif, the Plopiş Mountains almost entirely, most of the Zarand Mountains, the eastern periphery of Trascăului Mountains, the central ridge of Meseş Mountains, etc.;
- *intrusive rocks (granite, dacite, granodiorite, etc.)* are found in the western half of Metaliferi Mountains and the central platoon of their eastern half. Large areas are also found on the southern half (the Iadei Mountains) and the Eastern part of Vlădeasa Massif. Smaller spots can be identified in the extreme north of Bihor

Mountains, on central platoon of Muntele Mare, Găina Massif and in the central-southern platoon of Zarand ridge. Isolated on smaller areas intrusive rocks can be found in the north-western branch of Codru-Moma Mountains, in the Eastern side of Muntele Mare and in the South part of Trascău Mountains, etc.

Starting from these considerations, the main waterbodies identified in the Apuseni Mountains can be described in the following manner:

- *ROSO04* underground water body—*Bihor—Vlădeasa Mountains* is defined by a strong fragmentation of the Tribsic-Cretaceous carbonate deposits (limestone and dolomite), which have important water resources. Their discharge is done through karstic springs, with variable flows ranging from 0.04 to 550 l/s. From a hydrochemical point of view, the body's water is calcium and/or magnesium bicarbonate, having a slightly basic pH (7.15–7.86) and total mineralization oscillating between 125 and 529.7 mg/l [85, 87];
- the underground water body *ROMU06—Brădești—Trascău Mountains* has a mixed (phreatic + deep) layout, being located in Triasic-Jurasic age limestone deposits, which have an intense fragmentation, hydrogeologically reflected in a large number of karstic systems. The discharge of bicarbonate-calcium water to the surface is made through springs with flow rates between 0.2 and 234 l/s [86];
- underground water body *ROMU09—Poieni—Metaliferi Mountains* is characterized by Paleozoic crystalline limestone that generated mixed aquifers, that are eeding usually by rainfalls. From a chemical point of view, they are sodium bicarbonate water with discharge flow rates ranging from 0.1 to 322.5 l/s [86];
- the groundwater body *ROMU10—Abrud—Metaliferi Mountains* is of mixed type, accumulated in the Jurassic-Cretaceous deposits (limestone, sandstone, conglomerate, marlin, clayey shale) and is characterized by an effective infiltration between 220.5 and 315 mm/year [86];
- the *ROCR02 Zece Hotare, Pădurea Craiului Mountains* are formed in Triasic, Jurassic and Cretaceous deposits, represented by limestone and dolomites, intensely fractured and carstified. Local subterranean aquifers are fed both from precipitations and from surface waters, while the infiltration paths are represented by heavily fractured and cracked areas. Analysis of water chemistry has led to a classification of calcium bicarbonate or sodium bicarbonate [85];
- the underground water body *ROCR03 Dumbrăvița de Codru—Moneasa, Codru Moma Mountains* of karst-fisural type, corresponds to the Triasic-Cretaceous limestone and dolomites without covering layers. The flows of the springs are between 0.7 and 123 l/s, and the chemical classification shows the calcium-magnesian or calcic-sodium bicarbonate types [85];
- the underground water body *ROCR04 Clăptescu, Codru-Moma Mountains* is marked by intense underground water circulation, with springs discharges ranging from 0.01 to 80 l/s. The aquifer's feed is predominantly made from precipitation and subordinated from the sinking waters found on the non-karst slopes near Clăptescu area and infiltrated underground at the entrance to the carst. The discharge of the aquifer is done through springs as well as by feeding the phreatic layer from the Crișul Negru riverbed [85];

- *the underground water body ROCR05 Vaşcău, Codru-Moma Mountains* formed in limestone and Triassic dolomites, has significant quantitative resources, respectively evacuation flows between 0.05 and 139 l/s. The aquifer's feed is predominantly made from rainfall and subordinated from the gully-erosion waters from the non-karst slopes of the Vaşcău Plateau and infiltrated underground at the entrance to the carst. The hydrochemical typology inserts this waterbody into the bicarbonate-calcic-magnesian type [85].

16.3.2   Surface Water Resources

16.3.2.1   Elements of Average River Runoff

The rivers in the Apuseni Mountains have a mixed supply, with the predominance of superficial sources. Thus, the amount of water from rain and snow melting takes have proportions of 64 and 70% of the total, for the rivers influenced by the karst (Someşul Cald, Drăgan valley, the upper basin of Crişul Negru and Arieş river) and between 70 and 80% for the other watercourses. The area is defined by five types of water supply, which are differentiated, with specific features, correlated with their physico-geographic conditions.

(a)   The pluvio-nival type with moderate subterranean feeding includes most of the Apuseni Mountains, overlapping the northeastern, eastern and south-eastern slopes. Precipitation amounts have values between 36 and 45%, the one from snow melting 26–35%, while the underground is around 20–35%. This type is reflected in the regime of the river through large spring waters (April–May) and frequent floods in other seasons.

(b)   The rich underground plovio-nival type, which includes the Arieş basin, where the underground exceeds 35% due to the extensive carst area it drains.

(c)   The moderate pluvial type with moderate pluvial supplying amounts appears on the western slopes, with rich rainfall throughout the year. Rains participation is over 35%, snow is below 30%, and groundwater between 24 and 28%. The thickness of the snow layer and its shorter duration, as well as its discontinuity, due to the heating times, correlated with winter precipitation, determine a high frequency of winter floods and the formation of large spring waters starting with February.

(d)   The moderate pluvial type with rich underground feeding is associated with the upper basin of the Crişul Negru, where the underground karst contribution to total exceeds 35%.

(e)   The moderate underground supply of nivo-pluvial type has limited extensions to mountain areas with altitudes above 1700 m from Bihor and Muntele Mare massifs [31].

The run-off analysis for the watercourses in the Apuseni Mountains was based on the existing bibliographic data, information from hydrometric stations and water administration institutions, which manage the hydrographic resources in the analyzed area.

The distribution of run-off volumes during the year shows large variations in close connection with the evolution of climatic elements. As a rule, the largest volume of water flows in the spring season, with the melting of snow and the occurrence of specific rains (40–50% and the maximum flow is recorded during April). Minimum values are recorded usually in winter in specific conditions caused by negative temperatures involving the blocking of water volumes either in the form of snow or caught in the form of ice in the riverbed. In some situations, can be identified in a *small water period* even in autumn, especially in depression areas. Depending on the climatic factors, the altitude of the hydrographic basins taken into account and the position on the two sides of the mountains, several types of the regime can be highlighted: Western Pericarpathic, Transylvanian Pericarpathic, Western Carpathic and Transylvanian Carpathic. The classification and description of hydrological regimes are presented below after I. Ujvari, 1972 (Table 16.7).

Table 16.7 The average seasonal flow and the types of hydric regime in the main river basins of the Apuseni Mountains

River name	Hydrometric station	Seasonal mean runoff (%)				Regime type
		Winter	Spring	Summer	Autumn	
Someşul Cald	Someşul Cald	16.2	46.6	25.4	11.8	TC
Someşul Rece	Someşul Rece	14.9	42.6	28.9	13.6	TC
Someşul Mic	Cluj-Napoca	16.2	45.2	26.6	12.0	TC
Drăgan	Pârâul Crucii	17.5	44.2	24	14.3	WC
Iada	Leşu	24.9	39.9	21.5	13.7	WC
Crişul Repede	Oradea	26.9	41.1	21.0	11.0	WC + WPc
Crişul Pietros	Pietroasa	26.2	43.1	19.6	11.1	WC (carst)
Roşia	Pocola	38.4	35.9	15.2	10.5	WPc (carst)
Crişul Negru	Şuşti-Vaşcău	33.7	39.6	16.6	10.1	WC
Crişul Alb	Gurahonţ	36.1	42.8	13.7	7.4	WC + WPc
Dezna	Sebiş	38.7	37.2	15.5	8.6	WPc
Cigher	Mocrea	39.4	44.9	11.0	4.7	WPc
Arieş	Câmpeni	25.7	45.1	18.1	11.1	TC + WC
Arieş	Turda	18.0	43.0	25.2	13.8	TC
Iara	Iara	15.9	38.9	31.4	13.8	TC + OC
Ampoi	Zlatna	25.6	49.6	17.4	7.4	TC + WC
Galda	Benic	18.6	43.8	25.7	11.9	WPc
Geoagiu	Geoagiu	28.9	44.7	17.2	9.2	WPc

WC Western Carpathic, *WPc* Western Pericarpathic, *TC* Transylvanian Carpathic, *OC* Oriental Carpathic
Source Ujvari (1972), Ştef (1998)

Crişul Repede and Crişul Negru rivers belong to the western carpathian regime (WC). These rivers are characterized by the early appearance of the spring high waters (March–April), by flash floods in May–June due to the heavy rains that fall in this period, followed by a hydrological drought from June until September, with some disruptions due to the summer flash floods. The minimum flow occurs during autumn in the depression area and during the winter in the mountain area. In the depression area, it happens that the winter flow to overcome the spring flow (30–40%) due to the snow melting and winter rains. In these conditions, the highest monthly flow rates appear in February or March.

Crişul Alb river belongs on almost it entire length of western pericarpathic regime type (WPc). This is explained by the low altitude of it catchment area, (only 610 m at Crişcior, in the upper basin, and 437 m at the Bocsig, in the lower basin). It is noted the high instability of the winter regime. Due to the common heat waves that appear in the cold period of the year, massive and rapid melting of the snow layer takes place, so that the average winter flow is here approximately equal to the one recorded in spring (over 30%). Also, the early summer flash floods are very intense, with increased frequency, but of short duration. The minimum flow is observed in the summer.

Someşul Mic basin belongs to the transylvanian carpathian type of regime (TC). It is affected by the descent of air masses originating from the western north-western circulation, that makes the rainfall to drop to 1000–1100 mm/year even at altitudes of over 1 300 m. As a consequence, the values of the liquid average flow to stay under 700 mm. Due to low snow reserves in the Gilău Mountains and of convective summer rains in the Someşul Rece basin, the share of summer runoff exceeds 25% of the total runoff. The maximum flow is recorded especially in spring and summer, and the minimum values especially in the winter. The transylvanian carpathian type of regime is quite similar to the western carpathian one, only that it presents a much more stable winter regime, without the snow meltdowns episodes: winter nivo-pluvial flash floods have a frequency of only 10–20%.

The Arieş river has a regime with the transition character. Due to its large basin surface, the maximum flow rates result by cumulating the long duration rainfall or correlated rainfall with snowmelt periods. Upper Arieş and its tributaries belong to the transylvanian carpathian type of regime (TC), but with higher instability of the winter regime.

Its tributary, Iara, deviate from the regime to the rest of the tributaries, distinguishing himself through a stable winter regime with late snow melting, with rich flow from the heavy rains in May–July (it resembles much with the Someşul Rece regime). This makes the maximum monthly flow rate to be registered in May (nearly equal to April), and with high flow volumes, also, during the summer (31.4%).

In terms of *average flow*, were analysed the main parameters that describe this topic, with reference to the multiannual average values, maximum and minimum, respectively, to the values of the specific average flow and drained layer at the hydrometric stations (Table 16.8). The used data were obtained from various bibliographic sources, respectively from the statistical processing of the hydrological data strings, generated at the level of territorial structures for hydric environment monitoring.

Table 16.8 The distribution of water resources in the Apuseni Mountains (multiannual average values, extreme values, flow layer and specific average discharge)

No.	Major catchment	River	Hydrometric station	Monitoring period	F (km²)	H (m)	Q multiannual (m³/s)	Q max (m³/s)	Q min. (m³/s)	Y (mm)	q (l/s km²)
1	Arieş	Arieş	Scărişoara	1978–99	203	1126	5.58	275	0.3	867	27.5
2		Arieş	Câmpeni	1978–99	639	1020	12.2	735	0.756	603	19.1
3		Arieş	Baia de Arieş	1978–99	1189	965	18.9	860	1.260	502	15.9
4		Arieş	Buru	1978–99	1960	945	23.4	822	1.640	377	11.9
5		Arieş	Turda	1978–99	2358	892	24.9	950	1.400	333	10.6
6		Albac	Albac	1978–99	94	1110	1.66	53.7	0.076	557	17.7
7		Arieşul Mic	Ponorel	1978–99	133	1025	3.22	150	0.188	764	24.2
8		Abrud	Abrud	1965–99	99	891	1.42	84.5	0.025	453	14.3
9		Abrud	Câmpeni	1978–99	219	840	2.93	145	0.06	422	13.4
10		Valea Mare	Bistra	1973–99	69	1315	1.42	141	0.102	649	20.6
11		V. Muşcanilor	Muşca	1978–98	17	808	0.262	24.1	0.03	486	15.4
12		Poşaga	Poşaga	1978–99	106	1096	1.2	27	0.08	357	11.3
13		Ocoliş	Ocoliş	1978–99	63	903	0.526	9.5	0.018	263	8.3
14		Iara	Valea Ierii	1965–99	107	1359	0.678	86	0.036	200	6.3
15		Iara	Iara	1978–99	273	1120	0.877	77.4	0.053	101	3.2
16		Hăşdate	Petreştii de Jos	1978–99	189	629	0.591	64.8	0.03	99	3.1
17		Valea Largă	Viişoara	1978–99	200	395	0.31	27.1	0.02	49	1.6

(continued)

Table 16.8 (continued)

No.	Major catchment	River	Hydrometric station	Monitoring period	F (km²)	H (m)	Q multiannual (m³/s)	Q max (m³/s)	Q min. (m³/s)	Y (mm)	q (l/s km²)
18	Someşul Mic	Someşul Cald	Smida	1974–2018	103	1295	3.22	108	0.146	987	31.3
19		Beliş	Poiana Horea	1974–2018	85.1	1256	1.8	36.1	0.048	667	21.2
20		Someşul Rece	Someşu Rece-sat	1927–2018	328	1223	5.19	158	0.005	499	15.8
21		Someşul Mic	Gilău	1975–90	878	1143	13.5	272	0.156	485	15.4
22		Răcătău	Răcătău	1957–2018	101	1242	2.11	63.2	0.031	659	20.9
23	Crişul Alb	Crişul Alb	Blăjeni	1968–97	104	716	2	235	0.015	492	15.6
24		Crişul Alb	Crişcior	1968–97	333	637	3.66	254	0.032	347	11.0
25		Crişul Alb	Gurahonţ	1968–97	1581	515	15.09	544	0.267	301	9.5
28		Hălmagiu	Hălmagiu	1968–97	109	650	1.8	188	0.06	521	16.5
29		Hălmăgel	Hălmăgel	1968–97	67	746	1.17	126	0.03	551	17.5
30	Crişul Repede	Călata	Călata	1950–2004	74	890	1.01	4.81	0.019	431	13.6
31		Crişul Repede	Ciucea	1950–2004	814	904	11.9	50.2	0.673	461	14.6
32		Crişul Repede	Vadu Crişului	1950–2004	1328	821	20.5	98.9	1.35	486	15.4

(continued)

Table 16.8 (continued)

No.	Major catchment	River	Hydrometric station	Monitoring period	F (km²)	H (m)	Q multiannual (m³/s)	Q max (m³/s)	Q min. (m³/s)	Y (mm)	q (l/s km²)
33		Valea Drăganului	Valea Drăganului	1950–2004	228	1161	6.79	33.7	0.39	940	29.8
34		Iada	Leşu Amonte	1950–2004	67.5	1095	2.19	10.3	0.135	1024	32.4
35		Poicu	Vânători	1950–2004	37	638	0.384	2.26	0.022	328	10.4
36		Valea Stanciului	Răchiţele	1950–2004	55.6	1118	1.14	5	0.063	647	20.5
37		Săcuieu	Henţ	1950–2004	209	1006	3.28	14.5	0.194	495	15.7
38	Crişul Negru	Crişul Negru	Şuştiu	1950–2009	137	617	2.17	48	0.58	500	15.8
39		Crişul Negru	Beiuş	1950–2009	941	581	13.1	464	1.25	439	13.9
40		Crişul Pietros	Pietroasa	1950–2009	158	972	4.15	121	0.29	829	26.3
41		Crişul Băiţa	Ştei	1950–2009	65	796	1.23	175	0.13	597	18.9
42		Roşia	Pocola	1950–2009	267	427	3.4	145	0.04	402	12.7
43	Mureş	Aiud	Aiud	1970–2007	176	642	0.456	110	0.021	82	2.6
44		Geoagiu	Mogoş	1970–2007	30	1035	0.345	36	0.018	363	11.5
45		Geoagiu	Vl. Mănăstirii	1970–2007	135	916	1.23	82.3	0.089	288	9.1
46		Geoagiu	Teiuş	1970–2007	185	844	1.24	83	0.121	212	6.7
47		Galda	Benic	1970–2007	155	855	1.13	77.2	0.078	230	7.3

(continued)

Table 16.8 (continued)

No.	Major catchment	River	Hydrometric station	Monitoring period	F (km²)	H (m)	Q multiannual (m³/s)	Q max (m³/s)	Q min. (m³/s)	Y (mm)	q (l/s km²)
48		Ampoi	Zlatna	1970–2007	148	818	1.38	116	0.09	294	9.3
49	Middle Someş	Almaş	Almaşu	1950–2009	141	476	0.603	1.64	0.108	135	4.3
50		Almaş	Hida	1950–2009	556	242	1.83	5.42	0.463	104	3.3
51	Barcău	Bistra	Chiribiş	1950–2009	169	354	1.11	59	0.06	207	6.6

Source (Romanian Waters National Administration—Hydrographic Management National Plans of Mureş, Someş-Tisa and Crişuri; Ştef 2015; Sorocovschi 2006)

The available data revealed certain peculiarities of the flow formation, correlated with the physical-geographical factors influence: the trajectory of air masses from west sector movement, the petrographic and morphological mosaic related to the studied area.

The richest drainage is generated in the basin of the Arieş river, which is monitored through a number of 17 gauging stations. The main course, get through its tributaries, to a multiannual average value of about 25 m³/s (Turda hydrometric station). An important contribution, to this value, have the tributaries: Arieşul Mic (3.22 m³/s), Abrud (2.93 m³/s), Bistra, Poşaga and Iara. In Someşul Mic basin, the monitoring is done at the level of 5 hydrometric stations, which cumulated at the level of the Gilău closure section, a multiannual average flow of 13.5 m³/s. The most consistent volumes from the tributaries are collected by Someşul Rece (5.19 m³/s) and Răcătău with 2.11 m³/s (Table 16.8).

In Criş rivers basin, stands the hydrographic system of Crişul Repede river, which drains important volumes of water, exceeding the threshold of 20 m³/s, multiannual average value, at Vadu Crişului, gaging station. Lower values are associated with Crişul Negru (13.1 m³/s at Beiuş hydrometric station), respectively, on Crişul Alb (3.66 m³/s at Crişcior). The eastern slope and the southern is defined by a hydrographic network poorly evolved, which ensures quick drainage but insignificant. An exception to this statement could be Ampoi River and Geoagiu valley (Alba county) that have at the exit from mountain area, flow rates of more than 1 m³/s. The northern flank associated with middle Someş basin, respectively with Barcău creek, is represented by the watercourses which do not have high flow rates.

Maximum runoff is an important parameter for quantitative assessment of the water resource in terms of the effects they can generate on the anthropic component, in particular, related to the city infrastructure, land use, security of communication ways, etc. Expressed through the phases of high waters regime, respectively, episodes of flash flood, the maximum runoff in the Apuseni Mountains, is manifested in function of the specificities of the production and distribution in the territory of rainfall.

Thus, the long duration rains and those rich in quantity, showers, heavy snowfalls, or the sudden melting of accumulated snow can generate high flow rates. A special peculiarity is represented by winter floods, atypical phenomena in the mountain region in general. On the eastern flank, atypical episodes of warm waves, combined with rain led to the generation of dangerous phenomena (the basin of Arieş river, the basin of the Someşul Mic river—December 1995). The floods of winter can be multiplied also by the formation of the foehnale circulation to the east. High waters can be also the results of cumulated hydrological events (rich rainfall amounts which fall over the snow layer found in the process of melting), which may generate increased values of the flow rate at the beginning of spring. The floods of summer are determined in general by the convective rains, which can reach amplitudes very large, creating major floods (July 1975).

The drained layer at the level of the analysed area has values in a quite widely range, from over 1000 mm (in the north-central part, with maximum altitudes of the Bihor, Vlădeasa, Muntele Mare), to low values of under 100 mm, to the periphery of the mountain space (the southern flank, towards Mureş Corridor and western side to the golf type depressionary areas, which separates the mountains ridges of Zarand, Codru-Moma, Pădurea Craiului and Plopiş). An expression map of this territorial distribution, based on the information of ICPDR working group (International Commission for the Protection of the Danube River) is presented in Fig. 16.16.

The specific average flow shows high values (over 25 l/s km^2) in the high mountain area on the upper courses of Someşul Cald (Smida, 31.3 l/s km^2), Drăgan Valley (29.8 l/s km^2), Iadei (Leşu Amonte, 32.4 l/s km^2), Arieşului (Scărişoara, 27.5 l/s km^2) and Crişului Pietros (Pietroasa, 26.3 l/s km^2). The smallest values occur in the periphery of the mountainous area, associated with relatively short water courses with limited drainage (Aiud, Valea Largă, Hăşdate—on the eastern flank and Almaşul on the north).

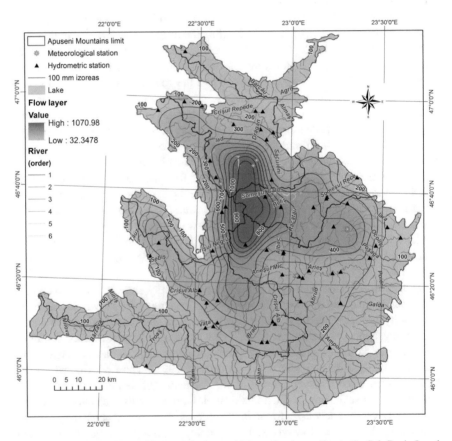

Fig. 16.16 The drained layer (Y—mm) in Apuseni Mountains (according to the Sub-Basin Level Flood Action Plan in the Tisza River Basin, 2009, compiled with information made in various studies by Basin Water Administration)

16.3.2.2   Storage Lakes as Water Reservoirs

The reservoirs from the upper basin of Someşul Mic river. Before 1968 there was in the specified area an artificial lake used for transporting logs to the lumber mill in the old village of Beliş. Subsequently, the authorities passed to the greath hydrotechnic improvement of the entire basin, that was made in two stages.

In the first stage (1968–1980) have been completed the greatest dams with storage lakes from the basin, where two built on the Someşul Cald river and one on the Someşul Mic River. The first functional dam with a reservoir from the system was Gilău I, in the year of 1972, followed by the Tarniţa, in 1973 and later Fântânele, in 1976.

Still, in the first stage, the building of the intakes and derivations from the Someşul Rece and Iara basins have started, some of them being given in service (Someşul Rece II).

In the second stage (1980–1990), have been given in service the rest of intakes and derivations mentioned above, bringing to the final the Someşul Cald river improvement, in 1983, with the given in service of the Someşul Cald dam and reservoir.

In the other hand, the Upper basin of Someşul Mic river includes 5 great lakes/reservoirs (Fântânele, Tarniţa, Someşul Cald, Gilău, Someşul Rece I— Fig. 16.13) whose total volume (334.16 million m^3) represents 72.0% of the total volume of the reservoirs from the Someş basin (464.32 million m^3). These storages lakes cumulate a surface of 1298 ha, which represents approximately 1% of the total area of the hydropower reservoirs from Romania [63].

These lakes occupy the areas of widening developed at the main course confluences with most important tributaries and upstream of narrowing imposed by the intrusions of hard rocks; in these conditions, the surface of the reservoir often extend over hundreds of hectares upstream of the dams (lakes Fântânele, Tarniţa— Table 16.9).

Table 16.9 Main morphometric parameters of the reservoirs from the Upper basin of Someşul Mic river (after Şerban 2007)

No.	River	Reservoir	N.L.R.[a] (m-B.S.)	Area (ha)	Length (km)	Maximum width (m)	L/W ratio
1.	Someşul Cald	Fântânele	991.00	815	19.13	748	25.57
2.	Someşul Cald	Tarniţa	521.50	220	8.40	597.8	14.05
3.	Someşul Cald	Someşul Cald	441.00	85	4.25	423.5	10.04
4.	Someşul Mic	Gilău	420.10	72	2.34	497.3	4.70
5.	Someşul Rece	Someşul Rece I	1020.5	8.9	1.12	145.8	7.68

[a]N.L.R. normal level of retention; m-B.S. meters towards the Black Sea

Table 16.10 Characteristic volumes of the reservoirs from the Upper basin of Someşul Mic river (after Şerban 2007)

No.	Reservoir	Characteristic volumes (mil m³)					
		Total	Raw	Useful	Backup vol.	Dead	Mitigation
1.	Fântânele	244.69	207.59	186.93	10.17	10.50	37.00
2.	Tarniţa	75.25	68.08	13.79	39.08	5.62	6.90
3.	Someşul Cald	8.45	6.45	0.86	3.41	2.18	1.99
4.	Gilău	4.10	2.90	0.80	1.60	0.50	1.21
5.	Someşul Rece I	1.67	1.34	0.98	–	–	0.33

The values of the reservoirs morphometric features are decreasing from upstream to downstream with their dimensions, as well as the influence that they exert on the hydric and morphologic processes and phenomena from the middle and lower sectors of the main course.

The characteristic volumes of reservoirs presented in Table 16.10, emphasize the existing differences between the lakes and highlight the numerous functions that they perform, as well as the existence of some processes that affect the storage capabilities of the hydraulic agent.

The characteristic volumes provide information on the total and useful capacity of reservoirs, as well as on the function of floods mitigation. The particularly low values of the unusable volume, in comparison with the total volume, highlight an accentuated stage of silting of that lake.

The functions of the improvements from the Upper basin of Someşul Mic river are multiple, often common to the four lakes—for example, the hydropower function. There are, however, specific functions of each retention, according to the purpose for which it was built [63].

Fântânele dam & lake has a complex function:

– energetic, respectively production of electricity in the Mărişelu underground station;
– multi-annual and annual liquid flow regularization on the course of the Someşul Cald river;
– mitigation of flash flood waves on the Someşul Cald river.

Tarniţa dam & lake also has a complex function:

– energetic, respectively production of electricity in the power plant located at the base of the dam;
– newer source for water supply of the very important network of settlements and economic units located in the downstream and neighbour areas: the municipality of Cluj-Napoca and its industrial units, older Căpuş and Aghireş mining interest regions, all the localities from Someşul Mic Corridor (includes the Gherla and Dej cities), the neighbour Sălaj county, using the pumping stations;
– annual liquid flow regularization on the course of the Someşul Cald river;
– mitigation of flash flood waves originating from the related catchment areas.

The *Someşul Cald dam & lake*, is a buffer lake between the Tarniţa and Gilău lakes; its functions are also important, namely:

- the production of electricity in the power plant located at the base of the dam;
- the recovery of water flows discharged from the Tarniţa hydropower plant;
- silt decanter for Agârbiciu tributary; in fact, this feature has been one of the most important reasoning in the design of this retention, just to mitigate the very high silting level of Gilău retention, situated immediately downstream;
- the mitigation of flash flood function is poorly represented, due to the reduced dedicated characteristic volume.

The *Gilău dam & lake*, although the smallest, has the complex function, which derives from the purpose for which it was made:

- older source for water supply of the same important network of settlements and economic units located in the downstream and neighbour areas: the municipality of Cluj-Napoca and its industrial units, older Căpuş and Aghireş mining interest regions, all the localities from Someşul Mic Corridor (includes the Gherla and Dej cities), the neighbour Sălaj county, using the pumping stations;
- compensation role for the liquid flow pulsations discharged from the peak hydro-electric power plants, located in the upstream;
- the production of electricity in the power plant located at the base of the dam;
- the mitigation of flash flood function is poorly represented, due to the reduced dedicated characteristic volume.

Someşul Rece I dam & reservoir represent a collector of Iara basin derivation and of Someşul Rece river, with the water drainage to the Fântânele reservoir; even the function of flash floods mitigation is, also, present, but limited by the low volume of this retention.

Not devoid of importance is, also, the tourist function, beeing specific for all areas around the reservoirs from the analyzed basin. This function is a complex one in the case of Fântânele and Gilău reservoirs (rest and recreation) and easy (leisure) in the case of the other lakes. This situation, in the first case, is explained by the possibilities of accommodation that they offer: the spaces around the Fântânele reservoir by comfortable hotels with a total capacity of 365 places, restaurants and bars and the spaces around the Gilău reservoir by the motel located on the left bank, equipped with an accommodation capacity of 114 places, restaurant and bar [55].

The dams with reservoirs from the Upper basin of Crişul Repede river. The Drăgan-Iada hydropower improvement has been made in order to capitalize in a superior way the water resources available in Drăgan, Iada and Săcuieu rivers (Fig. 16.14).

In order to satisfy the needs of industrial and irrigation water in Oradea municipality area, it was found necessary to stabilize the liquid flow of Crişul Repede river, by carrying in its basin of one or several reservoirs with a useful volume of water of 28 million m^3.

After analysing many location sites, has come to the conclusion that the optimal solution for the first stage was the onstruction of the Leşu reservoir on the Iada river [30].

Table 16.11 Characteristic volumes of the reservoirs from the Upper basin of Crişul Repede river

No.	Reservoir	Characteristic volumes (mil m^3)			
		Total	Raw	Useful	Mitigation
1.	Floroiu	124	112	100	6
2.	Leşu	32.8	28.3	–	–

After Horvath [30]

The hydrotechnical complex Drăgan-Iada can be divided in two components: *the Drăgan hyrotechnical assembly* with the Remeţi, Munteni I şi Munteni II hydropower plants, and on the other hand *the Leşu hydrotechnical assembly* with Leşu hydropower plant and its associated dam. The most important function is the capitalization of hydropower potential of Drăgan, Iada and Săcuieu rivers

Drăgan dam, was completed at the end of 1984, while Floroiu reservoir (290 ha) has reached the 84 m height of retention in May of 1985. The Leşu dam and reservoir (120 ha) are located on Iada river, at about 7 km upstream of the Remeţi village, the dam being made during the years 1969–1973.

The characteristic volumes, which gives the image of the functionality of the reservoirs from Drăgan-Iada hydrotehnical complex are shown in Table 16.11.

The dams from Arieş basin are represented by a single major reservoir, the Mihoieşti hydrotechnical facility which is located at the confluence of Arieşul Mare with Arieşul Mic rivers, 5 km upstream of Câmpeni town: Its main purposes are to ensure the water supply for several settlements, electricity generation and mitigation function for floods, generated in the upper basin. The reservoir total volume is 9 mil. m^3, with a surface area of about 123 ha, at the maximum exploitation level of 580 mBS. The reservoir has staretd to operate since 1987 and became the main structural measure for floods' mitigation on the Arieş River, the most important watercourse of the Apuseni Mountains.

16.4   Water Balance in the Apuseni Mountains

The making of the water balance in the Apuseni Mountains has been done following the quantitative relationships between the inputs and outputs of water in the area, at the level of the main collector basins (Table 16.12). Thus, the parameters used at the basis of this analysis are the average rainfall, the average flow layer and the evapotranspiration [62]. We also determined the index of aridity—K, as the ratio between evapotranspiration and precipitation.

The associated rainfall data indicates that on the whole area, the average value is about 832 mm, with maximum values in the Crişul Negru basin (924 mm), followed by the Crişul Alb basin and by the Someşul Mic basin. The lowest values are found in the northern extremity associated with the Plopiş and Meseş mountains, respectively in the Middle Someş (700.1 mm) and Barcău (709 mm) river basins.

Table 16.12 The structure of water balance in Apuseni Mountains (information provided by Water Basin Administrations of Mureş, Someş-Tisa, Crişuri)

No.	Collector basin/element	F (km²)	Precipitation (X)			Global flow (Y)			Evapotranspiration (Z)			Aridity index
			mm	mil. m³	%	mm	mil. m³	%	mm	mil. m³	%	
1	Mureş	5064.06	782.1	3960.70	41.18	223.3	1130.60	35.68	542.7	2748.34	43.57	0.69
2	Crişul Alb	2247.29	916.5	2059.56	21.41	239.0	537.09	16.95	596.9	1341.30	21.26	0.65
3	Crişul Repede	1444.83	892.2	1289.04	13.40	378.4	546.69	17.25	534.3	772.00	12.24	0.60
4	Crişul Negru	1081.16	924.2	999.22	10.39	327.8	354.39	11.18	565.3	611.15	9.69	0.61
5	Someşul Mic	988.50	901.6	891.26	9.27	543.0	536.79	16.94	483.8	478.22	7.58	0.54
6	Middle Someş	358.92	700.1	251.29	2.61	109.2	39.20	1.24	601.9	216.04	3.42	0.86
7	Barcău	235.40	709.5	167.01	1.74	103.7	24.40	0.77	598.9	140.98	2.23	0.84
8	Total Apuseni	11,420.16	832.3	9618.09	100.00	274.9	3169.15	100.00	560.5	6308.04	100.00	0.69

Regarding the liquid global flow, it has an average value of about 275 mm, with maximum in the Someşul Mic basin (543 mm) and minimum values in the North of Apuseni Mountains. The average value of the evapotranspiration reaches to 560 mm, with a maximum of about 600 mm in the North and minimum under 500 mm in the Someşul Mic basin (483 mm).

Regarding the Index of aridity, it has a global value of 0.69 on all the mountain area, with the lowest value in the Someşul Mic basin (0.54), and the highest in the area drained by the Middle Someş (0.86).

If we exclude the space of the lower mountains from the northern Apuseni Mountains, the study area is defined by important quantities associated with precipitation and runoff, combined with modest evapotranspiration. These elements make possible the framing of the studied space in the rich moisture area of Romanian territory (Ujvari 1972).

16.5   Conclusions

The Apuseni Mountains area has a significantly hydric budget, spatially expressed through a dense hydrographic network, with radial drainage, imposed by the high sector of central ridge of the Bihor-Vlădeasa and Muntele Mare mountains. The entire space is a tributary of the Tisa river, which receives the entire budget of the Apuseni Mountains through three major river systems: the Mureş river, the Crişuri rivers and the Someş river.

Having a very varied petrographic constitution, supplemented with a complicated morphology, with frequent intercalations of depression areas between the mountain chains, the Apuseni hydric budget is determined by the high rainfall quantities, brought by western air masses advection.

The development of foehn processes, on the eastern flank, induce an uneven distribution of water resources, in favour of the western flank. By the way, in this area the maximum rainfall at the national level recorded.

The hydrogeology of the Apuseni Mountains is marked by the development of the igneous structures, interleaved with metamorphic ones and on extensive areas with limestone surfaces. The latter are constituted in territories with complex drainage, expressed by numerous gorges and defiles sectors, but, also, by underground systems of caves, potholes, sinkholes and karst springs flow.

The physical-geographical factors contribute to the formation of a vigorous hydrographic network, through the fluviatile systems prism. The natural laccustric inventory is modest but complemented by multiple anthropogenic lakes with multiple functions: electric energy production, the water supply of the localities, mitigation of the flash flood. These anthropogenous lakes are found especially on the upper sector of the Someşul Mic and Crişul Repede rivers.

Underground resources present limited meanings from a therapeutic point of view, except a small area from the southern perimeter of Metaliferi Mountains, where it focuses more on hydro mineral and thermal sources.

Hydrological monitoring of Apuseni Mountains is made through a structure of hydrometric stations and representative hydrological basins, coordinated territorially by three water basin administrations: Someș-Tisa, Mureș and Crișuri.

From the analysis of the rainfall, of the petrographic conditions, edaphic and land-use patterns, it is apparent that at the level of the Apuseni Mountains water resources have provided favorable conditions for development, under the quantitative and qualitative report. The mountain areas with an altitude of over 1400 m, have significant reserves of water, regarding the map of liquid flow layer.

The chronologic approach of the flow put out in evidence areas with different types of hydric regime, but also in terms of the supply types of the rivers. Higher values of runoff are recorded on river basins related to the upper sectors of the Crișul Repede and Crișul Negru rivers and lower on water-courses, which drain the eastern flank of the Trascău and Metaliferi mountains: Valea Mănăstirii, Galda, Ampoi și Geoagiu.

A characteristic for the rivers from the Apuseni Mountains (in particular for those developed on the western flank) the flow is relative high in the winter months, between 20 and 35% of the annual flow, while the spring by over 40%, in comparison with the summer one, that does not exceed 15–25%, or with the autumn one with 10–15%.

16.6   Recommendations

Water resources of Apuseni Mountains can be better exploited in terms of floods mitigation or quality protection. The lack of wastewater treatment plant for many cities, led to frequent pollution episodes. The accelerated cutting of forests, especially in the upper hydrographic basins, will lead in time to a higher frequency of floods, but also to an increase of water turbidity. The closure of mining operations (in quarries or underground) does not guarantee the quality of river safety. Due to their high population density, compared with other mountain areas, the Apuseni Mountains are dealing with a great human pressure, over aquatic environment. Ocassionally, mostly in Spring, rivers changed into the most efficient sanitation agent, by taking over huge amounts of waste and residues, abandoned voluntarily or involuntarily around riverbeds. Future studies may emphasize the aspect mentioned above or can develop more analysis on run-off process and its spatial and temporal distribution.

References

1. Alboiu M, Nitulescu M, Paduraru A (1962) The drying up of the rivers from the Crișul Repede river basin (in Romanian), vol 3. Studii de hidrologie, București
2. Arghiuș V (2006) The study of flash floods from the Apuseni Mountains Eastern slope water courses and the associated risks (in Romanian), Casa Cărții de Știință, Cluj-Napoca, 251p

3. Aznar-Sánchez JA, Belmonte-Ureña LJ, Velasco-Muñoz JF, Manzano-Agugliaro F (2018) Economic analysis of sustainable water use: a review of worldwide research. J Clean Prod 198:1120–1132
4. Batinaş R, Sorocovschi V, Şerban Gh (2002) Hydrological risk phenomena induced by the flash floods in the lower basin of the Arieş river (in Romanian). Seminarul Geografic "Dimitrie Cantemir", nr 21–22, Iaşi
5. Bătinaş R (2010) The study of the surface waters quality in the Arieş river basin (in Romanian). Edit. Presa Universitară Clujeană, Cluj-Napoca
6. Bătinaş R, Şerban G, Sabău D, Rafan S (2016) Preliminary analysis on some physico-chemical rivers water features in Pricop-Huta-Certeze and Upper Tisa Natura 2000 Protected Areas. In: "Air and water—components of the environment" conference proceedings, Şerban Gh, Bătinaş R, Croitoru A, Holobâcă I, Horvath C, Tudose T (eds), 22–23 March, Babeş-Bolyai University, Faculty of Geography, Cluj-Napoca, România, Edit. Casa Cărţii de Ştiinţă, pp 314–319
7. Bird G, Brewer P, Macklin M, Balteanu D, Serban M, Zaharia S. (2003) The impact and significance of metal mining activities on the environmental quality of Romanian River systems. In: Proceedings of the first international conference on environmental research and assessment, Bucharest, March 23–27, pp 316–332
8. Blaney D (2014) Water resource management in a vulnerable world: the hydro-harzardscapes of climate change by Daanish Mustafa. Glob Environ Politics 14(1):138–139
9. Borsos B, Sendzimir J (2018) The Tisza River: Managing a Lowland River in the Carpathian Basin. https://doi.org/10.1007/978-3-319-73250-3_28 In book: Riverine Ecosystem Management
10. Buta I, Iacob E (1967) The alluvial runoff on the rivers from the north-west of Romania (in Romanian). Studia Univ, Babeş-Bolyai, Cluj-Napoca
11. Buz V (1976) Water balance in the Criş rivers basin (in Romanian). Studia Universitatis Geographia, Cluj-Napoca
12. Choiński A, Ilyin L, Marszelewski W, Ptak M (2008) Lakes supplied by Springs: selected examples. Limnol Rev 8(4):145–150
13. Cocean P (2000) Apuseni Mountains. Karstic processes and forms (in Romanian) Ed. Academiei Române, Bucureşti
14. Corduneanu F, Vintu V, Balan I, Crenganis L, Bucur D (2016) Impact of drought on water resources in North-Eastern Romania. Case study-the Prut River. Environ Eng Manag Jurnal (EEMJ) 15(6)
15. Costan C (2010) Natural and technological risks in the middle basin of the Arieş river (in Romanian). Ph.D. thesis—manuscript, Universitatea Babes-Bolyai, Facultatea de Stiinta Mediului, http://enviro.ubbcluj.ro/documente/teza%20cameia%20c.pdf
16. Crăciun A (2011) The indirect estimation, with the help of GIS, of the soil moisture, in the purpose of the rain floods modeling. Applications in Apuseni Mountains river (in Romanian). Ph.D. thesis—manuscript, Universitatea Babes-Bolyai, Facultatea de Geografie, Cluj-Napoca
17. Diaconu D (2008) The Siriu reservoir, Buzău river (România), Lakes, reservoirs and ponds. Romanian J Limnol, 1–2:141–149
18. Diaconu DC (2010) Management of storage lakes in Romania. Volumul Conferinţei Aerul şi apa - Componente ale mediului. Presa Universitara Clujeană, pp 149–155
19. Diaconu DC (2013) Water resources from Buzău river watershed (in Romanian). Editura Universitară, Bucureşti, 238p
20. Diaconu DC, Mailat E (2010) The management of Reservoir Silting in Romania. In: 10th international multidisciplinary scientific geoconference SGEM, vol II, pp 105–112
21. Drăgan M (2011). The resilience of the Apuseni Mountains regional system (in Romanian). Ph.D. thesis—manuscript, Universitatea Babes-Bolyai, Facultatea de Geografie, Cluj-Napoca
22. Duma S (1998) Geoecologic study of the mining exploitations from the southern area of the Apuseni Mountains, Poiana Ruscă Mountains and Mountains of Sebeş (in Romanian). Editura Dacia, Cluj-Napoca
23. Dume DM (2009) Hydrotechnical improvements from Crişul Repede basin and their impact on the liquid flow (in Romanian). Ph.D. thesis—manuscript, Universitatea din Oradea, Oradea

24. Forray FL (2002) Geochemistry of the environment in mining sites areas from the Arieş valley (Apuseni Mountains) (in Romanian). Ph.D. thesis—manuscript, Universitatea Babes-Bolyai, Catedra de Mineralogie, Cluj-Napoca, p 301
25. Gâştescu P (2003) Territorial distribution of water resources in Romania in terms of social-economic demand. Revue roumaine de géographie, tomes 47:48
26. Gâştescu P (2010) Water resources from Romania. Potential, quality, spatial distribution, management (in Romanian). În volumul „Resursele de apă din România – vulnerabilitate la presiunile antropice", Lucrările primului simpozion naţional de Limnogeografie, Editori Gâştescu, P., Breţcan, P., 11–13 iunie, Universitatea Valahia, Târgovişte, Edit. Transversal, pp 10–30
27. Gâştescu P (2012) Water resources in the Romanian Carpathians and their management. GEO-REVIEW: Scientific Annals of Stefan cel Mare University of Suceava. Geogr Ser 21(2):9–10
28. Gyori M-M (2013) The flash floods prediction in the conditions of limited data. Application to the small rivers from the Zărand and Săvârşin mountains (in Romanian). PhD. thesis—manuscript, Universitatea Babes-Bolyai, Facultatea de Geografie, Cluj-Napoca
29. Haidu I, Craciun AI, Bilasco St. (2007) The SCS-CN model assisted by GIS-alternative estimation of the hydric runoff in real time. Geographia Technica 1:1–7. ISSN 1842-5135
30. Horvath C (2008) The study of the reservoirs from the upper basin of the Crişul Repede river (in Romanian). Editura Casa Cărţii de Ştiinţă, Cluj-Napoca, p 208
31. Iacob E (1971) Apuseni Mountains, Hydrological Study (in Romanian), Ph.D. thesis—manuscript, Universitatea Babes-Bolyai, Facultatea de Biologie, Geografie şi Geologie, Cluj-Napoca
32. Konar M, Evans TP, Levy M, Scott CA, Troy TJ, Vörösmarty CJ, Sivapalan M (2016) Water resources sustainability in a globalizing world: who uses the water? Hydrol Process 30(18):3330–3336
33. Lakshmi V, Fayne J, Bolten J (2018) A comparative study of available water in the major river basins of the world. J Hydrol 567:510–532
34. Lal R (2015) World water resources and achieving water security. Agron J 107(4):1526–1532
35. Li P, Qian H (2018) Water resource development and protection in loess areas of the world: a summary to the thematic issue of water in loess. Environ Earth Sci 77(24):796
36. Linzer HG, Frisch W, Zweigel P, Girbacea R, Hann HP, Moser F (1998) Kinematic evolution of the Romanian Carpathians. Tectonophysics 297(1–4):133–156
37. Mac I (1998) Documentation of certification of the Fântânele tourist resort like as eco-ethnographic resort (in Romanian). Plan de Urbanism Zonal, S.C. FACIL SERVCOM S.R.L. Cluj-Napoca
38. Mălai M (1983) Hydropower improvement of Someşul Mic river downstream of Tarniţa dam (in Romanian). Hidrotehnica, 11, Bucureşti
39. Mirchi A, Watkins DW, Huckins CJ, Madani K, Hjorth P (2014) Water resources management in a homogenizing world: averting the growth and underinvestment trajectory. Water Resour Res 50(9):7515–7526
40. Morariu T, Savu Al (1956) The hydrographic regions of Transylvania (in Romanian), Bulletin Ştiintific, Sectia Geologie-Geografie I.3-4 Bucuresti
41. Mustafa D (2013) Water resource management in a vulnerable world: the hydro-hazardscapes of climate change. Philip Wilson Publishers
42. Orăşeanu I (2016) The hydrogeology of karst from Apuseni Mountains (in Romanian). Edit, Belvedere, Oradea
43. Pandi G (1997) The energy conception of the suspension alluvium formation and transport: application in Nord-West of Romania (in Romanian). Presa Universitară Clujeană, Cluj-Napoca, 229p
44. Pavel M (1975) Drăgan hydropower improvement on Iad river (in Romanian). Hidrotehnica 2, Bucuresti
45. Piasecki A, Marszelewski W (2014) Dynamics and consequences of water level fluctuations of selected lakes in the catchment of the Ostrowo-Gopło Channel. Limnological Review 14(4):187–194. https://doi.org/10.1515/limre-2015-0009

46. Pop Gh (1970) Fărcaş smoothing surface from Gilău Mountains (in Romanian). Ph.D. thesis (manuscript), Universitatea Babeş-Bolyai, Cluj-Napoca
47. Pop Gr. (1996) Romania. *Hydropower Geography (in Romanian)*, Edit. Presa Universitară Clujeana
48. Pop Gr (1992) Hydropower improvements from the basin of Crişul Repede river (in Romanian). Studia Univ. Babeş-Bolyai, 1–2, Cluj-Napoca
49. Popescu V, Florescu D (1976) Fântânele dam, solutions and technologies of building (in Romanian). Hidrotehnica, 6, Bucureşti
50. Posea A (1969) Winter phenomena in the basin of Crişul Repede river (in Romanian). Revista Terra, nr 1, Bucureşti
51. Posea A (1970) Hydrographic basin of Crişul Repede river (in Romanian) Ph.D. thesis— manuscript, Universitatea Babes-Bolyai, Facultatea de Biologie, Geografie şi Geologie, Cluj-Napoca, 225p
52. Romanescu G, Sandu I, Stoleriu C, Sandu IG (2014) Water resources in Romania and their quality in the main lacustrine basins. Rev Chim (Bucharest) 63(3):344–349
53. Rusu T (1988) Following undergound waters. The karst from Pădurea Craiului Mountains (in Romanian). Edit. Dacia, Cluj-Napoca
54. Savin C (1990) Water resources from Jiu river major riverbed (in Romanian). Editura Scrisul Românesc, Craiova
55. Schreiber WE, Idu PD, Sorocovschi V, Ciangă N, Maier A, Stoia Ileana (1987) Landschaftsbeeinflussung durch hydroenergetische anlagen im oberen einzugsbecken des Someşu Mic - flusses. Studia Univ. Babeş-Bolyai, Geol.-Geogr., XXXII, 3, Cluj-Napoca
56. Simionescu Al (1980a) The technology of excavation and concreting of the Mărişelu power plant cavern on Someşul Cald river (in Romanian). Hidrotehnica, 8, Bucureşti
57. Simionescu Al (1980b) Tarniţa dam. The technology of concreting (in Romanian). Hidrotehnica, 10, Bucureşti
58. Simionescu Al (1982) The organization and operation of the career from the Fântânele rocks dam (in Romanian). Hidrotehnica, 11, Bucureşti
59. Słyś D, Stec A, Zeleňáková M (2012) A LCC analysis of rainwater management variants. Ecol Chem Eng S 19(3):359–372
60. Sofronie C (2000) Hydrotechnical improvements in Someş-Tisa basin (in Romanian). Edit, Gloria, Cluj-Napoca
61. Sorocovschi V, Serban Gh, Batinas R (2002) Hydric risks in lower basin of Arieş river (in Romanian). Vol. "Riscuri si catastrofe", I, Editor V.Sorocovschi, Edit. Casa Cartii de Stiinta, Cluj-Napoca
62. Sorocovschi, V., Şerban, Gh (2012) Elements of climatology and hydrology. Part II—hydrology. ID education form (in Romanian). Edit. Casa Cărţii de Ştiinţă, Cluj-Napoca, 242p
63. Şerban Gh (2007) The reservoirs from Someşul Mic upper basin—hydrogeographic study (in Romanian). Presa Universitară Clujeană, Cluj-Napoca, Edit
64. Şerban Gh, Pandi G, Sima A (2012) The need for reservoir improvement in Vişeu river basin, with minimal impact on protected areas, in order to prevent flooding. Studia Univ. "Babeş-Bolyai", Geographia, LVII, nr.1, Cluj-Napoca, pp 71–80
65. Şerban Gh, Sabău A, Rafan S, Corpade C, Niţoaia A, Ponciş R (2016) Risks Induced by Maximum Flow with 1% Probability and Their Effect on Several Species and Habitats in Pricop-Huta-Certeze and Upper Tisa Natura 2000 Protected Areas. "Air and water—components of the environment" Conference Proceedings, Şerban Gh, Bătinaş R, Croitoru A, Holobâcă I, Horvath C, Tudose T (eds), 22–23 March, Babeş-Bolyai University, Faculty of Geography, Cluj-Napoca, România, Edit. Casa Cărţii de Ştiinţă, pp. 58–69
66. Teodosiu C, Barjoveanu G, Vinke-de Kruijf J (2013) Public participation in water resources management in Romania: issues, expectations and actual involvement. Environ Eng Manag J (EEMJ) 12(5)
67. Ujvári I (1972) The geography of Romanian waters (in Romanian), Edit. Ştiinţifică, Bucureşti, 578p

68. Woodward RT, Shaw WD (2008) Allocating resources in an uncertain world: water management and endangered species. Am J Agr Econ 90(3):593–605
69. Zaharia L (1999) Water resources in Putna catchment. A hydrological study (in Romanian). Editura Universităţii din Bucureşti, 305p
70. Zaharia L (2010) The Iron Gates reservoir—aspects concerning hydrological characteristics and water quality, Lakes, reservoirs and ponds. Romanian J Limnol 4(1–2):52–69
71. Zaharia L (2005) Study on water resources in Curvature Carpathians and Subcarpathians area to optimize their use for the population supply (in Romanian), in the volume „Lucrări şi rapoarte de cercetare. Centrul de cercetare „Degradarea terenurilor si Dinamica geomorfologica", vol. I, Ed. Universităţii Bucureşti, pp 137–171
72. Zeleňáková M, Hudáková G (2014) The concept of rainwater management in area of Košice region. Procedia Eng 89:1529–1536
73. Zeleňáková M, Markovič G, Kaposztásová D, Vranayová Z (2014) Rainwater management in compliance with sustainable design of buildings. Procedia Eng 89:1515–1521
74. *** (1968) Geologic Map of Romania (in Romanian), 1:200000. Geologic Institute of Romania, Bucharest
75. *** (1971–1979) Soil map of R.S. Romania (in Romanian), 1:200000. ICPA, Bucureşti
76. *** (1975) Romanian Groundwater Map (in Romanian). Geologic Institute of Romania, Bucharest.
77. *** (1978–1982) Topographic map of Romania (in Romanian), 1:25000. Military Topographic Direction, Bucharest
78. *** (1982) The Geographic Atlas of Romania (in Romanian). Ed. Didactică şi Pedagogică, Bucureşti
79. *** (1983) Geografia României, vol I, Edit. Academiei RSR, Bucuresti
80. *** (1987) Geografia României, vol III, Edit. Academiei RSR, Bucuresti
81. *** (1992) The Atlas of Water Cadastre of Romania (in Romanian). Ministry of Environment and Aquaproject S.A., Bucureşti, 683p
82. *** (2009) Sub-Basin Level Flood Action Plan Tisza River Basin. International Commission for the Protection of the Danube River—Flood Protection Expert Group, Hungary, Romania, Slovakia, Serbia, Ukraine
83. *** (2013) Guide for the activity of hydrometric stations on rivers (in Romanian). N.I.H.W.M.—The National Institute of Hydrology and Water Management, Bucharest
84. *** (2018) Corine land cover. https://land.copernicus.eu/pan-european/corine-land-cover/clc-2012
85. *** A.B.A.C. (2015) Planul de management al bazinului hidrografic Crişuri (C.W.B.A.—"Crişuri" Water Basin Administration—Hydrographic basin of Criş rivers management plan, in Romanian)
86. *** A.B.A.M. (2015) Planul de management al bazinului hidrografic Mureş (M.W.B.A.—"Mureş" Water Basin Administration—Hydrographic basin of Mureş river management plan, in Romanian)
87. *** A.B.A.S.T. (2015) Planul de management al spaţiului hidrografic Someş-Tisa (S.T.W.B.A.—"Someş-Tisa" Water Basin Administration—Someş-Tisa hydrographic space management plan, in Romanian)
88. *** Records of R.W.N.A. ("Romanian Waters" National Administration—in Romanian)
89. *** Records of S.T.W.B.A. ("Someş-Tisa" Water Basin Administration—in Romanian)
90. *** https://upload.wikimedia.org/wikipedia/commons/thumb/5/5a/Carpathians_dem.jpg/580px-Carpathians_dem.jpg
91. ***(2008) Romanian Climate. National Administration of Meteorology, Edit. Academiei Române, Bucureşti (in Romanian)
92. *** (1984) Popescu-Argeşel I, Arieş Valley, Edit. Sport-Turism, Bucureşti (in Romanian)

Part VII
Conclusion

Chapter 17
Update, Conclusions, and Recommendations for "Water Resources Management in Romania"

Ionut Minea, Abdelazim M. Negm and Martina Zeleňáková

Abstract This chapter highlights the update of the topic, main conclusions, and recommendations of the chapters presented in the book. Therefore, this chapter contains information on water resources management in Romania in the period of climate change. It focuses on hydrological extremes—droughts and flood—its assessment and protection, water quality in Romania and it is devoted to sustainable management of water resources. A set of recommendations for future research is pointed out to direct the future research towards sustainability of water resources management which is one of the strategic themes of the Romania.

Keywords Romania · Water resources management · Water quality · Climate change · Droughts · Floods · Hydrological extremes

17.1 Introduction

Water management, like the power industry, is not a sector per se, but it does secure access to water for all other sectors and for society as a whole according to need. However, unlike energy there are no alternative sources of water. And that is why for several years now we have considered water to be a strategic raw material. In addition to water provision, water management has another no less important task— protection from the undesired effects of hydrological extremes, such as drought

I. Minea
Department of Geography, Faculty of Geography and Geology, "Alexandru Ioan Cuza", University of Iasi, Iasi, Romania
e-mail: ionutminea1979@yahoo.com

A. M. Negm (✉)
Water and Water Structures Engineering Department, Faculty of Engineering, Zagazig University, Zagazig 44519, Egypt
e-mail: Amnegm@zu.edu.eg

M. Zeleňáková
Department of Environmental Engineering, Faculty of Civil Engineering, Technical University in Košice, Košice, Slovakia
e-mail: martina.zelenakova@tuke.sk

© Springer Nature Switzerland AG 2020
A. M. Negm et al. (eds.), *Water Resources Management in Romania*, Springer Water,
https://doi.org/10.1007/978-3-030-22320-5_17

and floods. Meteorology, climatology and hydrology in particular provide not only the marginal conditions but also direct input values into water management. For a long time, here in Romania and abroad, water management was determined based on the sources of water, the renewability of which was considered as a stationary process, whose central values and variance did not change over time. In considering of climate change, a phenomenon we are already confronting and which is primarily expressed in meteorological, climatological and hydrological processes, it is shown that these processes are non-stationary. This means that we identify trends in time-related climatic as well as in hydrological orders. Water resources may decrease or increase depending on the development of climate elements. In the past we were able to resolve annual or perennial fluctuations of available water sources either by using economic instruments or by creating water reserves in our conditions, with annual regulation.

This conclusion chapter presents a summary of the essential findings and conclusions of the studies on the hydrological hazards, water quality, and water quantity mainly from hydrological point of view completed by case studies. A set of recommendations extracted from all contributions are presented to help academicians, researchers, practitioners and decision makers to go forward towards reasonable and sustainable management of water resources in Romania.

17.2 Update

Assessing water resource management in a country is a problem that can be analyzed in many ways. First of all, we must take into account the social and economic development of the country in the last decades. In the case of Romania, the changes since 1990, through the post-communist period, have had an exponential impact on the ways of monitoring, evaluation and management of water resources. Moreover, by joining the European Union (starting with 2007), but also the pre-existing actions, required that the systems for the implementation of the water resource management plans, applied at the level of the administrative and territorial unit, be adapted to the new European requirements.

The transposition of the European Water Frameworks Directive (2000/60/EC for water resources management and 2007/60/EC for reducing the impact of hydrological risks associated with the maximum leakage) into national legislation has allowed a series of reassessments of monitoring, evaluation, analysis and resource management of water across the country [35, 36]. This was highlighted in this volume by Romanescu et al. and Paveluc et al. in the chapters related to *Implementation of EU Water Framework Directive (2000/60/CE) in Romania-European Qualitative Requirements* and *Monitoring and management of water in the Siret River Basin (Romania)*. Adopting European legislation on how to manage water resources has solved many problems. The first measure focused on how to report the distribution and the volume of water resources by replacing the administrative and territorial units (used in the communist period) with the natural aquatic units (drainage basins, underground water bodies) [9]. Obviously, this could not be achieved across the country, especially if we refer to cross-border hydrographic basins or to the Danube Delta and Black Sea area. In this respect, the Water Basin Administrations were established incorporated in the National Hydrological Administration or the Romanian Water Company. This allowed, firstly, a more accurate and real reevaluation of the spatial distribution and extent of all categories of water resources and identify vulnerable areas for that can be found effective measures to reduce water scarcity. In this respect, the two basic principles set out in the EU Water Framework Directive [10] have been integrated, the drainage basin principle and the unitary water management principle [38].

The next step, in line with European legislation translated into the Romanian one, was the identification of water pollution issues. Up to the 1990 level, information on water pollution was virtually censored. By adopting European laws and principles associated (like economic principle and the polluter pays principle) important steps have been taken in identifying situations related to water pollution [2, 6, 12, 25]. Thus are analyzed the causes and effects of water pollution in Romania (by Breabăn in the chapter *Causes and Effects of Water Pollution in Romania*) highlighting the reduction of the impact of industrial activities and the significant increase of the input of pollutants through the uncontrolled use of chemical fertilizers in agriculture (the lack of water treatment plants associated with human agglomerations) [5]. Secondly the environmental impact of surface water pollution was also assessed (in the chapter *Management of Surfaces Water Resources—Ecological Status of the Mureș Waterbody (Superior Mureș Sector), Romania* written by Morar și Rus).

Another important aspect associated with the effects of the adopted European legislation is the one related to the economic valorization of water resources and the associated problems [26] that arise in the case of water supply of the population in large urban agglomerations (Constantin and Niță in the chapter *Water Supply Challenges and Achievements in Constanta County*) and the quality of water used in the water supply systems of the population (in the chapter written by Bacotiu et al *Drinking Water Supply Systems—Evolution towards Efficiency*).

An important aspect was the identification of anthropogenic impacts on water resources through surface or underground water extraction [8, 19], by making hydro-technical constructions [22] or by reducing the woodland area across the country (Peptenatu et al. in the chapter *Deforestation and Frequency of Floods in Romania* and Minea partial in the chapter *The Vulnerability of Water Resources from Eastern Romania to Anthropic Impact and Climate Change*).

Also was analyzed the climate-surface water relation (by Briciu et al. in the chapter Assessment of Some Diurnal Stream water Profiles in Western and Northern Romania in Relation to Meteorological Data) and the impact of climate change on the evolution of water resources (by Zaharia et al. in the chapter Hydrological Impacts of Climate Changes in Romania and Minea partial in the chapter *The Vulnerability of Water Resources from Eastern Romania to Anthropic Impact and Climate Change*). Thus, local trends have been identified that integrate in regional trends (both at the climatic parameters and at the hydrological and hydrogeological parameters) [4, 20, 27, 39]. From the hydrological point of view, the most important trend identified is the one related to the trend of increasing the annual and seasonal average values for discharges for more than 50% of rivers [7]. It also highlighted the trend of increasing the groundwater level associated with the increase of surface water from precipitation or hydrographic network [20]. Associated with this type of analysis, the impact of climate change on the manifestation of extreme hydrological phenomena was highlighted (associated with either maximum leakage—like flash floods and floods, or minimum leakage spillage—like hydrological droughts) which in the last decades imposed additional pressure on the use of water resources for economic purposes (agricultural and energy) [1, 17, 23], and even to economic resounding bankruptcies with negative effect on the population (analyzed by Romanescu and Minea in the chapter *Drought and Insolvency: Case study of the Producer-Buyer Conflict: Romania, the period between the year 2011–2012*).

Obviously this type of analysis has been integrated with current situations related to the manifestation of extreme hydrological phenomena or the way of capitalizing the water resources in the cross-border basins which in the last decades have raised particular problems. We have in mind here the issues related to *Romanian Danube River Floodplain Functionality Assessment* (in the chapter written by Trifanov et al.), the *Particularities of Drain Liquid in the Small Wetland of Braila Natural Park* located in an intensely anthropic activity area over the last decades (in the chapter written by Diaconu) and Romanian Upper Tisa Watershed (*Between Water Resources and Hydrological Hazards*, in the chapter written by Șerban et al.).

On this line, the editing of the volume on Water Resources Management in Romania finds its place in the scientific approaches, by the importance of the methodologies used and the ways of assessing the anthropogenic impact on them. All water management issues that may occur in a country could not be covered, but important steps have been taken to align with European standards for monitoring, evaluation and management of this type of resource [3, 24, 28].

One of the main issues that remain unresolved is the reduction of water resource monitoring points, both in quantitative and qualitative terms. This will help to increase the evaluation of the anthropogenic impact on the quantity and quality of water resources by uncontrolled volumes of water both from the surface (rivers, lakes) or underground (hydrogeological drillings or individual wells), or by domestic waste water or agro-industrial origin intake [21, 29, 37].

In this respect, a number of recommendations are required which can be included in the future water resource management plans. The first recommendation is related to identifying the possibilities for increasing qualitative and quantitative monitoring points. This was done partially at the level of the various basin administrations by the introduction of automatic monitoring stations, which are insufficient and do not adequately cover the entire territory of the country. Many of these stations have been located in areas vulnerable to extreme hydrological events such as flash floods and floods, but supplementing these numbers can bring a surplus of vital information during the occurrence of these hydrological phenomena [13, 16]. At the same time, it is necessary to implement water quality monitoring systems especially around large urban agglomerations and to support investments in the refurbishment and construction of treatment plants to reduce the anthropic pollution degree [30, 31] In parallel, legalizing adjustments are needed to increase the integration of water resources in good and very good classes (currently only 60% of Romania's water resources can be integrated into this class). The main recommended measures to achieve a higher percentage are related to supporting investments in the implementation of a wastewater collection system, especially in rural areas, where surface and underground surface water pollution is accentuated by lack of septic tanks and garbage disposal sites waste. These measures should also be integrated with the investments to be made in the water supply systems and increasing the access to good quality water, especially in rural areas.

In the same sense, a series of measures can be taken to renaturation the main areas where the anthropic impact is significant. We are considering here increasing the terrestrial and aquatic areas that can be included in protected wetlands (currently at the level of Romania, only 19 sites with a total area of approximately 1.6 million hectares are included within protected areas under the RAMSAR Convention) and where the environmental impact has been significantly reduced by real measures to protect and reduce the level of anthropogenic pollution.

Another recommendation that must be included in future local or regional management plans is to reduce the impact of extreme hydrological phenomena on human communities and the economic activities. This can be achieved by concrete measures to relocate rural and urban sites, especially those severely affected in the last decades by floods [14, 34]. In parallel, work on hydro-technical structures designed to protect and defend against floods can be continued but adapted to new hydrological conditions where extreme hydrological phenomena such as flash floods and floods reach thresholds beyond the scenarios set out in the original plans [33]. Extreme hydrological phenomena associated with high temperatures and lack of rainfall that can generate hydrological droughts and which can be mitigated by investing in irrigation or water supply systems in water surplus areas (especially from mountain drainage basins) must not be forgotten [15, 32]. These actions can also reduce the economic effects of bankruptcy of energy-producing companies (like Hidroelectrica S.A.'s case) or a significant reduction in agricultural production. Also a series of non-structural measures can be taken that address the ways of evacuating the population affected by floods or real afforestation actions in the affected mountainous and sub-mountainous areas in recent decades due to intense grubbing-up legislative gaps.

A last recommendation and perhaps the most recent one is to integrate in water management plans the scenarios related to global and regional climate change. These changes, which at local level are unlikely to be felt, in time can generate devastating effects. Currently there are slight tendencies to increase the average air temperature with effects in increasing the quantities of evaporated water with effect in the drought phenomena that are becoming more and more frequent and severe. At the same time a slight concentration of atmospheric precipitation in the warm season of the year is observed. Exceptional rains with quantities exceeding 250–300 mm in only a few hours that generated severe floods and floods have become a hydrological normality [18]. That is why it is necessary to integrate all regional climate change scenarios into future management plans so that the most pragmatic measures are taken within a shorter time frame. At the same time, as proposed in the 2007 Framework Directive [11], it is necessary to carry out as much as possible flood hazard maps and flood risk maps in order to reduce the negative economic effects and, in particular, to reduce the number of people affected.

17.3 Conclusions

In the next sections, some of the conclusions and recommendations of the chapters in this volume of the Environmental Earth Science are presented. The following conclusions are mainly extracted the chapters presented in this volume:

1. The environmental objectives for the surface and groundwaters in Romania—in relation to the Framework Directive for Water 2000/60 of the European Union—have been achieved almost integrally. The status of waters is relatively good and good for about 60% water bodies. The point sources of pollution—belonging to old industrial units from the Communist period or to new and small ones, mainly diffuse sources of pollution—bring significant damage to the small streams.

2. Since joining the EU in 2007, Romania has significantly improved its environmental performance, but water quality problems are still persisting. The most important problems are related to the surface and groundwater within rural settlements because they are polluted by animal and human dejections that end up straight in the bodies of water. By connecting all the localities to the sewerage system and to the water supply system, pollution will cease almost automatically nationwide.

3. It is necessary to continue the process of developing the monitoring system to cover all the elements of quality (biological, hydro morphological, and physicochemical) and all investigation media (water, sediment, and biota) with a frequency that ensures high levels of confidentiality and precision in assessing the condition of water bodies.

4. The efficiency and low energy consumption are among the goals of the Romanian authorities, and the designers enrolled in the rehabilitation programmes, as the infrastructure for drinkable and irrigation water supply is aged. The environment protection has to be a priority for all decision-makers regarding the technology used in new construction sites, for maintaining or even restoring the natural habitats that give specific charm and beauty and, above all, assure a healthy living for the inhabitants.

5. There is a huge gap between urban areas and rural areas concerning the access to public water supply systems. Things are changing, but slowly, because urban water operators tend to expand their activity in the nearby rural areas and this will only be possible with big investments, based mainly on EU structural funds.

6. Anthropic activities have led to radical changes in the natural conditions under which the water bodies of Eastern Romania have been evolving. These changes require the rethinking of water management systems, both at the surface and

in the underground. Various climatic scenarios must take into account and formulate strategies for the adaptation to climate change through the introduction within water management plans of measures.

7. Scientific research can still bring important results in reconsideration of economic activities, although the technological solutions that can be applied immediately, exist and should be introduced without delay, with government support.

8. The role of forests on water resources and of the hydrological regime as well as the climate provides a powerful mitigation tool for the effects of global climate change. Cutting the forest to nearly 400,000 ha has caused major imbalances in ecosystems, changing the drainage of the water on the slopes being the immediate consequence for floods.

9. The most important climate changes with hydrological impacts detected in the last half-century (after 1960) are a general warming at annual and seasonal scale (excepting autumn) and increase of the precipitation amount in autumn and of the daily maximum rainfall (mainly in summer and autumn). Important changes also affected the snow-related indices (mean snow depth, number of days with snow cover and with snowfall, continuous snow cover duration), which had significant downward trends.

10. The water resources (surface and groundwater) in the management plan of the Siret, as well as in the other river basins in Romania, are monitored qualitatively and quantitatively for conservation and protection against pollution and floods. Following the collection of information, a database is being gathered, and it is analyzed then after the results are obtained, a series of measures are taken to improve or eliminate the negative effects.

11. The Romanian catchment of the Upper Tisa River indicate a surplus of water. The quantitative and qualitative monitoring network is very well developed in the basin, but not the hydrotechnical facilities, able to intervene in the maximum discharge phase and in the regularization of the waterflow, which are completely missing in the Vişeu basin.

12. From the point of view of the water resource necessary for fire extinguishing in the areas with greater fire risk (Small Braila Wetlands) there should be constructed reservoirs (water basins) from where water for extinguishing vegetation fires is supplied by pumping with individual groups of pumps and portable electrical generators.

13. The natural particularities of the monitoring sites can greatly impact the results of measurements. The evolution of air parameters can explain the evolution of the water parameters in natural areas, but it is only partly useful in areas with intense human activity, such as the areas where dams and hydroelectric plants exist.

14. The global climatic changes affect also Romania. For the past 25 years, drought periods are longer and more severe, though, the mean amount of precipitations has also increased. Unfortunately, heavy rains (100–300 mm/24 h) causing large amounts of water within a short time span have become more frequent and more significant; they lead inevitably, to catastrophic floods.

15. The Apuseni Mountains area has a significantly hydric budget, spatially expressed through a dense hydrographic network, with radial drainage, imposed by the high sector of central ridge of the Bihor-Vlădeasa and Muntele Mare mountains.

17.4 Recommendations

The following recommendations are mainly extracted from the chapters presented in this volume:

1. It is necessary to decreasing pollution by ecological measures like afforestation or erosion of tailings dumps that belong to the mining companies in the northern Carpathians, the Apuseni Mountains and Oltenia. It is also necessary to bloc emissions from abandoned quarries by coal, copper or other non-ferrous or ferrous ores in southern and western Romania.
2. The lack of septic tanks and the storage of rubbish especially at the rural areas make the water supply network seriously affected. It is necessary to urge the generalized water supply of all villages, to conduct sewage disposal and to educate the population.
3. Economic domains with the most significant contributions to the potential for pollution of wastewater are: (i) water capture and processing for the supply of the population; (ii) chemical processing; (iii) industries producing electricity or heat; (iv) extractive industry.
4. With regard to water quality, Romania needs to improve its policy according to the intervention logic of the Water Framework Directive. There are two aspects: (i) investments in urban waste water treatment plants; (ii) reducing pressures in agriculture by better identifying and defining both the compulsory measures that all farmers have to respect and the complementary ones for which they can receive funding.
5. The information on the implementation of relevant legislation on the website of competent European and national authorities should be more detailed.
6. The population should be educated to understand and address information on the quality of the environment and the rational use of natural resources and there should be greater transparency for the public in accessing financial instruments.
7. New wastewater treatment facilities have to be bulit. New investments in pressure control and automation equipment are needed to modernise the existing pumping stations.
8. The use of numerical simulation in all planning and designing activity. A sound programme for either initial or continuous education of the personnel involved in engineering design, construction and exploitation of the water systems.
9. Regarding water systems is necessary taking specific sectoral measures: increasing the number of customers connected to the centralized drinking water supply

system; extending the existing centralized drinking water supply system; permanently increasing the quality of water and services; improving operational and financial performances; avoiding an uncontrolled increase in the drinking water price.

10. Local or regional water operator should be permanently concerned to fulfil the following: reducing consumptions and network water losses; permanently improving customer relations; decreasing the duration of repairs; increasing the reaction speed to emergencies and complaints; permanently improving employees' skills.

11. It is necessary to include in the local and regional management plans of water resources the scenarios regarding the changes of climatic parameters (both temperature and precipitation) in order to identify the vulnerable areas.

12. In the context of increasing the demand for water for economic and social purposes, it is necessary either to make new large-scale hydro-technical works (such as reservoir lakes) or additions from adjacent water-surplus basins.

13. Planning policies depend on the spatial context in which they are implemented, and how they modify this context. The Ecological and Economic Resizing Program of the Romanian Danube Floodplain presented a spatial planning tool developed and built to design, analyze and evaluate long-term policies in a social, economic and ecological context.

14. The main component of the spatial planning tool should be a dynamic land use model applied to the entire territory of the Danube floodplain. To represent the processes that make and change the configuration of the Danube floodplain requires an overlayered model to represent processes on three geographical levels: national, regional and local.

15. It is necessary to carry out a spatial analysis of the deforested areas and the scale of the phenomenon, detailing the impact on changes on climatic factors, the degradation of the slopes, and last but not least on the local and regional economy.

16. A call to action on forests, water and climate is emerging on many fronts. Taking into account the effects of forests on water and climate, suggests that this call is urgent. Stimulating regional and continental approaches can contribute to the development of more appropriate governance, thus increasing the chances of success.

17. The development and updating of scientific knowledge on: water resources availability and requirements for different uses; future impacts of climate changes on rivers flow; water-related risks and their management are necessary. The scientific studies must be the basis for the adaptation measures. Better collaboration is needed between the research/academic institutions and stakeholders/authorities responsible for designing and implementing adaptation measures to climate changes (at different spatial scales).

18. It is necessary a cooperation between the research/academic institution and those holding hydro-climatic database, based on agreements, which facilitate the use of the database for scientific purposes. In this way it will be possible to use more effectively the hydro-climatic database nationwide, for studying

climate change and their effects. It is desirable that the research results on this topic be integrated into a national platform accessible to all those interested.

19. In order to obtain as accurate data as possible it is necessary to multiply the number of the observation points, such as: the hydrometric stations on the rivers, the evaporimetric stations both on the ground and on the lake, the automated rain gauges should also be multiplied. It is also imperative to use modern hydrometrical equipment (level sensors, water and air temperature sensors, automatic rain gauges).

20. Preventions, reduction and control of pollution on surface and groundwater and conservation of its living resources in accordance with generally accepted international rules and standards are necessary as well as conservation, improvement and rational use of surface water and groundwater and control of the hazards caused by accidents with pollutants, floods and frost.

21. A minimal and limited anthropogenic intervention in the mountainous (where the surface flow forms) are recommended, in order to maintain the high quality of water resource for the natural and anthropogenic environments.

22. The water sources and treatment plants must be constituted in the mountainous area and connected with the localities from the low altitude spaces through pipelines of large capacity. By this the clean water is assured, and the sustainable development can be made according regulations.

23. Although Romania, compared to Europe, does not face large-scale vegetation fires it is necessary to prepare an adequate infrastructure in order to prevent and mitigate its effects. Global climate change by emphasizing extreme phenomena and protecting natural areas is the premise of the new approaches.

24. The implementation of fire mitigation projects can be done gradually, especially within the areas that are difficult to access and which are in the fire risk. Staging development will offer the opportunity to optimize the technical solutions implemented.

25. The analysis of a physical, chemical or biological parameter of a river should take into account the diurnal cycle. The cycle may be useful for estimating a representative average of a day, for including a daily variation into analysis or for removing the diurnal oscillation/noise from the long term evolution of a time series.

26. Smoothing techniques may be applied to all mean profiles of river in order to obtain more abstract representations of diurnal evolution in every site.

27. One must consider reassessing the hydrological forecasts for hydropower production, taking into account, the minimum mean flows or other conditions accepted by all parties: producer, distributor, recipient.

28. Hidroelectrica S.A. must focus its financial resources on updating the technology of the great-capacity hydropower plants, which have been exploited for over 30 years.

29. The water resource is a particularly important, even decisive, component for the existence of natural and anthropic habitats.

30. Balanced or deficient water resource in a territory, it is obvious the necessity for its management, in the context of increasing vulnerability and the aggravation and chronicisation of dangerous meteorological and hydrological phenomena.

References

1. Adler MJ, Ungureanu V (2006) Multi-model technique for low flow forecasting. Climate variability and change—hydrological impacts. In: Proceedings of the fifth FRIEND world conference held at Havana, Cuba, November 2006, IAHS Publ. 308, pp 151–157
2. Bănăduc D, Pânzar C, Bogorin P, Hoza O, Curtean-Bănăduc A (2016a) Human impact on Tarnava Mare river and its effects on aquatic biodiversity. Acta Oecologica Carpatica IX:187–197
3. Bănăduc D, Rey S, Trichkova T, Lenhardt M, Curtean-Bănăduc A (2016) The lower Danube River-Danube Delta–North West Black Sea: a pivotal area of major interest for the past, present and future of its fish fauna—a short review. Sci Total Environ 545–546:137–151
4. Birsan MV, Zaharia L, Chendes V, Branescu E (2014) Seasonal trends in Romanian streamflow. Hydrol Process 28:4496–4505
5. Capatina C, Cirtina D (2017) Comparative study regarding heavy metals content in air from Targu Jiu and Rovinari. Rev. Chim. (Bucharest) 68(12):2839–2844
6. Cirtina D, Capatina C (2017) Quality issues regarding the watercourses from Moddle Basin of Jiu River. Rev Chim (Bucharest) 68(1):72–76
7. Croitoru AE, Minea I (2015) The impact of climate changes on rivers discharge in Eastern Romania. Theoret Appl Climatol 20(3–4):563–573. https://doi.org/10.1007/s00704-014-1194-z
8. Diaconu DC, Andronache I, Ahammer H, Ciobotaru AM, Zelenakova M, Dinescu R, Pozdnyakov AV, Chupikova SA (2017a) Fractal drainage model—a new approach to determinate the complexity of watershed. Acta Montanistica Slovaca 22(1):12–21
9. Diaconu DC, Peptenatu D, Simion A.G, Pintilii RD, Draghici CC, Teodorescu C, Grecu A, Gruia AK, Ilie AM (2017b) The restrictions imposed upon the urban development by the piezometric level. Case study: Otopeni-Tunari-Corbeanca. Urbanism Archit Constr 8(1):27–36
10. Directive 2000/60/EC of the European Parliament and of the council establishing a framework for the Community action in the field of water policy (2000) EU Water Framework Directive 2000/60/EC (2000)
11. Directive 2007/60/EC of the European Parliament and of the council on the assessment and management of flood risks (2007) EU Water Framework Directive 2007/60/EC (2007)
12. Enea A, Hapciuc OE, Iosub M, Minea I, Romanescu G (2017) Water quality assessment for the mountain region of Eastern Romania. Environ Eng Manag J 16(3):605–614
13. Hapciuc OE, Minea I, Romanescu G, Tomașciuc AI (2015) Flash flood risk managemenet for small basins in mountain-plateau transition zone. Case study for Sucevița catchment (Romania). In: International multidisciplinary scientific GeoConference—SGEM din Albena, Bulgaria, conference proceedings, hydrology and water resources, pp 301–308. https://doi.org/10.5593/sgem2015/b31/s12.039
14. Hapciuc OE, Romanescu G, Minea I, Iosub M, Enea A, Sandu I (2016) Flood susceptibility analysis of the cultural heritage in the Sucevita catchment (Romania). Int J Conserv Sci 7(2):501–510
15. Hapciuc OE, Iosub M, Tomașciuc AI, Minea I, Romanescu G (2016b) Identification of the potential risk areas regarding the floods occurrence within small mountain catchments. In: Geobalcanica 2nd international scientific conference, pp 177–183
16. Iordache I, Ursu A, Liviu M, Iosub M, Istrate V (2016) Using MODIS imagery for risk assessment in the cross-border area Romania-Republic of Moldova. In: 6th international multidisciplinary scientific GeoConference SGEM 2016, SGEM2016 conference proceedings, vol 2,

Photogrammetry and Remote Sensing, pp 1075–1082. https://doi.org/10.5593/sgem2016/b22/10.137

17. Iosub M, Iordache I, Enea A, Romanescu G, Minea I (2015) Spatial and temporal analysis of dry/wet conditions in Ozana drainage basin, Romania using the standardized precipitation index. In: International multidisciplinary scientific GeoConference—SGEM, Albena, Bulgaria, 585–592, https://doi.org/10.5593/sgem2015/b31/s12.075

18. Minea I (2012) Bahlui drainage river-hydrological study. Al.I.Cuza Press, Iași, Romania (in Romanian)

19. Minea I, Croitoru AE (2015) Climate changes and their impact on the variation of groundwater level in the Moldavian Plateau (Eastern Romania). International Multidisciplinary Scientific Geoconferences. In: SGEM 2015, 15th GeoConference on water resources, forest, marine and ocean ecosystems, conference proceedings, vol I, Hydrology and Water Resources, pp 137–145

20. Minea I, Croitoru AE (2017) Groundwater response to changes in precipitations in north-eastern Romania. Environ Eng Manag J 16(3):643–651

21. Mititelu LA (2010) Vidraru Reservoir, Romania. Environmental impact of the hydrotechnical constructions on the upper course of Arges River. Lakes, Reserv Ponds 4(1–2):152–166

22. Murarescu O, Druga M, Puscoi B (2008) The anthropic lakes from the hydrographic basin of upper Ialomita river (Romania). Lakes, Reserv Ponds 1–2:150–157

23. Murarescu O, Muratoreanu G, Frinculeasa M (2014) Agrometeorological drought in the Romanian plain within the sector delimited by the valleys of the Olt and Buzau Rivers. J Environ Health Sci Eng 12(1):152

24. Năvodaru I, Staras M, Cernisencu I (2008) The challenge of sustainable use of the Danube Delta Fisheries, Romania. Fish Manag Ecol 8(45):323–332

25. Omer I (2016) Water quality assessment of the groundwater body RODL01 from North Dobrogea. Rev Chim (Bucharest) 67(12):2405–2408

26. Pantea I, Ferechide D, Barbilian A, Lupusoru M, Lupusoru GE, Moga M, Vilcu ME, Ionescu T, Brezean I (2017) Drinking water quality assessment among rural areas supplied by a centralized water system in Brasov county. University Politehnica of Bucharest Scientific Bulletin, Series C-Electrical Engineering and Computer Science, Series B 79(1):61–70

27. Prăvălie R, Piticar A, Roșca B, Sfica L, Bandoc G, Tiscovschi A, Patriche C (2019) Spatio-temporal changes of the climatic water balance in Romania as a response to precipitation and reference evapotranspiration trends during 1961–2013. CATENA 172:295–312. https://doi.org/10.1016/j.catena.2018.08.028

28. Raischi MC, Oprea L, Deak G, Badilita A, Tudor M (2016) Comparative study on the use of new sturgeon migration monitoring systems on the lower Danube. Environ Eng Manag J 15(5):1081–1085

29. Romanescu G, Stoleriu C, Romanescu AM (2011) Water reservoirs and the risk of accidental flood occurrence. Case study: Stanca–Costesti reservoir and the historical floods of the Prut river in the period July–August 2008, Romania. Hydrol Process 25(13):2056–2070

30. Romanescu G, Iosub M, Sandu I, Minea I, Enea A, Dascălița D, Hapciuc OE (2016) Spatio-temporal analysis of the water quality of the Ozana river. Rev Chim (Bucharest) 67(1):42–47

31. Romanescu G, Hapciuc OE, Sandu I, Minea I, Dascalița D, Iosub M (2016) Quality indicators for Suceava river. Rev Chim (Bucharest) 67(2):245–249

32. Romanescu G, Tirnovan A, Cojoc GM, Sandu IG (2016c) Temporal variability of minimum liquid discharge in Suha basin. Secure water resources and preservation possibilities. Int J Conserv Sci 7(4):1135–1144

33. Romanescu G, Cimpianu CI, Mihu-Pintilie A, Stoleriu CC (2017) Historic flood events in NE Romania (post-1990). J Maps 13(2):787–798

34. Romanescu G, Hapciuc OE, Minea I, Iosub M (2018,) Flood vulnerability assessment in the mountain-plateau transition zone. Case study for Marginea village (Romania). J Flood risk Manag 5502–5013. https://doi.org/10.1111/jfr3.12249

35. Romanian Waters National Administration (2004) The hydrographical basins management plans. National Report 2004—Romania. Bucharest

36. Romanian Waters National Administration (2013) The national plan for the development of the hydrographical basins in Romania. Synthesis. Revised version—February 2013. Bucharest
37. Roșu L, Zagan R (2017) Management of drought and floods in Romania. In: Natural resources management: concepts, methodologies, tools, and applications. Information Resources Manag. Association, IGI Global, Hershey PA, USA 2063. https://doi.org/10.4018/978-1-5225-0803-8.ch002
38. Serban P, Galie A (2006) Water Management. European principles and regulations. Tipored Press, Bucharest (in Romanian)
39. Sfica L, Croitoru AE, Iordache I, Ciupertea AF (2017) Synoptic conditions generating heat waves and warm spells in Romania. Athmosfere 8(3):50. https://doi.org/10.3390/atmos8030050